Lecture Notes in Computer Science 6506

Commenced Publication in 1973
Founding and Former Series Editors:
Gerhard Goos, Juris Hartmanis, and Jan van Leeuwen

Otfried Cheong
Kyung-Yong Chwa
Kunsoo Park (Eds.)

Algorithms and Computation

21st International Symposium, ISAAC 2010
Jeju Island, Korea, December 15-17, 2010
Proceedings, Part I

 Springer

Volume Editors

Otfried Cheong
KAIST
Department of Computer Science
Daejeon 305-701, Korea
E-mail: otfried@kaist.edu

Kyung-Yong Chwa
KAIST
Department of Computer Science
Daejeon 305-701, Korea
E-mail: kychwa@jupiter.kaist.ac.kr

Kunsoo Park
Seoul National University
School of Computer Science and Engineering
Seoul 151-742, Korea
E-mail: kpark@snu.ac.kr

Library of Congress Control Number: 2010939852

CR Subject Classification (1998): F.2, I.3.5, E.1, C.2, G.2, F.1

LNCS Sublibrary: SL 1 – Theoretical Computer Science and General Issues

ISSN   0302-9743
ISBN-10   3-642-17516-3 Springer Berlin Heidelberg New York
ISBN-13   978-3-642-17516-9 Springer Berlin Heidelberg New York

springer.com

© Springer-Verlag Berlin Heidelberg 2010
Printed in Germany

Typesetting: Camera-ready by author, data conversion by Scientific Publishing Services, Chennai, India
Printed on acid-free paper   06/3180

Preface

This volume contains the proceedings of the 21st Annual International Symposium on Algorithms and Computations (ISAAC 2010), held in Jeju, Korea during December 15–17, 2010. Past editions have been held in Tokyo, Taipei, Nagoya, Hong Kong, Beijing, Cairns, Osaka, Singapore, Taejon, Chennai, Taipei, Christchurch, Vancouver, Kyoto, Hong Kong, Hainan, Kolkata, Sendai, Gold Coast, and Hawaii over the years 1990-2009.

ISAAC is an annual international symposium that covers the very wide range of topics in algorithms and computation. The main purpose of the symposium is to provide a forum for researchers working in algorithms and the theory of computation where they can exchange ideas in this active research community.

In response to the call for papers, ISAAC 2010 received 182 papers. Each submission was reviewed by at least three Program Committee members with the assistance of external referees. Since there were many high-quality papers, the Program Committee's task was extremely difficult. Through an extensive discussion, the Program Committee accepted 77 of the submissions to be presented at the conference. Two special issues, one of *Algorithmica* and one of the *International Journal of Computational Geometry and Applications*, were prepared with selected papers from ISAAC 2010.

The best paper award was given to "From Holant to #CSP and Back: Dichotomy for Holant^c Problems" by Jin-Yi Cai, Sangxia Huang and Pinyan Lu, and the best student paper award to "Satisfiability with Index Dependency" by Hongyu Liang and Jing He. Two eminent invited speakers, David Eppstein from University of California, Irvine, and Matt Franklin from University of California, Davis, also contributed to this volume.

We would like to thank all Program Committee members and external referees for their excellent work, especially given the demanding time constraints; they gave the conference its distinctive character. We thank all who submitted papers for consideration; they all contributed to the high quality of the conference. We also thank the Organizing Committee members for their dedicated contribution that made the conference possible and enjoyable. Finally, we thank our sponsor SIGTCS (Special Interest Group on the Theoretical Computer Science) of KIISE (The Korean Institute of Information Scientists and Engineers) for the assistance and support.

December 2010

Otfried Cheong
Kyung-Yong Chwa
Kunsoo Park

Organization

Program Chairs

Otfried Cheong   KAIST, Korea
Kyung-Yong Chwa   KAIST, Korea
Kunsoo Park   Seoul National University, Korea

Program Committee

Lars Arge   University of Aarhus, Denmark
Takao Asano   Chuo University, Japan
Danny Chen   University of Notre Dame, USA
Rudolf Fleischer   Fudan University, China
Satoshi Fujita   Hiroshima University, Japan
Mordecai Golin   Hong Kong UST, Hong Kong
Seok-Hee Hong   University of Sydney, Australia
Oscar Ibarra   University of California - Santa Barbara, USA
Giuseppe Italiano   University of Rome "Tor Vergata", Italy
Tao Jiang   UC Riverside, USA
Mihyun Kang   Humboldt University Berlin, Germany
Ming-Yang Kao   Northwestern University, USA
Tak-wah Lam   University of Hong Kong, Hong Kong
Gad Landau   University of Haifa, Israel
Peter Bro Miltersen   Aarhus University, Denmark
David Mount   University of Maryland, USA
Ian Munro   University of Waterloo, Canada
Yoshio Okamoto   Tokyo Institute of Technology, Japan
Frank Ruskey   University of Victoria, Canada
Kunihiko Sadakane   National Institute of Informatics, Japan
Steven Skiena   Stony Brook University, USA
Takeshi Tokuyama   Tohoku University, Japan
Ryuhei Uehara   Japan Advanced Institute of Science
     and Technology, Japan
Peter Widmayer   ETH Zurich, Switzerland
Chee Yap   Courant, NYU, USA
Hsu-Chun Yen   National Taiwan University, Taiwan
Afra Zomorodian   Dartmouth College, USA

Host Institute

KAIST (Korea Advanced Institute of Science and Technology)

Organizing Committee

Joon-Soo Choi	Kookmin University, Korea
Kyung-Yong Chwa	KAIST, Korea
Seung Bum Jo	KAIST, Korea
Hyunseob Lee	KAIST, Korea
Jung-Heum Park	The Catholic University of Korea, Korea

Referees

Mohammad Abam
Peyman Afshani
Hee-Kap Ahn
Laila El Aimani
Toru Araki
Abdullah Arslan
Ilia Averbouch
Laszlo Babai
Christian Bachmaier
Jeremy Barbay
Amir Barghi
Peter van Beek
Renévan Bevern
Binay Bhattacharya
Danny Breslauer
Gerth Stølting Brodal
Joshua Brody
Jonathan Buss
Ho-Leung Chan
Kun-Mao Chao
Ho-Lin Chen
Ming-Yang Chen
Minkyoung Cho
Marek Chrobak
David Cohen
Amin Coja-Oghlan
Maxime Crochemore
Pooya Davoodi
Yann Disser
Reza Dorrigiv
David Doty
Scot Drysdale
Vida Dujmovic
Robert Elsaesser
Guy Even

Arash Farzan
Sandor Fekete
Andreas Feldmann
Holger Flier
Mathew Francis
Robert Fraser
Sorelle Friedler
Hiroshi Fujiwara
Oliver Gableske
Joachim von zur Gathen
Petr Golovach
Martin Golumbic
Joachim Gudmundsson
Prosenjit Gupta
Gregory Gutin
Carsten Gutwenger
Michel Habib
Kristoffer Arnsfelt
    Hansen
Sariel Har-Peled
Masud Hasan
Mohammad Khairul
    Hasan
Meng He
Pinar Heggernes
Danny Hermelin
Tomio Hirata
Christian Hoffmann
Ming-Deh Huang
John Iacono
Keiko Imai
Toshimasa Ishii
Takehiro Ito
Kazuo Iwama
Li Jian

Jiongxin Jin
Shinhaeng Jo
Daniel Johannsen
Naoyuki Kamiyama
Tom Kamphans
Iyad Kanj
Bruce Kapron
Akinori Kawachi
Daniel Keren
Shuji Kijima
Hyo-Sil Kim
Jae-Hoon Kim
Masashi Kiyomi
Jan Willem Klop
Koji Kobayashi
Darek Kowalski
Kasper Dalgaard Larsen
Hyunseob Lee
Lap-Kei Lee
Mira Lee
Taehyung Lee
Avivit Levy
Chun-Cheng Lin
Maarten Loeffler
Daniel Lokshtanov
Alejandro Lopez-Ortiz
Vadim Lozin
Jun Luo
Thomas Mølhave
Khalegh Mamakani
Maurice Margenstern
Dimitri Marinakis
Tomomi Matsui
Yusuke Matsumoto
Yuichiro Miyamoto

Matthias Mnich
Morteza Monemizadeh
Petra Mutzel
Hiroshi Nagamochi
Shin-Ichi Nakano
C. Thach Nguyen
Patrick Nicholson
Takao Nishizeki
Martin Noellenburg
Katsuyuki Okeya
Hirotaka Ono
Yota Otachi
Konstantinos
  Panagiotou
Daniel Panario
Eunhui Park
Jeong-Hyeon Park
Jung-Heum Park
Anders Sune Pedersen
Benny Pinkas
Greg Plaxton
Sheung-Hung Poon
Sanguthevar
  Rajasekaran
Jörg Rambau

Bala Ravikumar
Iris Reinbacher
Daniel Roche
Juanjo Rué
Daniel Russel
Toshiki Saitoh
Jagan Sankaranarayanan
Ignasi Sau
Saket Saurabh
Joe Sawada
Dominik Scheder
Akiyoshi Shioura
Michiel Smid
Aaron Sterling
Mizuyo Takamatsu
Kenjiro Takazawa
Nicholas Tran
Xuehou Tan
Siamak Tazari
My Thai
Hing-Fung Ting
Alexander Tiskin
Etsuji Tomita
Lorenzo Traldi
Amitabh Trehan

Rahul Tripathi
Shi-Chun Tsai
Kostas Tsakalidas
Takeaki Uno
Yushi Uno
Leslie Valiant
Antoine Vigneron
Bow-Yaw Wang
Haitao Wang
Osamu Watanabe
Oren Weimann
Joel Wein
Renato Werneck
Aaron Williams
Peter Winkler
Alexander Wolff
Prudence Wong
Prudence W.H. Wong
Mutsunori Yagiura
Katsuhisa Yamanaka
Koichi Yamazaki
Deshi Ye
Xiao Zhou
Binhai Zhu
Anna Zych

Table of Contents – Part I

Session 2A. Data Structure and Algorithm I

Session 2B. Combinatorial Optimization

Session 3A. Graph Algorithm I

Session 3B. Complexity II

Session 4A. Computational Geometry I

Session 4B. Graph Coloring I

Session 5A. Fixed Parameter Tractability

Session 5B. Optimization

Table of Contents – Part II

Session 10B. Computational Geometry III

Regular Labelings and Geometric Structures

David Eppstein

Computer Science Department, University of California, Irvine

Abstract. Three types of geometric structure—grid triangulations, rectangular subdivisions, and orthogonal polyhedra—can each be described combinatorially by a *regular labeling*: an assignment of colors and orientations to the edges of an associated maximal or near-maximal planar graph. We briefly survey the connections and analogies between these three kinds of labelings, and their uses in designing efficient geometric algorithms.

O. Cheong, K.-Y. Chwa, and K. Park (Eds.): ISAAC 2010, Part I, LNCS 6506, p. 1, 2010.

Algorithmic Aspects of Secure Computation and Communication

Matt Franklin (U.C. Davis)

Abstract. We survey some recent progress in the design of efficient protocols for secure computation and communication, in a variety of cryptographic settings. The common thread is the usefulness of interesting algorithmic methods originally developed for non-cryptographic applications. We also present some intriguing open problems for which new algorithmic ideas may be needed.

O. Cheong, K.-Y. Chwa, and K. Park (Eds.): ISAAC 2010, Part I, LNCS 6506, p. 2, 2010.

Faster Algorithms for Feedback Arc Set Tournament, Kemeny Rank Aggregation and Betweenness Tournament[*]

Marek Karpinski[1],[**] and Warren Schudy[2],[***]

[1] University of Bonn
marek@cs.uni-bonn.de
[2] IBM T.J. Watson
ws@cs.brown.edu

Abstract. We study fixed parameter algorithms for three problems: Kemeny rank aggregation, feedback arc set tournament, and betweenness tournament. For Kemeny rank aggregation we give an algorithm with runtime $O^*(2^{O(\sqrt{OPT})})$, where n is the number of candidates, $OPT \leq \binom{n}{2}$ is the cost of the optimal ranking, and $O^*(\cdot)$ hides polynomial factors. This is a dramatic improvement on the previously best known runtime of $O^*(2^{O(OPT)})$. For feedback arc set tournament we give an algorithm with runtime $O^*(2^{O(\sqrt{OPT})})$, an improvement on the previously best known $O^*(OPT^{O(\sqrt{OPT})})$ [4]. For betweenness tournament we give an algorithm with runtime $O^*(2^{O(\sqrt{OPT/n})})$, where n is the number of vertices and $OPT \leq \binom{n}{3}$ is the optimal cost. This improves on the previously known $O^*(OPT^{O(OPT^{1/3})})$ [28], especially when OPT is small. Unusually we can solve instances with OPT as large as $n(\log n)^2$ in polynomial time!

Keywords: Kemeny rank aggregation, Feedback arc set tournament, Fixed parameter tractability, Betweenness tournament.

1 Introduction

Suppose you ran a chess tournament, everybody played everybody (a.k.a. round robin) and you wanted to use the results to rank everybody. Unless you were really lucky, the results would not be acyclic, so you could not just sort the players by who beat whom. A natural objective is to find a ranking that minimizes the number of upsets, where an upset is a pair of players where the player ranked lower in the ranking beat the player ranked higher. Minimizing the number of upsets is called *feedback arc set problem* on tournaments (FAST). The complementary problem of *maximizing* the number of pairs that are *not upsets* is

---

[*] A preliminary version of this work appeared in version 1 of the arXiv preprint Karpinski and Schudy [23].
[**] Parts of this work done while visiting Microsoft Research.
[***] Parts of this work done while a student at Brown U. and while visiting U. of Bonn.

O. Cheong, K.-Y. Chwa, and K. Park (Eds.): ISAAC 2010, Part I, LNCS 6506, pp. 3–14, 2010.
© Springer-Verlag Berlin Heidelberg 2010

called the *maximum acyclic subgraph problem* on tournaments. These problems are NP-hard [2, 3, 11, 15], but a polynomial-time approximation scheme (PTAS) [26] is known.

In statistics and psychology, one motivation is *ranking by paired comparisons* [29]: here, you wish to sort some set by some objective but you do not have access to the objective, only a way to compare a pair and see which is greater; for example, determining people's preferences for types of food. This problem attracted computational attention as early as 1961 [29]. Feedback arc set tournament and closely related problems have also been used in machine learning [14, 2].

The FAST problem can be generalized to a problem we call *weighted FAST*, sometimes known as *feedback arc set with probability constraints*. The input is a complete directed graph with arc weights $\{w_{uv}\}_{u,v}$. The weights satisfy $w_{uv} + w_{vu} = 1$ for every pair of vertices u, v. The objective is to minimize the total weight of the backwards arcs.

We study the *parameterized complexity* of this problem, in particular the parameter OPT, the cost of an optimal ranking. We use the notation $O^*(\cdot)$ to hide factors that are polynomial in the input size, that is $f(I) \in O^*(g(I))$ iff $f(I) \leq g(I)|I|^c$ for some $c > 0$ and all sufficiently large inputs I. The first fixed-parameter algorithms for feedback arc set tournament had runtime $O^*(2^{O(OPT)})$ and later algorithms made a dramatic improvement to $O^*(OPT^{O(\sqrt{OPT})})$ [4]. We improve this to $O^*(2^{O(\sqrt{OPT})})$.

Theorem 1. *There exists a deterministic parameterized subexponential algorithm for weighted FAST with runtime $2^{O(\sqrt{OPT})} + n^{O(1)}$. A variant of the algorithm uses $OPT^{O(\sqrt{OPT})} + n^{O(1)}$ time and $n^{O(1)}$ space.*

The *Exponential time hypothesis* (ETH) [21] is that 3-SAT cannot be solved in time $2^{o(\text{number of variables})}$. We also give a matching lower bound assuming the ETH:

Theorem 2. *There does not exist a parameterized algorithm for weighted FAST with runtime $O^*(2^{o(\sqrt{OPT})})$ unless the exponential time hypothesis [21] is false.*

We leave open the possibility that *unweighted* FAST may admit an exact algorithm with runtime $O^*(2^{o(\sqrt{OPT})})$.

We note that independently of this work Feige [19] gave an unrelated algorithm for *unweighted* FAST matching Theorem 1.

An important application of weighted feedback arc set tournament is *rank aggregation*. Frequently, one has access to several rankings of objects of some sort, such as search engine outputs [18], and desires to aggregate the input rankings into a single output ranking that is similar to all of the input rankings: it should have minimum average distance from the input rankings, for some notion of distance. A natural notion of distance is the number of pairs of vertices that are in different orders, which is known as the Kendall-Tau distance. This defines the *Kemeny rank aggregation* problem (KRA) [24, 25, 30]. This problem is NP-hard [5], even with only four voters [18], and has a PTAS [26].

We denote the *average* distance between the optimal ranking and the input rankings by $OPT \leq \binom{n}{2}$. Two parameters have attacted the bulk of the study:

OPT and the average Kendall-Tau distance between the input rankings. It is easy to see (triangle inequality) that these two parameters are within a constant factor of each other, so these parameters give equivalent runtimes up to constants in the exponent. All previous work give algorithms with runtime $O^*(2^{O(OPT)})$ [8]. There is a standard reduction from KRA to weighted FAST [2, 16, 26], so we improve the best known parameterized algorithm for KRA dramatically to $O^*(2^{O(\sqrt{OPT})})$ as a corollary of our Theorem 1.

Corollary 3. *Let n be the number of candidates and $OPT \leq \binom{n}{2}$ the optimum value. There exists a deterministic parameterized subexponential algorithm for Kemeny Rank Aggregation with runtime and space $2^{O(\sqrt{OPT})} + n^{O(1)}$. A variant uses $OPT^{O(\sqrt{OPT})} + n^{O(1)}$ time and $n^{O(1)}$ space.*

Some other paramters have attracted attention. The parameter of maximum Kendall-Tau distance has been studied but yield bounds no tighter (up to constants in the exponent) than is known for the average Kendall-Tau distance [8]. Another parameter is the maximum $r_{\max}$, over candidates c and pairs of voters v_1, v_2, of the absolute difference between the rank of c in v_1 and v_2. The best runtime known is $O^*(2^{O(r_{\max})})$ [8].

In the Betweenness problem we are given a ground set of *vertices* and a set of *betweenness constraints* involving 3 vertices and a *designated* vertex among them. The objective function of a ranking of the elements is the number of betweenness constraints for which the designated vertex is not between the other two vertices. The goal is to *minimize* the objective function. For the status of the general Betweenness problem, see e.g. Opatrny [27], Chor and Sudan [13], Ailon and Alon [1], Charikar et al. [12]. We refer to the Betweenness problem in tournaments, that is in instances with a constraint for every triple of vertices, as the BETWEENNESSTOUR problem (see Ailon and Alon [1]). This problem is NP-hard [1] and has a recently discovered polynomial-time approximation scheme [23]. We study its parameterized complexity.

Theorem 4. *There exists a randomized parameterized subexponential algorithm for BETWEENNESSTOUR with runtime and space $2^{O(\sqrt{OPT/n})} \cdot n^{O(1)}$, where n is the number of vertices and OPT is the cost of the optimal ranking. It succeeds with constant probability.*

The previously best known runtime was $O^*(2^{O(OPT^{1/3} \log OPT)})$ [28]. Our result is better by a logarithmic factor in the exponent for the largest possible $OPT = \Theta(n^3)$ and even better for smaller OPT. Interestingly we can solve all instances with $OPT = O(n \log^2 n)$ in polynomial time!

Our results easily generalize to all fully dense ranking CSPs of arity three with *fragile* constraints as introduced by Karpinski and Schudy [23]. For simplicity we limit ourselves to the well-known problems discussed above.

We now outline the organization of our paper. Section 2 discusses weighted feedback arc set tournament, including our algorithm (Section 2.1), analysis (2.2), and lower bound (2.3). Section 3 discusses our results for betweenness tournament, including our algorithm (Section 3.1) and analysis (3.2).

2   Feedback Arc Set Tournament

2.1   Algorithm

We now outline some of our key techniques. Firstly any two low-cost rankings for a FAST problem are nearby in Kendall-Tau distance. Secondly two rankings that are Kendall-Tau distance D apart are equivalent to within additive $O(\sqrt{D})$ in how good each position for each a vertex is (Lemma 6). Thirdly most vertices (in a low-cost instance) have a vee-shaped cost versus position curve and optimal rankings are locally optimal so we know that each vertex belongs at the bottom of its curve. The uncertainty in this curve by $\sqrt{D}$ causes an uncertainty in the optimal position also around $\sqrt{D}$ (Lemmas 7 and 8). Our algorithm simply computes uncertainties $r(v)$ in the positions of all of the vertices v and solves a dynamic program for the optimal ranking that is near a particular constant-factor approximate ranking. We remark that Braverman and Mossel [9] and Betzler et al. [7, 8] previously applied dynamic programming to FAST and KRA.

First we state some core notation. Throughout this paper let V refer to the set of objects (vertices) being ranked and n denote $|V|$. Our $O(\cdot)$ hides absolute constants only. Our $O^*(\cdot)$ hides a polynomial in n. A *ranking* is a bijective mapping from a set $S \subseteq V$ to $\{1, 2, 3, \ldots, |S|\}$. We call $\pi(v)$ the position of v in the ranking π. We let $d(\pi, \pi')$ denote the Kendall-Tau distance between rankings π and π', i.e. the number of pairs of vertices in different orders in the two rankings. An *ordering* is an injection from S into $\mathbb{R}$. We use π and σ (with superscripts) to denote rankings and orderings respectively.

The input to weighted FAST is a set V of vertices and arc weights $\{w_{uv}\}_{u,v\in V}$ such that $w_{uv} + w_{vu} = 1$ for all $u, v \in V$. The FAST objective function is the weight of the backwards arcs $C(\pi) = \sum_{u,v\in V:\pi(v)>\pi(u)} w_{vu}$. For ranking π, vertex $v \in V$ and $p \in \mathbb{R}$ (with $\pi(u) \neq p$ for all $u \neq v$) we define $b(\pi, v, p) = \sum_{u\neq v} \begin{cases} w_{vu} & \text{if } p > \pi(u) \\ w_{uv} & \text{if } p < \pi(u) \end{cases}$, i.e. the cost of the arcs incident to v in the ordering formed by moving v to position p in π. Let π^* denote an optimal ranking and $OPT = C(\pi^*)$ its cost.

Before running our main Algorithm 1 we compute a small *kernel*, that is a smaller instance with the same optimal cost as the input instance (up to a known shift). This preliminary step allows us to separate the dependence on n and OPT in the runtime, yielding the runtime stated in Theorem 1.

Algorithm 1. Exact algorithm for FAST. If dynamic programming is used in the last line the runtime and space are both $n^{O(1)}2^{O(\sqrt{OPT})}$. If divide-and-conquer is used the runtime is $n^{O(\sqrt{OPT})}$ and the space is $n^{O(1)}$.

Input: Vertex set V, arc weights $\{w_{uv}\}_{u,v\in V}$.

1: Sort by weighted indegree Coppersmith et al. [16], yielding ranking π^1 of V.
2: Set $r(v) = 4\sqrt{2C(\pi^1)} + 2b(\pi^1, v, \pi^1(v))$ for all $v \in V$.
3: Use dynamic programming or divide-and-conquer (Details: Lemma 9) to find the optimal ranking π^2 with $|\pi^2(v) - \pi^1(v)| \leq r(v)$ for all v.

Dom et al. [17] give an algorithm for computing kernels of *unweighted* FAST instances with $O(OPT^2)$ vertices. This was later improved to $O(OPT)$ vertices by Bessy et al. [6]. There is a kernelization algorithm for Kemeny rank aggregation in Betzler et al. [8], but it produces an instance of size $O((\text{Number of voters}) \cdot OPT)$, not the desired $OPT^{O(1)}$. To get the desired kernel for general weighted FAST we consider a slight variant of the algorithm from Dom et al. [17].

Lemma 5. *There is polynomial-time computable $O(OPT^2)$-vertex kernel for weighted FAST.*

Proof (sketch). Let $OPT \leq U \leq 5OPT$ be the cost of a 5-approximate ranking [16].

We say that an arc is a *majority arc* if it has greater weight than its reverse, with ties broken arbitrarily. A majority arc clearly has weight at least $1/2$. The *majority tournament* [2] is the unweighted directed graph with vertex set V and arc set equal to the majority arcs.

Our kernelization algorithm is simple: we apply the following two reduction rules, which are extensions of two reduction rules in Dom et al. [17], as often as possible.

The first reduction rule is eliminating a vertex that is part of no cycles of three arcs in the majority tournament (henceforth triangles). Consider some such vertex v. It is easy to see that there exists an optimal ranking that puts every predecessor of v (in the majority tournament) before v and every successor of v after v, while implies the validity of this rule.

The second reduction rule concerns an arc (u, v) of the majority tournament that is in more than $2U$ triangles. Any feedback arc set not not paying for such an arc must pay for more than $2OPT$ other arcs of the majority tournament, each of cost at least $1/2$, and hence cannot be optimal. Therefore we record that we must pay w_{uv}, then set weight w_{uv} to zero and w_{vu} to one.

Now we argue that the resulting instance after these two rules are exhaustively applied has $O(OPT^2)$ vertices. An optimal feedback arc set, which necessarily has cost OPT, can include at most $2OPT$ majority arcs. Each such majority arc is in at most $10OPT$ triangles by the second rule, so there are at most $20OPT^2$ triangles. Finally by the first rule every vertex is in a triangle, so there are at most $60OPT^2$ vertices.

We have not investigated whether or not the $O(OPT)$ vertex kernel for unweighted FAST [6] can be extended to weighted FAST.

2.2 Analysis

Variants of the following Lemma are given in Mathieu and Schudy [26] and Karpinski and Schudy [23]. We give a simplified proof here for completeness.

Lemma 6 (Mathieu and Schudy [26], Karpinski and Schudy [23]). *Let π and π' be rankings over V. It follows that $|b(\pi, v, p) - b(\pi', v, p)| \leq 2\sqrt{d(\pi, \pi')}$ for all $v \in V$ and $p \in \mathbb{R} \setminus \mathbb{Z}$.*

Proof. Fix $v \in V$, $p \in \mathbb{R} \setminus \mathbb{Z}$ and rankings π, π'. Consider the sets of vertices $L = \{ u \in V \setminus \{v\} : \pi(u) < p < \pi'(u) \}$ and $R = \{ u \in V \setminus \{v\} : \pi'(u) < p < \pi(u) \}$. Intuitively these are the vertices that cross p from left to right (resp. right to left) when going from π to π'. It follows easily from the definition of b that $|b(\pi, v, p) - b(\pi', v, p)| \leq |L| + |R|$, so we now proceed to bound $|L|$ and $|R|$.

The bijective nature of π and π' implies that $|L| = |R|$. Observe that all vertices in L are before all vertices in R in π, and vice versa for π', hence $d(\pi, \pi') \geq |L||R|$. Putting these facts together proves the Lemma.

Lemma 7. *In Algorithm 1 we have $|\pi^*(v) - \pi^1(v)| \leq r(v)$ for all $v \in V$ and any optimal ranking π^* of V.*

Proof. The weight of an arc and its reverse sum to one so $d(\pi^*, \pi^1) \leq C(\pi^*) + C(\pi^1) \leq 2C(\pi^1)$. By Lemma 6 therefore

$$|b(\pi^*, v, j + 1/2) - b(\pi^1, v, j + 1/2)| \leq 2\sqrt{2C(\pi^1)} \tag{1}$$

for any $j \in \mathbb{Z}$.

Now fix $v \in V$. We conclude

$$
\begin{aligned}
&|\pi^*(v) - \pi^1(v)| \\
&\leq b(\pi^1, v, \pi^*(v)) + b(\pi^1, v, \pi^1(v)) && (w_{uv} + w_{vu} = 1 \text{ for all } u) \\
&= b(\pi^1, v, \pi^*(v) + 1/2) + b(\pi^1, v, \pi^1(v) + 1/2) && (\pi^1 \text{ is integral}) \\
&\leq b(\pi^*, v, \pi^*(v) + 1/2) + 2\sqrt{2C(\pi^1)} + b(\pi^1, v, \pi^1(v) + 1/2) && (\text{By } (1)) \\
&\leq b(\pi^*, v, \pi^1(v) + 1/2) + 2\sqrt{2C(\pi^1)} + b(\pi^1, v, \pi^1(v) + 1/2) && (\text{Optimality of } \pi^*) \\
&\leq 4\sqrt{2C(\pi^1)} + 2b(\pi^1, v, \pi^1(v) + 1/2) && (\text{By } (1)) \\
&= r(v) && (\text{Definition of } r(v)).
\end{aligned}
$$

Lemma 8. *In Algorithm 1 we have $\max_{j \in \mathbb{Z}} |\{ v \in V : |\pi^1(v) - j| \leq r(v) \}| = O(\sqrt{OPT})$.*

Proof. Fix $j \in \mathbb{Z}$. Let $R = \{ v \in V : |\pi^1(v) - j| \leq r(v) \}$, the cardinality of which we are trying to bound. We say $v \in V$ is *pricey* if $2b(\pi^1, v, \pi^1(v)) > \sqrt{2C(\pi^1)}$. Clearly $2C(\pi^1) = \sum_v b(\pi^1, v, \pi^1(v)) \geq (\text{number pricey})\frac{1}{2}\sqrt{2C(\pi^1)}$ hence the number of pricey vertices is at most $\frac{2C(\pi^1)}{(1/2)\sqrt{2C(\pi^1)}} = 2\sqrt{2C(\pi^1)}$. All non-pricey vertices in R have $|\pi^1(v) - j| \leq r(v) \leq 5\sqrt{2C(\pi^1)}$, so at most $10\sqrt{2C(\pi^1)} + 1$ non-pricey vertices are in R. We conclude $|R| \leq 12\sqrt{2C(\pi^1)} + 1 = O(\sqrt{OPT})$ since π^1 is a 5-approximation [16].

Lemma 9. *There is a dynamic program for FAST that finds the optimal ranking π^2 with $|\pi^2(v) - \pi^1(v)| \leq r(v)$ for all v using space and runtime $O(|V|^2)2^\psi$, where $\psi = \max_j |\{ v \in V : |\pi^1(v) - j| \leq r(v) \}|$. A divide and conquer variant uses $|V|^{O(\psi)}$ time and $|V|^{O(1)}$ space.*

Proof. Say that a set $S \subseteq V$ is *valid* if it contains all vertices v with $\pi^1(v) \leq |S| - r(v)$ and no vertex v with $\pi^1(v) > |S| + r(v)$. Observe that for any $s \in \mathbb{N}$ all valid sets of size s agree except for the presence or absence of ψ vertices. Therefore there are at most $n2^\psi$ valid sets.

We say that a ranking π of valid set S is *valid* if $\{ v : \pi(v) \leq j \}$ is a valid set for all $0 \leq j \leq |S|$. It is easy to see that a ranking π is valid if and only if satisfies $|\pi(v) - \pi^1(v)| \leq r(v)$ for all v.

One can easily see the following optimal substructure property: prefixes of an optimal valid ranking are optimal valid rankings themselves.

For any valid set S let $\bar{C}(S)$ denote the cost of the optimal valid ranking of S. The recurrence relation is

$$\bar{C}(S) = \min_{v \in S : S \setminus \{v\} \text{ is valid}} \left[\bar{C}(S \setminus \{v\}) + \sum_{u \in S \setminus \{v\}} w_{vu} \right].$$

The space-efficient variant evaluates $\bar{C}$ using divide and conquer instead of dynamic programming, similar to Dom et al. [17]. Details deferred.

Now we put the pieces together and prove Theorem 1.

Proof (of Theorem 1). The kernelization algorithm of Lemma 5 allows us to assume without loss of generality that $n = O(OPT^2)$. Algorithm 1 returns an optimal ranking by Lemmas 7 and 9. Lemmas 8 and 9 allow us to bound the runtime and space requirements of the dynamic program.

2.3   Lower Bound

Proof (of Theorem 2). For sake of contradiction suppose we have an algorithm for weighted FAST with runtime $2^{o(\sqrt{OPT})}$. We present a series of reductions which converts such an algorithm into a subexpontial-time algorithm for vertex cover, the existence of which is known to contradict the ETH Flum and Grohe [20].

Let an instance of vertex cover with n vertices be given. Applying Karp's reduction from vertex cover to feedback arc set [22] produces a feedback arc set instance with $2n$ vertices. Finally one can reduce this to a weighted FAST instance with the same number of vertices by representing incomparable pairs of vertices by opposite arcs of weight $1/2$. The result is an weighted FAST instance with $2n$ vertices that is equivalent to the original vertex cover instance. The optimal cost for this instance is at most its number of arcs, which is $O(n^2)$, so the hypothesized algorithm has runtime $2^{o(\sqrt{OPT})} = 2^{o(n)}$. This runtime is subexponential, contradicting the ETH.

3   Betweenness Tournament

3.1   Algorithm

We now introduce some new notation for the betweenness problem. We let $\binom{n}{k}$ (for example) denote the standard binomial coefficient and $\binom{V}{k}$ denote the set of subsets of set V of size k. For any ordering σ let $\mathrm{Ranking}(\sigma)$ denote the ranking naturally associated with σ.

Let $v \mapsto p$ denote the ordering over $\{v\}$ which maps v to p. For set Q of vertices and ordering σ with domain including Q let $Q \mapsto \sigma$ denote the ordering over Q which maps $u \in Q$ to $\sigma(u)$, i.e. the restriction of σ to Q. For orderings σ^1 and σ^2 with disjoint domains let $\sigma^1|\sigma^2$ denote the natural combined ordering over $Domain(\sigma^1) \cup Domain(\sigma^2)$. For example of our notations, $Q \mapsto \sigma|v \mapsto p$ denotes the ordering over $Q \cup \{v\}$ that maps v to p and $u \in Q$ to $\sigma(u)$.

A ranking 3-CSP consists of a ground set V of *vertices* and a *constraint system* c, where c is a function from rankings of 3 vertices to $[0,1]$. For brevity we henceforth abuse notation and and write $c(\mathrm{Ranking}(\sigma))$ by $c(\sigma)$. The objective of a ranking CSP is to find an ordering σ (w.l.o.g. a ranking) minimizing $C(\sigma) = \sum_{S \in \binom{\mathrm{Domain}(\sigma)}{k}} c(S \mapsto \sigma)$. We will only ever deal with one constraint system c at a time, so we leave the dependence of C on c implicit in our notations. Abusing notation we sometimes refer to $S \subseteq V$ as a *constraint*, when we really are referring to $c(S \mapsto \cdot)$. Clearly one can model BETWEENNESSTOUR as a ranking 3-CSP.

Let $b(\sigma, v, p) = \sum_{Q:\ldots} c(Q \mapsto \sigma|v \mapsto p)$, where the sum is over sets $Q \subseteq Domain(\sigma) \setminus \{v\}$ of size 2. Note that this definition is valid regardless of whether or not v is in $\mathrm{Domain}(\sigma)$. The only requirement is that the range of σ excluding $\sigma(v)$ must not contain p. This ensures that the argument to $c(\cdot)$ is an ordering (injective).

Our algorithm and analysis for BETWEENNESSTOUR are analogous to our results for FAST with two major differences. Firstly no kernel for betweenness tournament is known, which hurts the runtime somewhat. Secondly we use a more complicated approach to get the preliminary constant-factor approximation ranking π^1. Our analysis requires not only that π^1 be of *cost* comparable to π^* but also that it be close in *Kendall-Tau distance*. Fortunately the known PTAS [23] (with an appropriate error parameter to get a 2-approximation) is known to also produce a ranking close to the optimum in Kendall-Tau distance.

Theorem 10 (Theorem 4 in Karpinski and Schudy [23]). *There exists a polynomial-time algorithm for* BETWEENNESSTOUR *that produces a set Π of $O(1)$ rankings. With constant probability one of the rankings $\pi \in \Pi$ satisfies $d(\pi, \pi^*) = O(OPT/n)$ and has cost at most $2C(\pi^*)$, where π^* is some optimal ranking.*

3.2   Analysis

The following two lemmas are given in Karpinski and Schudy [23] in more generality. We sketch proofs here for completeness.

Lemma 11 (Karpinski and Schudy [23]). *For any rankings π and π' over vertex set V, vertex $v \in V$ and $p \in \mathbb{R}$ we have*

$$|b(\pi, v, p) - b(\pi', v, p)| \le 3(n-1)\sqrt{d(\pi, \pi')}.$$

Algorithm 2. Our algorithm for BETWEENNESSTOUR. The runtime is $n^{O(1)}2^{O(\sqrt{OPT/n})}$.

Input: Vertex set V

1: Use the Algorithm from Theorem 10 to construct a set of rankings Π
2: Let π^{good} be the ranking from Π with lowest cost
3: **for** each $\pi^1 \in \Pi$ **do**
4:     **if** $C(\pi^1) \leq 2C(\pi^{good})$ **then**
5:         Set $r(v) = \alpha_1\sqrt{C(\pi^1)/n} + \alpha_2 b(\pi^1, v, \pi^1(v))/n$ for all $v \in V$, where α_1 and α_2 are absolute constants.
6:         Use dynamic programming (see Lemma 9) to find the optimal ranking π^2 with $|\pi^2(v) - \pi^1(v)| \leq r(v)$ for all v.
7:     **end if**
8: **end for**
9: Return the best of the π^2 rankings.

Proof (Sketch). We use L and R from the proof of Lemma 6. A constraint $\{u, u', v\}$ contributes identically to $b(\pi, v, p)$ and $b(\pi', v, p)$ unless either:

1. $\{u, u'\}$ and $(L \cup R)$ have a non-empty intersection (or)
2. $\mathbb{1}(\pi(u) < \pi(u')) \neq \mathbb{1}(\pi'(u) < \pi'(u'))$.

To finish the proof we use the bound $|L| = |R| \leq \sqrt{d(\pi, \pi')}$ from the proof of Lemma 6 and the trivial bound $d(\pi, \pi') \leq n(n-1)/2$.

Lemma 12 (Karpinski and Schudy [23]). *Let π be a ranking of V, $|V| = n$, $v \in V$ be a vertex and $p, p' \in \mathbb{R}$. Let B be the set of vertices (excluding v) between p and p' in π. Then $b(\pi, v, p) + b(\pi, v, p') \geq \frac{(n-2)|B|}{2}$.*

Proof (Sketch). By definition

$$b(\pi, v, p) + b(\pi, v, p') = \sum_{Q:\cdots} [c(Q \mapsto \pi | v \mapsto p) + c(Q \mapsto \pi | v \mapsto p')] \qquad (2)$$

where the sum is over sets $Q \subseteq V \setminus \{v\}$ of 2 vertices. The quantity in brackets in (2) is at least 1 for every Q that has at least one vertex between p and p' in π. There are at least $|B|(n-2)/2$ such sets.

Lemma 13. *During the iteration of Algorithm 2 that considers the ranking with $d(\pi^1, \pi^*) = O(OPT/n)$ and $C(\pi^1) \leq 2C(\pi^*)$ guaranteed by Theorem 10 we have $|\pi^*(v) - \pi^1(v)| \leq r(v)$ for all $v \in V$.*

Proof. By Lemma 11 and Theorem 10 we have

$$|b(\pi^*, v, j + 1/2) - b(\pi^1, v, j + 1/2)| = O(n\sqrt{OPT/n}) \qquad (3)$$

for any $j \in \mathbb{Z}$.
    Fix $v \in V$. We conclude

$$|\pi^*(v) - \pi^1(v)|\frac{n-2}{2}$$

$$\leq b(\pi^1, v, \pi^1(v) + 1/2) + b(\pi^1, v, \pi^*(v) + 1/2) \qquad \text{(Lemma 12)}$$

$$\leq b(\pi^*, v, \pi^*(v) + 1/2) + O(\sqrt{nOPT}) + b(\pi^1, v, \pi^1(v) + 1/2) \quad \text{(By (3))}$$

$$\leq b(\pi^*, v, \pi^1(v) + 1/2) + O(\sqrt{nOPT}) + b(\pi^1, v, \pi^1(v) + 1/2) \quad \text{(Optimality of } \pi^*)$$

$$\leq O(\sqrt{nOPT}) + 2b(\pi^1, v, \pi^1(v) + 1/2) \qquad \text{(By (3))}$$

$$= r(v)\frac{n-2}{2} \qquad \text{(Definition of } r(v)).$$

Lemma 14. *In Algorithm 2 we have* $\max_{j \in \mathbb{Z}} |\{ v \in V : |\pi^1(v) - j| \leq r(v) \}| = O(\sqrt{C(\pi^1)/n})$.

Proof. We proceed analogously to the proof of Lemma 8. Fix j. Let $R = \{ v \in V : |\pi^1(v) - j| \leq r(v) \}$, whose cardinality we are trying to bound. We say $v \in V$ is *pricey* if $b(\pi^1, v, \pi^1(v))/n > \sqrt{2C(\pi^1)/n}$. Clearly $3C(\pi^1) = \sum_v b(\pi^1, v, \pi^1(v)) \geq$ (number pricey)$n\sqrt{2C(\pi^1)/n}$ hence the number of pricey vertices is at most $3C(\pi^1)/(\sqrt{2nC(\pi^1)}) = O(\sqrt{C(\pi^1)/n})$. All non-pricey vertices in R have $|\pi^1(v) - j| = O(\sqrt{C(\pi^1)/n})$, so $O(\sqrt{C(\pi^1)/n})$ non-pricey vertices are in R. We conclude $|R| = O(\sqrt{C(\pi^1)/n})$.

Lemma 15. *There is a dynamic program for betweenness that finds the optimal ranking* π^2 *with* $|\pi^2(v) - \pi^1(v)| \leq r(v)$ *for all* v, *with space and runtime* $O(|V|^3 2^\psi)$ *where* $\psi = \max_j |\{ v \in V : |\pi^1(v) - j| \leq r(v) \}|$. *A divide and conquer variant uses* $|V|^{O(\psi)}$ *time and* $|V|^{O(1)}$ *space.*

Proof (Sketch). We adopt the definition of valid sets and rankings from the proof of Lemma 9. For any ranking π over S let $C'(\pi)$ denote the portion of the cost shared by all orderings with prefix π. That is, the cost of all constraints with at most 1 vertex outside S. We use the natural recurrence relation for the C' cost of the optimal (w.r.t. C') valid ranking of S.

Proof (of Theorem 4). Lemmas 15 and 14, plus the test of the "if" in Algorithm 2, allow us to bound the runtime and space requirements of the dynamic program used by Algorithm 2 by $n^{O(1)} 2^{O(\sqrt{C(\pi^{good})/n})}$, which is of the correct order since $C(\pi^{good}) \leq 2C(\pi^*)$. The "for" loop is over a constant number of options and hence does not impact the runtime.

For correctness we focus on the iteration of Algorithm 2 that considers the $\pi^1 \in \Pi$ with $d(\pi^1, \pi^*) = O(\sqrt{C(\pi^*)/n})$ and $C(\pi^1) \leq 2C(\pi^*)$ as guaranteed by Theorem 10. Theorem 10 ensures $C(\pi^1) \leq 2C(\pi^*) \leq 2C(\pi^{good})$ and hence the "if" is passed. By Lemma 13 π^* is among the orders the dynamic program considers.

Acknowledgements

We would like to thank Venkat Guruswami, Claire Mathieu, Prasad Raghavendra and Alex Samorodnitsky for interesting remarks and discussions.

References

[1] Ailon, N., Alon, N.: Hardness of fully dense problems. Inf. Comput. 205(8), 1117–1129 (2007)

[2] Ailon, N., Charikar, M., Newman, A.: Aggregating inconsistent information: Ranking and clustering. Journal of the ACM 55(5), Article No. 23 (2008)

[3] Alon, N.: Ranking tournaments. SIAM J. Discrete Math. 20(1), 137–142 (2006)

[4] Alon, N., Lokshtanov, D., Saurabh, S.: Fast FAST. In: Albers, S., Marchetti-Spaccamela, A., Matias, Y., Nikoletseas, S., Thomas, W. (eds.) ICALP 2009. LNCS, vol. 5555, pp. 49–58. Springer, Heidelberg (2009)

[5] Bartholdi III, J., Tovey, C., Trick, M.: Voting schemes for which it can be difficult to tell who won the election. Social Choice and Welfare 6, 157–165 (1989)

[6] Bessy, S., Fomin, F.V., Gaspers, S., Paul, C., Perez, A., Saurabh, S., Thomassé, S.: Kernels for feedback arc set in tournaments. In: FSTTCS 2009: 29th Foundations of Software Technology and Theoretical Computer Science (2009)

[7] Betzler, N., Fellows, M.R., Guo, J., Niedermeier, R., Rosamond, F.A.: Fixed-parameter algorithms for Kemeny scores. In: Fleischer, R., Xu, J. (eds.) AAIM 2008. LNCS, vol. 5034, pp. 60–71. Springer, Heidelberg (2008)

[8] Betzler, N., Fellows, M.R., Guo, J., Niedermeier, R., Rosamond, F.A.: How similarity helps to efficiently compute Kemeny rankings. In: AAMAS 2009: 8th International Conference on Autonomous Agents and Multiagent Systems, pp. 657–664 (2009); Journal version in Theoretical Computer Science 410, 4554–4570 (2009)

[9] Braverman, M., Mossel, E.: Noisy sorting without resampling. In: Procs. 19th ACM-SIAM SODA, pp. 268–276 (2008)

[10] Bredereck, R.: Fixed-parameter algorithms for computing Kemeny scores—theory and practice. Technical report, Studienarbeit, Institut für Informatik, Friedrich-Schiller-Universität Jena, Germany (2009) arXiv:1001.4003v1

[11] Charbit, P., Thomasse, S., Yeo, A.: The minimum feedback arc set problem is NP-hard for tournaments. Combinatorics, Probability and Computing 16, 1–4 (2007)

[12] Charikar, M., Guruswami, V., Manokaran, R.: Every Permutation CSP of Arity 3 is Approximation Resistant. In: 24th IEEE CCC (2009)

[13] Chor, B., Sudan, M.: A geometric approach to betweenness. SIAM J. Discrete Math. 11(4), 511–523 (1998)

[14] Cohen, W.W., Schapire, R.E., Singer, Y.: Learning to order things. J. Artificial Intelligence Research 10, 243–270 (1999)

[15] Conitzer, V.: Computing Slater rankings using similarities among candidates. In: Procs. 21st AAAI, pp. 613–619 (2006)

[16] Coppersmith, D., Fleischer, L., Rudra, A.: Ordering by weighted number of wins gives a good ranking for weighted tournaments. In: Procs. 17th ACM-SIAM SODA, pp. 776–782 (2006)

[17] Dom, M., Guo, J., Hüffner, F., Niedermeier, R., Truß, A.: Fixed-Parameter Tractability Results for Feedback Set Problems in Tournaments. In: Calamoneri, T., Finocchi, I., Italiano, G.F. (eds.) CIAC 2006. LNCS, vol. 3998, pp. 320–331. Springer, Heidelberg (2006)

[18] Dwork, C., Kumar, R., Naor, M., Sivakumar, D.: Rank aggegation methods for the web. In: Procs. 10th WWW, pp. 613–622 (2001), The NP-hardness proof is in the online-only appendix available from http://www10.org/cdrom/papers/577/

[19] Feige, U.: Faster fast (feedback arc set in tournaments), arXiv:0911.5094 (2009)

[20] Flum, J., Grohe, M.: Parameterized Complexity Theory. Springer, Heidelberg (2006)

[21] Impagliazzo, R., Paturi, R.: Which problems have strongly exponential complexity? J. Computer and System Sciences 63, 512–530 (2001)

[22] Karp, R.: Reducibility among combinatorial problems. In: Procs. Complexity of Computer Computations, pp. 85–103 (1972)

[23] Karpinski, M., Schudy, W.: Approximation schemes for the betweenness problem in tournaments and related ranking problems. arXiv:0911.2214 (2009)

[24] Kemeny, J.: Mathematics without numbers. Daedalus 88, 571–591 (1959)

[25] Kemeny, J., Snell, J.: Mathematical models in the social sciences. Blaisdell, New York (1962); Reprinted by MIT press, Cambridge (1972)

[26] Mathieu, C., Schudy, W.: How to Rank with Few Errors. In: 39th ACM STOC, pp. 95–103 (2009), In Submission,
http://www.cs.brown.edu/~ws/papers/fast_journal.pdf

[27] Opatrny, J.: Total ordering problems. SIAM J. Comput. 8(1), 111–114 (1979)

[28] Saurabh, S.: Chromatic coding and universal (hyper-) graph coloring families. In: Parameterized Complexity News, pp. 3–4 (June 2009),
http://mrfellows.net/Newsletters/2009June_FPT_News.pdf

[29] Slater, P.: Inconsistencies in a schedule of paired comparisons. Biometrika 48, 303–312 (1961)

[30] Young, P.: Optimal voting rules. The Journal of Economic Perspectives 9(1), 51–64 (1995)

A 3/2-Approximation Algorithm for Generalized Steiner Trees in Complete Graphs with Edge Lengths 1 and 2

Piotr Berman[1,*], Marek Karpinski[2,**], and Alexander Zelikovsky[3,* * *]

[1] Department of Computer Science & Engineering, Pennsylvania State University,
University Park, PA 16802
berman@cse.psu.edu
[2] Department of Computer Science, University of Bonn, 53117 Bonn
marek@cs.uni-bonn.de
[3] Department of Computer Science, Georgia State University, Atlanta, GA 30303
alexz@cs.gsu.edu

Abstract. Given a graph with edge lengths and a set of pairs of vertices which should be connected (requirements) the Generalized Steiner Tree Problem (GSTP) asks for a minimum length subgraph that connects every requirement. For the Generalized Steiner Tree Problem restricted to complete graphs with edge lengths 1 and 2, we provide a 1.5-approximation algorithm. It is the first algorithm with the approximation ratio significantly better than 2 for a class of graphs for which GSTP is MAX SNP-hard.

Keywords: Steiner trees, generalized Steiner trees, approximation algorithms.

1 Introduction

Given a graph with lengths on edges, the Generalized Steiner Tree Problem asks for the minimum length subgraph that connects given set of pairs of vertices. The first nontrivial approximation ratio of 2 has been given in [1] and for two decades there were no improvements of this ratio for any sufficiently wide subclass of graphs for which the problem is MAX SNP-hard. Note that as the Steiner tree problem, geometric cases admit PTAS [5].

The class of complete graphs with edge length 1 and 2 has attracted considerable attention following publication of [3] culminating in recent 1.25-approximation of Steiner trees in such graphs [2]. In this paper we generalize "potential technique" introduced in [2] to obtain a better factor for Generalized Steiner tree Problem. The resulted 3/2-approximation algorithm is fairly simple but requires an elaborate nontrivial analysis.

In the next section we formulate the Generalized Steiner tree problem and introduce necessary notations. In Section 3 we describe our approximation algorithm and in Section 4 we prove the approximation ratio of 3/2.

---

* Research partially done while visiting Department of Computer Science, University of Bonn and supported by DFG grant Bo 56/174-1.
** Supported in part by DFG grants, Procope grant 31022, and the Hausdorff Center grant EXC59-1.
* * * Supported in part by NSF grant IIS-0916948.

O. Cheong, K.-Y. Chwa, and K. Park (Eds.): ISAAC 2010, Part I, LNCS 6506, pp. 15–24, 2010.

2 Definitions and Notation

Let $G = (V, E, d)$ be a graph with edge lengths $d : E \to \mathcal{R}^+$. Let $R \subseteq V^2$ be a set of *requirements*, i.e., pairs of vertices which are required to be connected. We now can formulate the following

Generalized Steiner Tree Problem (GSTP). Given a graph $G = (V, E, d)$ and a set of requirements R, find a minimum length subset of edges $F \subseteq E$, such that each pair $r \in R$ is contained in a connected component of (V, F).

Clearly, any minimal feasible solution has no cycles, so we will refer to it as a Steiner forest. We say that a vertex is a *terminal* if it belongs to a at least one requirement. Let $T \subseteq V$ be the set of all terminals. We define *requirement components* as connected components of (T, R). A required component is *degenerate* if it consists of a single terminal. Since no edge is needed to satisfy a requirement forming degenerate components, a required component will further refer only to non-degenerate one. When there exists a single required component, then the GSTP becomes a famous Steiner Tree Problem (STP) requiring connection of all terminals. While a solution for GSTP is a Steiner forest, a solution for STP is a Steiner tree.

It has been noticed that it is useful to split a Steiner tree into *full components* which are maximal subtrees whose terminals coincide with its leaves. Full components are building blocks for all recent Steiner tree approximations (see, e.g., [7,4]).

We will further assume that the graph G is complete and each edge has length either 1 or 2. We will refer to these edges as 1-edges and 2-edges, and the sets of all 1-edges and 2-edges will be denoted E_1 and E_2, respectively. The corresponding GSTP and STP will be referred as GSTP[1,2] and STP[1,2], respectively. Our algorithm uses three types of full components:

   (i) a 1-edge connecting two terminals,
  (ii) an s-star consisting of a non-terminal c (referred as the center) connected to s
       terminals $t_1, \ldots, t_s$ with 1-edges $(c, t_1), \ldots, (c, t_s)$, and
 (iii) a 2-edge connecting two terminals.

The algorithms select edges (full components) that will be used in the solution and transform the instance by collapsing (contracting) them, i.e., replacing the endpoints by a single vertex. We use the term *selection* since we may need to unselect certain edges later. Let A be a set of selected edges. Note that these edges are selected a such way that A cannot contain a cycle. The set A defines a partition of V into connected components of (V, A), and we use V/A to denote the collection of these components.

In turn, we define graph $(V/A, E/A)$ where $(\mathbf{u}, \mathbf{v}) \in E/A$ if $(u, v) \in E$ for some $u \in \mathbf{u}, v \in \mathbf{v}$. We say that (u, v) is a representative of $(\mathbf{u}, \mathbf{v})$. We define $d(\mathbf{u}, \mathbf{v})$ as the minimum $d(u, v)$ such that (u, v) is a representative of $(\mathbf{u}, \mathbf{v})$. In exactly same manner A induces the set of requirements R/A on the set of terminals T/A.

In our algorithms, when we select some "building block" C (a star or another set of edges) then A is augmented with the representatives of the edges in C, and this changes the residual graph $(V/A, E/A)$ in which we make our next selection. For this reason, we use terms "select" and "collapse" as synonyms. Further, instead of referring to V/A, E/A, T/A, and R/A, we will simply refer to the "current" V, E, T, and R.

> 1. Repeatedly select 1-edges between terminals.
> 2. Repeatedly select s-stars with the largest s, $s \geq 3$.
> 3. Repeatedly select 2-edges between terminals.

Fig. 1. Rayward-Smith's Heuristic for STP[1,2]

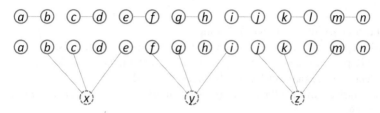

Fig. 2. Blind selection of inter-required full components may lead to over-merging

3   The Algorithm

Our algorithm for GSTP[1,2] is a generalization of the following Rayward-Smith's heuristic [6] analyzed in [3].

In order to apply the above algorithm to the GSTP[1,2] we need to distinguish the case when connection between terminals is required (i.e., *intra-required full components* connecting terminals from the same required component) from the case when it is not required (i.e., *inter-required full components* connecting terminals from at least 2 distinct required components).

The intra-required full components can be treated the same way as in Rayward-Smith's heuristic. But the inter-required full components cannot be blindly used as seen from the following example (see Fig. 2). The required pairs are $\{a, a\}, \{b, b\}, ..., \{n, n\}$, 1-edges connect terminals $(a, b), \ldots, (m, n)$ and 3-star centers x, y, and z to terminals $(x, b), \ldots, (z, m)$. The optimal solution would connect each required pair with cost 2 summing up to the total length of 28. Blindly selected 1-edges and 3-stars of total length 16 merge all pairs to a larger whole thus making the ratio $1 + 16/28 > 1.5$. Although greedy "pruning" at the end would replace these bad choices, it is not easy to estimate the effect of such pruning.

On the other hand, one cannot completely avoid using inter-required full components as seen from the following example (see Fig. 3). Suppose that we have a chain of inter-required 1-edges in which we have required sets of size 5 such that within the same requirement all vertices have the same label, e.g., a. The optimal solution would use length 1 per terminal and 5 per required component, but the solution that blindly

Fig. 3. Blind avoiding of inter-required full components may lead to under-merging

avoids inter-required full components would cost 8 per required component. Thus blind avoidance of such components would have approximation ratio of $8/5 > 1.5$.

We now describe our Generalized Rayward-Smith's Heuristic (see Fig. 4). Phase P1 selects intra-required full components the same way as the original Rayward-Smith's heuristic. Phases P3 and P4 are also the same as Rayward-Smith's steps 2 and 3, respectively. The difference is in Phase P2 whose main goal is to select inter-required 1-edges and prune some of them afterwards. The details of the Phase P2 are described below.

P1. Selection of intra-required full components
   (a) Repeatedly select intra-required 1-edges between terminals.
   (b) Repeatedly select intra-required s-stars with the largest s, $s \geq 3$.
P2. Selection of remaining 1-edges between terminals
   (a) Mark as unsafe all required pairs and mark as safe all other required components
   (b) Repeatedly
      (i) select a 1-edge (u, v) between terminals
      (ii) if (u, v) is an intra-required 1-edge,
           then mark as safe the required component containing u and v
           else if both merged required components are unsafe,
                then mark as unsafe the resulted required component
                otherwise, mark as safe the resulted required component
   (c) Prune (unselect) all edges that form unsafe requirement components
   (d) Select all intra-required 2-edges between terminals of remaining required pairs
P3. Repeatedly select inter- and intra-required s-stars with the largest s, $s \geq 3$.
P4. Repeatedly select intra-required 2-edges between terminals.

Fig. 4. The Generalized Rayward-Smith's Heuristic for GSTP[1,2]

Note that over-merging can harm only when we connect *required pairs*, i.e., requirement components of size 2. So Phase P2 treats required pairs differently from larger "safe" required components – it prunes some inter-required 1-edges between them and finally connects them with intra-required 2-edges. In the selection step P2(b), intra-required 1-edges may reappear after selection of an inter-required 1-edge (see Fig. 5). Such 1-edges are useful and will be never pruned.

Pruning of unsafe components in the step P2(c) is followed by connecting of required pairs in the step P2(d). These two steps are necessary to avoid over-merging that can otherwise happen in Phase P3. Indeed, the unsafe components (if unpruned) will be not connected in P2(d) awaiting to be connected in Phase P3 resulting in violation of 3/2 ratio (see Fig. 6). Assume that (a_i, a_i') and (b_i, b_i'), $i = 1, \ldots, n$, are required pairs connected with the inter-required 2-edges. The 1-edges connect b_i' with a_i and also 3-star centers with b_i's. Each unsafe component formed by 1-edges (b_i', a_i) will be pruned in step P2(c) and required pairs (a_i, a_i') and (b_i, b_i'), $i = 1, \ldots, n$, will be connected

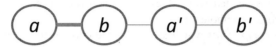

Fig. 5. For two requirements $\{a, a'\}$ and $\{b, b'\}$, selection of the bold inter-required edge (a, b) makes the other two 1-edges intra-required

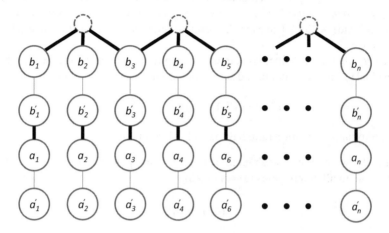

Fig. 6. An instance of the GSTP with $2n$ required pairs (a_i, a'_i) and (b_i, b'_i), $i = 1, \ldots, n$, and $n - 1$ Steiner points (dashed circles). The thick solid edges have length 1 and thin edges have length 2. The Generalized Rayward-Smith's Heuristic finds optimal solution of length $4n$ but will find solution of length $6.5n - 0.5$ if unsafe components are not pruned in the step P2(c).

by 2-edges in step P2(d). As a result, the algorithm will find the optimal solution of length $4n$. If unsafe components are not pruned in the step P2(c) then the algorithm will choose besides 2-edges connecting required pairs also all 1-edges resulting in the cost of $6.5n - 0.5$ and the ratio exceeds 3/2. Note that although greedy "pruning" at the end would replace these useless 1-edges, it is not easy to estimate the effect of such pruning.

In the next section we will prove the following main result of the paper.

Theorem 1. *The generalized Rayward-Smith's Heuristic for* GSP[1,2] *has an approximation ratio of at most* $3/2$.

4   The Analysis of the Generalized Rayward-Smith's Heuristic

The proof of the approximation ratio is based on so called "potential method" developed in [2]. While following the phases P1-P4 of the algorithm we will maintain the following forests and values:

- the set of selected edges A which is empty in the beginning and forms a feasible solution at the end,
- the *residual reference solution* F_{ref} which is initialized as an optimum solution F^* and is gradually reduced to an empty set while more and more edges are selected and added to A,

- the *potentials* associated with edges of residual reference solution F_{ref} and the following types of components:
  (a) *reference components* which are connected components of the residual reference solution F_{ref};
  (b) *1-components* which are connected components with at least **three vertices** of $(V, F_{ref} \cap E_1)$, the residual reference solution F_{ref} restricted to 1-edges;
  (c) *full components* of $(V, F_{ref} \cap E_1)$ which are maximal subtrees whose terminals coincide with its leaves in the residual reference solution F_{ref} restricted to 1-edges
- the *estimated cost* defined as sum of the length of selected edges, the length of the residual reference solution, and sum of potentials assigned to edges and components

$$Cost = d(A) + d(F_{ref}) + \sum P(X)$$

The proof of approximation ratio has the following scheme:

1. Initially, we assign to each edge of $e \in F_{ref}$ the potential $P(e) = \frac{1}{2}d(e)$ and 0 potential to all components (a)-(c) making

$$Cost = \frac{3}{2}d(F^*)$$

2. Potential change after each selection in Phases P1-P4 as well as major potential redistribution after Phases P1 and P2 preserve the following two properties (invariants): estimated cost *Cost* never increases and the total potential is non-negative, $\Sigma P \geq 0$
3. When the algorithm terminates, A is a feasible solution.

4.1   Preserving the Invariants during Phase P1

The first potential				
reference component	1-component	full component	1-edge	2-edge
0	0	0	1/2	1

Phase P1 is amortized with the initial distribution of the potential. If we select an edge with cost 1 that is contained in a requirement component, we remove/collapse an edge in F_{ref}, so $d(A)$ increases by 1, $d(F_{ref})$ decreases by 1 and ΣP decreases as well. If we select a star with s terminals, $s \geq 3$, $d(A)$ increases by s, we also remove/collapse $s - 1$ edges in F_{ref}, hence $d(F_{ref})$ decreases by at least $s - 1$, and ΣP also decreases, by at least $(s - 1)/2 \geq 1$. In either case, *Cost* does not increase.

4.2   The Potential Redistribution

We redistribute the potential and restructure F_{ref} to achieve two effects:

- A selection can merge reference components, and in that case we do not decrease $d(F_{ref})$; to avoid an increase of *Cost* we decrease the sum of potentials given to reference components; this requires that reference components have positive potential.

- When we select required pairs at the end of Phase P2, we may remove, say, p required pairs that belong to some reference component S. We increase $d(A)$ by $2p$ and we remove $2p$ vertices from S, we will make sure that this decreases $d(F_{ref})$ by $2p$. However, this may be false if F_{ref} is the optimum solution, so we transform it using Break rules, to a solution with a largest cost, but for which it is true. To prevent *Cost* from increasing, we give negative potential to 1-components.

When we define the second potential, we use a fixed formula for potential of edges and 1-components (full components have potential 0 at this stage), and we define rules how to alter the potential of reference components. Later we can estimate the potential $P(S)$ for a reference component S with some properties (see Lemmas 1 and 3 below).

We start the redistribution by transferring some potential from edges to their reference components: suppose $e \in F_{ref}, e \subset S$ and S is a reference component. If $d(e) = 1$ we decrease the potential of e from $\frac{1}{2}$ to $\frac{1}{3}$ and we increase the potential of S by $\frac{1}{6}$. If $d(e) = 2$ we decrease the potential of e from 1 to 0 and we increase the potential of S by 1.

Next, we transfer some potential from 1-components to reference components, namely if 1-component C is contained in reference component S, we decrease the potential of C from 0 to $-\frac{2}{3}$, and we increase the potential of S by $\frac{2}{3}$.

The second potential				
reference component	1-component	full component	1-edge	2-edge
variable	$-2/3$	0	$1/3$	0

Lemma 1. *Each reference component S gets potential $P(S) \geq 1$, and $P(S) = 1$ only if S is a required pair.*

Proof. If S contains an edge of cost 2, this edge alone transfers 1 to $P(S)$. So it suffices to consider the case when S is a single component of edges with cost 1.

If S contains just one edge, then this is a terminal edge inside an requirement component, a contradiction because we would select this edge in Phase P1. If S contains just two edges, then either it contains three terminals, which must form a single requirement component, and we have the same contradiction, or those two edges could be replaced with a single edge of cost 2. Thus S forms a 1-component with at least 3 edges, so at least $\frac{2}{3} + \frac{3}{6}$ is transferred to $P(S)$.

4.3 Bipartization of F_{ref} and Preserving the Invariants in Phase P2

Consider selection of an edge e in the step P2(b). If e connects nodes from two different reference components, say S_0 and S_1, then we merge them into a single reference component S with $P(S) = P(S_0) + P(S_1) - 1$.

If both ends of e belong to the same reference component S, we modify F_{ref} by removing an edge of S, say e', which formed cycle with e. If e' is a 2-edge then we increase $P(S)$ by $d(e') - d(e) = 1$, and if e' is a 1-edge, we increase $P(S)$ by $p(c') = \frac{1}{3}$.

One can see that the potential of a reference component equals 1 if and only if it forms an unsafe requirement component. Moreover, after the step P2(b) we do not have terminal edges.

Before the step P2(c) we alter F_{ref} without changing the potential of reference components. The resulting F_{ref} becomes bipartite, i.e., 1-edges will connect only terminals with non-terminals and no edge will connect two non-terminals. In particular, we break 1-components using 3 rules applied as long as possible. Recall that a 1-component C is a component of edges from E_1, i.e. with cost 1, with at least 2 such edges. We consider removing an edge e, which would break C into C_0 and C_1.

Break large 1-components: remove e if both C_0 and C_1 have at least two edges. *Cost* does not change, because we replace edge with cost 1 and potential $\frac{1}{3}$ with an edge that has cost 2 and we have another 1-component with potential $-\frac{2}{3}$.

A *Steiner point* of F_{ref} is a non-terminal u that is incident to at least one edge of cost 1 that belongs to F_{ref}; the number of such edge is the *degree* of u.

Break dead ends: remove e if it is incident to a Steiner point of degree 1. This decreases $d(F_{ref})$ by 1. It can also change ΣP: we subtract $P(e)$, and we may also subtract $P(C)$ if u was in a 1-component with just two edges, afterwards this component of $F_{ref} \cap E_1$ has only one edge, so it is not classified as a 1-component. If so, we increase ΣP by $\frac{2}{3} - \frac{1}{3}$, but because of the reduction in $d(F_{ref})$ we decrease *Cost*.

Break bridges: remove e if it is incident to a a Steiner point u of degree 2. This rule removes a pair of edges of cost 1 and replaces them with a single edge with cost 2, the sum of edge potentials decreases by $\frac{2}{3}$, and the sum of 1-component potentials may increase, but by at most $\frac{2}{3}$, as at most one 1-component can vanish.

Lemma 2. *When Break rules do not apply, a 1-component C with a non-terminal edge has this structure: a Steiner vertex c connected to at least 3 terminals $t_1, \ldots, t_k$, $k \geq 3$.*

Now consider an unsafe requirement component S that was formed from some p required pairs by selecting $p - 1$ edges. If it forms a reference component, the sum of its components of *Cost* equal $3p$: p edges of S that have cost 2, $p - 1$ selected edges that have cost 1 and $p(S) = 1$, so selecting required pairs of S instead reduces the contribution to *Cost* to $2p$. If it is a part of a larger reference component S', we remove $p + 1$ vertices from S', and $p - 1$ edges from A (by de-selecting them), so it suffices to show that for each vertex we remove from A we reduce *Cost* by at least 1 (and then we can use $p - 1 + p + 1 = 2p$ to connect required pairs of S).

If a removed vertex u is "isolated", we remove edge of cost 2 from S'.

If u is in a 1-component of S' with at least 4 terminals, we simply remove the edge e that connects u with its star center, and we can also use $p(e)$.

If u is in a 1-component of S' with exactly 3 terminals, we remove that star and connect its other two terminals of with an edge of cost 2; this removes from S' 3 edges of cost 1 and a 1-component, so the reduction in *Cost* is $3 + \frac{3}{3} - \frac{2}{3} - 2 = 1 + \frac{1}{3}$.

Thus the selection/collapse of required pairs that follows the de-selection of unsafe requirement components does not increase *Cost*.

Lemma 3. *When the step P2(c) terminates, a reference component with i 1-components has potential at least $1 + \frac{1}{2} + \frac{1}{6}i$.*

Proof. Consider reference component S. Because Break rules do not change $P(S)$ we consider the situation before they were applied. Because there are no required pairs when the step P2(c) terminates, each requirement component has at least 3 terminals and thus the same holds for reference components.

First consider the case of $i = 0$. If S contains two edges with cost 2 then $P(S) \geq 2$. If S contains one edge with cost 2, then the claim fails only if it contains no 1-component (before the Break rules), so S can contain at most 2 edges of cost 1, hence at most 4 vertices. Because there are no required pairs, this means that all terminals of S are in the same R-comp, while the edges of cost 1 in S are not terminal edges. This is not possible. Thus consider the case when S contains no edges of cost 2, hence, a single 1-component. This 1-component transfers $\frac{2}{3}$ to $P(S)$, therefore the claim fails only if S has at most 4 edges. Because there are no terminal edges, there must be a Steiner vertex, hence S has at most 4 terminals, and thus, a single requirement component. Therefore S cannot have more than two terminals connected to a Steiner vertex, so S must have at least 2 Steiner vertices and at most 3 terminals, not really possible.

Next, consider the case of $i = 1$. Because the "surviving" 1-component is a star, this star must have at least 3 terminals, and at most 2 of them from the same requirement component, hence there has to be at least 2 requirement components, hence at least 6 terminals, and at least 2 of these terminals not adjacent to the star center. Thus before the breaking S had at least 2 Steiner vertices, hence, at least 8 vertices, hence, at least 7 edges, so at least $\frac{2}{3} + \frac{7}{6}$ was transferred to $P(S)$.

Now, consider the case of $i > 1$. One can see that the least possible transfer to $P(S)$ occurs if S started as a single 1-component that consists of i stars, each with 3 terminals, and connected by $i - 1$ edges between the Steiner vertices, for the total of at least $4i - 1$ edges. Thus the minimum transfer to $P(S)$ is $\frac{2}{3} + \frac{4i-1}{6} \geq \frac{3}{2} + \frac{i}{6}$ (the last inequality is equivalent to $i \geq 2$).

4.4   Potential Redistribution and Preserving the Invariants in Phase P3

When Break Rules terminate, each 1-component is also an full component. By Lemma 3 we can redistribute the potential as follows: assign the potential of each "surviving" 1-component S to its full component, and transfer $\frac{1}{6}$ from the reference components that contains S to $P(S)$. This reduces the potential of reference components, but they are still at least $\frac{3}{2}$, and it increases the potential of S to $-\frac{1}{2}$. Thus we do not increase ΣP be defining the potential as follows:

The third potential				
reference component	1-component	full component	1-edge	2-edge
3/2	0	−1/2	1/3	0

Now we can follow steps of Phase P3.

When we select a star with $k + 1$ terminals, say $t_0, \ldots, t_k$, we view it as creating k connections between terminals, say $\{t_0, t_i\}$ for $i = 1, \ldots, k$. Each connection has cost $1 + 1/k$. For each connection we modify F_{ref}.

If t_0 and t_i are in different reference components, we merge these two reference components. this does not change $d(F_{ref})$, and we replace two reference components potentials with one, hence ΣP drops by $\frac{3}{2}$. Because $k \geq 2$, Cost does not increase.

If t_0 and t_i are in the same reference component but in different 1-components, we merge these 1-components and we remove from F_{ref} an edge with cost 2. The potential does not change, while $d(A) + d(F_{ref})$ drops by $2 - 1 - 1/k$.

If t_0 and t_i are in the same 1-component, we remove an edge e from this 1-component, say, the edge that connects t_i with the center of some full component B (which is a star). If B has at least 4 terminals, we know that we have selected a star with at least 4 terminals, i.e. $k \geq 3$. While we add at most $1 + \frac{1}{3}$ to $d(A)$, $d(F_{ref})$ drops by 1 and ΣP drops by $P(e) = \frac{1}{3}$.

If B has 3 terminals, after removing e we replace the remaining two edges of B, say e_1, e_2, with a single edge with cost 2. Thus the change in $Cost$ is $1 + 1/k - 1 - P(B) - P(e) - P(e_1) - P(e_2) = 1/k - 1/2$.

Thus in no case $Cost$ increases.

4.5   Preserving the Invariants in Phase P4

In Phase P4 we select cost 2 edges that are contained in requirement components, and thus, in reference components. Note that no full component can survive Phase P3, otherwise it would be a star that the algorithm could select. Thus a selection in Phase P4 increases $d(A)$ by 2, and decreases $d(F_{ref})$ also by 2. The only changes in ΣP occur when we connect the last two vertices of some reference component, because then the potential of that reference component drops to 0. Thus $Cost$ does not increase.

Clearly, at the end of Phase P4 edge set A is a valid solution. Thus we have proved Theorem 1.

References

1. Agrawal, A., Klein, P.N., Ravi, R.: When trees collide: An approximation algorithm for the generalized Steiner tree problem on networks. In: STOC, pp. 134–144 (1991)
2. Berman, P., Karpinski, M., Zelikovsky, A.: 1.25-Approximation Algorithm for Steiner Tree Problem with Distances 1 and 2. In: Dehne, F., et al. (eds.) WADS 2009. LNCS, vol. 5664, pp. 86–97. Springer, Heidelberg (2009)
3. Bern, M., Plassmann, P.: The Steiner problem with edge lengths 1 and 2. Information Processing letters 32, 171–176 (1989)
4. Byrka, J., Grandoni, F., Rothvoß, T., Sanità, L.: An improved LP-based approximation for Steiner tree. In: STOC (to appear, 2010)
5. Czumaj, A., Lingas, A., Zhao, H.: Polynomial-time approximation schemes for the Euclidean survivable network design problem. In: Widmayer, P., Triguero, F., Morales, R., Hennessy, M., Eidenbenz, S., Conejo, R. (eds.) ICALP 2002. LNCS, vol. 2380, pp. 973–984. Springer, Heidelberg (2002)
6. Rayward-Smith, V.J.: The computation of nearly minimal Steiner trees in graphs. Internat. J. Math. Educ. Sci. Tech. 14, 15–23 (1983)
7. Robins, A.Z.: Tighter Bounds for Graph Steiner Tree Approximation. SIAM Journal on Discrete Mathematics 19(1), 122–134 (2005); Preliminary version appeared in Proc. SODA 2000, pp. 770–779 (2000)

Approximate Periodicity

Amihood Amir[1,2,*], Estrella Eisenberg[1], and Avivit Levy[3,4]

[1] Department of Computer Science, Bar-Ilan University, Ramat-Gan 52900, Israel
amir@cs.biu.ac.il
[2] Department of Computer Science, Johns Hopkins University, Baltimore, MD 21218
[3] Department of Software Engineering, Shenkar College, 12 Anna Frank,
Ramat-Gan, Israel
avivitlevy@shenkar.ac.il
[4] CRI, Haifa University, Mount Carmel, Haifa 31905, Israel

Abstract. We consider the question of finding an approximate period in a given string S of length n. Let S' be a periodic string closest to S under some distance metric. We consider this distance the error of the periodic string, and seek the smallest period that generates a string with this distance to S. In this paper we consider the *Hamming* and *swap* distance metrics. In particular, if S is the given string, and S' is the closest periodic string to S under the Hamming distance, and if that distance is k, we develop an $O(nk \log \log n)$ algorithm that constructs the smallest period that defines such a periodic string S'. We call that string the approximate period of S under the Hamming distance. We further develop an $O(n^2)$ algorithm that constructs the approximate period under the swap distance. Finally, we show an $O(n \log n)$ algorithm for finite alphabets, and $O(n \log^3 n)$ algorithm for infinite alphabets, that approximates the number of mismatches in the approximate period of the string.

1   Introduction

String Periodicity is a classic topic in Computer Science. It has been extensively studied over the years [16] and linear time algorithms for exploring the periodic nature of a string were suggested (e.g. [5]). Multidimensional periodicity [1] and periodicity in parameterized strings [3] was also explored. In addition, periodicity has played a role in efficient parallel string algorithms e.g. [7,2,4].

Nevertheless, periodicity models cyclic natural phenomena. Such phenomena abound in such diverse areas as Astronomy, Geology, Earth Science, Oceanography, Meteorology, Biological Systems, the Genome, Economics, and more. Realistic data may contain errors. Such errors may be caused by the process of gathering the data which might be prone to transient errors. Moreover, errors can also be an inherent part of the data because the periodic nature of the data represented by the string may be inexact. However, it is still valuable to detect and utilize the underlying periodicity. This calls for the notion of *approximate periodicity*. To our knowledge, although there has been quite an extensive research

---

* Partly supported by NSF grant CCR-09-04581 and ISF grant 347/09.

O. Cheong, K.-Y. Chwa, and K. Park (Eds.): ISAAC 2010, Part I, LNCS 6506, pp. 25–36, 2010.

on the related notion of approximate multiple tandem repeats as discussed below in the related work subsection, all previous work on (full) periodicity dealt with exact periodicity. Approximate tandem repeats deals with different metrics and differs from the natural problem of approximate periodicity that we define.

A natural way of handling errors is the following. Since we may not be confident of our measurement or suspect the periodic process to be inexact, then we may be interested in finding the current *approximate periodic* nature of the string, i.e., what is the smallest period that defines the given string with the smallest number of errors. It is natural to ask if such an approximate period can be found efficiently. The error cause varies with the different phenomena. This is formalized by considering different distance metrics. In this paper we study approximate periodicity under two metrics: the Hamming distance and the swap distance. It may also be interesting to get an approximation of the number of errors, assuming approximate periodicity, even if the period or the exact number of errors are unknown. Such a number can either indicate that any cyclic phenomena is not interesting because it has too many errors, or may identify a small number of errors that may indeed define periodicity in the data. Can such an approximation be achieved quickly?

Results. In this paper these questions are studied for the *Hamming Distance*, where strings are possibly corrupted with *substitution* errors, i.e., a character may be substituted by a different character, and for the *swap distance*, where the errors are the exchange of two adjacent symbols (with no symbol participating in more than one exchange). Let S be a string of length n over alphabet Σ. We prove the following:

Approximate Period under the Hamming Distance. The approximate period of S under Hamming distance can be found in time $O(nk \log \log n)$, where k is the number of errors in S (Theorem 2).

Approximate Period under the Swap Distance. The approximate period of S under the swap distance can be found in time $O(n^2)$ (Theorem 3).

Fast Approximation of the Error Bound For Hamming Distance. The number of mismatches in the approximate period of S under Hamming distance can be approximated to within a factor of 2 in time $O(|\Sigma|n \log n)$ (Theorem 4). For infinite alphabets, for every $\epsilon > 0$, the number of errors in the approximate period of S under Hamming distance can be approximated to a factor of $2(1 \pm \epsilon)$ in time $O(\frac{1}{\epsilon} \cdot n \log^3 n)$ (Theorem 5). Omitted proofs will appear in the full version of the paper.

Related Work. The notion of approximate periodicity is related to the known and studied notion of approximate tandem repeats. A *perfect single tandem repeat* is defined as a nonempty string that can be divided into two identical sub-strings, e.g., *abcabc*. It is a well-studied problem. Main and Lorentz [17] present an $O(n \log n)$ algorithm, which reports all perfect tandem repeats. The motivation for studying tandem repeats came from research in formal languages. Repeats also occur frequently in biological sequences, yet they are seldom exact.

This motivated the study of approximate tandem repeats. An *approximate single repeat* is a nonempty string that can be divided into two similar substrings. The distance between the two substrings must be less than a given threshold k, in order for the two parts to be considered similar. Common studied distances are the Hamming and the (possibly weighted) edit distances.

The problem of finding all approximate single repeats is a sub-problem of finding all approximate multiple repeats in a string. A *perfect multiple repeat* is a nonempty string that can be divided into a number of identical adjacent substrings, or periods. The last period of the repeat can be partial. Note that the related problem of finding the (exact) period of a given string is a simpler problem, because this period should repeat from beginning to end. It is, therefore, expected that better algorithms can be found for this problem. An *approximate multiple repeat* is a multiple repeat in which the periods of the repeat are approximate. The definition of the approximate multiple repeat, when trying to measure the distance among several strings, is not naturally apparent. Unfortunately, the computation of many of the most intuitive measures is $\mathcal{NP}$-complete. A number of different definitions were studied, even for variations that are all based on the Hamming distance (e.g. [15,18,13]). All these definitions attempt to capture the biologically relevant relationships between biosequences, and have, in some sense, a local approach to defining the "approximate repetitions" (e.g., bounding the number of mismatches between every two adjacent repeats). This paper, however, focuses on the natural and simpler problem of approximate (full) periodicity. Our definition of approximate periodicity is, therefore, different from the notion of approximate multiple tandem repeats in all papers we are aware of, even for those definitions that are based on the Hamming distance. Moreover, it is not even clear that our approach yields a precise definition when attempting to use it for defining approximate multiple tandem repeats. It should also be noted that we are not aware of any work considering the swap distance as a measure of similarity when defining approximate multiple tandem repeats.

Paper Contribution. The contributions of the paper are three-fold:

1) Giving a first simple and natural definition of the approximate periodicity problem. Although the definition is simple, it precludes a straightforward use of standard tools (e.g., convolution, kangaroo jumps) and necessitates a careful properties study for providing a solution.
2) Providing efficient solutions to the problem under two different common metrics: the Hamming and the swap metrics.
3) Showing that it is possible to efficiently approximate the error bound under the Hamming distance. A surprising and important feature of our algorithm is the independence of the actual bound, i.e., its complexity is (almost) linear even if many errors are needed in order to assume periodicity in the input string. We were also able to extend the result for general alphabets.

Open Problems. Some remaining open questions are:

1) Can the approximate period under other metrics, e.g., the edit distance, be found efficiently?

2) How hard is it to find a periodic string whose distance to the given string is within the factor 2 or the factor (1+epsilon) of optimal?

These questions should be addressed in a future research.

2   Preliminaries

In this section we give basic definitions of periodicity and approximate periodicity as well as a formal definition of the problem.

Definition 1. *Let S be a string. Denote by $|S|$ the length of a string S, and let $|S| = n$. S is called* periodic *if $S = P^i pref(P)$, where $i \in \mathbb{N}$, $i \geq 2$, P is a substring of S such that $|P| \leq n/2$, P^i is the concatenation of P to itself i times, and $pref(P)$ is a prefix of P. The smallest such substring P is called the* period *of S. If S is not periodic it is called* aperiodic.

Remark. Throughout the paper we use p to denote a period length and P the period string, i.e., $|P| = p$.

Definition 2. *Let P be a string of length p. Let $n \in \mathbb{N}$ such that $2 \cdot p \leq n$. The string S_P is defined to be a periodic string of length n with period P, i.e., $S_P = P^{\lfloor \frac{n}{p} \rfloor} pref(P)$, where $pref(P)$ is the prefix of P of length $n - \lfloor \frac{n}{p} \rfloor \cdot p$.*

Definition 3. *Let S be n-long string over alphabet Σ. Let d be a metric defined on strings. S is called* periodic with k errors *if there exists a string P over Σ, $p \in \mathbb{N}$, $p \leq n/2$, such that $d(S_P, S) = k$. The string P is called a k-error period of S.*

Definition 4. *Let S be a string of length n over alphabet Σ. A string P over alphabet Σ is called the* approximate period *of S if:*

1) P is a k-error period of S for some $k \geq 1$.
2) for every k'-error period of S, $k' \geq k$.
3) for every P', a k-error period of S, $p' \geq p$.

Definition 5. *Given a string metric d, the* Approximate Period Problem under the metric d, *is the following:*

INPUT: *String S of length n over alphabet Σ.*
OUTPUT: *The approximate period of S under the metric d, P, and k such that P is a k-error period of S under d.*

Definition 6. *Given a string metric d, let P be the string of length p such that $d(S, S_P)$ is minimal over all possible strings of length p. We call P the* approximate-p-length-period *under d. Note that for some p and S, P does not exist.*

Example: Given the string $S = ACB\ BAC\ BAC\ BAC\ BAC\ BAC\ BAC\ BA$. $P = BAC$ is the approximate period under the Hamming distance with $k = 3$ errors. The closest periodic string under the Hamming distance is, therefore, $S_P =$ **BAC** $BAC\ BAC\ BAC\ BAC\ BAC\ BAC\ BA$, $H(S_P, S) = 3$. For length $p = 3$, the approximate-3-length-period under the swap distance is $P = ABC$ with $k = 8$ swap errors $S = ACB$ **BAC BAC BAC BAC BAC BAC BA**. However, the approximate period under the swap distance is $P = ABCBACBAC$ of length 9 with $k = 3$ errors. The closest periodic string under the swap distance is, therefore, $S_P =$ **ABC**$BACBAC$ **ABC**$BACBAC$ **ABC**BA, where $d_{swap}(S_P, S) = 3$.

Remark. Note that the number of errors of the approximate period under swap cannot be obtained by dividing by 2 the number of mismatches between the closest periodic string under the Hamming distance and S. Also, the approximate period under the swap distance may even have a different size than that of the approximate period under the Hamming distance. Thus, the algorithm designed for solving the problem under Hamming distance cannot be used in order to solve the problem under the swap distance, and special algorithms should be designed for the latter problem.

3   The Approximate Period under the Hamming Distance

In this section we show that, given a string S of length n and k the upper bound of error number, it is possible to efficiently find the approximate period under the Hamming distance, such that $H(S_P, S) \leq k$. We begin by giving some basic definitions and tools used in this section.

Definition 7. *Let S be a string of length n over alphabet Σ, and let $p \in \mathbb{N}$, $p \leq n/2$. For every i, $0 \leq i \leq p - 1$, the i-th frequent element with regard to p, denoted F_i^p, is the symbol $\sigma \in \Sigma$ which appears the largest number of times in locations $i + r \cdot p$ in S, where $r \in [0..\lfloor n/p \rfloor]$, and $i + r \cdot p \leq n$.*[1]

The main tool we use is the *kangaroo* method of Landau-Vishkin [14] and Galil-Giancarlo [8]:

Theorem 1. *[Landau-Vishkin 86, Galil-Giancarlo 86] Let S be a string of length n over alphabet Σ. Then S can be preprocessed in time $O(n \log \min\{n, |\Sigma|\})$ allowing subsequent constant-time answers to queries of the form:*

Let S_i, S_j be the suffixes of S starting at locations i and j, respectively, i.e. $S_i = S[i]S[i+1] \cdots S[n]$, and $S_j = S[j]S[j+1] \cdots S[n]$. Return $LCP(S_i, S_j) = \ell$, the length of the longest common prefix of S_i and S_j.

The original implementation of the kangaroo method was via suffix trees [20] and Lowest Common Ancestor (LCA) queries [9]. It can also be implemented

---

[1] For simplicity of exposition we will henceforth refer to positive integers of the form $i + r \cdot p$, $r \in [0..\lfloor n/p \rfloor]$ for some $p \leq n/2$ as locations in S without mentioning explicitly the condition $i + r \cdot p \leq n$. It is always assumed that only valid positions in S are referred to.

using suffix arrays [10] and Longest Common Prefix computations [12]. Using these last constructions, the preprocessing time for alphabet $\Sigma = \{1, ..., n\}$ is linear-time.

Definition 8. *Let S_i and S_j be two suffixes. Call $LCP(S_i, S_j)$ the first kangaroo jump. Assume $LCP(S_i, S_j) = \ell_1$. If $\ell_1 = \min\{|S_i|, |S_j|\}$, we say that there is one possible kangaroo jump between S_i and S_j. Otherwise, call $LCP(S_{i+\ell_1+1}, S_{j+\ell_1+1}) = \ell_2$ the second kangaroo jump. Again, if $\ell_2 = \min\{|S_{i+\ell_1+1}|, |S_{j+\ell_1+1}|\}$, we say that there are two possible kangaroo jumps between S_i and S_j.*

In general, Call $LCP(S_{i+\ell_1+\ell_2+\cdots\ell_{k-1}+k-1}, S_{j+\ell_1+\ell_2+\cdots\ell_{k-1}+k-1}) = \ell_k$ the k-th kangaroo jump. If $\ell_k = \min\{|S_{i+\ell_1+\ell_2+\cdots\ell_{k-1}+k-1}|, |S_{j+\ell_1+\ell_2+\cdots\ell_{k-1}+k-1}|\}$, we say that there are k possible kangaroo jumps between S_i and S_j.

Algorithm's Idea. The algorithm checks for every length p, $1 \leq p \leq n/2$, if there exists an approximate p-length-period with at most k mismatch errors. This is done by allowing at most $2 \cdot k$ LCP jumps, where Lemma 1 assures that this upper bound indeed suffices. Each such jump identifies a mismatch, but also enables to compute the current frequency of the symbols that participate in the mismatch, as explained in the proof of Lemma 2. The final frequencies of the symbols allow to decide what is the approximate p-length period, if it exists (if at most $2 \cdot k$ LCP jumps were enough), by using the majority criterion. Finally, the algorithm chooses the smallest such period with the smallest number of errors. For completeness, a detailed description of the algorithm is given in the full version of the paper.

Analysis. We are now ready for the analysis of the algorithm. Lemma 1 and 2 provide the main basis for the correctness and complexity of the algorithm. Theorem 2, then follows.

Lemma 1. *Let S be a string of length n over alphabet Σ, and let $p \in \mathbb{N}$, $p \leq n/2$. Assume that S is periodic with k errors for a period of length p. Let ℓ be the number of possible jumps between S and S_p. Then $\ell \leq 2 \cdot k$.*

Lemma 2. *Let S be a string of length n over alphabet $\Sigma = \{1, \ldots, n\}$, and let $p \in \mathbb{N}$, $p \leq n/2$. Assume that S is periodic with k errors for a period of length p. Let $P = F_1^p F_2^p \cdots F_p^p$ be the k-error period of S. There are at most k elements of P for which $F_i^p \neq S[i + r \cdot p]$. It is possible to identify these elements of the k-error period of S in time $O(k \log \log n)$.*

Proof. The main idea of the proof is noting that a jump means that all the positions in the period are unchanged until the jump. In particular, it lets us know, for the location in the period where there is a mismatch, how many times the symbol appeared repeatedly within this position – since the last jump at this period location.

Formally, let ℓ_1 be the result of the first jump ($LCP(S_p, S_0)$), ℓ_1 is the first error text location. Let $loc = (\ell_1 \mod p)$, the error location in the period, and assume that $S[\ell_1] = a$ and $S[\ell_1 + p] = b$. This means that $S[loc + r \cdot p] = a, \forall r =$

$0, \ldots, (\lceil \ell_1/p \rceil - 1)$, and $S[loc + \lceil \ell_1/p \rceil \cdot p] = b$. We introduce variable V_a^{loc} and give it value $\lceil \ell_1/p \rceil$. This indicates the number of times the symbol a appears in location $loc + r \cdot p$ so far.

Let ℓ_j be the result of the j-th jump, $((\sum_{i=1}^{j}(\ell_i+1))-1)$ is the j-th error text location. Let $loc = (((\sum_{i=1}^{j}(\ell_i+1))-1) \bmod p)$, the error location in the period, and assume that $S[\sum_{i=1}^{j}(\ell_i + 1) - 1] = a$ and $S[(\sum_{i=1}^{i}(\ell_i + 1)) - 1 + p] = b$. If V_a^{loc} was already defined, then add to its value $\lceil \ell_j/p \rceil$. If V_a^{loc} had not been defined yet, define it and give it value $\lceil \ell_j/p \rceil$.

After the $2k$-th jump, all values $loc \in \{0, 1, \ldots, p-1\}$ that have no defined variables have $F_{loc}^p = S[loc]$. All others have $F_{loc}^p = a$, where a is symbol for which V_a^{loc} is maximal. Because of Lemma 1 the $2k$ jumps are sufficient. The only operation that needs more than constant time is checking whether V_a^{loc} is defined. However, for alphabet $\{1, \ldots, n\}$, a data structure such as van Emde Boas [19] can be used, making the search time $O(\log \log n)$ for a total time of $O(k \log \log n)$.   □

Theorem 2. *Let S be a string of length n. Then the approximate period of S, P, can be found in time $O(nk \log \log n)$, where k is the number of errors in P.*

4   The Approximate Period under the Swap Distance

In this section we present an algorithm that finds the approximate period of the string under the swap distance, if it exists.

Algorithm's Idea. Let S be a string of length n. At the onset of the algorithm, there is no knowledge of the length of the minimum period. Thus, the algorithm actually assumes all possible lengths p, $1 \le p \le \lfloor \frac{n}{2} \rfloor$. For each such length it tries to create a periodic string of period length p and calculate its swap distance from S. The best result, if such a result exists, is chosen and returned by the algorithm. Note that unlike the Hamming distance, where there is always an answer not exceeding $\frac{n}{2}$, with the swap metric there may not be any periodic string whose distance from S is finite. To see this, consider for example, the string $S = ABCABCABD$. Under the Hamming distance, S is periodic with one error, however, under the swap distance there is no periodic string whose distance from S is finite (i.e., S cannot be 'fixed' to be periodic by using swaps). Having fixed period length p, partition S to $\lceil \frac{n}{p} \rceil$ consecutive substrings, $P_1, \ldots, P_{\lceil \frac{n}{p} \rceil}$, each having length p. For completeness, a detailed description of the algorithm is given in the full version of the paper.

Definition 9. *Let P_i, P_{i+1} be two adjacent substrings. A swap between two characters within P_i or P_{i+1} is called an internal swap. A swap between the rightmost character of Pi and the leftmost character of P_{i+1} is called an external swap.*

Suppose that the approximate period exists and its length is p. Let $Hist_p$ be the histogram of the approximate period, i.e., the list of symbols that appear in the period, with the number of times each symbol occurs in the period. For every symbol σ that appear in S, we count the number of times that the symbol appears in S, denoted by $\#_\sigma$. This number is then divided by the number of times that the period appears in S, $\frac{n}{p}$, and the integer value of the result is then taken. We call the resulting number *the frequency of σ in $Hist_p$*, denoted by f_σ^p. If there exists an approximate period of length p, then $\sum_\sigma f_\sigma^p = p$. Note that if $\sum_\sigma f_\sigma^p \neq p$, there is no approximate period of length p. For example, if $S = ABC\ ABC\ ABC\ ABC$ there is no approximate period for length $p = 2$ or $p = 4$.

Lemma 3. *Let S be a string of length n. Let p be a period length, $1 \leq p \leq \lfloor \frac{n}{2} \rfloor$, such that string S_P generated by a period of length p has a finite swap distance to S. Then $Hist_p$ can be constructed from S in time $O(n)$.*

Example: Take $S = BACDEFACBDEFBACDEAFCBEDFAB$, as a running example. For $p = 6$, the split to substrings gives:
$S = BACDEF\ ACBDEF\ BACDEA\ FCBEDF\ AB.\ Hist_6 = \{B, A, C, D, E, F\}$

The algorithm performs all external swaps, so that the histogram of each period appearance will equal $Hist_p$. The external swaps are done sequentially from left to right. We begin with the leftmost substring P_{i_0} whose histogram does not equal $Hist_p$. If a swap of its rightmost element with the leftmost element of P_{i_0+1} adjusts its histogram to equal $Hist_p$, and proceed to substring P_{i_0+1}. If such a swap does not fix the histogram of P_i, then there is no period of length P for which $d_{swap}(S, S_P) < \infty$. When the process completes, either there is no approximate period, or all substrings have the same histogram, $Hist_p$. In our example $S' = BACDEF\ ACBDEF\ BACDE\mathbf{F}\ \mathbf{A}CBEDF\ AB$.

We now need to construct the approximate period. This is done via internal swaps since each substring now has the appropriate histogram. For ease of exposition of the intuition behind the algorithm, we will now assume that all characters of the approximate period P are distinct. In Lemma 4 we prove that this assumption does not detract from the generality of the algorithm.

We construct the approximate period as follows. Every character in the period (and they are assumed to be distinct) appears in a certain location in each of the substrings P_i. For each symbol a, consider the set of indices Loc_a in which it appears in the various substrings. If Loc_a has two elements ℓ, j for which $|\ell - j| > 2$, it immediately implies that there is no swap match. The reason is that a swap can only change the position of a character by one, a character whose position in the period P is j can only appear in locations $j - 1, j$, and $j + 1$ of each of the P_i's. We can assume, then, that our sets Loc_a all have either one, two, or three elements. The elements in sets of size 2 are either adjacent or of the form $j - 1, j + 1$. The elements of sets of size three are of the form $j - 1, j, j + 1$.

Our algorithm constructs P as follows. For a character a that appears in all the substrings in the same location j, the algorithm sets $P[j]$ to have value a. This follows from the desire to minimize the number of swaps, because there is no point is swapping all a's in all substrings with an element b that must also be in the same location in all substrings. In our example, $P[5]$ is set to F. When

$Loc_a = \{j-1, j+1\}$ or when $Loc_a = \{j-1, j, j+1\}$, the algorithm sets $P[j]$ to value a. In our example, B appears in locations $Loc_B = \{0, 1, 2\}$ therefore $P[1]$ gets value B.

At this point, the only sets left are those that have adjacent locations. In our example, $Loc_A = \{0, 1\}$, $Loc_C = \{1, 2\}$, $Loc_D = \{3, 4\}$, $Loc_E = \{3, 4\}$. For every such character a, if $Loc_a = \{j, j+1\}$ and if $P[j]$ (or, symmetrically, $P[j+1]$) is already set to another character, then the algorithm sets $P[j+1]$ (or, $P[j]$, in the corresponding case) to the value a. In our example, $P[0]$ is set to A and $P[2]$ is set to C, because $P[1]$ already has the value B.

Now we are left with pairs of equal sets. In our example, Loc_D and Loc_E are such a pair because $Loc_D = Loc_E = \{3, 4\}$. Since both locations in P are not set, the algorithm sets the characters in this locations according to majority criterion, to minimize the number of swap errors. In our example, $P[3] = D$, $P[4] = E$.

Analysis. We are now ready to prove the correctness of the algorithm. We begin by justifying the assumption that all characters of the period are distinct, is valid. This is done in Lemma 4. For this purpose we give formal definitions to the appropriate problems in Definitions 10 and 11.

Definition 10. *The* Approximate-p-Length-Period Problem *is the following:*
INPUT: String S of length n over alphabet Σ, period length p.
OUTPUT: A string P of length p for which the value $d_{swap}(S, S_P)$ is smallest. Notify if there is no P with a finite $d_{swap}(S, S_P)$.

Definition 11. *The* Approximate-p-Length-Period Problem with distinct characters *is the following:*
INPUT: String S of length n over alphabet Σ, where $|\Sigma| = p$.
OUTPUT: The approximate-p-length-period of S.

Lemma 4. *The approximate-p-length-period problem is reducible in linear time to the approximate period problem with distinct characters.*

The following next lemmas completes the proof of the algorithm's correctness. Theorem 3 then follows.

Lemma 5. *Let S be a string of length n. Let p be a period length, $1 \le p \le \lfloor \frac{n}{2} \rfloor$, such that swap-approximate-p-length-period exists in S. Every swap that the External Swap Procedure preforms is necessary.*

Lemma 6. *If the histogram of substring P_i is $Hist_p$, then P_i should not initiate an external swap.*

Lemma 7. *Given an input string S, the string S_P defined for the period P, returned by Algorithm Swap-k-Error-Period, has a swap match with S.*

Lemma 8. *Let S be a string. If an approximate period P with k swap errors exists in S, then given the input S algorithm Swap-k-Error-Period returns P and k.*

Theorem 3. *Let S be a string of length n. The approximate period of S, P, can be found in time $O(n^2)$.*

5   Approximating the Hamming Error Bound

Theorem 2 enables determining the approximate period as well as the error bound k of the number of mismatch errors in S. However, the time complexity depends on both n and k. Therefore, if S has many errors, it does not give a fast way of knowing this. In this section, we show that the number of mismatches in S can be approximated to a constant in $\tilde{O}(n)$ time, no matter what the value of k is. To this end, we use another tool – the *self-convolution vector*.

Definition 12. *Let S be a string of length n over alphabet Σ, and let $\bar{S}$ be the string S concatenated with n \$'s (where \$ $\notin \Sigma$). The* self-convolution vector *of S, v, is defined for every i, $0 \le i \le n/2 - 1$, $v[i] = \sum_{j=0}^{n-1} f(\bar{S}[i+j], S[j])$, where*

$$f(\bar{S}[i+j], S[j]) = \begin{cases} 1, \text{ if } \bar{S}[i+j] \ne S[j] \text{ and } \bar{S}[i+j] \ne \$; \\ 0, \text{ otherwise.} \end{cases}$$

Lemma 9 follows from standard FFT techniques. Lemmas 10 and 11 give the background for Theorems 4 and 5.

Lemma 9. [6] *The self-convolution vector of a length n string S over alphabet Σ can be computed in time $O(|\Sigma|n \log n)$.*

Lemma 10. *Let S be a string of length n with P a k-error period of S. Let $a \in \mathbb{N}$, $a \le \frac{n}{2p}$ and let P^a be a length $a \cdot p$ d-error period of S, then $d \le k$.*

Corollary 1. *Let S be a string of length n with P the approximate period of S, and let k be such that P is a k-error period of S. Then there exists P' a k-error period of S such that $n/4 < p' \le n/2$.*

Lemma 11. *Let S be a string of length n with P the approximate period of S, and let k be such that P is a k-error period of S. Let min be the minimum value in the range $[n/4 + 1, n/2]$ of the self-convolution vector of S. Then $\frac{1}{2} \cdot min \le k \le 2 \cdot min$.*

Lemma 11 suggests a very simple algorithm for approximating the number of errors in the approximate period of S: Compute v, the self-convolution vector of S in time $O(|\Sigma|n \log n)$ (by Lemma 9) and then in $O(n)$ time find the minimum value of v in the range $[n/4 + 1, n/2]$. Theorem 4 follows.

Theorem 4. *Let S be a string of length n over alphabet Σ. Then the minimum number of errors in the approximate period of S can be approximated to a factor of 2 in time $O(|\Sigma|n \log n)$.*

Theorem 4 gives a fast approximation for bounded alphabet. For unbounded alphabet, we do not compute the self-convolution vector exactly but rather an *approximated self-convolution vector*. Note that, the self-convolution vector of a string S gives at every location j the *Hamming distance* between the suffix of S starting at j and the prefix of S that ends at $j-1$. Thus, Karloff's algorithm [11] for approximating the Hamming distance between a pattern and a text can be used to compute an approximated self-convolution vector, by taking the text to be $\bar{S}$ (i.e. S concatenated with n \$ signs) and the pattern to be S. The resulting vector can be corrected in linear time by subtracting from each position the number of \$'s that are encountered in the alignment with S. Karloff's algorithm runs in time $O(\frac{1}{\epsilon} \cdot \log^3 n)$ and gives an approximation of $1 \pm \epsilon$ to the Hamming distance. Combining this with Lemma 11 we get Theorem 5.

Theorem 5. *Let S be a n-long string. Then for every $\epsilon > 0$, the number of errors in the approximate period P of S can be approximated to a factor of $2(1 \pm \epsilon)$ in time $O(\frac{1}{\epsilon} \cdot n \log^3 n)$.*

References

1. Amir, A., Benson, G.: Two-dimensional periodicity and its application. SIAM J. Comp. 27(1), 90–106 (1998)
2. Amir, A., Benson, G., Farach, M.: Optimal parallel two dimensional text searching on a crew pram. Information and Computation 144(1), 1–17 (1998)
3. Apostolico, A., Giancarlo, R.: Periodicity and repetitions in parameterized strings. Discrete Appl. Math. 156(9), 1389–1398 (2008)
4. Cole, R., Crochemore, M., Galil, Z., Gąsieniec, L., Harihan, R., Muthukrishnan, S., Park, K., Rytter, W.: Optimally fast parallel algorithms for preprocessing and pattern matching in one and two dimensions. In: Proc. 34th IEEE FOCS, pp. 248–258 (1993)
5. Crochemore, M.: An optimal algorithm for computing the repetitions in a word. Information Processing Letters 12(5), 244–250 (1981)
6. Fischer, M.J., Paterson, M.S.: String matching and other products, Complexity of Computation. In: Karp, R.M. (ed.) Complexity of Computation. SIAM-AMS Proceedings, vol. 7, pp. 113–125 (1974)
7. Galil, Z.: Optimal parallel algorithms for string matching. In: Proc. 16th ACM Symposium on Theory of Computing, vol. 67, pp. 144–157 (1984)
8. Galil, Z., Giancarlo, R.: Improved string matching with k mismatches. SIGACT News 17(4), 52–54 (1986)
9. Harel, D., Tarjan, R.E.: Fast algorithms for finding nearest common ancestor. Computer and System Science 13, 338–355 (1984)
10. Kärkkäinen, J., Sanders, P.: Simple linear work suffix array construction. In: Baeten, J.C.M., Lenstra, J.K., Parrow, J., Woeginger, G.J. (eds.) ICALP 2003. LNCS, vol. 2719, pp. 943–955. Springer, Heidelberg (2003)
11. Karloff, H.: Fast algorithms for approximately counting mismatches. Information Processing Letters 48(2), 53–60 (1993)
12. Kasai, T., Lee, G., Arimura, H., Arikawa, S., Park, K.: Linear-time longest-common-prefix computation in suffix arrays and its applications. In: Amir, A., Landau, G.M. (eds.) CPM 2001. LNCS, vol. 2089, pp. 181–192. Springer, Heidelberg (2001)
13. Kolpakov, R.M., Kucherov, G.: Finding Approximate Repetitions under Hamming Distance. In: Meyer auf der Heide, F. (ed.) ESA 2001. LNCS, vol. 2161, pp. 170–181. Springer, Heidelberg (2001)
14. Landau, G.M., Vishkin, U.: Efficient string matching with k mismatches. Theoretical Computer Science 43, 239–249 (1986)
15. Landau, G.M., Schmidt, J.P., Sokol, D.: An algorithm for approximate tandem repeats. Journal of Computational Biology 8(1), 1–18 (2001)

16. Lothaire, M.: Combinatorics on words. Addison-Wesley, Reading (1983)
17. Main, M.G., Lorentz, R.J.: An $o(n \log n)$ algorithm for finding all repetitions in a string. Journal of Algorithms 5, 422–432 (1984)
18. Sim, J.S., Park, K., Iliopoulos, C.S., Smyth, W.F.: Approximate periods of strings. In: Crochemore, M., Paterson, M. (eds.) CPM 1999. LNCS, vol. 1645, pp. 123–133. Springer, Heidelberg (1999)
19. van Emde Boas, P., Kaas, R., Zijlstra, E.: Design and implementation of an efficient priority queue. Mathematical systems Theory 10, 99–127 (1977)
20. Weiner, P.: Linear pattern matching algorithm. In: Proc. 14 IEEE Symposium on Switching and Automata Theory, pp. 1–11 (1973)

Approximating the Average Stretch Factor of Geometric Graphs*

Siu-Wing Cheng[1], Christian Knauer[2], Stefan Langerman[3,**], and Michiel Smid[4]

[1] Department of Computer Science and Engineering, HKUST, Hong Kong
[2] Institute of Computer Science, Universität Bayreuth
[3] Département d'Informatique, Université Libre de Bruxelles
[4] School of Computer Science, Carleton University, Ottawa

Abstract. Let G be a geometric graph whose vertex set S is a set of n points in $\mathbb{R}^d$. The stretch factor of two distinct points p and q in S is the ratio of their shortest-path distance in G and their Euclidean distance. We consider the problem of approximating the sum of all $\binom{n}{2}$ stretch factors determined by all pairs of points in S. We show that for paths, cycles, and trees, this sum can be approximated, within a factor of $1 + \epsilon$, in $O(n\, polylog(n))$ time. For plane graphs, we present a $(2+\epsilon)$-approximation algorithm with running time $O(n^{5/3} polylog(n))$, and a $(4 + \epsilon)$-approximation algorithm with running time $O(n^{3/2} polylog(n))$.

1 Introduction

Let S be a set of n points in $\mathbb{R}^d$ and let G be a connected graph with vertex set S in which the weight of any edge (p, q) is equal to the Euclidean distance $|pq|$ between p and q. The length of a path in G is defined to be the sum of the weights of the edges on the path. For any two points p and q of S, we denote by $|pq|_G$ the minimum length of any path in G between p and q. If $p \neq q$, then the *stretch factor* of p and q is defined to be $|pq|_G/|pq|$. If $t \geq 1$ is a real number such that each pair of distinct points in S has stretch factor at most t, then we say that G is a *t-spanner* of S. The smallest value of t such that G is a t-spanner of S is called the *stretch factor* of G.

The problem of computing, given any set S of points in $\mathbb{R}^d$ and any $t > 1$, a t-spanner of S, has been well-studied; see the book by Narasimhan and Smid [8].

For the related problem of computing, or approximating, the stretch factor of a given geometric graph, much less is known. Narasimhan and Smid [7] show that the problem of approximating the stretch factor of any geometric graph on n vertices can be reduced to performing approximate shortest-path queries for $O(n)$ pairs of points. Agarwal *et al.* [1] show that the exact stretch factor of a geometric path, tree, and cycle on n points in the plane can be computed in $O(n \log n)$, $O(n \log^2 n)$, and $O(n\sqrt{n} \log n)$ expected time, respectively. They also present algorithms for the three-dimensional versions of these problems. Klein *et al.* [6] consider the problem of reporting all pairs of vertices

---
* Research of Cheng was supported by Research Grant Council, Hong Kong, China (project no. 612107). Research of Smid was supported by NSERC.
** Maître de Recherches du F.R.S.-FNRS.

O. Cheong, K.-Y. Chwa, and K. Park (Eds.): ISAAC 2010, Part I, LNCS 6506, pp. 37–48, 2010.

whose stretch factor is at least some given value t; they present efficient algorithms for the cases when the input graph is a geometric path, tree, or cycle.

Given a method to compute the stretch factor of a graph, a natural question is whether the graph connectivity can be adjusted to lower the stretch factor. For instance, this would be helpful in reducing the maximum commute time in a road network and related problems have been considered (e.g. [4]). However, the stretch factor can be high just because the stretch factors of a few pairs of points are high while the stretch factors of the other pairs are low. A more robust measure is the *average* stretch factor which we define as follows. Let $SSF(G)$ denote the sum of all stretch factors, i.e.,

$$SSF(G) = \sum_{\{p,q\} \in \mathcal{P}_2(S)} \frac{|pq|_G}{|pq|},$$

where $\mathcal{P}_2(S)$ denotes the set of all $\binom{n}{2}$ unordered pairs of distinct elements in S. The value $SSF(G)/\binom{n}{2}$ is equal to the *average* stretch factor of the graph G.

To the best of our knowledge, even for a simple graph G such as a path, it is not known if $SSF(G)$ can be computed in $o(n^2)$ time. We remark that Wulff-Nilsen shows in [9] that the related problem of computing the Wiener index (i.e., $\sum_{\{p,q\} \in \mathcal{P}_2(S)} |pq|_G$) of an unweighted planar graph can be solved in $O(n^2 \log \log n / \log n)$ time.

In this paper, we consider the problem of approximating $SSF(G)$. We start in Section 2 by showing that, not surprisingly, the well-separated pair decomposition (WSPD) of Callahan and Kosaraju [3] can be used to approximate $SSF(G)$. In Section 3, we apply this general approach to compute a $(1+\epsilon)$-approximation to $SSF(G)$ in $O(n \log^2 n)$ time, for the cases when G is a path or a cycle. In Section 4, we modify the general approach of Section 2 and show how to compute a $(1 + \epsilon)$-approximation to $SSF(G)$ in $O(n \log^2 n / \log \log n)$ time for the case when G is a tree. Finally, in Section 5, we consider plane graphs. We further modify the general approach of Section 2 and obtain a $(2+\epsilon)$-approximation to $SSF(G)$ in $O((n \log n)^{5/3})$ time, and a $(4+\epsilon)$-approximation in $O(n^{3/2} \log^2 n)$ time.

2   The General Approach Using Well-Separated Pairs

Let $s > 0$ be a real number, called the *separation ratio*. We say that two point sets A and B in $\mathbb{R}^d$ are *well-separated* with respect to s, if there exist two balls, one containing A and the other containing B, of the same radius, say ρ, which are at least $s\rho$ apart. If A and B are well-separated, a and a' are points in A, and b and b' are points in B, then it is easy to verify that

$$|ab| \leq (1 + 4/s)|a'b'|. \tag{1}$$

Let S be a set of n points in $\mathbb{R}^d$. A *well-separated pair decomposition (WSPD)* of S is a sequence $\{A_1, B_1\}, \ldots, \{A_m, B_m\}$ of well-separated pairs of subsets of S, such that, for any two distinct points p and q in S, there is a unique index i such that $p \in A_i$ and $q \in B_i$ or $p \in B_i$ and $q \in A_i$.

Callahan and Kosaraju [3] have shown that a WSPD can be obtained from the *split-tree* $T(S)$ of the point set S. This tree is defined as follows: If $n = 1$, then $T(S)$

consists of one single node storing the only element of S. Assume that $n \geq 2$. Consider the bounding box $\mathcal{B}$ of S. By splitting the longest edge of $\mathcal{B}$ into two parts of equal size, we obtain two boxes $\mathcal{B}_1$ and $\mathcal{B}_2$. The split tree $T(S)$ consists of a root having two subtrees, which are recursively defined split trees $T(S_1)$ and $T(S_2)$ for the point sets $S_1 = S \cap \mathcal{B}_1$ and $S_2 = S \cap \mathcal{B}_2$, respectively.

Given a separation ratio $s > 0$, the split tree $T(S)$ can be used to compute a WSPD of S, where each subset A_i (and each subset B_i) corresponds to a node v of the split-tree: A_i equals the set S_v of all points that are stored at the leaves of the subtree rooted at v.

Theorem 1 (Callahan and Kosaraju [3]). *Let S be a set of n points in $\mathbb{R}^d$ and let $s > 0$ be a real constant. In $O(n \log n)$ time, the split tree $T(S)$ and a corresponding WSPD $\{A_1, B_1\}, \ldots, \{A_m, B_m\}$ of S can be computed, such that $m = O(n)$ and $\sum_{i=1}^{m} \min(|A_i|, |B_i|) = O(n \log n)$.*

If G is a connected graph with vertex set S, then

$$SSF(G) = \sum_{i=1}^{m} \sum_{p \in A_i, q \in B_i} \frac{|pq|_G}{|pq|}.$$

Let $s = 4/\epsilon$. By (1), all distances $|pq|$, where $p \in A_i$ and $q \in B_i$, are within a factor of $1 + \epsilon$ of each other. For each i, choose an arbitrary point x_i in A_i and an arbitrary point y_i in B_i, and consider the summation

$$SSF'(G) = \sum_{i=1}^{m} \frac{1}{|x_i y_i|} \sum_{p \in A_i, q \in B_i} |pq|_G.$$

Then $1/(1 + \epsilon) \leq SSF'(G)/SSF(G) \leq 1 + \epsilon$. In order to compute $SSF'(G)$, we need to compute the values

$$\sum_{p \in A_i, q \in B_i} |pq|_G. \tag{2}$$

3 Paths and Cycles

Assume that the graph G is a path $(p_1, p_2, \ldots, p_n)$ on the points of the set S. For two indices i and j with $1 \leq i < j \leq n$, we say that p_i is to the *left* of p_j in G, and p_j is to the *right* of p_i in G.

Before we present the algorithm that approximates $SSF(G)$, we describe the main idea. Consider a pair $\{A_i, B_i\}$ of the WSPD. Let p be an arbitrary point in A_i, let $b_1, \ldots, b_k$ be the points in B_i that are to the left of p in G, and let $b'_1, \ldots, b'_{k'}$ be the points in B_i that are to the right of p in G. Then

$$\sum_{q \in B_i} |pq|_G = (k - k')|p_1 p|_G + \sum_{j=1}^{k'} |p_1 b'_j|_G - \sum_{j=1}^{k} |p_1 b_j|_G. \tag{3}$$

Let v be the node in the split tree such that $B_i = S_v$, i.e., B_i is the subset of S that is stored in the subtree rooted at v. Assume that we have a balanced binary search tree $\mathcal{T}_v$

storing the points of S_v at its leaves, sorted according to their indices in the path G. Also assume that each node u of this tree stores (i) the number of points stored in the subtree of u and (ii) the sum of the path lengths $|p_1 q|_G$, where q ranges over all points stored in the subtree of u. Then by searching in $\mathcal{T}_v$ for p, we obtain, in $O(\log |B_i|) = O(\log n)$ time, (i) a partition, into $O(\log n)$ canonical subsets, of all points in B_i that are to the left of p in G, and (ii) a partition, into $O(\log n)$ canonical subsets, of all points in B_i that are to the right of p in G. From the information stored at the canonical nodes, we can compute the summation in (3) in $O(\log n)$ time.

Based on this discussion, we obtain the following algorithm.

Step 1: Compute the split tree $T(S)$ and the corresponding WSPD $\{A_1, B_1\}, \ldots,$ $\{A_m, B_m\}$ of Theorem 1, with separation ratio $s = 4/\epsilon$. Assume that $|A_i| \leq |B_i|$ for all $1 \leq i \leq m$.

Step 2: Traverse the path G and store with each point p_i $(1 \leq i \leq n)$ the path length $|p_1 p_i|_G$.

Step 3: Traverse the split tree $T(S)$ in post-order, maintaining the following invariant: After having just visited node v, this node contains a pointer to the above data structure $\mathcal{T}_v$ storing the set S_v. Let v be the node of $T(S)$ that is currently visited.

1. If v is a leaf of $T(S)$, then initialize $\mathcal{T}_v$ such that it contains only the point stored at v. Otherwise, let v_1 and v_2 be the two children of v. If the size of $\mathcal{T}_{v_1}$ is at most that of $\mathcal{T}_{v_2}$, then insert all elements of $\mathcal{T}_{v_1}$ into $\mathcal{T}_{v_2}$, discard $\mathcal{T}_{v_1}$, and rename $\mathcal{T}_{v_2}$ as $\mathcal{T}_v$. Otherwise, insert all elements of $\mathcal{T}_{v_2}$ into $\mathcal{T}_{v_1}$, discard $\mathcal{T}_{v_2}$, and rename $\mathcal{T}_{v_1}$ as $\mathcal{T}_v$.
2. For each pair $\{A_i, B_i\}$ in the WSPD for which $B_i = S_v$, do the following: Let w be the node of the split tree such that $A_i = S_w$. Traverse the subtree rooted at w and for each point p stored in this subtree, use $\mathcal{T}_v$ to compute the value in (3). The sum of all these values (over all p in A_i) gives the summation in (2).

Theorem 2. *Let G be a path on n points in $\mathbb{R}^d$ and let $\epsilon > 0$ be a real constant. In $O(n \log^2 n)$ time, we can compute a real number that lies between $SSF(G)/(1 + \epsilon)$ and $(1 + \epsilon)SSF(G)$.*

By using a slight modification of the above algorithm, we can prove the following result:

Theorem 3. *Let G be a cycle on n points in $\mathbb{R}^d$ and let $\epsilon > 0$ be a real constant. In $O(n \log^2 n)$ time, we can compute a real number that lies between $SSF(G)/(1 + \epsilon)$ and $(1 + \epsilon)SSF(G)$.*

4   Trees

Let S be a set of n points in $\mathbb{R}^d$ and let G be a spanning tree of S. Assume that $n \geq 3$. Let c be a *centroid* of G, i.e., c is a node whose removal from G (together with its incident edges) results in two forests G'_1 and G'_2, each one having size at most $2n/3$. It is well known that such a centroid always exists and can be computed in $O(n)$ time. Let G_1 be the tree obtained by adding c to G'_1, together with the edges of G between c and G'_1. Define G_2 similarly with respect to G'_2. We have

$$SSF(G) = SSF(G_1) + SSF(G_2) + \sum_{p \in G_1'} \sum_{q \in G_2'} \frac{|pq|_G}{|pq|}. \qquad (4)$$

We will show that the summation in (4) can be approximated in $O(n \log n)$ time. Therefore, by recursively approximating the values $SSF(G_1)$ and $SSF(G_2)$, we obtain an approximation of $SSF(G)$ in $O(n \log^2 n)$ time.

We color each point of G_1' red and each point of G_2' blue. The centroid c does not get a color. Consider the split tree $T(S)$ and the corresponding WSPD of Theorem 1, where $s = 4/\epsilon$. For each i, let A_i^r and A_i^b be the set of red and blue points in A_i, respectively, and let B_i^r and B_i^b be the set of red and blue points in B_i, respectively. Then (4) is equal to

$$\sum_{i=1}^{m} \left(\sum_{p \in A_i^r} \sum_{q \in B_i^b} \frac{|pq|_G}{|pq|} + \sum_{p \in B_i^r} \sum_{q \in A_i^b} \frac{|pq|_G}{|pq|} \right).$$

For each i, fix $x_i \in A_i$ and $y_i \in B_i$. Then

$$\sum_{i=1}^{m} \left(\frac{1}{|x_i y_i|} \left(\sum_{p \in A_i^r} \sum_{q \in B_i^b} |pq|_G + \sum_{p \in B_i^r} \sum_{q \in A_i^b} |pq|_G \right) \right)$$

approximates the summation in (4) within a factor of $1 + \epsilon$. Observe that

$$\sum_{p \in A_i^r} \sum_{q \in B_i^b} |pq|_G = |B_i^b| \sum_{p \in A_i^r} |pc|_G + |A_i^r| \sum_{q \in B_i^b} |cq|_G$$

and

$$\sum_{p \in B_i^r} \sum_{q \in A_i^b} |pq|_G = |A_i^b| \sum_{p \in B_i^r} |pc|_G + |B_i^r| \sum_{q \in A_i^b} |cq|_G.$$

This leads to the following algorithm for approximating the summation in (4):

Traverse the tree G in postorder (assuming it is rooted at the centroid c) and store with each point p the path length $|pc|_G$.

Traverse the split tree $T(S)$ in postorder and store with each node v the number of red points in S_v and the number of blue points in S_v.

For each leaf v of the split tree $T(S)$, do the following: Let p be the point stored at v. If p is red, then set $redsum(v) = |pc|_G$ and $bluesum(v) = 0$. If p is blue, then set $redsum(v) = 0$ and $bluesum(v) = |pc|_G$. If p is the centroid, then set $redsum(v) = 0$ and $bluesum(v) = 0$.

Traverse the split tree $T(S)$ in postorder. For each internal v, with children v_1 and v_2, set $redsum(v) = redsum(v_1) + redsum(v_2)$ and $bluesum(v) = bluesum(v_1) + bluesum(v_2)$.

Consider a pair $\{A_i, B_i\}$ in the WSPD, and let v and w be the nodes in the split tree such that $A_i = S_v$ and $B_i = S_w$. Node v stores the values $|A_i^r|$ and $|A_i^b|$. Also, the values of $redsum(v)$ and $bluesum(v)$ are equal to $\sum_{p \in A_i^r} |pc|_G$ and $\sum_{q \in A_i^b} |cq|_G$, respectively. Similarly, from the information stored at w, we obtain the values of $|B_i^r|$, $|B_i^b|$, $\sum_{p \in B_i^r} |pc|_G$, and $\sum_{q \in B_i^b} |cq|_G$.

Theorem 4. *Let G be a tree on n points in $\mathbb{R}^d$ and let $\epsilon > 0$ be a real constant. In $O(n \log^2 n)$ time, we can compute a real number that lies between $SSF(G)/(1 + \epsilon)$ and $(1 + \epsilon)SSF(G)$.*

We now show how the running time can be improved by a doubly-logarithmic factor. Consider the recursion tree of the above divide-and-conquer algorithm, and consider a node in this tree. Let S' be the set of points in S that are involved in the call at this node, and let n' be the size of S'. The total time spent at this node is equal to the sum of (i) $O(n' \log n')$, which is the time to compute the split tree and the WSPD of S', and (ii) $O(n')$, which is the time for the rest of the algorithm at this node of the recursion tree. Assume that, at this node, we do not compute the split tree and the WSPD of S', but use the split tree and the WSPD for the entire point set S. Consider a centroid c' of the subtree of G that corresponds to S'. This centroid splits the set S' into two subsets, which we color red and blue, whereas the centroid c' does not get a color. Also, no point of $S \setminus S'$ gets a color. Now we can use the split tree $T(S)$ to compute an approximation of the summation in (4) in $O(n)$ time.

Let h be a positive integer such that $h = O(\log n)$. By using the split tree $T(S)$ and the corresponding WSPD of the entire set S at the levels $0, 1, \dots, h - 1$ of the recursion tree, the total time spent at these levels is $O(n \log n + 2^h n)$. At each node at level h of the recursion tree, we compute the split tree and the WSPD for the points involved in the recursive call at this node. In this way, the total time of our algorithm is $O((n \log n + 2^h n)\frac{\log n}{h})$. For $h = \log \log n$, this gives the following result:

Theorem 5. *Let G be a tree on n points in $\mathbb{R}^d$ and let $\epsilon > 0$ be a real constant. In $O(n \log^2 n / \log \log n)$ time, we can compute a real number that lies between $SSF(G)/(1 + \epsilon)$ and $(1 + \epsilon)SSF(G)$.*

5   Plane Graphs

Let G be a plane connected graph whose vertex set is a set S of n points in $\mathbb{R}^d$, and let C be a *separator* of G. That is, C is a subset of the point set S, such that the following is true: By removing the points of C (together with their incident edges) from G, we obtain two graphs, with vertex sets, say, A and B, such that G does not contain any edge joining some point of A with some point of B.

For any point p in $S \setminus C$, let p' be a point of C for which $|pp'|_G$ is minimum. The following lemma appears in Arikati *et al.* [2].

Lemma 1. *Let p be a point in A, let q be a point in B, and assume that $|pp'|_G \leq |qq'|_G$. Then*

$$|pp'|_G + |p'q|_G \leq 2|pq|_G.$$

The following notions were introduced by Frederickson [5]. A *division* of G is a sequence $R_1, \dots, R_k$ of subsets of S (called *regions*), for some $k \geq 1$, such that $\cup_{i=1}^k R_i = S$ and for each i and each p in R_i,

1. either p is an *interior* point of R_i, i.e., (i) p is not contained in any other subset in the sequence and (ii) for every edge (p, q) in G, the point q also belongs to R_i
2. or p is a *boundary* point of R_i, i.e., p is contained in at least one other subset in the sequence.

A division $R_1, \ldots, R_k$ is called an *r-division*, if $k = O(n/r)$ and each region R_i contains at most r points and $O(\sqrt{r})$ boundary points. Frederickson [5] has shown that such an *r*-division can be computed in $O(n \log n)$ time. The total number of boundary points in an *r*-division is $k \cdot O(\sqrt{r}) = O(n/\sqrt{r})$. Also, for any i, the boundary points of R_i form a separator of the graph G.

5.1 Approximating $SSF(G)$ within $2 + \epsilon$

Consider an *r*-division $R_1, \ldots, R_k$ of G. It partitions the pairs in $\mathcal{P}_2(S)$ into two groups: The pair $\{p, q\}$ is of *type 1*, if p and q belong to the same region; otherwise, $\{p, q\}$ is of *type 2*. For $j \in \{1, 2\}$, we define

$$SSF_j(G) = \sum_{\{p,q\} \text{ of type } j} \frac{|pq|_G}{|pq|}.$$

Then $SSF(G) = SSF_1(G) + SSF_2(G)$. We start by showing how a 2-approximation of $SSF_1(G)$ can be computed. Using the algorithm of [2], we preprocess the graph G in $O(n^{3/2})$ time, after which, for any two points p and q, a 2-approximation of $|pq|_G$ can be computed in $O(\log n)$ time. Since the total number of pairs of type 1 is at most $kr^2 = O(rn)$, this leads to a 2-approximation of $SSF_1(G)$ in $O(n^{3/2} + rn \log n)$ time. In the rest of this section, we will show how to compute a $(2 + \epsilon)$-approximation of $SSF_2(G)$ in time

$$O\left(\frac{n^2 \log^2 n}{\sqrt{r}}\right). \tag{5}$$

By choosing $r = (n \log n)^{2/3}$, we obtain the following result:

Theorem 6. *Let G be a plane graph on n points in $\mathbb{R}^d$ and let $\epsilon > 0$ be a real constant. In $O((n \log n)^{5/3})$ time, we can compute a real number that lies between $SSF(G)/(2 + \epsilon)$ and $(2 + \epsilon)SSF(G)$.*

A $(2+\epsilon)$-approximation of $SSF_2(G)$ is obtained in the following way. For each boundary point p, we run Dijkstra's shortest-path algorithm with source p. In this way, we obtain the shortest-path lengths for all pairs p, q of points, where p ranges over all boundary points and q ranges over all points of S. We store the values $|pq|_G$ in a table so that we can access any one of them in $O(1)$ time. This part of the algorithm takes $O((n^2 \log n)/\sqrt{r})$ time, which is within the time bound in (5).

Nest, we compute the split tree $T(S)$ and the corresponding WSPD $\{A_1, B_1\}, \ldots,$ $\{A_m, B_m\}$ of Theorem 1, with separation ratio $s = 8/\epsilon$. Assume that $|A_i| \leq |B_i|$ for all $1 \leq i \leq m$.

We repeat the following for each region R in the *r*-division $R_1, \ldots, R_k$. We color each point of R *red* and color each point of $S \setminus R$ *blue*. For each i with $1 \leq i \leq m$, let A_i^r and A_i^b be the set of red and blue points in A_i, respectively, and let B_i^r and B_i^b be the set of red and blue points in B_i, respectively.

As in Section 4, in order to obtain a $(2+\epsilon)$-approximation of $SSF_2(G)$, it is sufficient to compute a 2-approximation of the values

$$\sum_{p \in A_i^r} \sum_{q \in B_i^b} |pq|_G \text{ and } \sum_{p \in B_i^r} \sum_{q \in A_i^b} |pq|_G, \tag{6}$$

for all i with $1 \leq i \leq m$. We will show how a 2-approximation of the first summation in (6) can be computed. The second summation can be approximated in a symmetric way.

Let $b_1, \ldots, b_\ell$ be the boundary points of the region R. Recall that $\ell = O(\sqrt{r})$. For each point p of S, let p' be a point in $\{b_1, \ldots, b_\ell\}$ for which $|pp'|_G$ is minimum. Observe that, since we have run Dijkstra's algorithm from every boundary point, each point p "knows" the closest boundary point p'.

Consider a pair $\{A_i, B_i\}$ in the WSPD, let v be the node in the split tree $T(S)$ such that $B_i = S_v$, and let S_v^b be the set of blue points in S_v. Assume that v stores the following information:

1. Balanced binary search trees $\mathcal{T}_{v,j}$ for $1 \leq j \leq \ell$. Each such tree $\mathcal{T}_{v,j}$ stores the points q of S_v^b at its leaves, sorted according to the values $|qq'|_G$. Moreover, each node u in this tree stores
   (a) the number of leaves in the subtree of u and
   (b) the sum of the values $|qb_j|_G$, where q ranges over all points in the substree of u.
2. Balanced binary search trees $\mathcal{T}'_{v,j}$ for $1 \leq j \leq \ell$. Each such tree $\mathcal{T}'_{v,j}$ stores the points q of $\{q \in S_v^b : q' = b_j\}$ at its leaves, sorted according to the values $|qq'|_G$. Each node u in this tree stores
   (a) the number of leaves in the subtree of u and
   (b) the sum of the values $|qb_j|_G$, where q ranges over all points in the substree of u.

Let us see how these trees can be used to obtain a 2-approximation of the summation in (6). Let p be a point in A_i^r and let q be a point in B_i^b.

1. Assume that $|pp'|_G \leq |qq'|_G$. By Lemma 1, $|pp'|_G + |p'q|_G$ is a 2-approximation of $|pq|_G$. Recall that we know the value $|pp'|_G$. If j is the index such that $p' = b_j$, then the value $|p'q|_G = |b_jq|_G$ is stored in the tree $\mathcal{T}_{v,j}$.
2. Assume that $|pp'|_G > |qq'|_G$. By Lemma 1, $|qq'|_G + |q'p|_G$ is a 2-approximation of $|pq|_G$. We know the value $|q'p|_G$. If j' is the index such that $q' = b_{j'}$, then the value $|qq'|_G = |qb_{j'}|_G$ is stored in the tree $\mathcal{T}_{v,j'}$.

Based on this, we do the following, for each point p in A_i^r: Let j be the index such that $p' = b_j$. By searching in $\mathcal{T}_{v,j}$ with the value $|pp'|_G$, we compute, in $O(\log n)$ time,

1. the number N of points q in B_i^b for which $|pp'|_G \leq |qq'|_G$,
2. the summation

$$X = \sum_{q \in B_i^b, |pp'|_G \leq |qq'|_G} |p'q|_G,$$

3. the value $N|pp'|_G + X$.

Observe that

$$N|pp'|_G + X = \sum_{q \in B_i^b, |pp'|_G \leq |qq'|_G} (|pp'|_G + |p'q|_G).$$

Next, for all j' with $1 \leq j' \leq \ell$, by searching in the trees $\mathcal{T}'_{v,j'}$ with the value $|pp'|_G$, we compute, in $O(\ell \log n) = O(\sqrt{r} \log n)$ total time,

1. the numbers $N_{j'}$ of points q in $\{q \in B_i^b : q' = b_{j'}\}$ for which $|pp'|_G > |qq'|_G$,
2. the summations

$$X_{j'} = \sum_{q \in B_i^b, q' = b_{j'}, |pp'|_G > |qq'|_G} |qq'|_G,$$

3. the summation

$$\sum_{j'=1}^{\ell} (N_{j'}|pq'|_G + X_{j'}).$$

Observe that this last summation is equal to

$$\sum_{q \in B_i^b, |pp'|_G > |qq'|_G} (|pq'|_G + |q'q|_G).$$

Thus, for a fixed point p in A_i^r, we have computed, in $O(\sqrt{r} \log n)$ time, a 2-approximation of the summation $\sum_{q \in B_i^b} |pq|_G$. Therefore, in $O(|A_i^r|\sqrt{r} \log n)$ time, we have computed a 2-approximation of the summation in (6). (Recall that this assumes that we have the trees $\mathcal{T}_{v,j}$ and $\mathcal{T}'_{v,j}$ for all j with $1 \leq j \leq \ell$.)

By traversing the split tree $T(S)$ in post-order, as we did in Step 3 of the algorithm in Section 3, we obtain 2-approximations of the summations in (6), for all i with $1 \leq i \leq m$, in total time which is the sum of

1. $O(\sqrt{rn} \log^2 n)$: This is the total time to compute all binary search trees $\mathcal{T}_{v,j}$ and $\mathcal{T}'_{v,j}$.
2. $O(\sqrt{rn} \log^2 n)$: This is the total time to search in all these binary search trees.

Recall that we repeat this algorithm for each of the $O(n/r)$ regions R in the r-division. It follows that the total time used to compute a $(2 + \epsilon)$-approximation of $SSF_2(G)$ is within the time bound in (5). This completes the proof of Theorem 6.

5.2 Approximating $SSF(G)$ within $4 + \epsilon$

In this section, we improve the running time in Theorem 6, while increasing the approximation factor to $4 + \epsilon$. Since the algorithm is recursive, a generic call solves a more general problem.

Let R be a subset of S, which we can think of to be a region in a division of the graph G. Recall that a point p of R is an interior point, if for every edge (p, q) in G, the point q is also in R. All other points of R are boundary points. We denote the sets of interior and boundary points of R by $int(R)$ and ∂R, respectively. The subgraph of G that is induced by R is denoted by $G[R]$.

The input for the algorithm consists of a subset R of S such that $|int(R)| = r$ and $|\partial R| = O(\sqrt{r})$. The output will be a real number that is between $SSF(R)/(4 + \epsilon)$ and $(4 + \epsilon)SSF(R)$, where

$$SSF(R) = \sum_{\{p,q\} \in \mathcal{P}_2(int(R))} \frac{|pq|_G}{|pq|}.$$

By running this algorithm with $R = S$ (in which case $int(R) = S$ and $\partial R = \emptyset$), we obtain a $(4 + \epsilon)$-approximation of $SSF(G)$.

We assume that the entire graph G has been preprocessed using the algorithm of Arikati *et al.* [2]. Recall that this preprocessing takes $O(n^{3/2})$ time, after which, for any two points p and q, a 2-approximation of $|pq|_G$ can be computed in $O(\log n)$ time.

If r is less than some constant, we use the data structure of [2] to compute a 2-approximation of $SSF(R)$ in $O(\log n)$ time. For a large value of r, the algorithm does the following.

Let $r' = r/2$. Use the algorithm of Frederickson [5] to compute an r'-division of the graph $G[R]$. Since $G[R]$ has size $O(r)$, this takes $O(r \log r)$ time and produces $k = O(r/r') = O(1)$ regions, each one having size at most $r' = r/2$ and $O(\sqrt{r'}) = O(\sqrt{r})$ boundary points. Thus, the total number of boundary points in the r'-division is $O(\sqrt{r})$.

The r'-division partitions the pairs in $\mathcal{P}_2(int(R))$ into three groups:

1. The pair $\{p, q\}$ is of *type 0*, if at least one of p and q is a boundary point of some region.
2. The pair $\{p, q\}$ is of *type 1*, if p and q are interior points of the same region.
3. The pair $\{p, q\}$ is of *type 2*, if p and q are interior points of different regions.

For $j \in \{0, 1, 2\}$, we define

$$SSF_j(R) = \sum_{\{p,q\} \text{ of type } j} \frac{|pq|_G}{|pq|},$$

so that

$$SSF(R) = SSF_0(R) + SSF_1(R) + SSF_2(R).$$

We obtain a 2-approximation of $SSF_0(R)$ by querying the data structure of [2] with each pair of type 0. Since the number of such pairs is $O(r^{3/2})$, this takes time

$$O\left(r^{3/2} \log n\right). \tag{7}$$

We obtain a $(4 + \epsilon)$-approximation of $SSF_1(R)$, by running the algorithm recursively on each region in the r'-division. Recall that k denotes the number of regions in the r'-division. For $1 \leq i \leq k$, we denote by r_i the number of interior points of the i-th region and by $\mathcal{T}(r_i)$ the running time of the recursive call on the i-th region. Then, the total time to approximate $SSF_1(R)$ is

$$O(r) + \sum_{i=1}^{k} \mathcal{T}(r_i). \tag{8}$$

Observe that $r_1 + \ldots + r_k \leq r$ and each value r_i is at most $r/2$.

It remains to show how to approximate $SSF_2(R)$. We first do the following for each region R' in the r'-division. Consider the graph $G[R']$. We add a dummy vertex and connect it by an edge to every point of $\partial R'$; each such edge gets weight, say, one. Then we run Dijkstra's algorithm on the resulting graph with the source being the dummy vertex. This gives, for each point p in $int(R')$ a point p' of $\partial R'$ such that

$|pp'|_{G[R']}$ is minimum, together with the shortest-path length $|pp'|_{G[R']}$. Observe that $|pp'|_{G[R']} = |pp'|_G$. Since the number of regions R' is $O(1)$ and each one has size $O(r)$, this part of the algorithm takes $O(r \log r)$ time.

The value of $SSF_2(R)$ is approximated, by doing the following for each pair R', R'' of distinct regions in the r'-division. We compute the split tree and the corresponding WSPD $\{A_1, B_1\}, \ldots, \{A_m, B_m\}$ of Theorem 1 for the point set $int(R') \cup int(R'')$, with separation ratio $s = 16/\epsilon$. We color the points of $int(R')$ and $int(R'')$ red and blue, respectively. For each pair $\{A_i, B_i\}$, we define A_i^r, A_i^b, B_i^r, and B_i^b as in Section 5.1. As before, we want to approximate the values

$$\sum_{p \in A_i^r} \sum_{q \in B_i^b} |pq|_G \quad \text{and} \quad \sum_{p \in B_i^r} \sum_{q \in A_i^b} |pq|_G,$$

for all i with $1 \leq i \leq m$.

Recall that, during the approximation of $SSF_0(R)$, we have computed a 2-approximation $\delta(b, p)$ of $|bp|_G$, for each b in $\partial R' \cup \partial R''$ and each p in $int(R') \cup int(R'')$. Consider a point p in $int(R')$ and a point q in $int(R'')$. Since any path in G between p and q passes through a point of $\partial R' \cup \partial R''$, we can use Lemma 1 to approximate $|pq|_G$:

1. If $|pp'|_G \leq |qq'|_G$, then $|pp'|_G + \delta(p', q)$ is a 4-approximation of $|pq|_G$.
2. If $|pp'|_G > |qq'|_G$, then $|qq'|_G + \delta(q', p)$ is a 4-approximation of $|pq|_G$.

Let $b_1, \ldots, b_\ell$ be the elements of $\partial R' \cup \partial R''$. We now use the binary search trees $\mathcal{T}_{v,j}$ and $\mathcal{T}'_{v,j}$ as in Section 5.1. The only difference is that each node u in any of these trees stores the sum of the values $\delta(b_j, q)$, where q ranges over all points in the subtree of u. By using the same algorithm as in Section 5.1, we obtain 4-approximations of the summations $\sum_{p \in A_i^r} \sum_{q \in B_i^b} |pq|_G$ and $\sum_{p \in B_i^r} \sum_{q \in A_i^b} |pq|_G$ in total time $O(r^{3/2} \log^2 r)$.

Thus, since the number of pairs of distinct regions is $O(1)$, the total time for computing a $(4 + \epsilon)$-approximation of $SSF_2(R)$ is

$$O\left(r^{3/2} \log^2 r\right). \tag{9}$$

If we denote the total running time of the algorithm by $\mathcal{T}(r)$, then it follows from (7), (8), and (9) that

$$\mathcal{T}(r) = O\left(r^{3/2}(\log n + \log^2 r)\right) + \sum_{i=1}^{k} \mathcal{T}(r_i).$$

Recall that $r_1 + \ldots + r_k \leq r$ and each value r_i is at most $r/2$. A straightforward inductive proof shows that

$$\mathcal{T}(r) = O\left(r^{3/2}(\log n + \log^2 r)\right).$$

As mentioned before, we obtain a $(4 + \epsilon)$-approximation of $SSF(G)$, by running this algorithm with $R = S$. We have proved the following result:

Theorem 7. *Let G be a plane graph on n points in $\mathbb{R}^d$ and let $\epsilon > 0$ be a real constant. In $O(n^{3/2} \log^2 n)$ time, we can compute a real number that lies between $SSF(G)/(4 + \epsilon)$ and $(4 + \epsilon)SSF(G)$.*

References

1. Agarwal, P.K., Klein, R., Knauer, C., Langerman, S., Morin, P., Sharir, M., Soss, M.: Computing the detour and spanning ratio of paths, trees, and cycles in 2D and 3D. Discrete & Computational Geometry 39, 17–37 (2008)
2. Arikati, S., Chen, D.Z., Chew, L.P., Das, G., Smid, M., Zaroliagis, C.D.: Planar spanners and approximate shortest path queries among obstacles in the plane. In: Díaz, J. (ed.) ESA 1996. LNCS, vol. 1136, pp. 514–528. Springer, Heidelberg (1996)
3. Callahan, P.B., Kosaraju, S.R.: A decomposition of multidimensional point sets with applications to k-nearest-neighbors and n-body potential fields. Journal of the ACM 42, 67–90 (1995)
4. Farshi, M., Giannopoulos, P., Gudmundsson, J.: Improving the stretch factor of a geometric graph by edge augmentation. SIAM Journal on Computing 38, 226–240 (2008)
5. Frederickson, G.N.: Fast algorithms for shortest paths in planar graphs, with applications. SIAM Journal on Computing 16, 1004–1022 (1987)
6. Klein, R., Knauer, C., Narasimhan, G., Smid, M.: On the dilation spectrum of paths, cycles, and trees. Computational Geometry: Theory and Applications 42, 923–933 (2009)
7. Narasimhan, G., Smid, M.: Approximating the stretch factor of Euclidean graphs. SIAM Journal on Computing 30, 978–989 (2000)
8. Narasimhan, G., Smid, M.: Geometric Spanner Networks. Cambridge University Press, Cambridge (2007)
9. Wulff-Nilsen, C.: Wiener index and diameter of a planar graph in subquadratic time. In: EuroCG, pp. 25–28 (2009)

Satisfiability with Index Dependency[*]

Hongyu Liang and Jing He

Institute for Theoretical Computer Science,
Tsinghua National Laboratory for Information Science and Technology (TNList),
Tsinghua University, Beijing, China
{hongyuliang86,hejing2929}@gmail.com

Abstract. We study the Boolean Satisfiability Problem (SAT) restricted on input formulas for which there are linear arithmetic constraints imposed on *the indices of variables occurring in the same clause*. This can be seen as a structural counterpart of Schaefer's dichotomy theorem which studies the SAT problem with additional constraints on *the assigned values of variables in the same clause*. More precisely, let k-SAT$(m, \mathcal{A})$ denote the SAT problem restricted on instances of k-CNF formulas, in every clause of which the indices of the last $k - m$ variables are totally decided by the first m ones through some linear equations chosen from $\mathcal{A}$. For example, if $\mathcal{A}$ contains $i_3 = i_1 + 2i_2$ and $i_4 = i_2 - i_1 + 1$, then a clause of the input to 4-SAT$(2, \mathcal{A})$ has the form $y_{i_1} \vee y_{i_2} \vee y_{i_1+2i_2} \vee y_{i_2-i_1+1}$, with y_i being x_i or $\overline{x_i}$. We obtain the following results:

1. If $m \geq 2$, then for any set $\mathcal{A}$ of linear constraints, the restricted problem k-SAT$(m, \mathcal{A})$ is either in P or NP-complete assuming $P \neq NP$. Moreover, the corresponding #SAT problem is always #P-complete, and the MAX-SAT problem does not allow a polynomial time approximation scheme assuming $P \neq NP$.
2. $m = 1$, that is, in every clause only one index can be chosen freely. In this case, we develop a general framework together with some techniques for designing polynomial-time algorithms for the restricted SAT problems. Using these, we prove that for any $\mathcal{A}$, #2-SAT$(1, \mathcal{A})$ and MAX-2-SAT$(1, \mathcal{A})$ are both polynomial-time solvable, which is in sharp contrast with the hardness results of general #2-SAT and MAX-2-SAT. For fixed $k \geq 3$, we obtain a large class of non-trivial constraints $\mathcal{A}$, under which the problems k-SAT$(1, \mathcal{A})$, #k-SAT$(1, \mathcal{A})$ and MAX-k-SAT$(1, \mathcal{A})$ can all be solved in polynomial time or quasi-polynomial time.

Keywords: Boolean Satisfiability Problem, index-dependency, index-width, bandwidth, dichotomy.

1 Introduction

The Boolean Satisfiability Problem (SAT) has received extensive studies since its first proposed as a natural NP-complete problem in [7]. It is also well known that

[*] This work was supported in part by the National Natural Science Foundation of China Grant 60553001, 61073174, 61033001 and the National Basic Research Program of China Grant 2007CB807900, 2007CB807901.

O. Cheong, K.-Y. Chwa, and K. Park (Eds.): ISAAC 2010, Part I, LNCS 6506, pp. 49–60, 2010.

3-SAT is still NP-complete [7] while 2-SAT is linear-time tractable [3], where by k-SAT we mean the the problem of deciding whether a given CNF formula with exactly k literals in each clause is satisfiable. The counting class $\#P$ is introduced by Valiant [18], and he proved in [19] that #SAT, the problem of counting the number of satisfying assignments of a given formula, is $\#P$-complete even on instances of monotone 2-CNFs. The optimization version MAX-3SAT, and even MAX-2SAT, have been shown to be APX-hard [2,10].

People thus became interested in examining the complexity of certain restricted versions of SAT, such as Horn-SAT [11,20] and 1-in-3-SAT [16], where the former is polynomial-time decidable while the latter is NP-complete. The most significant step in this line of research is Schaefer's dichotomy theorem [16], which we will briefly review as follows. In most cases, the restrictions associated with the SAT problem can be seen as a set of logical relations $S = \{R_1, \ldots, R_m\}$, where for all $i \in \{1, \ldots, m\}$, $R_i \subseteq \{0, 1\}^k$ for some integer $k \geq 1$. An R-clause, written as $R(x_1, \ldots, x_k)$, is said to be satisfied iff $(b_1, \ldots, b_k) \in R$ where b_i is the assigned value to the variable x_i. Any R-clause with $R \in S$ is called an S-clause. The problem SAT(S) is then defined to be the problem of deciding whether a given set of S-clauses are simultaneously satisfiable. It is easy to see that many variants of SAT, including the canonical k-SAT, Horn SAT and 1-in-3 SAT, fall into this classification. Schaefer's dichotomy theorem states that for every such S, SAT(S) is polynomial-time decidable or NP-complete. Allender et al. [1] proposed a refinement of Schaefer's theorem considering AC^0 isomorphisms instead of polynomial-time reductions.

Roughly speaking, Schaefer's dichotomy theorem accounts for the variants of SAT problems for which certain constraints are imposed on the *assigned values of variables appearing in the same clause*, while it does not directly consider the *structure of variables in the input formula* itself. For illustration, the fact that 3-SAT remains NP-complete on instances in which every variable occurs at most 4 times [17], cannot be derived from Schaefer's theorem. An interesting point is that 3-SAT with each variable occuring at most 3 times is always satisfiable [17]. This "phase transition" phenomenon is generalized to k-SAT by Kratochvíl, Savický and Tuza [12]. (See [8] for a recent result for determining this threshold.) Other examples include the NP-completeness of the PLANAR-3-SAT problem [13], which is to decide whether a given 3-CNF formula with a planar incidence graph is satisfiable.

In this paper, we initiate the study of SAT problems restricted on formulas for which there are some pre-known constraints on the *indices of variables occurring in the same clause*. This type of "index-dependency constraints" can be seen as a structural counterpart of what is considered by Schaefer [16]. We focus on linear arithmetic constraints, since they are the most natural ones we may encounter in various scenes.

Besides its pure theoretical interest as being a comparison to Schaefer's results, another motivation for studying this type of constraints on the clauses comes from the generation of random CNF formulas for testing the behavior of algorithms or other purposes. In practice, we usually use $O(km \log n)$ random bits, obtained by a pseudorandom number generator, to produce a random k-CNF formula of n variables and m clauses. A dilemma is that, when the number of formulas wanted is large, we cannot hope to get "good" random formulas if the pseudorandom number generator is too simple, while it costs much more time using a complex (but close to "true randomness") generator. A natural question is that to what extent we can relax the requirement on the quality of randomness and still get good (or, hard) enough formulas. Can we still get hard instances of SAT if the pseudorandom number generator is based on simple linear arithmetic? Can we get hard formulas by first using only a few bits obtained from a complex generator, and then performing some simple deterministic steps? One of our results provides evidence to a positive answer of these questions, which essentially says that in a k-CNF formula only the indices of the first two (or arbitrarily two, since the order doesn't matter) variables in any clause are "important", and others can just be generated by simple linear arithmetic operations (Theorem 5). It is thus interesting to study the complexity of these types of restricted SAT problems.

We now briefly explain the model we use. More rigorous definitions can be found in Section 2. Let $f = \bigwedge_{i=1}^{t} \bigvee_{j=1}^{k} y_{a_{i,j}}$ be a k-CNF formula, where y_l denotes either x_l or $\overline{x_l}$. A simple way for demonstrating the linear constraints on indices is to introduce a matrix $\mathcal{A}$ and a vector b, and requiring that $\mathcal{A}(a_{i,1}, \ldots, a_{i,k})^{\mathrm{T}} = b$ for all $i \in \{1, \ldots, t\}$. But for convenience of our proof, and without loss of generality, we adopt a more strict requirement that in any clause, the first m indices can be set freely, while the last $k - m$ indices are totally decidable by the first m ones through some linear equations. More precisely, there exists an integer m and a matrix $\mathcal{A} \in \mathbb{Z}^{(k-m)\times(m+1)}$ such that

$$\forall i \in \{1, 2, \ldots, t\}, (a_{i,m+1}, \ldots, a_{i,k})^{\mathrm{T}} = \mathcal{A}(a_{i,1}, \ldots, a_{i,m}, 1)^{\mathrm{T}}.$$

For instance, when $\mathcal{A} = \begin{pmatrix} 1 & 1 & 0 \\ 2 & -1 & -3 \end{pmatrix}$, every clause has the form $(y_i \vee y_j \vee y_{i+j} \vee y_{2i-j-3})$, with y_l denoting x_l or $\overline{x_l}$. Since subtraction is allowed to define an arithmetic constraint, we assume that the variables can be indexed by any integer, not necessarily positive.

We refer to the problem of deciding whether a given k-CNF formula of this form is satisfiable as D-k-SAT$(m, \mathcal{A})$. When $m = k$ it degenerates to the classical k-SAT problem. Other variants like MAX-k-SAT$(m, \mathcal{A})$ and #k-SAT$(m, \mathcal{A})$ can be defined similarly.

Our Contributions. We systematically study the complexity of such classes of SAT problems. In the first part of this article (Section 3), we prove a dichotomy theorem for the case $m \geq 2$. More precisely, for any $k \geq m \geq 2$ and $\mathcal{A} \in \mathbb{Z}^{(k-m)\times(m+1)}$, D-$k$-SAT$(m, \mathcal{A})$ is either in P or NP-complete, assuming $P \neq NP$. We also prove that for any $k \geq m \geq 2$ and $\mathcal{A} \in \mathbb{Z}^{(k-m)\times(m+1)}$, #$k$-SAT$(m, \mathcal{A})$ is #P-complete and MAX-k-SAT$(m, \mathcal{A})$ does not have a polynomial time approximation scheme, unless $P = NP$.

The second part (Section 4) of this paper is devoted to the case $m = 1$, where things become subtler. We develop a framework for designing polynomial-time algorithms for the SAT problems restricted on this case. First, we define a new parameter on the formulas, which we call the *index-width* (see Section 2 for the definition). Roughly speaking, it captures how "far away" two variables in the same clause can be from each other, under some ordering of the variables. We compare it to the existing definition of *bandwidth* [14] (also referred to as the *diameter* in [9]). We show that small index-width implies small bandwidth if the considered formula has large "index-dependency" (or, in our notation, $m = 1$). Moreover, a "layout" of the formula achieving a good (small) bandwidth can be constructed efficiently if its index-width is small. (In fact, we are able to prove that the two measures have the same magnitude up to some constant factor when $m = 1$. However, we will not present the proof in this version.) This guarantees that the algorithm in [9] for solving variants of SAT, which takes the bandwidth as a parameter, runs in polynomial time. With a slight modification, we can apply it to the weighted maximization version as well.

The reason that we define a new width parameter, but not directly use bandwidth, is that this new concept is easier to handle. To get an intuitive idea, the bandwidth cares a lot about the clauses of a formula, while the index-width deals mainly with the variables themselves. Although two arbitrary formulas on n Boolean variables may have very different structures of clauses, their variables can, without loss of generality, be regarded as the same set $\{x_1, x_2, \ldots, x_n\}$. By arguing on the index-width, we are able to handle a large class of formulas at the same time. Another reason is that this definition is well related to the index-dependency, since they both consider the indices, or more generally, the integers. We are then able to use tools from number theory and combinatorics to obtain desired results.

Keeping the ideas in mind and the techniques at hand, we first show that for any $\mathcal{A} \in \mathbb{Z}^{1\times 2}$, WMAX-2-SAT$(1, \mathcal{A})$ (the weighted version of MAX-SAT) and #2-SAT$(1, \mathcal{A})$ are both polynomial-time solvable. This is in sharp contrast with the hardness results of general #2-SAT [19] and MAX-2SAT [10]. We then consider the case where $m = 1$ and $k = 3$. We show that there is a large (and surely non-trivial) class of $\mathcal{A} \in \mathbb{Z}^{2\times 2}$ for which both WMAX-k-SAT$(1, \mathcal{A})$ and #k-SAT$(1, \mathcal{A})$ can be solved in polynomial time. This is harder to prove than the previous case $k = 2$. We then generalize the results to the case $k > 3$, showing that a large and natural class of $\mathcal{A}$ makes the problems solvable in DTIME$(n^{\text{polylog}(n)})$, which is generally believed not to contain any NP-hard problem. We also believe that, besides the obtained results, the concept of index-width and the techniques used are of their own interest.

Due to space limitations, several proofs are omitted and will appear in the full version of this paper.

2   Preliminaries

In this paper $[a, b]$ denotes the integer set $\{\lceil a \rceil, \lceil a \rceil + 1, \ldots, \lfloor b \rfloor\}$. We use $\log x$ to denote the logarithm base 2 of x. $\mathbb{Z}$ is the set of all integers, $\mathbb{Z}^+$ is the set of all positive integers and $\mathbb{Z}^{a\times b}$ is the set of all $a \times b$ integer matrices. Let $\mathcal{A}[i, j]$

denote the element of a matrix $\mathcal{A}$ at the crossing of the i-th row and the j-th column.

Let $X = \{x_i \mid i \in \mathbb{Z}\}$ be an infinite set of Boolean variables indexed by all integers. Without loss of generality, all Boolean variables considered in this paper are chosen from X. For all $i \in \mathbb{Z}$, we write y_i to represent a literal chosen from $\{x_i, \overline{x_i}\}$. Define an index function $\mathcal{I}$ as $\mathcal{I}(y_i) = i$ for all $i \in \mathbb{Z}$.

A clause is the disjunction of literals, which can be seen as an ordered tuple. A CNF formula (or simply CNF) is the conjunction of clauses, and a k-CNF is a CNF in which all clauses have exactly k literals. For a clause c, let $\mathcal{I}(c) = \{\mathcal{I}(x) \mid x \in c\}$ be the set of indices occurring in c, and $\mathcal{I}(c, i)$ be the index of the i-th literal of c. For a CNF f, let $\mathcal{I}(f) = \bigcup_{c \in f} \mathcal{I}(c)$. Define the *index-width* of f as:

$$\mathcal{W}(f) = \max_{c \in f} \max_{i,j \in \mathcal{I}(c)} |i - j|.$$

Observe that the index-width of f is just the maximum difference between any two indices in the same clause of f. This concept is a dual analog of the bandwidth of a formula [14], whose definition will be restated below in a more helpful form. Given a CNF formula $f = \bigwedge_{i=1}^{m} c_i$, a *layout* of f is an injective function $l : \{c_1, \ldots, c_m\} \to \mathbb{Z}$ (or, a relabeling of the clauses). The *bandwidth* of f under layout l is defined as the minimum integer k s.t. whenever a variable or its negation appears in two clauses c_i and c_j, the inequality $|l(c_i) - l(c_j)| \leq k$ holds. The bandwidth of f is the minimum bandwidth of f under all possible layouts.

Definition 1. *Let $k \geq m \geq 1$ be two positive integers and $\mathcal{A} \in \mathbb{Z}^{(k-m) \times (m+1)}$. Define k-CNF$(m, \mathcal{A})$ to be the collection of all k-CNF f such that for any clause $c \in f$, it holds that $(\mathcal{I}(c, m+1), \ldots, \mathcal{I}(c, k))^T = \mathcal{A}(\mathcal{I}(c, 1), \ldots, \mathcal{I}(c, m), 1)^T$. With a little abuse of notation, we also use k-CNF$(m, \mathcal{A})$ to denote a formula chosen from this set.[1]*

Definition 2. *Let $k \geq 2, m \in [1, k]$ and $\mathcal{A} \in \mathbb{Z}^{(k-m) \times (m+1)}$. An instance of k-SAT$(m, \mathcal{A})$ consists of a string 1^n as well as a formula $f \in k$-CNF$(m, \mathcal{A})$ with $\mathcal{I}(f) \subseteq [-n, n]$. The goal is to find an assignment to $\{x_i \mid i \in [-n, n]\}$ that satisfies f, or correctly report that no such assignment exists. Define D-k-CNF$(m, \mathcal{A})$, MAX-k-CNF$(m, \mathcal{A})$, WMAX-k-CNF$(m, \mathcal{A})$ and #k-CNF$(m, \mathcal{A})$ to be the decision version, the optimization version, the weighted optimization version and the counting version of k-SAT$(m, \mathcal{A})$, respectively.[2]*

3   Dichotomy Results for k-SAT$(m, \mathcal{A})$ When $m \geq 2$

3.1   Dichotomy Theorem for 3-SAT(2,$\mathcal{A}$)

We examine the complexity of variants of 3-SAT(2,$\mathcal{A}$). In this case $\mathcal{A}$ is just a 3-dimensional integer vector (a, b, c), and every clause of an instance has the form

---

[1] For example, any clause of a formula $f \in 4$-CNF$\left(2, \begin{pmatrix} 1 & 1 & 0 \\ 2 & -1 & -3 \end{pmatrix}\right)$ has the form $y_i \vee$

$y_j \vee y_{i+j} \vee y_{2i-j-3}$.

[2] In our notation, D-k-SAT$(k, \emptyset)$, MAX-k-SAT$(k, \emptyset)$ and #k-SAT$(k, \emptyset)$ correspond to the canonical problems k-SAT, MAX-k-SAT and #k-SAT, respectively.

$y_i \vee y_j \vee y_{ai+bj+c}$. It is easy to find some special vectors $\mathcal{A}$ for which 3-SAT(2,$\mathcal{A}$) degenerates to 2-SAT: When $\mathcal{A} = (0, 1, 0)$ or $(1, 0, 0)$ it is obvious, and when $\mathcal{A} = (0, 0, c)$ it suffices to examine the two 2-SAT instances where x_c is set to 0 and 1 respectively. Therefore 3-SAT(2,$\mathcal{A}$) can be solved in polynomial time under these choices of $\mathcal{A}$.

It is somehow surprising that they are the only possibilities that 3-SAT(2,$\mathcal{A}$) is polynomial-time solvable, assuming $P \neq NP$. For convenience we define $Easy\mathcal{A} = \{(0,1,0),(1,0,0)\} \cup \{(0,0,c) \mid c \in \mathbb{Z}\}$ which consists of all easy cases.

Theorem 1. *Let $\mathcal{A} \in \mathbb{Z}^{1 \times 3}$. Assuming $P \neq NP$, D-3-SAT(2,$\mathcal{A}$) is in P if $\mathcal{A} \in Easy\mathcal{A}$, and is NP-complete otherwise.*

We also obtain the following inapproximability results of the corresponding optimization problems.

Theorem 2. *Let $\epsilon > 0$ be any positive constant. Then, for $\mathcal{A} \in \mathbb{Z}^{1 \times 3}$, it is NP-hard to approximate MAX-3-SAT(2,$\mathcal{A}$) better than*

- $21/22 + \epsilon$ *if* $A \in \{(0, 1, 0), (1, 0, 0)\}$;
- $43/44 + \epsilon$ *if* $\mathcal{A} \in \{(0, 0, c) \mid c \in \mathbb{Z}\}$;
- $23/24 + \epsilon$ *if* $\mathcal{A} \in \mathbb{Z}^{1 \times 3} \setminus Easy\mathcal{A}$.

Finally, we have the following hardness result for the counting version.

Theorem 3. *#3-SAT(2,$\mathcal{A}$) is #P-complete for any $\mathcal{A} \in \mathbb{Z}^{1 \times 3}$.*

3.2   Results for $k > 3$ and $m \geq 2$

In this subsection we consider the cases where $k \geq 4$ and $m \geq 2$. The following two theorems are not hard to prove given Theorem 1.

Theorem 4. *Let $k \geq m \geq 3$ be two integers and $\mathcal{A} \in \mathbb{Z}^{(k-m) \times (m+1)}$. Then D-$k$-SAT$(m, \mathcal{A})$ is NP-complete.*

Theorem 5. *Let $k \geq 3$ and $\mathcal{A} \in \mathbb{Z}^{(k-2) \times 3}$. Assuming $P \neq NP$, D-k-SAT$(2, \mathcal{A})$ is in P if all row vectors of $\mathcal{A}$ are in $Easy\mathcal{A}$, and is NP-complete otherwise.*

Combining the above statements and noting that the preceding hardness results also carry over to the case $k > 3$, we obtain:

Corollary 1. *Let $k \geq m \geq 2$ and $\mathcal{A} \in \mathbb{Z}^{(k-m) \times (m+1)}$. Then,*

- D-k-SAT$(m, \mathcal{A})$ *is either in P or NP-complete, assuming $P \neq NP$;*
- #k-SAT$(m, \mathcal{A})$ *is always #P-complete;*
- MAX-SAT$(m, \mathcal{A})$ *does not admit a polynomial time approximation scheme (PTAS) unless $P = NP$.*

4   Tractability Results for $m = 1$

We now turn to the case $m = 1$. A little surprisingly, we will show that for any $\mathcal{A} \in \mathbb{Z}^{1 \times 2}$, all the variants of 2-SAT(1,$\mathcal{A}$) defined previously are polynomial-time solvable. For $k \geq 3$, there is a large class of non-trivial choices of $\mathcal{A} \in \mathbb{Z}^{(k-1)\times 2}$ for which the variants of k-SAT$(1,\mathcal{A})$ can all be solved in polynomial time or quasi-polynomial time (DTIME($n^{\mathrm{polylog}(n)}$)). In this short version, we will focus on the case $k \leq 3$. We also assume without loss of generality that every input formula has pairwise different clauses, since otherwise we can easily detect and remove redundant clauses (and add the weight to another clause when dealing with MAX-SAT). We first define the following subsets of $\mathbb{Z}^{1\times 2}$:

$$\mathscr{A}_1 := \left\{ \begin{pmatrix} 0 & b_1 \\ a_2 & b_2 \end{pmatrix} \in \mathbb{Z}^{2\times 2} \right\} \cup \left\{ \begin{pmatrix} a_1 & b_1 \\ 0 & b_2 \end{pmatrix} \in \mathbb{Z}^{2\times 2} \right\} \cup \left\{ \begin{pmatrix} 1 & 0 \\ a_2 & b_2 \end{pmatrix} \in \mathbb{Z}^{2\times 2} \right\}$$

$$\cup \left\{ \begin{pmatrix} a_1 & b_1 \\ 1 & 0 \end{pmatrix} \in \mathbb{Z}^{2\times 2} \right\} \cup \left\{ \begin{pmatrix} a_1 & b_1 \\ a_1 & b_1 \end{pmatrix} \in \mathbb{Z}^{2\times 2} \right\};$$

$$\mathscr{A}_2 := \left\{ \begin{pmatrix} 1 & b_1 \\ 1 & b_2 \end{pmatrix} \in \mathbb{Z}^{2\times 2} \right\}; \quad \mathscr{A}_3 := \left\{ \begin{pmatrix} a_1 & 0 \\ a_2 & 0 \end{pmatrix} \in \mathbb{Z}^{2\times 2} \right\};$$

$$\mathscr{A}_4 := \left\{ \begin{pmatrix} a_1 & b_1 \\ a_2 & b_2 \end{pmatrix} \in \mathbb{Z}^{2\times 2} \mid b_1 b_2 \neq 0 \text{ and } (a_1 - 1)/b_1 = (a_2 - 1)/b_2 \right\};$$

$$\mathscr{A} := \mathscr{A}_1 \cup \mathscr{A}_2 \cup \mathscr{A}_3 \cup \mathscr{A}_4.$$

Theorem 6. *For any $\mathcal{A} \in \mathbb{Z}^{1\times 2}$, the problems* 2-SAT$(1,\mathcal{A})$, WMAX-2-SAT $(1,\mathcal{A})$ *and* #2-SAT$(1,\mathcal{A})$ *are all polynomial-time solvable.*

Theorem 7. *For any $\mathcal{A} \in \mathscr{A}$, the problems* 3-SAT$(1,\mathcal{A})$, WMAX-3-SAT$(1,\mathcal{A})$ *and* #3-SAT$(1,\mathcal{A})$ *are all polynomial-time solvable.*

We will make use of the index-width of f to prove the two theorems. The following lemma is needed.

Lemma 1. *Fix $k \geq 3$ and $\mathcal{A} \in \mathbb{Z}^{(k-1)\times 2}$. Given any formula $f \in k$-CNF$(1, \mathcal{A})$, we can construct a layout for f in polynomial time under which f has bandwidth $O(\mathcal{W}(f))$.*

Proof. Let $f \in k$-CNF$(1, \mathcal{A})$. We construct a layout l for f as follows. For any ordered clause $c = (y_{i_1} \vee \ldots \vee y_{i_k}) \in f$, let $l(c) = 2^k i_1 + r$ where $r \in [0, 2^k - 1]$ depends on the polarity of literals of c in the obvious way (2^k possible cases in total; recall that $i_2, \ldots, i_k$ are fixed after i_1 is chosen). This is a valid layout since f contains no duplicated clauses. Furthermore, suppose x_i or its negation appear in both c_1 and c_2. Let $j = \mathcal{I}(c_1, 1)$ and $k = \mathcal{I}(c_2, 1)$. By definition, we have $|j - i| \leq \mathcal{W}(f)$ and $|k - i| \leq \mathcal{W}(f)$, and therefore $|l(c_1) - l(c_2)| \leq 2^k |j - k| + 2^k < 2^{k+1}(\mathcal{W}(f) + 1)$. Thus, f has bandwidth $O(\mathcal{W}(f))$ under layout l. □

Given any k-CNF formula f, we can first apply Lemma 1 to get an equivalent formula f' with bandwidth $O(\mathcal{W}(f))$, and then use the algorithms in [9] (see Theorem 3 in [9]) to solve SAT, MAX-SAT and #SAT on f. Furthermore, their algorithm can be modified to solve the weighted version WMAX-SAT. Thus we obtain the following lemma.

Lemma 2. SAT, WMAX-SAT *and* #SAT *can be solved in time $n^{O(1)} 2^{O(\mathcal{W}(f))}$ on any input formula f with $\mathcal{I}(f) \subseteq [-n, n]$.*

Lemma 2 implies that we can solve variants of SAT efficiently on instances with $\mathcal{W}(f) = O(\log n)$. Since for any $\mathcal{A} = \begin{pmatrix} 1 & b_1 \\ 1 & b_2 \end{pmatrix}$ and $f \in$ 3-CNF(1,$\mathcal{A}$) it holds that $\mathcal{W}(f) \leq \max\{|b_1|, |b_2|, |b_1 - b_2|\} = O(1)$, we have:

Corollary 2. *The result of Theorem 7 holds for all $\mathcal{A} \in \mathscr{A}_2$.*

Given Lemma 2, a natural way for efficiently solving SAT on f is to find a family of injective mappings $\{\mathcal{F}_n\}$ where $\mathcal{F}_n : [-n, n] \to [-n', n']$ for some $n' = n^{O(1)}$, such that the new formula f', obtained by replacing all indices in f with their images under $\mathcal{F}_n$, obeys $\mathcal{W}(f') = O(\log n)$. Notice that the injectivity of $\mathcal{F}_n$ is important since it ensures the SAT-equivalence of f and f'. Now we prove Theorem 6 using this idea.

Proof (of Theorem 6). Let $\mathcal{A} = (a, b)$ and $f \in$ 2-CNF(1,$\mathcal{A}$) with $\mathcal{I}(f) \subseteq [-n, n]$. Construct a directed graph $G = (V, E)$ with $V = \{v_i \mid i \in [-n, n]\}$ and $E = \{(v_i, v_j) \mid j = ai + b, j \neq i\}$. Every vertex in V has in-degree at most 1 and out-degree at most 1. Now we prove that every connected component is a chain (a single vertex is regarded as a chain of length 0). If this is not the case, then there exists a connected component that is a cycle, which indicates the existence of t distinct integers $i_1, i_2, \ldots, i_t, t \geq 2$, such that $i_{l+1} = ai_l + b$ for all $l \in [1, t-1]$ and $i_1 = ai_t + b$. Note that in this case $a \neq 1$. Simple calculations show that $i_1 = b/(1 - a)$. We then reach a contradiction, since $i_2 = ai_1 + b = i_1$.

Thus, E is a collection of disjoint chains. We give an arbitrary ordering on the chains. For every $i \in [-n, n]$, let $q(i), r(i)$ be the integers such that i is the $r(i)$-th node on the $q(i)$-th chain. For every edge $(v_i, v_j) \in E$, we have $q(i) = q(j)$ and $|r(i) - r(j)| = 1$. Now define a function $\mathcal{F}_n$ as $\mathcal{F}_n(i) = (2n + 2)q(i) + r(i)$ for all $i \in [-n, n]$. This is an injective function since $r(i) \leq 2n + 1$. We construct f' by replacing all literals y_i in f with $y_{\mathcal{F}_n(i)}$. It is easy to see that $\mathcal{W}(f') \leq 1$, and by Lemma 2 we are done. $\qquad\square$

Lemma 3. *The result of Theorem 7 holds for all $\mathcal{A} \in \mathscr{A}_1$.*

Proof. It is obvious that for any $\mathcal{A} \in \mathscr{A}_1$, 3-SAT(1,$\mathcal{A}$) can be reduced to 2-SAT(1,$\mathcal{A}'$) for some $\mathcal{A}' \in \mathbb{Z}^{1 \times 2}$. Other variants can be treated analogously. $\qquad\square$

Next we deal with $\mathcal{A} = \begin{pmatrix} a_1 & 0 \\ a_2 & 0 \end{pmatrix} \in \mathscr{A}_3$, trying to find a family of injective mappings $\{\mathcal{F}_n\}$ as suggested above. There is a simple construction of $\{\mathcal{F}_n\}$ when $|a_1|, |a_2| \geq 2$ and a_1 and a_2 are co-prime, in which case every integer m can be uniquely represented by an integer-tuple (m_1, m_2, m_3) such that $m = a_1^{m_1} a_2^{m_2} m_3$, $a_1 \nmid m_3$ and $a_2 \nmid m_3$. For all $m \in [-n, n]$, let $\mathcal{F}_n(m) = m_3 n + m_1(\lceil \log n \rceil + 1) + m_2$. Since $|m_1|, |m_2| \leq \lceil \log n \rceil$, this is an injective mapping for large enough n. The corresponding tuple associated with $a_1 m$ is $(m_1 + 1, m_2, m_3)$, so we have $|\mathcal{F}_n(a_1 m) - \mathcal{F}(m)| = O(\log n)$ and similarly $|\mathcal{F}_n(a_2 m) - \mathcal{F}(m)| = O(1)$, giving us the desired result. However, this method fails if a_1 and a_2 are not co-prime. In the following, we will attack the general case using a different approach.

Lemma 4. *The result of Theorem 7 holds for all $\mathcal{A} \in \mathscr{A}_3$.*

Proof. Fix $a_1, a_2 \in \mathbb{Z} \setminus \{0\}$. Let $f \in 3\text{-SAT}(1, \begin{pmatrix} a_1 & 0 \\ a_2 & 0 \end{pmatrix})$ with $\mathcal{I}(f) \subseteq [-n, n], n \geq 2$. We construct an undirected graph $G = (V, E)$ (note that in the proof of Theorem 6 we use a directed graph) as follows: Let $V = \{v_i \mid i \in [-n, n]\}$, and E be the set of all (v_i, v_j) such that i and j can appear in the same clause of a $3\text{-CNF}(1, \begin{pmatrix} a_1 & 0 \\ a_2 & 0 \end{pmatrix})$ formula. By our construction, $(v_i, v_j) \in E$ indicates $i = \lambda j$ for $\lambda \in \{a_1, a_2, 1/a_1, 1/a_2, a_1/a_2, a_2/a_1\}$. Notice that the graph G actually only depends on a_1, a_2 and n.

Suppose G has connected components $\{G_i = (V_i, E_i) \mid i \in [1, c^*]\}, c^* \geq 1$. Inside each component G_i, we pick an arbitrary vertex v (which we call the *root* of G_i) and define $l(v')$ for all $v' \in V_i$ to be the the length of the shortest path between v' and v (note $l(v) = 0$). Then for all possible values t and all $v' \in V_i$ with $l(v') = t$, say, $l(v_{i_1}) = \ldots = l(v_{i_m}) = t$, we define $r(v_{i_q}) = q$ for all $q \in [1, m]$ (notice that the order of vertices in this list can be arbitrary). Finally let $c(v') = i$ for all $v' \in V_i$. In this way we have defined three functions $c(\cdot), l(\cdot)$ and $r(\cdot)$ on V, which can be computed in polynomial time using simple breadth-first-search algorithms. Intuitively, v is the $r(v)$-th node with distance $l(v)$ to the root of $G_{c(v)}$, and $(c(\cdot), l(\cdot), r(\cdot))$ can be seen as a new indexing on the vertices in that $(c(v), l(v), r(v)) = (c(v'), l(v'), r(v'))$ implies $v = v'$. We next prove two lemmas bounding $l(v)$ and $r(v)$ from above.

Lemma 5. *There exists $c_1 > 0$, which is independent of n, such that $l(v) \leq c_1 \log n$ for all $v \in V$.*

Proof. For any $v_i \in V$, let v_j be the root of $G_{c(v_i)}$ and P is (one of) the shortest path(s) between v_j and v_i. For any edge $(v_{t_1}, v_{t_2}) \in P$, we have $t_2 = \lambda t_1$ where $\lambda \in \{a_1, a_2, 1/a_1, 1/a_2, a_1/a_2, a_2/a_1\}$, implying that $i = j a_1^{u_1} a_2^{u_2}$ for some $u_1, u_2 \in \mathbb{Z}$ and obviously $l(v_i) \leq |u_1| + |u_2|$. Let $\{p_1, \ldots, p_t\}$ be the set of all prime divisors of a_1 or a_2. Due to the fundamental theorem of arithmetic (see e.g. Chapter 3.5 in [15]), we can assume $a_1 = (-1)^{r_0} \prod_{l=1}^t p_l^{r_l}$ and $a_2 = (-1)^{r_0'} \prod_{l=1}^t p_l^{r_l'}$ where $r_0, r_0' \in \{0, 1\}$ and $r_1, \ldots, r_t, r_1', \ldots, r_t' \in \mathbb{Z}^+ \cup \{0\}$. Note that $p_1, \ldots, p_t, r_0, \ldots, r_t, r_0', \ldots, r_t'$ are all constants that only depend on a_1 and a_2. Suppose $i/j = (-1)^{s_0} \prod_{l=1}^t p_l^{s_l}$ where $s_0 \in \{0, 1\}$ and $s_l \in \mathbb{Z}$ for any $l \in [1, t]$. We have $|s_l| \leq c \log n$ for some constant c because $p_l \geq 2$. Substituting them into $i = j a_1^{u_1} a_2^{u_2}$ gives

$$r_0 u_1 \oplus r_0' u_2 = s_0 \text{ and } \forall l \in [1, t] : r_l u_1 + r_l' u_2 = s_l ,$$

where "$\oplus$" denotes the modulo-2 addition. We know that there exists a solution $\mathbf{u} = (u_1, u_2)$ to these equations, and $l(v_i) \leq |u_1| + |u_2|$ for any solution (u_1, u_2). If there exist $l_1, l_2 \in [1, t]$ s.t. $\begin{vmatrix} r_{l_1} & r_{l_1}' \\ r_{l_2} & r_{l_2}' \end{vmatrix} \neq 0$, we will get a unique solution of $(u_1, u_2) = (\begin{vmatrix} s_{l_1} & r_{l_1}' \\ s_{l_2} & r_{l_2}' \end{vmatrix} / \begin{vmatrix} r_{l_1} & r_{l_1}' \\ r_{l_2} & r_{l_2}' \end{vmatrix}, \begin{vmatrix} r_{l_1} & s_{l_1} \\ r_{l_2} & s_{l_2} \end{vmatrix} / \begin{vmatrix} r_{l_1} & r_{l_1}' \\ r_{l_2} & r_{l_2}' \end{vmatrix})$ and hence

$|u_1|, |u_2| \leq O(|s_{l_1}| + |s_{l_2}|) \leq c' \log n$ for some constant c'. Otherwise all equations but the first one (which only reflects whether the number is positive or negative) become equivalent. From the property and general formulas of linear diophantine equations over two variables (see, e.g. Chapter 3.7 in [15]), we know that there exists one solution for which $|u_1|, |u_2| \leq O(s_1) \leq c'' \log n$ for some constant c''. Therefore $l(v_i) \leq 2\max\{c', c''\} \log n$. $\qquad\square$

Lemma 6. *There exists $c_2 > 0$, which is independent of n, such that $r(v) \leq c_2(\log n + 1)$ for all $v \in V$.*

Proof. Let $v_i \in V$ and v_j be the root of $G_{c(v_i)}$. From the proof of Lemma 5, for any edge $(v_{t_1}, v_{t_2}) \in E$ where $t_1/j = a_1^{e_1} a_2^{e_2}$, we have $t_2/j = a_1^{e_1'} a_2^{e_2'}$ for some $(e_1', e_2') \in \{(e_1+1, e_2), (e_1-1, e_2), (e_1, e_2+1), (e_1, e_2-1), (e_1+1, e_2-1), (e_1-1, e_2+1)\}$, from which follows $|e_1'| \leq |e_1| + 1, |e_2'| \leq |e_2| + 1$ and $|e_1' + e_2'| \leq |e_1 + e_2| + 1$. Define $h(v_i) = \min_{i/j = a_1^{u_1} a_2^{u_2}; u_1, u_2 \in \mathbb{Z}} \max\{|u_1|, |u_2|, |u_1 + u_2|\}$. Since $h(v_j) = 0$ and $h(v_{t_2}) \leq h(v_{t_1}) + 1$ for any edge $(v_{t_1}, v_{t_2}) \in E$, we have $h(v_i) \leq l(v_i)$.

On the other hand we prove $l(v_i) \leq h(v_i)$ as follows. Assume (u_1^*, u_2^*) witnesses $h(v_i)$; that is, $i/j = a_1^{u_1^*} a_2^{u_2^*}$ and $h(v_i) = \max\{|u_1^*|, |u_2^*|, |u_1^* + u_2^*|\}$. We have either (1) $u_1^* u_2^* \geq 0$, and obviously $l(v_i) \leq |u_1^*| + |u_2^*| = |u_1^* + u_2^*|$, or (2) $u_1^* u_2^* < 0$, in which case $l(v_i) \leq \max\{|u_1^*|, |u_2^*|\}$ (for example, if $u_1^* > 0, u_2^* < 0$ and $|u_1^*| \geq |u_2^*|$, we have $a_1^{u_1^*} a_2^{u_2^*} = a_1^{|u_1^*| - |u_2^*|}(a_1/a_2)^{|u_2^*|}$ and thus $|u_1^*|$ edges suffice to connect v_i and v_j; other cases can be proven similarly). So $l(v_i) = h(v_i)$.

Note that $r(v_i)$ is at most the number of vertices at the same level with v_i. For any fixed value $l(v_i)$, the number of pairs (u_1, u_2) satisfying $l(v_i) = \max\{|u_1|, |u_2|, |u_1 + u_2|\}$ is at most $2(2l(v_i) + 1) + 2(l(v_i) + 1) = 6l(v_i) + 4$. Thus $r(v_i) \leq 6l(v_i) + 4$ and the result is straightforward by Lemma 5. $\qquad\square$

We continue the proof of Lemma 4. Define $\mathcal{F}_n : [-n, n] \to \mathbb{Z}$ as $\mathcal{F}_n(i) = c(v_i) \cdot n + l(v_i) \cdot \lceil c_2 \rceil (\lceil \log n \rceil + 2) + r(v_i)$ for any $i \in [-n, n]$, where c_2 is the constant ensured by Lemma 6. By Lemmas 5, 6, there exists $N_0 > 0$ such that for all $n > N_0$, we have $l(v_i) \cdot \lceil c_2 \rceil (\lceil \log n \rceil + 2) + r(v_i) < n$ and $\mathcal{F}(i) \leq n^3$. Using the idea of division with remainders, we have $\mathcal{F}_n(i) = \mathcal{F}_n(j) \Leftrightarrow i = j$, showing the injectivity of $\mathcal{F}_n$ for $n > N_0$. Furthermore, for any two indices i, j appearing in the same clause of f, we have $c(v_i) = c(v_j), |l(v_i) - l(v_j)| \leq 1, |r(v_i) - r(v_j)| \leq O(\log n)$ and thus $|\mathcal{F}_n(i) - \mathcal{F}_n(j)| \leq O(\log n)$ holds. Lemma 2 then again gives us a polynomial-time algorithm. Finally, we can just perform the exhaustive search when $n \leq N_0$, since N_0 is a constant. $\qquad\square$

By a simple reduction to 3-SAT$\left(1, \begin{pmatrix} a_1 & 0 \\ a_2 & 0 \end{pmatrix}\right)$, we can also prove that Theorem 7 holds for $\mathcal{A} \in \mathscr{A}_4$. Hence, combining with Lemmas 3,4 and Corollary 2, Theorem 7 follows.

We note that the above proofs rely on the existence of $\{\mathcal{F}_n\}$, whereas in some cases the non-existence of such mapping can be proved. Such an example is given by setting $\begin{pmatrix} a_1 & b_1 \\ a_2 & b_2 \end{pmatrix} = \begin{pmatrix} 1 & 1 \\ 2 & 0 \end{pmatrix}$, for which it is provable that the formulas have index-width $\Omega(n/\log n)$ under any relabeling of the variables. We omit the details.

Finally, we mention the following theorem for the case $k > 3$, the proof of which is omitted from this short version. It basically follows a similar line as above, with some more careful (and more complicated) analysis in dealing with the parameters.

Theorem 8. *Let $\mathcal{A} \in \mathbb{Z}^{(k-1)\times 2}$ with $k \geq 4$. Then the problems k-SAT$(1,\mathcal{A})$, WMAX-k-SAT$(1,\mathcal{A})$ and $\#k$-SAT$(1,\mathcal{A})$ are*

- *solvable in polynomial time if $A[1,1] = A[2,1] = \ldots = A[k-1,1] = 1$;*
- *solvable in time $n^{O(\log^{k-3}(n))}$ if $A[1,2] = A[2,2] = \ldots = A[k-1,2] = 0$.*

5   Discussions and Open Problems

We list several interesting questions that are left for future work.

- Can we decide the complexity of 3-SAT$(1,\mathcal{A})$ for every $\mathcal{A} \in \mathbb{Z}^{1\times 2}$? It is tempting to conjecture that 3-SAT$(1,\mathcal{A})$ is polynomial-time solvable for every $\mathcal{A}$. However, proving this requires new techniques and insights into the structure of the instances. More ambitiously, can we characterize the complexity of k-SAT$(1,\mathcal{A})$ for every $k \geq 3$ and $\mathcal{A} \in \mathbb{Z}^{(k-1)\times 2}$?
- Can we design polynomial-time approximation algorithms for 3-SAT$(2,\mathcal{A})$ for some $\mathcal{A} \in \mathbb{Z}^{1\times 3} \setminus Easy\mathcal{A}$ with performance guarantee better than 7/8 (the tight approximation threshold for 3-SAT)? Can we achieve faster exact algorithms for the index-dependent SAT variants?
- Can we prove threshold behaviors for random instances of k-SAT$(m, \mathcal{A})$ similar to that of random k-SAT? (Here the "random instances" should be defined carefully.)

Acknowledgements. The authors are grateful to the anonymous referees for their helpful comments and suggestions on an early version of this paper.

References

1. Allender, E., Bauland, M., Immerman, N., Schnoor, H., Vollmer, H.: The complexity of satisfiability problems: refining Schaefer's theorem. J. Comput. System Sci. 75(4), 245–254 (2009)
2. Arora, S., Lund, C., Motwani, R., Sudan, M., Szegedy, M.: Proof verification and the hardness of approximation problems. J. ACM 45(3), 501–555 (1998)
3. Aspvall, B., Plass, M.F., Tarjan, R.E.: A linear-time algorithm for testing the truth of certain quantified boolean formulas. Inf. Process. Lett. 8(3), 121–123 (1979)
4. Borosh, I., Flahive, M., Rubin, D., Treybig, B.: A Sharp Bound for Solutions of Linear Diophantine Equations. Proceedings of the American Mathematical Society 105(4), 844–846 (1989)
5. Borosh, I., Flahive, M., Treybig, B.: Small solution of linear Diophantine equations. Discrete Mathematics 58(3), 215–220 (1986)

6. Bradley, G.H.: Algorithms for Hermite and Smith Normal Matrices and Linear Diophantine Equations. Mathematics of Computation 25(116), 897–907 (1971)
7. Cook, S.A.: The complexity of theorem proving procedures. In: Proceedings of the 3rd ACM STOC, pp. 151–158 (1971)
8. Gebauer, H., Szabó, T., Tardos, G.: The local lemma is tight for SAT. CoRR (Computing Research Repository), arXiv:1006.0744 (2010)
9. Georgiou, K., Papakonstantinou, P.A.: Complexity and algorithms for well-structured k-SAT instances. In: Kleine Büning, H., Zhao, X. (eds.) SAT 2008. LNCS, vol. 4996, pp. 105–118. Springer, Heidelberg (2008)
10. Håstad, J.: Some optimal inapproximability results. J. ACM 48(4), 798–859 (2001)
11. Henschen, L., Wos, L.: Unit refutations and Horn sets. J. ACM 21(4), 590–605 (1974)
12. Kratochvíl, J., Savický, P., Tuza, Z.: One more occurrence of variables makes satisfiability jump from trivial to NP-complete. SIAM J. Comput. 22(1), 203–210 (1993)
13. Lichtenstein, D.: Planar formulae and their uses. SIAM J. Comput. 11(2), 329–343 (1982)
14. Monien, B., Sudborough, I.H.: Bandwidth constrained NP-complete problems. In: Proceedings of the 13th ACM STOC, pp. 207–217 (1981)
15. Rosen, K.H.: Elementary number theory and its applications, 5th edn. Addison-Wesley, Reading (2005)
16. Schaefer, T.J.: The complexity of satisfiability problems. In: Proceedings of the 10th ACM STOC, pp. 216–226 (1978)
17. Tovey, C.A.: A simplified satisfiability problem. Discrete Appl. Math. 8(1), 85–89 (1984)
18. Valiant, L.G.: The complexity of computing the permanent. Theoret. Comput. Sci. 8(2), 189–201 (1979)
19. Valiant, L.G.: The complexity of enumeration and reliability Problems. SIAM J. Comput. 8(3), 410–421 (1979)
20. Yamasaki, S., Doshita, S.: The satisfiability problem for the class consisting of Horn sentences and some non-Horn sentences in propositional logic. Infor. Control 59(1-3), 1–12 (1983)

Anonymous Fuzzy Identity-Based Encryption for Similarity Search[*]

David W. Cheung, Nikos Mamoulis, W.K. Wong, S.M. Yiu, and Ye Zhang

Department of Computer Science, University of Hong Kong, Hong Kong
{dcheung,nikos,wkwong2,smyiu,yzhang4}@cs.hku.hk

Abstract. In this paper, we consider the problem of predicate encryption and focus on the predicate for testing whether the Hamming distance between the attribute X of a data item and a target V is equal to (or less than) a threshold t where X and V are of length m. Existing solutions either do not provide attribute protection or produce a big ciphertext of size $O(2^m)$. For the equality version of the problem, we provide a scheme which is match-concealing (MC) secure and the sizes of the ciphertext and token are both $O(m)$. For the inequality version of the problem, we give a practical scheme, also achieving MC security, which produces a ciphertext with size $O(m^{t_{max}})$ if the maximum value of t, t_{max}, is known in advance and is a constant. We also show how to update the ciphertext if the user wants to increase t_{max} without constructing the ciphertext from scratch.

Keywords: predicate encryption, anonymous fuzzy IBE, inner-product encryption.

1   Introduction

It is getting more popular for a data owner to take advantage of the storage and computing resources of a data center to hold the data in encrypted form. Depending on the access right of a user, only authorized records can be retrieved. Due to privacy and security concerns, the data should not be decrypted at the data center and checked against the criteria. Thus computation should be carried out on encrypted data directly. Usually, users are given a token (by the owner) and based on this token, only authorized records are selected and later decrypted on the user site. Examples of outsourcing applications which are based on this model are retrieval of encrypted documents by keyword matching, selection of encrypted audit logs using multi-dimensional range queries on authorized IP addresses or port numbers, and Hamming distance based similarity search on encrypted DNA sequence data. The problem, in fact, has received much attention from both database [8,17] and cryptography [15,2,14,6,10] communities.

In general, the problem can be stated as follows. For each data item M, there is an associated predicate attribute value X (X may or may not be part of the

---

[*] This work was supported by Grant HKU 715509E from Hong Kong RGC.

O. Cheong, K.-Y. Chwa, and K. Park (Eds.): ISAAC 2010, Part I, LNCS 6506, pp. 61–72, 2010.

record M). Let f be a predicate on X representing the computation we want to carry out so that the data item M can be successfully decrypted if and only if $f(X) = 1$. Authorized users will obtain a token generated by the owner in order to perform the predicate evaluation. A different token can be generated for different users with different access power. For example, consider a database of medical records. Each record (M) can be encrypted based on a selected region of the DNA sequence (X) of the person. Note that X is the associated predicate attribute of M and needs not be part of the record M. When a research team is authorized to investigate the relationship between a certain DNA sequence V with diseases, this team would acquire a token which corresponds to the predicate f such that $f(X) = 1$ if and only if $HammingDist(X, V) \leq t$, say $t = 5$. By using the token, the research team would decrypt all medical records for which the corresponding DNA sequence is similar to V. In the above motivating example, it is obvious that the research team should not infer any information on records for which the corresponding attribute X is far away from V (i.e. $HammingDist(X, V) > 5$) since they are not authorized to do so. In addition, it is desirable that the ciphertext $E(pk, I, M)$, where pk is the public key of the data owner and E is the encryption algorithm, is the same for different V and t such that the encryption of data items needs only to be done once. This emerging branch of encryption schemes are referred to as *predicate encryption*.

Here we focus on the predicate f that tests whether the Hamming distance between V and X is equal to (or less than) a certain threshold t, where V and X are assumed to be bit vectors of equal length $m \in \mathbb{N}$. Hamming distance is an important searching criterion for record retrieval, with many interesting applications in databases, bioinformatics, and other areas. Note that V and t can vary and will be given to the owner for the generation of a token independent of the ciphertext $E(pk, I, M)$.

The security of predicate encryption [10] can be classified into (1) protecting the data items only; and (2) protecting both the data items and the associated predicate attributes. Attribute protection is usually referred to as *anonymous* in general and can be further classified into two levels: *match-revealing (MR)* [14] and *match-concealing (MC)* [6] (or say *attribute-hiding* in [10]). The difference between MR and MC is that predicate attributes will remain hidden in MC level even if they satisfy the predicate. While in MR level, if attribute X satisfies the predicate f (i.e. $f(X) = 1$), some more information on X (or even the whole X) other than the information of $f(X) = 1$ may be known. In our "medical record" example, we sometimes require the encryption scheme to be anonymous in order for the DNA sequence to be protected since the DNA sequence may contain genetic disorder information which should be kept private for individuals. It depends on applications whether we require MC or MR level of security. For example, if attribute X is part of data item M, when X satisfies the predicate, data item M will be properly decrypted and therefore people can see the entire X anyway. In such case, MR security seems to be a proper choice. So far, the predicate encryption scheme supporting this Hamming distance predicate is the one in [12], called "Fuzzy Identity-Based Encryption". However, this scheme does not provide the property of anonymity (i.e., attribute protection). In this paper, we propose "Anonymous Fuzzy Identity-Based Encryption" schemes to handle both the equality (i.e., $HammingDist(X, V) = t$) and inequality (i.e., $HammingDist(X, V) \leq t$) threshold versions of the predicate.

It is not trivial how to make the scheme in [12] anonymous. On the other hand, there is a generic solution [6] (see Appendix A) that can support the predicate we study with the property of anonymity and it is MC secure. This general construction supports for any polynomially computable predicate. However, it embeds (pre-computes for) every possible value of $V \in \{0, 1\}^m$ and t in the ciphertext even for the equality threshold version of the problem (the same applies to the inequality version), thus the size of each ciphertext is $O(t2^m)$ which is impractical even for moderate m although the token size is constant.

1.1 Our Contributions

For the equality threshold version, we provide an anonymous fuzzy identity-based encryption scheme achieving the MC level of security with both the size of ciphertext and token equal to $O(m)$. The core idea of our scheme comes from [10] which provides an inner-product encryption scheme. We represent the Hamming distance computation as an inner product such that X and V can be separated into the ciphertext and the token, respectively, so that V can be given only when the token is needed to be generated.

For the inequality threshold version, we provide a practical scheme to solve the problem. In many applications (e.g., in bioinformatics applications), $t << m$. Even assuming that we know the maximum value of t (t_{max}) in advance and is a constant, the size of the ciphertext produced by the solution based on [6] is still $O(2^m)$. In our scheme, also achieving the MC security level, the size of ciphertext is only $O(m^{t_{max}})$ (precisely, $\sum_{i=0}^{t_{max}+1} \binom{m}{i}$) which is much smaller than $O(2^m)$ if $t_{max} << m$. The core of this scheme is to come up with an inner product expression with a total number of $\sum_{i=0}^{t+1} \binom{m}{i}$ terms to express whether $HammingDist(X, V) \leq t$ and modifying the scheme in [10] to a new primitive to support our encryption scheme. We also show how to update the ciphertext to increase the value of t_{max} without recomputing the ciphertext from scratch.

1.2 Related Work

The predicate that was studied in the beginning is "exact keyword matching". That is, whether the value associated with the token is equal to the attribute value hidden in the ciphertext. Schemes that only provide data item security are basically "Identity-Based Encryption" [3]. Schemes protecting both the data item and the attributes were initiated by Song et al. [15] in the private-key setting and by Boneh et al. [2] in the public-key setting. The relationship between [2] and "Anonymous Identity-Based Encryption" [7] was revisited in [1].

Later, range predicates were also considered. Boneh et al. devised an Augmented Broadcast Encryption [5] which allows checking if the attribute value falls within a range on encrypted data. Their scheme also provides attribute protection. Then, Boneh and Waters [6] extended it for multi-dimensional range queries. Shi *et al.* [14] devised a more efficient scheme, but it is MR secure.

The predicate investigated in this paper was initiated by [12] where their scheme only protects the data item. However, there is no practical scheme supporting this predicate with attribute protection in a public-key setting. Park *et al.* [11] investigated this problem in the private-key setting and their solution is IND2-CKA secure. Liesdonk [16] proposed a public-key setting for this problem. However, the scheme requires the threshold value t to be fixed in the setup time.

Our work is using [10] as a framework. [10] provided schemes for handling predicates represented as inner products. We show how to represent the Hamming distance computation as inner products, and then derive a slightly different encryption scheme for better performance when considering the inequality case. In our work, we consider the problem of attribute protection in a public-key setting. In some applications, people may also want to provide protection to predicate ("the token"), which is inherently unachievable in public-key setting. A predicate encryption supporting inner product in private-key setting has been devised in [13] which can provide predicate privacy.

In this paper, we only consider a non-interactive solution (i.e., an encryption scheme). We should note that there are interactive solutions (e.g., [9]) for the same problem.

1.3   Paper Organization

The rest of this paper is organized as follows. Section 2 introduces the framework of the encryption scheme and the MC security model. Section 3 presents the scheme for the equality threshold version (i.e., $HammingDist(V, X) = t$) of the problem and Section 4 deals with the inequality threshold version (i.e., $HammingDist(V, X) \leq t$) of the problem.

2   Preliminaries

We assume that the attribute X is represented as a bit vector of length $m \in \mathbb{N}$. The attribute V (referred to as the *target attribute*) provided by the user to generate the token is also a bit vector of the same length as X. In the rest of the paper, for simplicity, we focus on predicate-only encryption, that is, we assume that we only have X without the data item M. So, the scheme will output "1" to indicate the decryption is successful ($f(X) = 1$) and "0" otherwise. Note that extending solutions for predicate-only encryption to include the data item M can be done easily (e.g., [10]). Also, there exist applications that we only need to encrypt the attribute X and based on the decryption result to retrieve the corresponding records separately.

Let $\mathcal{G}$ be a group generator which takes security parameter $n \in \mathbb{N}$ as input and (randomly) outputs $(p, q, r, \mathbb{G}, \mathbb{G}_T, \hat{e})$, where $\hat{e} : \mathbb{G} \times \mathbb{G} \to \mathbb{G}_T$ is a bilinear map which can be computed efficiently. $\mathbb{G}$ and $\mathbb{G}_T$ are cyclic with the same composite order $N = pqr$ where p, q and r are three distinct large primes. Let $\mathbb{G}_p$, $\mathbb{G}_q$ and $\mathbb{G}_r$ be the cyclic subgroups of $\mathbb{G}$ with order of p, q and r separately.

2.1   Framework

An anonymous fuzzy identity-based encryption scheme Π consists of the following four probabilistic polynomial-time (PPT) algorithms.

- Setup(1^n): On an unary string input 1^n where $n \in \mathbb{N}$ is a security parameter, it produces the public-private key pair (pk, sk).
- Encrypt(pk, X): On the public key pk and attribute vector X, it outputs the ciphertext C.
- GenTK(pk, sk, V, t): The token generation algorithm takes the public key pk, private key sk, outputs the token TK for the vector V and threshold t.
- Test(pk, TK, C): On the ciphertext C, the token TK and the public key pk, it outputs "1" if the Hamming distance between the vector X associated with C and the vector V associated with TK is equal to t associated with TK (is less than or equal to t for the inequality version); "0" otherwise.

2.2   The MC Security Model

We define the MC security in the selective model [6,14,10] as follows.

Definition 1. *An anonymous fuzzy identity-based encryption scheme Π is selectively MC secure if for any PPT adversary $\mathcal{A}$, the advantage of $\mathcal{A}$ in the following game is negligible.*

Setup: The adversary $\mathcal{A}(1^n)$ outputs two possible equal-length vectors X_0 and X_1 to the challenger $\mathcal{C}$. $\mathcal{C}$ runs Setup(1^n) and gives pk to $\mathcal{A}$.

Challenge: The challenger $\mathcal{C}$ picks a random bit $b \in \{0,1\}$ and encrypts X_b under pk. The ciphertext C^* is given to $\mathcal{A}$.

Phase 1: The adversary $\mathcal{A}$ may adaptively request a polynomially bounded number of tokens for any (V_i, t_i), with the restriction that $t_i = HammingDist(V_i, X_j)$ for both $j = 0, 1$ or $t_i \neq HammingDist(V_i, X_j)$ for both $j = 0, 1$. (For inequality threshold: $t_i < HammingDist(V_i, X_j)$ for both $j = 0, 1$ or $t_i \geq HammingDist(V_i, X_j)$ for both $j = 0, 1$)

Guess: $\mathcal{A}$ outputs a guess bit b'. The advantage of $\mathcal{A}$ is defined as $\left| \Pr[b' = b] - \frac{1}{2} \right|$.

The intuition behind Definition 1 is that if the encryption scheme is MC secure that C^* leaks no information on the attribute X_0 and X_1, then the adversary $\mathcal{A}$ cannot distinguish X_0 from X_1 to output a proper guess bit $b' = b$. To restrict $t_i = HammingDist(V_i, X_j)$ for both $j = 0, 1$ or $t_i \neq HammingDist(V_i, X_j)$ for both $j = 0, 1$ prevents $\mathcal{A}$ trivially distinguishes X_0 from X_1 because the only information allowed to leak in the MC security is whether t_i is equal to $HammingDist(V_i, X_j)$ or not. A similar restriction is applied to the inequality threshold case as well.

3   Scheme for Equality Threshold

In this section, we describe our scheme for handling the equality threshold version of the Hamming distance predicate. Recall that both the target attribute V and threshold t will only be known when the user wants to obtain a token. We need to produce a ciphertext based on attribute X; a token based on V and t even after X is encrypted. The Test() combines the ciphertext and token together to compute Hamming distance $HammingDist(X,V)$. To the best of our knowledge, we are aware that only bilinear map can provide such ability while not being too powerful to break the security. Intuitively, given g^a and g^b, bilinear map combines a and b by computing $\hat{e}(g^a, g^b) = \hat{e}(g,g)^{ab}$. More specifically, if we encrypt attribute X as ciphertext $C = g^{f(X)}$ and generate token $TK = g^{y(V,t)}$ for target attribute V and threshold t, we can construct Test(C,TK) as $\hat{e}(C,TK) = \hat{e}(g^{f(X)}, g^{y(V,t)}) = \hat{e}(g,g)^{f(X)\cdot y(V,t)}$. If we can find $f(X)$ and $y(V,t)$ such that $f(X)y(V,t) = HammingDist(X,V)$, Test$(C,TK)$ will function correctly. More generally, $f(X)$ and $y(V,t)$ would output a vector. This is because given two vector $(g^{a_1},\ldots,g^{a_m})$ and $(g^{b_1},\ldots,g^{b_m})$, we would combine $\boldsymbol{a} = (a_1,\ldots,a_m)$ and $\boldsymbol{b} = (b_1,\ldots,b_m)$ by computing $\prod_{i=1}^m \hat{e}(g^{a_i}, g^{b_i}) = \hat{e}(g,g)^{\sum_{i=1}^m a_i,b_i} = \hat{e}(g,g)^{\boldsymbol{a}\cdot\boldsymbol{b}}$ where $\boldsymbol{a} \cdot \boldsymbol{b}$ denotes the inner product [10,4] of $\boldsymbol{a}$ and $\boldsymbol{b}$.

Lemma 1. *Given two bit vectors X and V of equal length m, $HammingDist$ (X,V) equals $\sum_{i=1}^m x_i(1-2v_i) + 1 \times \sum_{i=1}^m v_i$, where $X = x_1\ldots x_m$ and $V = v_1\ldots v_m$.*

The encryption scheme [10] allows us to generate a ciphertext C based on $\boldsymbol{a} = (a_1,\ldots,a_n)$ and a token TK based on $\boldsymbol{b} = (b_1,\ldots,b_n)$ such that given C and TK, we can compute $e(g,g)^{s[\sum_{i=1}^n a_i b_i]}$, where s is a random number, which gives $1_{\mathbb{G}_T}$ only when the inner product $\sum_{i=1}^n a_i b_i = 0$, or a random number otherwise. [10] is MC secure for the above inner product predicate which allows us devising encryption schemes based on the inner product expression which will be also MC secure. To evaluate whether $HammingDist(X,V) = t$, according to Lemma 1, we can check whether $e(g,g)^{s[\sum x_i(1-2v_i)+1\times(\sum v_i-t)]}$ equals $1_{\mathbb{G}_T}$ or not. Equivalently, we construct $f(X) = \boldsymbol{a} = (x_1,\ldots,x_m,1)$ and $y(V,t) = \boldsymbol{b} = (1-2v_1,\ldots,1-2v_m,\sum v_i - t)$.

The sizes of both ciphertext and token are $O(n)$ in [10] provided that $\boldsymbol{a}$ and $\boldsymbol{b}$ are n-length vectors, meaning that the sizes of both ciphertext and token in the above scheme are $O(m)$.

Security analysis: Our encryption scheme is MC secure. The proof is based on a reduction. Assume that there exits an adversary $\mathcal{A}_1$ that can win the MC game of our scheme with non-negligible advantage, we can use $\mathcal{A}_1$ as a subroutine to construct an adversary $\mathcal{A}_2$ that can win the MC game of the scheme in [10] with non-negligible advantage: When $\mathcal{A}_1$ outputs two vectors X_0 and X_1 to be challenged, $\mathcal{A}_2$ forwards $(X_0,1)$ and $(X_1,1)$ to the challenger. When $\mathcal{A}_1$ asks for a token query for (V,t) to $\mathcal{A}_2$, since $HammingDist(X,V) = t$ (or $\neq t$) corresponds to $\sum x_i(1-2v_i) + 1\times(\sum v_i - t) = 0$ (or $\neq 0$), $\mathcal{A}_2$ is able to answer the query by asking the challenger a token query for $(1-2v_1,\ldots,1-2v_m,\sum v_i - t)$. The detailed proof is omitted in this paper.

4   Scheme for Inequality Threshold

We borrow an idea from [6] to construct a generic solution for the case of having an inequality threshold. This solution is MC secure. The details of this generic solution are given in Appendix A. The ciphertext size of this solution is $O(t2^m)$ although the token size is constant which is not practical. In the following, we provide a practical scheme to handle the inequality threshold version.

If we can know the maximum value for the threshold t, t_{max}, in advance, we can have a scheme which is better than the generic solution. The size of the ciphertext can be reduced to $O(\sum_{i=0}^{t_{max}+1} \binom{m}{i})$. In some applications, $t_{max} << m$ and is a constant. In that case, the size becomes $O(m^{t_{max}})$. The restriction on setting t_{max} seems to be quite stringent. At the end of this section, we show how one can update the ciphertext if the user decides to increase t_{max} without computing ciphertext from scratch. We first present the scheme for known t_{max}.

The idea behind our construction is based on the observation that for Hamming distance H, $H \leq t$ if and only if $H(H-1)\ldots(H-t) = 0$. Then, if we evaluate $\hat{e}(g,g)^{sH(H-1)\ldots(H-t)}$ as Test() result where s is random, when $H \leq t$, Test() will be $1_{\mathbb{G}_T}$ (no information is leaked except from the fact that $H \leq t$); when $H > t$, $H(H-1)\ldots(H-t) \neq 0$, Test() will output a random number (still no information is leaked except from the fact $H > t$ since Test() $\neq 1_{\mathbb{G}_T}$ computationally). Note that although evaluating $H(H-1)\ldots(H-t)$ seems trivial in performance, it helps to ensure no information can be leaked which is required in the MC security.

Since the formula $H(H-1)\ldots(H-t)$ where $H = \sum x_i(1-2v_i)+\sum v_i$ contains both information from ciphertext (i.e. knowledge of x_i) and token (i.e. knowledge of v_i and t) which cannot be available at the same time, we need to split the formula to these two parts (ciphertext and token). Recall that as we discussed in Section 3, we can split the formula to $f(X)$ and $y(V,t)$ whose inner product $f(X) \cdot y(V,t)$ provides the result for $H(H-1) \cdot \ldots \cdot (H-t)$. The following lemma expands $H(H-1) \cdot \ldots \cdot (H-t)$ so that we can find $f(X)$ and $y(V,t)$. We let

$$H(H-1) \cdot \ldots \cdot (H-t) = a_{t+1}H^{t+1} + a_t H^t + \ldots + a_1 H \qquad (1)$$

where we assume a_k $(k = 1, \ldots, t+1)$ can be efficiently determined.

Lemma 2. *Given attribute $X = (x_1, \ldots, x_m)$, target attribute $V = (v_1, \ldots, v_m)$ and threshold t, we denote H as the Hamming distance $HammingDist(X,V)$ and define $a_0 = 0$ and b_j $(j = 0, \ldots, t+1)$ as*

$$b_j = a_{t+1}\binom{t+1}{t+1-j}(\sum v_i)^{t+1-j} + a_t\binom{t}{t-j}(\sum v_i)^{t-j} + \ldots + a_j\binom{j}{0}. \qquad (2)$$

Then, we have $H(H-1) \cdot \ldots \cdot (H-t)$

$$= \sum_{j=0}^{t+1} b_j \Big(\sum_{k_1+\ldots+k_m=j} \frac{j!}{k_1! \cdot \ldots \cdot k_m!}(1-2v_1)^{k_1} \cdot \ldots \cdot (1-2v_m)^{k_m} x_1^{k_1} \cdot \ldots \cdot x_m^{k_m}\Big). \qquad (3)$$

Now, $H(H - 1) \cdot \ldots \cdot (H - t)$ can be represented as inner product of $f(X) \cdot y(V,t) = \sum f_i(X) \cdot y_i(V,t)$. This is the key idea to our construction for inequality threshold. However, notice that $x_i \in \{0, 1\}$, we would future reduce the number of items in Eq. (3) based on the observation that $x_1^{k_1} \cdot \ldots \cdot x_m^{k_m} = \prod_{\{i:k_i>0\}} x_i$ if $x_i \in \{0, 1\}$. Then, Eq. (3) can be refined as: $H(H - 1) \cdot \ldots \cdot (H - t)$

$$
\begin{aligned}
&= b_0 + \sum_{1 \le j \le m} \left[\sum_{1 \le k_1 \le t+1} b_{k_1} (1 - 2v_j)^{k_1} \right] x_j \\
&+ \sum_{1 \le j_1 < j_2 \le m} \left[\sum_{k_1 + k_2 \le t+1; k_i \ge 1} \frac{(k_1 + k_2)!}{k_1! k_2!} b_{k_1 + k_2} (1 - 2v_{j_1})^{k_1} (1 - 2v_{j_2})^{k_2} \right] x_{j_1} x_{j_2} \\
&+ \ldots \\
&+ \sum_{1 \le j_1 < \ldots < j_l \le m} \left[\sum_{\substack{k_1 + \ldots + k_l \le t+1; \\ k_i \ge 1}} \frac{(k_1 + \ldots + k_l)!}{k_1! \ldots k_l!} b_{k_1 + \ldots + k_l} (1 - 2v_{j_1})^{k_1} \ldots (1 - 2_{j_l})^{k_l} \right] x_{j_1} \ldots x_{j_l} \\
&+ \ldots \\
&+ \sum_{1 \le j_1 < \ldots < j_{t+1} \le m} \left[(t+1)! b_{t+1} (1 - 2v_{j_1}) \ldots (1 - 2v_{j_{t+1}}) \right] x_{j_1} \ldots x_{j_{t+1}}.
\end{aligned}
$$
(4)

For simplicity, we can denote Eq. (4) as below $H(H - 1) \cdot \ldots \cdot (H - t)$:

$$
\begin{aligned}
&= B_0 + B_1 x_1 + B_2 x_2 + \ldots + B_m x_m \\
&+ B_{m+1} x_1 x_2 + \ldots + B_{m + \binom{m}{2}} x_{m-1} x_m \\
&+ \ldots \\
&+ B_{m + \binom{m}{2} + \ldots + \binom{m}{t} + 1} x_1 x_2 \ldots x_{t+1} + \ldots + B_{m + \binom{m}{2} + \ldots + \binom{m}{t+1}} x_{m-t} \ldots x_m.
\end{aligned}
$$
(5)

The number of items in Eq. (5) is $1 + \binom{m}{1} + \ldots + \binom{m}{t+1} = \sum_{i=0}^{t+1} \binom{m}{i}$. We now describe a construction based on [10] and Eq. (5) whose ciphertext and token size are both $O(\sum_{i=0}^{t_{max}+1} \binom{m}{i})$.

Recall $(\texttt{Setup}, \texttt{Enc}, \texttt{GenKey}, \texttt{Dec})$ in [10] can support n-dimension vectors $\boldsymbol{a}$ and $\boldsymbol{b}$ such that $C \xleftarrow{\$} \texttt{Enc}(\boldsymbol{a})$ and $TK \xleftarrow{\$} \texttt{GenKey}(\boldsymbol{b})$ where $\texttt{Dec}(C, TK) = 1$ if and only if the inner product $\boldsymbol{a} \cdot \boldsymbol{b} = 0$. In our construction, we let n be $\sum_{i=0}^{t_{max}+1} \binom{m}{i}$. Encryption algorithm $\texttt{Encrypt}(X = x_1, \ldots, x_m)$ in our construction calls $\texttt{Enc}()$ with input vector[1]:

$$
\boldsymbol{a} = (1, x_1, \ldots, x_m, x_1 x_2, \ldots, x_{m-1} x_m, \ldots, x_{m - t_{max}} \cdot \ldots \cdot x_m).
$$
(6)

Token for V and t is generated by calling $\texttt{GenKey}()$ with input vector:

$$
\boldsymbol{b} = (B_0, \ldots, B_{m + \ldots + \binom{m}{t} + 1}, \ldots, B_{m + \ldots + \binom{m}{t+1}}, 0 \ldots, 0).
$$
(7)

---

[1] Note that although there exists x_1, x_2 and $x_1 x_2$ in $\boldsymbol{a}$, given ciphertext for x_1 and x_2, we cannot reuse x_1 and x_2 to compute $x_1 x_2$. This is because bilinear map is able to do only one multiplication (on encrypted data), however, we have used this ability to combine ciphertext and token, therefore, such redundancy in $\boldsymbol{a}$ seems to be necessary.

Note that a and b are constructed according to Eq. (5) and therefore, the inner product of $a \cdot b = H(H-1) \cdot \ldots \cdot (H-t)$.

This construction has ciphertext and token both of size $O(n) = O(\sum_{i=0}^{t_{max}+1} \binom{m}{i})$, however, some items in the token are in fact "0" since t may be less than t_{max}; more specifically, $H(H-1)\ldots(H-t)$ is $t+1$ degree and items in a whose degree larger than $t+1$ (i.e. $x_1 x_2 \ldots x_{t+2}, \ldots, x_{m-t_{max}} \ldots x_m$) will have coefficient "0" in b (Eq. (7)). This allows us to further reduce the token size. To do so, we devise an encryption scheme slightly different from [10] such that a is still n-dimensional while b can be any n'-dimensional ($n' \le n$) and decryption will output "1" if and only if the inner product $\sum_{i=1}^{n'} a_i b_i = 0$. We describe this construction as follows:

- $\mathtt{Setup}(1^n)$. This algorithm is the same as $\mathtt{Setup}(1^n)$ in [10].
- $\mathtt{Encrypt}(pk, X = x_1 \ldots x_n)$. It is the same as $\mathtt{Enc}(pk, X = x_1 \ldots x_n)$ in [10].
- $\mathtt{GenTK}(pk, sk, V = v_1 \ldots v_{n'})$. Note that $n' \le n$. It randomly selects $\{r_{1,i}, r_{2,i}\}_{i \in [1,n']}$ and f_1, f_2 from $\mathbb{Z}_N$. Then, it randomly selects Q'' from $\mathbb{G}_q$ and R'' from $\mathbb{G}_r$. It outputs token TK:

$$\left\{ \begin{array}{c} K_0 = Q''R'' \prod_{i=1}^{\boxed{n'}} h_{1,i}^{-r_{1,i}} h_{2,i}^{-r_{2,i}} \\ \left[K_{1,i} = g_p^{r_{1,i}} g_q^{f_1 v_i}, K_{2,i} = g_p^{r_{2,i}} g_q^{f_2 v_i} \right]_{i \in \boxed{[1,n']}} \end{array} \right\}.$$

- $\mathtt{Test}(pk, TK, C)$. It computes $r = \hat{e}(C_0, K_0) \prod_{i=1}^{\boxed{n'}} \hat{e}(C_{1,i}, K_{1,i}) \hat{e}(C_{2,i}, K_{2,i})$. If $r = 1_{\mathbb{G}_T}$, it will output "1"; otherwise it outputs "0".

The above encryption scheme is MC secure provided that Assumption 1 in [10] holds. Although we cannot directly reduce (i.e., by a black-box way) the security of [10] to the security of the above scheme because K_0 in [10] is fixed for n-length (rather than for any $n' < n$ in our case), we are able to prove the security of the above scheme from scratch, following a similar idea as [10].

Applying the above encryption scheme instead of the original scheme of [10] to Eq. (6) and (7), we obtain the final construction Π_1. Note that the above scheme also makes our security analysis of the final construction much easier (see the full paper). Π_1 is described as follows.

- $\mathtt{Setup}(1^n)$: The algorithm first runs $\mathcal{G}(1^n)$ to obtain $(p, q, r, \mathbb{G}, \mathbb{G}_T, \hat{e})$. Then, it randomly selects g_p from $\mathbb{G}_p$, g_q from $\mathbb{G}_q$ and g_r from $\mathbb{G}_r$. It also randomly selects $\{h_{1,l,i}, h_{2,l,i}\}_{l \in [1,t_{max}+1], i \in [1,\binom{m}{l})]}$ from $\mathbb{G}_p$. Then it randomly selects h_3, h_4 from $\mathbb{G}_p$. It also randomly selects R, R_3, R_4 and $\{R_{1,l,i}, R_{2,l,i}\}_{l \in [1,t_{max}+1], i \in [1,\binom{m}{l})]}$ from $\mathbb{G}_r$. It outputs

$$pk = \left\{ \begin{array}{c} g_p, g_r, Q = g_q R, \\ [H_{1,l,i} = h_{1,l,i} R_{1,l,i}, H_{2,l,i} = h_{2,l,i} R_{2,l,i}]_{l \in [1,t_{max}+1], i \in [1,\binom{m}{l})]}, \\ H_3 = h_3 R_3, H_4 = h_4 R_4 \end{array} \right\}$$

and

$$sk = \left\{ \begin{array}{c} p, q, r, g_q, \\ [h_{1,l,i}, h_{2,l,i}]_{l \in [1,t_{max}+1], i \in [1,\binom{m}{l})]}, \\ h_3, h_4 \end{array} \right\}.$$

- $\texttt{Encrypt}(pk, X = x_1 \ldots x_m)$: Encryption algorithm first randomly selects s, α, β from $\mathbb{Z}_N$ and $\{R'_{1,l,i}, R'_{2,l,i}\}_{l \in [1, t_{max}+1], i \in [1, \binom{m}{i})]}, R'_3, R'_4$ from $\mathbb{G}_r$. Then it outputs ciphertext C:

$$\left\{ \begin{array}{c} C_0 = g_p^s, \\ \left[C_{1,l,i} = H_{1,l,i}^s Q^{\alpha x_{j_1} \cdots x_{j_l}} R'_{1,l,i}, C_{2,l,i} = H_{2,l,i}^s Q^{\beta x_{j_1} \cdots x_{j_l}} R'_{2,l,i} \right]_{1 \le j_1 < \ldots < j_l \le m}^{l \in [1, t_{max}+1];} \\ C_3 = H_3^s Q^\alpha R'_3, C_4 = H_4^s Q^\beta R'_4 \end{array} \right\}.$$

- $\texttt{GenTK}(pk, sk, V = v_1 \ldots v_m, t)$: It randomly selects $\{r_{1,l,i}, r_{2,l,i}\}_{l \in [1, t+1], i \in [1, \binom{m}{i})]}$, r_3, r_4 and f_1, f_2 from $\mathbb{Z}_N$. Then, it randomly selects Q'' from $\mathbb{G}_q$ and R'' from $\mathbb{G}_r$. It outputs token TK:

$$\left\{ \begin{array}{cc} K_0 = Q'' R'' h_3^{-r_3} h_4^{-r_4} \prod_{l=1}^{t+1} \prod_{i=1}^{\binom{m}{i}} h_{1,l,i}^{-r_{1,l,i}} h_{2,l,i}^{-r_{2,l,i}}, & \\ K_{1,1,1} = g_p^{r_{1,1,1}} g_q^{f_1 B_1}, & K_{2,1,1} = g_p^{r_{2,1,1}} g_q^{f_2 B_1} \\ \ldots & \ldots \\ K_{1,1,m} = g_p^{r_{1,1,m}} g_q^{f_1 B_m}, & K_{2,1,m} = g_p^{r_{2,1,m}} g_q^{f_2 B_m} \\ K_{1,2,1} = g_p^{r_{1,2,1}} g_q^{f_1 B_{m+1}}, & K_{2,2,1} = g_p^{r_{2,2,1}} g_q^{f_2 B_{m+1}} \\ \ldots & \ldots \\ K_{1,2,\binom{m}{2}} = g_p^{r_{1,2,\binom{m}{2}}} g_q^{f_1 B_{m+\binom{m}{2}}}, & K_{2,2,\binom{m}{2}} = g_p^{r_{2,2,\binom{m}{2}}} g_q^{f_2 B_{m+\binom{m}{2}}} \\ \ldots & \ldots \\ K_{1,t+1,1} = g_p^{r_{1,t+1,1}} g_q^{f_1 B_{m+\ldots+\binom{m}{t}+1}}, & K_{2,t+1,1} = g_p^{r_{2,t+1,1}} g_q^{f_2 B_{m+\ldots+\binom{m}{t}+1}} \\ \ldots & \ldots \\ K_{1,t+1,\binom{m}{t+1}} = g_p^{r_{1,t+1,\binom{m}{t+1}}} g_q^{f_1 B_{m+\ldots+\binom{m}{t+1}}}, & K_{2,t+1,\binom{m}{t+1}} = g_p^{r_{2,t+1,\binom{m}{t+1}}} g_q^{f_2 B_{m+\ldots+\binom{m}{t+1}}} \\ \multicolumn{2}{c}{K_3 = g_p^{r_3} g_q^{f_1 B_0}, K_4 = g_p^{r_4} g_q^{f_2 B_0}} \end{array} \right\}$$

- $\texttt{Test}(pk, sk, TK, C)$: It outputs "1" if $r = 1_{\mathbb{G}_T}$ and "0" otherwise, where

$$r = \hat{e}(K_0, C_0) \hat{e}(K_3, C_3) \hat{e}(K_4, C_4) \prod_{l=1}^{t+1} \prod_{i=1}^{\binom{m}{i}} \hat{e}(K_{1,l,i}, C_{1,l,i}) \hat{e}(K_{2,l,i}, C_{2,l,i}).$$

The size of ciphertext is still $O(\sum_{i=0}^{t_{max}+1} \binom{m}{i})$ but the size of token is now $O(\sum_{i=0}^{t+1} \binom{m}{i})$ for threshold t. The security of the scheme is stated in Theorem 1 and proved in the full paper.

Theorem 1. *Our construction Π_1 in Section 4 is Selectively MC secure provided that Assumption 1 in [10] holds.*

Lastly, to show that it is feasible to compute the coefficients a_k ($k = 1, \ldots, t+1$) in Eq. (1), we have implemented an algorithm which, in fact, can calculate all elements of vector $\boldsymbol{b}$ in Eq. (7). For example, with input $m = 100$ and $t = 3$, it took about 16 seconds to calculate all elements on an Intel Core 2 Due E6750 2.66GHz CPU platform.

Increasing t_{max}: If the value α, β and s generated in $\texttt{Encrypt}()$ are kept by that user, the user can update the ciphertext to increase t_{max} without producing the ciphertext from scratch. When the maximum threshold is updated from t_{max} to T', the corresponding vector $\boldsymbol{a}$ in Eq. (6) also needs to be updated as:

$$\boldsymbol{a}' = (\boldsymbol{a}, x_1 \cdot \ldots \cdot x_{t_{max}+2}, \ldots, x_{m-T'} \cdot \ldots \cdot x_m). \tag{8}$$

Recall that when the maximum threshold is t_{max}, $\boldsymbol{a}$ contains all items whose degree $l \leq t_{max} + 1$. Thus, when the maximum threshold becomes T', we need to produce those items whose degree is within $t_{max} + 2$ to $T' + 1$, namely $x_{j_1}...x_{j_l}$ where $t_{max} + 2 \leq l \leq T' + 1$ and $1 \leq j_1 < ... < j_l \leq m$ in Eq. (8). This can be done without from scratch by calculating $C_{1,i} = H_{1,i}^s Q^{\alpha a_i''} R_{1,i}$ and $C_{2,i} = H_{2,i}^s Q^{\beta a_i''} R_{2,i}$ for $i = 1, \ldots, k$, given α, β and s. Where we denote vector $\boldsymbol{a}'' = (a_1'', \ldots, a_k'')$ such that $\boldsymbol{a}' = (\boldsymbol{a}, \boldsymbol{a}'')$.

The above update procedure can be shown to be MC secure. Intuitively, when $x_1, \ldots, x_m$ in $\boldsymbol{a}$ are determined, all items in $\boldsymbol{a}$ (also in $\boldsymbol{a}'$) have been determined since they are the multiplication of two or more items in $\{x_1, \ldots, x_m\}$. For any possible $t_{max} \geq 0$, $\boldsymbol{a}$ (and therefore $\boldsymbol{a}'$) contains $(x_1, \ldots, x_m)$ for sure. That means all terms including the one to be generated due to the increase in t_{max} has been fixed although they are not computed yet. Therefore, an adaptive attack will not work since it has no way to adaptively modify how the missing items are generated. The detailed proof is omitted in this paper.

In the worst case, $t_{max} = m$, the size of the ciphertext (and token) becomes $O(2^m)$. Although it is better than $O(m2^m)$ (since $t = m$) for the generic solution in Appendix A, it is not practical. So, this scheme should be used when t_{max} is small.

References

1. Abdalla, M., Bellare, M., Catalano, D., Kiltz, E., Kohno, T., Lange, T., Malone-Lee, J., Neven, G., Paillier, P., Shi, H.: Searchable encryption revisited: Consistency properties, relation to anonymous IBE, and extensions. In: Shoup, V. (ed.) CRYPTO 2005. LNCS, vol. 3621, pp. 205–222. Springer, Heidelberg (2005)
2. Boneh, D., Crescenzo, G.D., Ostrovsky, R., Persiano, G.: Public key encryption with keyword search. In: Cachin, C., Camenisch, J.L. (eds.) EUROCRYPT 2004. LNCS, vol. 3027, pp. 506–522. Springer, Heidelberg (2004)
3. Boneh, D., Franklin, M.K.: Identity-based encryption from the weil pairing. In: Kilian, J. (ed.) CRYPTO 2001. LNCS, vol. 2139, pp. 213–229. Springer, Heidelberg (2001)
4. Boneh, D., Goh, E.-J., Nissim, K.: Evaluating 2-DNF formulas on ciphertexts. In: Kilian, J. (ed.) TCC 2005. LNCS, vol. 3378, pp. 325–341. Springer, Heidelberg (2005)
5. Boneh, D., Waters, B.: A fully collusion resistant broadcast, trace, and revoke system. In: CCS (2006)
6. Boneh, D., Waters, B.: Conjunctive, subset, and range queries on encrypted data. In: Vadhan, S.P. (ed.) TCC 2007. LNCS, vol. 4392, pp. 535–554. Springer, Heidelberg (2007)
7. Boyen, X., Waters, B.: Anonymous hierarchical identity-based encryption (without random oracles). In: Dwork, C. (ed.) CRYPTO 2006. LNCS, vol. 4117, pp. 290–307. Springer, Heidelberg (2006)
8. Hacıgümüş, H., Iyer, B., Li, C., Mehrotra, S.: Executing SQL over encrypted data in the database-service-provider model. In: SIGMOD (2002)
9. Jarrous, A., Pinkas, B.: Secure hamming distance based computation and its applications. In: Abdalla, M., Pointcheval, D., Fouque, P.-A., Vergnaud, D. (eds.) ACNS 2009. LNCS, vol. 5536, pp. 107–124. Springer, Heidelberg (2009)

10. Katz, J., Sahai, A., Waters, B.: Predicate encryption supporting disjunctions, polynomial equations, and inner products. In: Smart, N.P. (ed.) EUROCRYPT 2008. LNCS, vol. 4965, pp. 146–162. Springer, Heidelberg (2008)
11. Park, H.-A., Kim, B.H., Lee, D.H., Chung, Y.D., Zhan, J.: Secure similarity search. In: GRC (2007)
12. Sahai, A., Waters, B.: Fuzzy identity-based encryption. In: Cramer, R. (ed.) EUROCRYPT 2005. LNCS, vol. 3494, pp. 457–473. Springer, Heidelberg (2005)
13. Shen, E., Shi, E., Waters, B.: Predicate privacy in encryption systems. In: Reingold, O. (ed.) TCC 2009. LNCS, vol. 5444, pp. 457–473. Springer, Heidelberg (2009)
14. Shi, E., Bethencourt, J., Chan, T.-H.H., Song, D., Perrig, A.: Multi-dimensional range query over encrypted data. In: IEEE Symposium on Security and Privacy (2007)
15. Song, D.X., Wagner, D., Perrig, A.: Practical techniques for searches on encrypted data. In: IEEE Symposium on Security and Privacy (2000)
16. van Liesdonk, P.: Anonymous and fuzzy identity-based encryption. Master's thesis, Eindhoven University (2007)
17. Wong, W.K., Cheung, D.W., Kao, B., Mamoulis, N.: Secure kNN computation on encrypted databases. In: SIGMOD (2009)

A   A Generic Construction from [6]

The main idea is that we can generate a ciphertext vector $(C_0, C_1, \ldots, C_{t_{max}})$ for each possible $V \in \{0,1\}^m$. Given $j \in [0, t_{max}]$, if $HammingDist(X, V) \leq j$, C_j is an encryption of the message "1"; otherwise, C_j will be an encryption of "0". When we Test() for a certain (V, t), we find the ciphertext vector of V and then decrypt the t-th element C_t in the vector. If $HammingDist(X, V) \leq t$, the decryption result should be "1". More specifically, Let $(\mathsf{G}, \mathsf{E}, \mathsf{D})$ be an IND-CPA secure encryption scheme.

- Setup(1^n) : Run $\mathsf{G}(1^n)$ to generate $\{(pk_{l,j}, sk_{l,j})\}_{l \in \{0,1\}^m, j \in [0, t_{max}]}$ for $(t_{max} + 1)2^m$ times. Return the public-key pk as $\{pk_{l,j}\}_{l \in \{0,1\}^m, j \in [0, t_{max}]}$ and the secret key sk as $\{sk_{l,j}\}_{l \in \{0,1\}^m, j \in [0, t_{max}]}$.
- Encrypt$(pk, X = x_1 \ldots x_m)$: For each $l \in \{0,1\}^m$, return $(C_0, C_1, \ldots, C_{t_{max}})_l$ where

$$C_j = \begin{cases} \mathsf{E}_{pk_{l,j}}(\text{"1"}) & \text{if } HammingDist(X, l) \leq j; \\ \mathsf{E}_{pk_{l,j}}(\text{"0"}) & \text{otherwise.} \end{cases}$$

- GenTK(pk, sk, V, t): Return $sk_{V,t}$ as the token TK.
- Test(pk, TK, C): It first finds $(C_0, C_1, \ldots, C_m)_V$ and returns $\mathsf{D}_{TK}(C_t)$.

The security of the above solution comes from the IND-CPA secure encryption scheme, see appendix A of [6] for more details.

Improved Randomized Algorithms for 3-SAT

Kazuo Iwama[1], Kazuhisa Seto[1], Tadashi Takai[2], and Suguru Tamaki[1]

[1] Kyoto University, Yoshida Honmachi, Sakyo-ku, Kyoto 606-8501, Japan
{iwama,seto,tamak}@kuis.kyoto-u.ac.jp
[2] East Japan Railway Company, Japan

Abstract. This pager gives a new randomized algorithm which solves 3-SAT in time $O(1.32113^n)$. The previous best bound is $O(1.32216^n)$ due to Rolf (J. SAT, 2006). The new algorithm uses the same approach as Iwama and Tamaki (SODA 2004), but exploits the non-uniform initial assignment due to Hofmeister et al. (STACS 2002) against the Schöning's local search (FOCS 1999).

Keywords: 3-SAT, exponential-time algorithms, randomized algorithms.

1 Introduction

3-SAT, the CNF satisfiability problem with clauses of size three, is often called *a benchmark problem* in the field of design and analysis of combinatorial algorithms. As shown in Table 1 (not exhaustive), many researchers have been involved in this problem, i.e., "breaking the world record" at that time or succeeded in decreasing the value of constant $c < 2$ such that their algorithms run in time $O(c^n)$, where n is the number of Boolean variables in the formula.

Not only improving the performance itself, many of these results include important new ideas being appreciated in their own right. For example, Paturi, Pudlák, Saks and Zane gave a backtrack approach, denoted by PPSZ in this paper, that works well for instances with strong constraint as well as how to increase constraints of the given instance automatically [PPZ97, PPSZ98]. Schöning (SCH) provided a simple and beautiful probabilistic analysis for the performance of the basic local search algorithm in [Sch99]. Hofmeister, Schöning, Schuler and Watanabe (HSSW) found that the distribution of initial assignments for the local search plays an important role and gave an optimal (non-uniform) distribution for the Schöning's local search [HSSW02]. Most recently Iwama and Tamaki (IT) proposed a nontrivial combination of the PPSZ's backtrack and the SCH's local search [IT04]. As one can see from the table, improvement has been getting harder and harder (the amount of reduction in c has become smaller and smaller) recently and in fact we have had no improvement at all for more than five years.

In this paper, we give a new upper bound, $O(1.32113^n)$, using the same approach as IT. As mentioned above, (i) IT is a combination of PPSZ and SCH, (ii) HSSW is exactly the same as SCH except the initial assignment distribution, and most importantly (iii) HSSW is faster than SCH. So, it is very natural to try

O. Cheong, K.-Y. Chwa, and K. Park (Eds.): ISAAC 2010, Part I, LNCS 6506, pp. 73–84, 2010.

Table 1. Worst case upper bounds for 3-SAT

c	type	ref.	c	type	ref.
1.839	det.	[MS79]	1.362	rand.	[PPSZ98]
1.769	det.	[Dan83]	1.334	rand.	[Sch99]
1.618	det.	[MS85]	1.3302	rand.	[HSSW02]
1.579	det.	[Sch92]	1.32971	rand.	[Rol03a]
1.505	det.	[Kul99]	1.3290	rand.	[BS03]
1.497	det.	[Sch96]	1.32793	rand.	[Rol03b]
1.481	det.	[DGH+02]	1.3238	rand.	[IT04]
1.476	det.	[Rod96]	1.32267	rand.	[IT04]+[PPSZ05]
1.473	det.	[BK04]	1.32216	rand.	[Rol06]
1.465	det.	[Sch08]	1.32113	rand.	This paper

to combine PPSZ and HSSW instead of PPSZ and SCH for better performance. Unfortunately, it immediately turns out that this naive idea does not work for the following reason: For an improved performance, different distributions are necessary for initial assignments of the two algorithms. This is easy if we can use two different initial assignments for the two algorithms, but the key thing of IT is that we must use a *single* initial assignment, the same one for both PPSZ and SCH in each repetition of the algorithm. Thus the obvious main issue of this paper is how to cope with this difficulty, which might be interesting in its own. Our answer is given in Section 3, after more detailed explanation of those previous algorithms.

2   Preliminaries

2.1   Definitions and Notations

An input formula uses n variables $\{x_1, x_2, \ldots, x_n\}$. A *literal* is a Boolean variable x_i or its negation $\bar{x}_i$. A *CNF formula* is a conjunction of clauses, where each clause is a disjunction of literals. If each clause contains at most k literals, we say the formula is k-CNF. A *satisfying assignment* is a 0/1 assignment to each variable, such that every clause contains at least one true literal under the assignment. For any finite set S, $|S|$ denote the cardinality of S. Let $[n]$ denote $\{1, 2, \ldots, n\}$. The *Hamming distance* $d(y, z)$ between assignments y and z in $\{0, 1\}^n$ is defined as $|\{i \mid y_i \neq z_i\}|$.

A *subcube* with respect to an assignment $z \in \{0, 1\}^n$ and $I \subseteq [n]$ is $\mathbf{C}(z, I) \equiv \{y \in \{0, 1\}^n \mid y_i = z_i \text{ for any } i \in I\}$ and then $\{x_i \mid i \in I\}$ is called *defining variables* of $\mathbf{C}(z, I)$. For example, let $z = (1, 0, 1, 0, \ldots, 0, 1)$ and $I = \{1, 3, 4\}$, then $\mathbf{C}(z, I) = \{y \in \{0, 1\}^n \mid y_1 = 1, y_3 = 1, y_4 = 0\}$ and defining variables are $\{x_1, x_3, x_4\}$. Note that, given a nonempty subset $S \subseteq \{0, 1\}^n$, the whole space $\{0, 1\}^n$ can be partitioned into a family $\{\mathbf{C}(z, I_z) \mid z \in S\}$ of disjoint subcubes so that $\mathbf{C}(z, I_z)$ contains the assignment $z \in S$ but no other $z' \in S - \{z\}$ (see [PPSZ05] for proof). We call such a partition (not unique in general) a *subcube partition* of $\{0,1\}^n$ with respect to S. We sometimes use the notation x_I to denote a vector $\{x_i\}_{i \in I}$ (ordered in a natural way).

In this paper, we always deal with probabilistic algorithms that have a probability of $2^{-\Theta(n)}$. Thus, we sometimes omit a factor of the form $2^{o(n)}$ in the calculation.

2.2 Algorithms PPSZ and SCH

Both PPSZ and SCH repeat an exponential number of tries. Each try of PPSZ consists of

(1) Generate a random assignment y.
(2) Generate a random permutation π of $[n]$ for the order of variables.
(3) Execute Davis-Putnam procedure with respect to π and y in n steps.

We use the following result stated in terms of subcube partitions. Given a formula F with a set S of satisfying assignments and subcube partition $\{\mathbf{C}(z, I_z) \mid z \in S\}$, $\tau(F_s, z \mid \mathbf{C}(z, I_z))$ is defined as the probability that a single try of PPSZ finds the assignment z under the condition that the initial assignment y is in $\mathbf{C}(z, I_z)$.

Lemma 1 ([Rol06]). *For any satisfiable 3CNF formula F and any partition $\mathbf{C}(z, I_z)$ described above, if $y \in \mathbf{C}(z, I_z)$ is chosen uniformly at random, then the value $\tau(F_s, z \mid \mathbf{C}(z, I_z))$ is bounded as follows:*

$$\tau(F_s, z \mid \mathbf{C}(z, I_z)) \geq 2^{-(1-\gamma)n + (1-\gamma-\beta)|I_z|},$$

where $\gamma = 0.6122939734$ and $\beta = 0.9062404894$.

Each try of SCH consists of

(1) Generate a random assignment y.
(2) Execute a local search for $3n$ steps starting from y.

Lemma 2 ([Sch99]). *Let F be a 3CNF formula and z be a satisfying assignment for F. For each assignment y, the probability that a single try of SCH starting from y finds z is at least $(1/2)^{d(y,z)}$.*

The most technical part of this paper is to bound the quantity $\mathbf{E}[(1/2)^{d(y,z)}]$ where y is generated with a certain probability distribution. Specifically, we are interested in the case when variables are partitioned into several groups $\{x_j\}_{j \in I_1} \cup \{x_j\}_{j \in I_2} \cup \cdots$ where $[n] = I_1 \cup I_2 \cup \cdots$ such that assignments in each group can be generated independently from the others. According to such a partition $\{I_i\}_i$ of $[n]$, an assignment y is decomposed into the set of substrings $\{y_{I_i}\}_i$. The following lemma is frequently used in the literature:

Lemma 3. *Under the above condition, the probability that a satisfying assignment is found by a single try of SCH starting from random y is at least*

$$\prod_i \mathop{\mathbf{E}}_{y_{I_i}} \left[\left(\frac{1}{2}\right)^{d(y_{I_i}, z_{I_i})} \right].$$

Proof. Straightforward due to the facts that $c^{v+w} = c^v \cdot c^w$, $d(y, z) = \sum_i d(y_{I_i}, z_{I_i})$ and $\mathbf{E}[XY] = \mathbf{E}[X]\,\mathbf{E}[Y]$ for independent random variables X, Y.   □

Now, given a formula F with a set S of satisfying assignments and a subcube partition $\{\mathbf{C}(z, I_z) \mid z \in S\}$, $\sigma(F, z|\mathbf{C}(z, I_z))$ is defined as the probability (averaged over y) that a single try of SCH finds the assignment z under the condition that the initial assignment $y \in \mathbf{C}(z, I_z)$.

Lemma 4 ([IT04]). *For any satisfiable 3CNF formula F and any partition $\mathbf{C}(z, I_z)$ described above, if $y \in \mathbf{C}(z, I_z)$ is chosen uniformly at random, then the value $\sigma(F, z|\mathbf{C}(z, I_z))$ is bounded as follows:*

$$\sigma(F, z|\mathbf{C}(z, I_z)) \geq \left(\frac{3}{4}\right)^{n - |I_z|}.$$

2.3   Algorithm BF

In this subsection, we recall the algorithm BF of [BS03]. Two clauses C and C' are called *independent* if they have no variable in common. For a formula F, a *maximal independent clause set* (MICS for short) $\mathcal{C}$ is a subset of the clauses of F such that all clauses in $\mathcal{C}$ are (mutually) independent and no clause in $F \setminus \mathcal{C}$ can be added to $\mathcal{C}$ without destroying this property. Note that the size of a MICS of 3CNF-formula is at most $n/3$. As the following lemma shows, BF that runs in polynomial time finds either a satisfying assignment or a MICS of relatively a large size.

Lemma 5 ([BS03]). *For any satisfiable 3-CNF formula F and any positive integer m, BF returns either a satisfying assignment of F, or a MICS of size $\geq m$, with probability at least 6^{-m}. (For example, if $m > n/3$, then BF always provides a satisfying assignment with probability 6^{-m} since there are no such MICS.)*

3   Main Results

3.1   Basic Ideas

It is well-known that PPSZ runs faster for less satisfying assignments, while SCH for more satisfying assignments. To exploit this complementary nature is a natural idea, but it does not work sufficiently well if we simply use the number of satisfying assignments as a trade-off parameter. Instead, IT [IT04] uses the size of the subcube the initial assignment of PPSZ and SCH falls into as a trade-off parameter (and hence both must use the same assignment). Now a simple calculation is enough to obtain the desired bound ($O(1.32266^n)$ and $O(1.32216^n)$ due to the improvement of the journal version [PPSZ05] of [PPSZ98]) from Lemmas 1 and 4.

To obtain better upper bounds, we would like to improve the bound of Lemma 4. One promising way is to use the idea of [HSSW02] to improve the initial assignment to SCH. Their idea is as follows: Let $C = (u \vee v \vee w)$ be some clause of satisfiable 3CNF-formula F. Since $(u, v, w) = (0, 0, 0)$ cannot be a satisfying assignment, it seems inefficient to generate $(0, 0, 0)$ as an initial assignment in SCH. If one assigns other values with more probability, the success probability seems to get higher. To keep the consistency of the entire assignment, the modification of initial assignments should be done for a clause set which only contains mutually independent clauses, i.e., for a MICS.

To adapt the above idea in the combination of PPSZ and SCH, there is one problem, that is, if we modify the distribution for generating initial assignments, then we can no longer use Lemma 1 since it assumes uniform initial assignments. Our trick is as follows: First we pick a uniformly random assignment. This is fed to PPSZ. Then we modify it using some randomized procedure (called Reassign) so that the distribution will be a desired one.

3.2   New Algorithm

We describe our algorithm formally. We begin with Reassign, the following randomized procedure that takes a MICS $\mathcal{C}$ and an assignment y as inputs and returns a modified assignment.

Reassign(MICS $\mathcal{C}$, assignment y)
for each $C = (x'_a \vee x'_b \vee x'_c) \in \mathcal{C}$ ($x'_a = x_a$ or $\bar{x}_a$, and similarly for x'_b and x'_c),
modify y by reassigning (y'_a, y'_b, y'_c) according to Fig. 1
where $y'_a = y_a$ if $x'_a = x_a$ and $1 - y_a$ otherwise, and similarly for y'_b and y'_c;
end for each
return y;

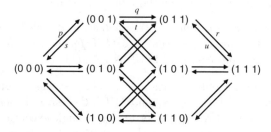

Fig. 1. Probability of changing assignments

Fig. 1 shows the probability that each assignment is changed. For instance, $(0, 0, 0)$ is changed into $(0, 0, 1)$ with probability p, $(0, 0, 1)$ into $(0, 1, 1)$ with probability q, and so on. We assume that optimal values for those probabilities satisfy the following two conditions: (i) Any change should occur between states whose Hamming distances differ by one. (ii) Optimal value of these probabilities should be symmetric, i.e., for example, the probability from $(0, 1, 0)$ to $(0, 1, 1)$

must be the same as the one from $(0, 1, 0)$ to $(1, 1, 0)$. (We do not have a formal proof that these assumptions do not lose of generality, but it seems obvious.) Note that $p, q, r, s, t, u \geq 0$ and $3p, 2q + s, r + 2t, 3u \leq 1$. Now we are ready to present our main algorithm:

ISTT(3CNF-formula F, integer m, T)
(01) **repeat** T times
(02) $(z, \mathcal{C}) = \mathbf{BF}(F)$;
(03)   **if** z satisfies F **then** output('Satisfiable'), exit;
(04)   **if** $|\mathcal{C}| \geq m$ **then break**; /* BF found a MICS of a good size. */
(05) **end repeat**
(06) **if** $|\mathcal{C}| < m$ **then** output('Unsatisfiable'), exit; /* BF found neither satisfying assignments nor MICSs of a good size after sufficiently many repetitions. */
(07) **repeat** T times
(08) $y = $ uniformly random vector $\in \{0, 1\}^n$;
(09)   Only (2) of **PPSZ**;
(10)   **if** a satisfying assignment is found **then** output('Satisfiable'), exit;
(11) $y' = \mathbf{Reassign}(\mathcal{C}, y)$;
(12)   Only (2) of **SCH** using y';
(13)   **if** a satisfying assignment is found **then** output('Satisfiable'), exit;
(14) **end repeat**
(15) output('Unsatisfiable'), exit;

---

4   Success Probability of ISTT

4.1   Easier Bounds

For better exposition, we first prove the following bound:

Theorem 1. *There exists a choice of parameters, m in* ISTT *and (p, q, r, s, t, u) in* Reassign, *such that for any satisfiable 3-CNF formula F, the success probability of $T = 1.32146^n$ repetitions of* ISTT *is $1 - o(1)$.*

Proof. Given a 3CNF-formula F with a set S of satisfying assignments and the subcube partition $\{\mathbf{C}(z, I_z) \mid z \in S\}$ and MICS $\mathcal{C}$ of size m, $\rho(F, z|\mathbf{C}(z, I_z))$ is defined as the probability (averaged over y) that a single execution of (8)(11)(12) in ISTT finds the assignment z under the condition that the initial assignment $y \in \mathbf{C}(z, I_z)$. We use the following lemma proven later:

Lemma 6. *For any $z \in S$, the value $\rho(F, z|\mathbf{C}(z, I_z))$ is bounded as*

$$\rho(F, z|\mathbf{C}(z, I_z)) \geq 1.012795^m \cdot 1.2845745^{|I_z|} \cdot \left(\frac{3}{4}\right)^n.$$

We see how Theorem 1 is proved with Lemmas 1, 5 and 6. Set m_0 to satisfy $6^{m_0} \approx 1.32146^n$. By Lemma 5, we can find a satisfying assignment or MICS $\mathcal{C}$ of size at least m_0 with probability $1 - o(1)$ after $T = 1.32146^n$ repetition of (2). If the former happens, we are done. Thus, assume the latter happens ($m \geq m_0$). The success probability α of one execution of steps (8)(9)(11)(12) in ISTT is written as:

$$\alpha \geq \sum_{z \in S} \mathbf{Pr}[y \in \mathbf{C}(z, I_z)] \max\{\tau(F, z|\mathbf{C}(z, I_z)), \rho(F, z|\mathbf{C}(z, I_z))\}.$$

Note that

$$\max\{\tau(F, z|\mathbf{C}(z, I_z)), \rho(F, z|\mathbf{C}(z, I_z))\}$$
$$\geq \min_{m \geq m_0, 0 \leq |I_z| \leq n} \max\{\tau(F, z|\mathbf{C}(z, I_z)), \rho(F, z|\mathbf{C}(z, I_z))\}$$
$$\geq \Omega(1.32146^{-n}),$$

and hence $\alpha \geq \Omega(1.32146^{-n})$, where the equality above holds when $m = m_0$ and $|I_z| = 0.0278212n$, which complete the proof. □

Proof. (of Lemma 6) From now on, we will fix an (arbitrary) subcube partition $\{\mathbf{C}(z, I_z) \mid z \in S\}$ and will show the bound on $\rho(F, z|\mathbf{C}(z, I_z))$ with respect to any satisfying assignment z and its corresponding subcube $\mathbf{C}(z, I_z)$.

Consider the distribution of an assignment y' obtained in step (11) of ISTT. To bound $\rho(F, z|\mathbf{C}(z, I_z))$, all we have to do is to apply Lemma 3 given a random assignment y' generated by Reassign from y under the condition $y \in \mathbf{C}(z, I_z)$. The set of variables is decomposed into those for each clause in MICS and the set of the remaining variables. Since the remaining variables are not reassigned, we can compute the value $\mathbf{E}[(1/2)^{d(y'_I, z_I)}]$ easily for the set I of indices of variables outside the MICS. That is, $\mathbf{E}[(1/2)^{d(y'_I, z_I)}] = 1^{|I'|}(3/4)^{|I''|}$, where $I' = I \cap I_z$, $I'' = I \setminus I'$. For the variables contained in an independent clause C, we have to do case analysis.

For example, consider $C = (x_1 \vee x_2 \vee x_3)$ and $\mathbf{C}(z, I_z)$ such that $I_z = \{1, 3\}$, $z_1 = 1$ and $z_3 = 0$. In this case, the distribution of (y_1, y_2, y_3) and (y'_1, y'_2, y'_3) is shown as in Table 2. With respect to this $\mathbf{C}(z, I_z)$, the satisfying assignment z should satisfy either $(z_1, z_2, z_3) = (1, 0, 0)$ or $(z_1, z_2, z_3) = (1, 1, 0)$. We can calculate the value $\mathbf{E}[(1/2)^{d((y'_1, y'_2, y'_3), (z_1, z_2, z_3))}]$ for each case and consider worst values as a lower bound of it. For example, assume $(z_1, z_2, z_3) = (1, 0, 0)$.

Table 2. Conditional distribution

(y_1, y_2, y_3)	prob.	(y_1, y_2, y_3)	prob.	(y'_1, y'_2, y'_3)	prob.	(y'_1, y'_2, y'_3)	prob.
(0,0,0)	0	(1,0,0)	$\frac{1}{2}$	(0,0,0)	$\frac{s}{2}$	(1,0,0)	$\frac{(1-2q-s+t)}{2}$
(0,0,1)	0	(1,0,1)	0	(0,0,1)	0	(1,0,1)	$\frac{q}{2}$
(0,1,0)	0	(1,1,0)	$\frac{1}{2}$	(0,1,0)	$\frac{t}{2}$	(1,1,0)	$\frac{(1+q-r-2t)}{2}$
(0,1,1)	0	(1,1,1)	0	(0,1,1)	0	(1,1,1)	$\frac{r}{2}$

Then

$$\begin{aligned}
\mathbf{E}[(1/2)^{d(y',z)}] &= \mathbf{Pr}[d(y',z)=0](1/2)^0 + \mathbf{Pr}[d(y',z)=1](1/2)^1 \\
&\quad + \mathbf{Pr}[d(y',z)=2](1/2)^2 + \mathbf{Pr}[d(y',z)=3](1/2)^3 \\
&= \frac{(1-2q-s+t)}{2} \times 1 + \left(\frac{s}{2}+\frac{q}{2}+\frac{(1+q-r-2t)}{2}\right) \times 1/2 \\
&\quad + \left(\frac{t}{2}+\frac{r}{2}\right) \times 1/4 + 0 \times 1/8 = \frac{3}{4}-\frac{q}{4}-\frac{r}{8}-\frac{s}{4}+\frac{t}{2}.
\end{aligned}$$

This is the essence of the analysis. We can classify each independent clause into nine different types shown in Table 3. The above example matches type 21, (= clause form $(0,1,*)$) since we have two fixed values 0 and 1 for z_3 and z_1 whose subscripts are in I_z, respectively, and one free value of z_2 whose subscript is not in I_z). Note that we ignore the ordering of variables in this classification. The values of $\mathbf{E}[(1/2)^{d(a,a')}]$ denoted by †, ‡ in the third and fourth columns correspond to two specific parameter values, $(p,q,r,s,t,u) = (1/3,0,1/21,0,2/21,0)$ and $(p,q,r,s,t,u) = (1/3,0,0.006061,0,0.060606,0)$, respectively. Note that these two examples play important roles later. In general, we denote by P_j the value $\mathbf{E}[(1/2)^{d(y',z)}]$ of a type j clause as a function of (p,q,r,s,t,u). It should be noted that each P_j is in the form of $c_1 p + c_2 q + c_3 r + c_4 s + c_5 t + c_6 u + c_7$, where each c_i is a constant depending on type j. We can easily prove that $c_1 \geq 0$, $c_4 \leq 0$ for any j. Hence, setting $p = 1/3$, $s = 0$ is always best to achieve maximum success probability.

Table 3. Clause type and $\mathbf{E}[(1/2)^{d(y',z)}]$

type ID	form of clause	$\mathbf{E}[(1/2)^{d(y',z)}]$†	$\mathbf{E}[(1/2)^{d(y',z)}]$‡
0	$(*,*,*)$	$\frac{3}{7}$	0.427273
10	$(0,*,*)$	$\frac{379}{672}$	0.565909
11	$(1,*,*)$	$\frac{181}{336}$	0.548864
20	$(0,0,*)$	$\frac{3}{4}$	0.750000
21	$(0,1,*)$	$\frac{29}{42}$	0.718182
22	$(1,1,*)$	$\frac{29}{42}$	0.718182
31	$(0,0,1)$	1	1
32	$(0,1,1)$	$\frac{37}{42}$	0.936363
33	$(1,1,1)$	1	1

Given (p,q,r,s,t,u), we can obtain the value $\mathbf{E}[(1/2)^{d(y'_I, z_I^*)}]$ for each I corresponding to $C \in \mathcal{C}$. Let t_j be the number of clauses of type j in MICS and d' be the number of defining variables not included in MICS, then we have

$$\rho(F, z | \mathbf{C}(z, I_z)) \geq \left(\prod_j P_j{}^{t_j}\right) \cdot 1^{d'} \cdot \left(\frac{3}{4}\right)^{n-3(t_0+t_{10}+t_{11}+t_{20}+t_{21}+t_{22}+t_{31}+t_{32}+t_{33})-d'}$$

$$= \left\{\prod_j \left(\frac{64 P_j}{27}\right)^{t_j}\right\} \cdot \left(\frac{4}{3}\right)^{d'} \cdot \left(\frac{3}{4}\right)^n \tag{1}$$

where

$$d' \geq 0, \forall j : t_j \geq 0, m = t_0 + t_{10} + t_{11} + t_{20} + t_{21} + t_{22} + t_{31} + t_{32} + t_{33}, \quad (2)$$

$$|I_z| = (t_{10} + t_{11}) + 2(t_{20} + t_{21} + t_{22}) + 3(t_{31} + t_{32} + t_{33}) + d'. \quad (3)$$

If we fix the Reassign parameters (p, q, r, s, t, u) (only (q, r, t, u) actually since we assumed $p = 1/3$ and $s = 0$), the values P_j are fixed. If we can calculate the smallest value $\rho(F, z | \mathbf{C}(z, I_z))$ under the condition (2) and (3), it would bound from below our success probability under the current value of (q, r, t, u). Thus our task is to find (q, r, t, u) that maximize the minimum value of $\rho(F, z | \mathbf{C}(z, I_z))$. For this purpose we conducted a numerical computation for (q, r, t, u) with precision up to 10^{-6} and LP formalization for minimizing $\rho(F, z | \mathbf{C}(z, I_z))$ (notice that if we take logarithm of (1), it would become a linear equation). As a result, it turned out that under the values of $(q, r, t, u) = (0, 0.006061, 0.060606, 0)$, we can have $\rho(F, z | \mathbf{C}(z, I_z)) \geq 1.012795^m \cdot 1.2845745^{|I_z|} \cdot \left(\frac{3}{4}\right)^n$. Note that the equality (the minimum value) is achieved for $t_{11} = |I_z|$, $t_0 = m - t_{11}$, $d' = 0$ and $t_j = 0$ for all $j \neq 0, 11$. $\qquad\square$

4.2   Better Bounds

We are now ready to prove our main result:

Theorem 2. *There exists a choice of parameters, m in ISTT and (p, q, r, s, t, u) in Reassign, such that for any satisfiable 3-CNF formula F, the success probability of $T = 1.32113^n$ repetitions of ISTT is $1 - o(1)$.*

Proof. In the previous subsection, the worst case means that all the independent clauses become type 0 or 11 and type 11 occurs a linear number of times for the initial random assignment y. However, we can easily see that the probability that this happens is exponentially small.

For example, see Fig. 2: Here the formula has three independent clauses $(x_1 \vee x_2 \vee x_3), (x_4 \vee x_5 \vee x_6)$ and $(x_7 \vee x_8 \vee x_9)$ and the figure shows the structure of the subcube partition consisting of seven subcubes such as the one determined by $x_1 = 1, x_7 = 0$ and $x_9 = 0$ (the third leaf node from the right in the figure). Suppose that $x_1 = 1$ in the initial assignment y. Then we have no choice of values for x_2 or x_3 to determine y's subcube. Thus first clause becomes type 11, a bad one in the previous analysis.

However in the random y, x_1 also becomes 0 with the same probability, namely, the first clause becomes type 11 with probability $1/2$ and type 10 with probability $1/2$. This is exactly same for $(x_7 \vee x_8 \vee x_9)$ after $x_1 = 1$: This clause becomes type 11, 20, and 21 with probabilities $1/2, 1/4$ and $1/4$, respectively. Now it turns out that there are 15 different patterns for how a single clause becomes different types. We show patterns 1 and 2 (corresponding to $(x_1 \vee x_2 \vee x_3)$ and $(x_7 \vee x_8 \vee x_9)$ mentioned above) in Fig. 2 (the remaining 13 one's are omitted due to the page limitation).

We next introduce new parameters random variables $t_{i,j}$ for $i = 1, \ldots, 15$ and $j \in \{0, 10, 11, 20, 21, 21, 31, 32, 33\}$, meaning that the initial y creates $t_{i,j}$

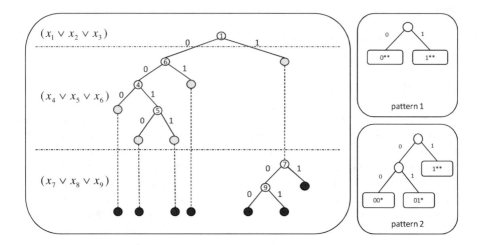

Fig. 2. Decision trees of subcube partition

clauses of type j and pattern i. For example, $x_1 = 1$, $x_7 = 0$ and $x_9 = 0$ in the previous example creates a clause of (type 11, pattern 1) and a clause of (type 20, pattern 2). Define random variables t_i' for $i = 1, \ldots, 15$ and t_j for $j \in \{0, 10, 11, 20, 21, 21, 31, 32, 33\}$ as

$$t_i' = \sum_j t_{i,j} \text{ and } t_j = \sum_i t_{i,j}.$$

Intuitively, t_i' denotes the number of clauses of pattern i created by y and t_j the number of clauses of type i (and any pattern) created by y.

Now we have an important observation: Recall that $t_1' = t_{1,10} + t_{1,11}$. As mentioned before, when we go through a pattern 1 clause, it will become type 10 or 11 with an equal probability. So, let us consider new random variables $X_1(\ell)$ for the value of $|t_{1,10} - t_{1,11}|$ after we have gone through ℓ clauses of pattern 1. Then we obviously have

$$\mathbf{E}[(X_1(\ell) \mid (X_1(1), \ldots, X_1(\ell-1)] = X_1(\ell-1),$$

and therefore $X_1(\ell)$ is martingale. Then we can use the Azuma's bound to have that for any constant $\epsilon_1 > 0$, the probability that

$$|t_{1,10} - t_{1,11}| > \epsilon_1 m$$

holds is exponentially small (note that $\ell \leq m$). Thus we can conclude that with high probability

$$|t_{1,10} - t_1'/2| \leq \epsilon_1 m \text{ and } |t_{1,10} - t_1'/2| \leq \epsilon_1 m.$$

Using all such inequalities for t_1' through t_{15}', we can rewrite equation (1) as

$$\rho(F, z | \mathbf{C}(z, I_z)) \geq \left(\frac{64P_j}{27}\right)^{t_0} \left(\prod_i Q_i^{t_i'}\right) \cdot \left(\frac{4}{3}\right)^{d'} \cdot \left(\frac{3}{4}\right)^n \cdot 2^{-\epsilon m},$$

where

$$Q_i^{t_i'} = \prod_{j \neq 0} \left(\frac{64P_j}{27}\right)^{t_{i,j}},$$

and when

$$d' \geq 0, t_0 \geq 0, \forall i : t_i' \geq 0 \text{ and } t_0 + \sum_{i=1}^{15} t_i' = m$$

with high probability for any constant $\epsilon > 0$. Here ϵ is the sum of ϵ_1 through ϵ_{15} introduced above. Now we can repeat exactly the same numerical computation as in Section 4.1 and obtain the desired bound, $O(1.321128^{-n})$, for the values of $(p, q, r, s, t, u) = (1/3, 0, 1/21, 0, 2/21, 0)$, where we can neglect the term $2^{-\epsilon m}$ since $\epsilon > 0$ can be any small constant. It might be interesting that this Reassign parameter values differ a lot from those in the previous subsection.   □

References

[AS08]   Alon, N., Spencer, J.H.: The probabilistic method, 3rd edn. Wiley Inter-science, Hoboken (2008)

[BS03]   Baumer, S., Schuler, R.: Improving a probabilistic 3-SAT algorithm by dynamic search and independent clause pairs. In: Giunchiglia, E., Tacchella, A. (eds.) SAT 2003. LNCS, vol. 2919, pp. 150–161. Springer, Heidelberg (2004)

[BK04]   Brüeggemann, T., Kern, W.: An improved deterministic local search algorithm for 3-SAT. Theoretical Computer Science 329(1-3), 303–313 (2004)

[Dan83]   Dantsin, E.: Two systems for proving tautologies based on the split method. Journal of Mathematical Science 22, 1293–1305 (1983)

[DGH+02]   Dantsin, E., Goerdt, A., Hirsch, E.A., Kannan, R., Kleinberg, J., Papadimitriou, C., Raghavan, P., Schöning, U.: A deterministic $2 - \frac{2}{k+1}$ algorithm for k-SAT based on local search. Theoretical Computer Science 289(1), 69–83 (2002)

[HSSW02]   Hofmeister, T., Schöning, U., Schuler, R., Watanabe, O.: A probabilistic 3-SAT algorithm further improved. In: Alt, H., Ferreira, A. (eds.) STACS 2002. LNCS, vol. 2285, pp. 192–202. Springer, Heidelberg (2002)

[IT04]   Iwama, K., Tamaki, S.: Improved upper bounds for 3-SAT. In: Proceedings of the 15th Annual ACM-SIAM Symposium on Discrete Algorithms, pp. 328–329 (2004)

[Kul99]   Kullmann, O.: New methods for 3-SAT decision and worst-case analysis. Theoretical Computer Science 223(1-2), 1–72 (1999)

[Luc84]   Luckhardt, H.: Obere Komplexitätsschranken für TAUT-Entscheidungen. In: Proceedings of Frege Conference 1984, pp. 331–337 (1984)

[MS79]   Monien, B., Speckenmeyer, E.: 3-Satisfiability is testable in $O(1.62^r)$ steps. Technical Report Beicht Nr. 3/1979, Reihe Theoretische Informatik, Universität-Gesamthochschule-Paderborn (1979)

[MS85]   Monien, B., Speckenmeyer, E.: Solving satisfiability less than 2^n steps. Discrete Applied Mathematics 10, 287–295 (1985)

[PPSZ98]   Paturi, R., Pudlák, P., Saks, M.E., Zane, F.: An improved exponential-time algorithm for k-SAT. In: Proceedings of the 39th Annual IEEE Symposium on Foundations of Computer Science (FOCS), pp. 628–637 (1998)

[PPSZ05]   Paturi, R., Pudlák, P., Saks, M.E., Zane, F.: An improved exponential-time algorithm for k-SAT. Journal of the ACM 52(3), 337–364 (2005)

[PPZ97]   Paturi, R., Pudlák, P., Zane, F.: Satisfiability coding lemma. In: Proceedings of the 38th Annual IEEE Symposium on Foundations of Computer Science (FOCS), pp. 566–574 (1997)

[Rod96]   Rodosek, R.: A new approach on solving 3-Satisfiability. In: Pfalzgraf, J., Calmet, J., Campbell, J. (eds.) AISMC 1996. LNCS, vol. 1138, pp. 197–212. Springer, Heidelberg (1996)

[Rol03a]   Rolf, D.: 3-SAT $\in$ RTIME($O(1.32971^n)$). Diploma thesis, Department Of Computer Science, Humboldt University Berlin, Germany (Janury 2003)

[Rol03b]   Rolf, D.: 3-SAT $\in$ RTIME(O(1.32793^n)). Electronic Colloquium on Computational Complexity, TR03-054 (2003)

[Rol06]   Rolf, D.: Improved bound for the PPSZ/Schöning-algorithm for 3-SAT. Journal on Satisfiability, Boolean Modeling and Computation 1, 111–122 (2006)

[Sch08]   Scheder, D.: Guided search and a faster deterministic algorithm for 3-SAT. In: Laber, E.S., Bornstein, C., Nogueira, L.T., Faria, L. (eds.) LATIN 2008. LNCS, vol. 4957, pp. 60–71. Springer, Heidelberg (2008)

[Sch92]   Schiermeyer, I.: Solving 3-Satisfiability in less than 1.579^n steps. In: Martini, S., Börger, E., Kleine Büning, H., Jäger, G., Richter, M.M. (eds.) CSL 1992. LNCS, vol. 702, pp. 379–394. Springer, Heidelberg (1993)

[Sch96]   Schiermeyer, I.: Pure literal look ahead: An $O(1.497^n)$ 3-Satisfiability algorithm. In: Proceedings of the Workshop on the Satisfiability Problem, pp. 127–136 (1996)

[Sch99]   Schöning, U.: A Probabilistic algorithm for k-SAT and constraint satisfaction problems. In: Proceedings of the 40th Annual IEEE Symposium on Foundations of Computer Science (FOCS), pp. 410–414 (1999)

Quantum Counterfeit Coin Problems

Kazuo Iwama[1,*], Harumichi Nishimura[2,**],
Rudy Raymond[3], and Junichi Teruyama[1]

[1] School of Informatics, Kyoto University, Japan
{iwama, teruyama}@kuis.kyoto-u.ac.jp
[2] School of Science, Osaka Prefecture University, Japan
hnishimura@mi.s.osakafu-u.ac.jp
[3] IBM Research – Tokyo, Japan
raymond@jp.ibm.com

Abstract. The counterfeit coin problem requires us to find all false coins from a given bunch of coins using a balance scale. We assume that the balance scale gives us only "balanced" or "tilted" information and that we know the number k of false coins in advance. The balance scale can be modeled by a certain type of oracle and its query complexity is a measure for the cost of weighing algorithms (the number of weighings). In this paper, we study the quantum query complexity for this problem. Let $Q(k, N)$ be the quantum query complexity of finding all k false coins from the N given coins. We show that for any k and N such that $k < N/2$, $Q(k, N) = O(k^{1/4})$, contrasting with the classical query complexity, $\Omega(k \log(N/k))$, that depends on N. So our quantum algorithm achieves a *quartic* speed-up for this problem. We do not have a matching lower bound, but we show some evidence that the upper bound is tight: any algorithm, including our algorithm, that satisfies certain properties needs $\Omega(k^{1/4})$ queries.

1 Introduction

Exponential speed-ups by quantum algorithms have been highly celebrated, but their specific examples are not too many. In contrast, almost every unstructured search problem can be sped up simply by using amplitude amplification [8,5,6], providing a huge number of combinatorial problems for which quantum algorithms are *quadratically* faster than classical ones. Interestingly there are few examples in between. (For instance, [7] provides a cubic speed-up while their classical lower bound is not known.) The reason is probably that the amplitude amplification is too general to combine with other methods appropriately. In fact we know few such cases including the one by [15] where they improved a simple Grover search algorithm for triangle finding by using clever combinatorial ideas (but unfortunately still less than quadratically compared to the best classical algorithm). This paper achieves a *quartic* speed-up for a well-known combinatorial problem.

* Research supported in part by KAKENHI (19200001,22240001).
** Research supported in part by KAKENHI (21244007,22700014).

O. Cheong, K.-Y. Chwa, and K. Park (Eds.): ISAAC 2010, Part I, LNCS 6506, pp. 85–96, 2010.
© Springer-Verlag Berlin Heidelberg 2010

The counterfeit coin problem is a mathematical puzzle whose origin dates back to 1945; in the American Mathematical Monthly, 52, p. 46, E. Schell posed the following question which is probably one of the oldest questions about the complexity of algorithms: "You have eight similar coins and a beam balance. At most one coin is counterfeit and hence underweight. How can you detect whether there is an underweight coin, and if so, which one, using the balance only twice?" The puzzle immediately fascinated many people and since then there have been several different versions and extensions in the literature (see e.g., [16,9,10,14]).

This paper considers the quantum version of this problem, which, a bit surprisingly, has not appeared in the literature. To make our model simple, we assume that we cannot obtain information on which side is heavier when the scale is tilted. So, the balance scale gives us only binary information, *balanced* (i.e., two sets of coins on the two pans are equal in weight) or *tilted* (different in weight). Our goal is to detect the false coin with a minimum number of weighings. The problem is naturally extended to the case that there are two or more (= k that is known in advance) false coins with equal weight. For the simplest case that $k = 1$, the following easy (classic) algorithm exists: We put (approximately) $N/4$ coins on both pans. If the scale is tilted, then we know the false coin is in those $N/2$ coins and if it is balanced, then the false one should be in the remaining $N/2$ ones. Also, it is easy to see that two weighings are enough for $N = 4$. Thus $\lceil \log N \rceil$ weighings are enough for $k = 1$ and this is also an information theoretic lower bound. (The original version of the problem assumes ternary outputs from the balance, left-heavy, right-heavy and balanced, and that the false coin is always underweight. As one can see easily, however, the same idea allows us to obtain the tight upper bound of $\lceil \log_3 N \rceil$.)

Our model of a balance scale is a so-called *oracle*. A *balance oracle* or simply a *B-oracle* is an N-bit register, which includes (originally unknown) N bits, $x_1 x_2 \cdots x_N \in \{0,1\}^N$. In order to retrieve these values, we can make *a query* with *a query string* $q_1 q_2 \cdots q_N \in \{0, 1, -1\}^N$ including the same number (= l) of 1's and −1's. Then the oracle returns a one-bit answer χ defined as:

$$\chi = 0 \text{ if } x_1 q_1 + \cdots + x_N q_N = 0 \text{ and } \chi = 1 \text{ otherwise.}$$

Consider $x_1, \cdots, x_N$ as N coins where 0 means a fair coin and 1 a false one. Then, $q_i = 1$ means we place coin x_i on the left pan and $q_i = -1$ on the right pan. Since we must have the same number of 1's and −1's, the answer χ correctly simulates the balance scale, i.e., $\chi = 0$ means it is balanced and $\chi = 1$ tilted. The number of weighings needed to retrieve x_1 through x_N (or to identify all the false coins) is called *query complexity*.

The main purpose of this paper is to obtain *quantum* query complexity for the counterfeit coin problems. Observe that if we know in advance that an even-cardinality set X includes *at most one* false coin, then by using the balance for any equal-size partition of X we can get the parity of X, i.e., the parity of the number (zero or one, now) of false coins in X. This means that for strings including at most one 1, the B-oracle is equivalent to the so-called IP oracle [4]. Therefore, by Bernstein-Vazirani algorithm [4], we need only one weighing to detect the false coin. (Note that this observation was essentially done by Terhal and

Smolin [20].) This already allows us to design the following quantum algorithm for general k: Recall that we know k in advance. So, if we sample N/k coins at random, then they include exactly one false coin with high probability and we can find it using the B-oracle just once as mentioned above. Thus, by using the standard amplitude amplification [6] (together with a bit careful consideration for the answer-confirmation procedure), we need $O(k)$ weighings to find all k false coins. For a small k, this is already much better than $\Omega(k \log(N/k))$ that is an information theoretic lower bound for the classical case.

Our Contribution. This paper shows that this complexity can be furthermore improved quartically, namely, our new algorithm needs $O(k^{1/4})$ weighings. Note that the above idea, the one exploiting Bernstein-Vazirani, already breaks down for $k = 2$, since the scale tilts even if the pans hold two (even) false coins if they both go to a same pan. Moreover, if k grows, say as large as linear in N, the balance will be tilted almost always for randomly selected equal partitions. Nevertheless, Bernstein-Vazirani is useful since it essentially reduces our problem (identifying false coins) to the problem of deciding the parity of the number of the false coins that turns out to be an easier task for B-oracles. By this we can get a single quadratic speed-up and another quadratic speed-up by amplitude amplification.

We conjecture that this bound is tight, but unfortunately, we cannot prove it at this moment. The main difficulty is that we have a lot of freedom on "the size of the pans" (= the number of coins placed on the two pans of the scale), which makes it hard to design a single weight scheme of the adversary method [1]. However, we do have a proof claiming that we cannot do better unless we can remove the two fundamental properties of our algorithm. These properties are (i) the big-pan property and (ii) the random-partition property. We have considered several possibilities for escaping from them, but not successful for even one of them.

Related Work. Query complexities have been studied almost always for the standard *index oracle*, which accepts an index i and returns the value of x_i. Other than this oracle, we know few ones including the IP oracle [4] mentioned before and the even more powerful one that returns the number (not the parity) of 1's in the string [20]. Also, [20] presented a single-query quantum algorithm for the binary search problem under the IP oracle, which is essentially based on the same idea as the $k = 1$ case of our problem mentioned above.

The quantum adversary method, which is used for B-oracles in this paper, was first introduced by Ambainis [1] for the standard oracle. Many variants have followed including weighted adversary methods [2,21], spectral adversary method [3], Kolmogorov complexity method [13], all of which were shown to be equivalent [19]. After Høyer et al. [11] introduced a stronger quantum adversary method called the negative adversary method, Reichardt [17,18] showed that this method is "optimal" for any Boolean function.

Models. A *B-oracle* is a binary string $x = x_1 \cdots x_N$ where $x_i = 1$ (resp. $= 0$) means that the i-th coin is false (resp. fair). For instance, the string 0001 for

$N = 4$ means that the fourth coin is a unique false coin. A query to the oracle is given as a string $q = q_1 \cdots q_N \in \{0, 1, -1\}^N$ that must be in the set $Q^{(B)} = \bigcup_{l=0}^{\lfloor N/2 \rfloor} Q_l$ where Q_l is the set of strings q such that q has exactly l 1's and l -1's. Here, 1 (or -1, resp.) in the i-th component means that we place the i-th coin on the left pan (on the right pan, resp.) and 0 means that the i-th coin is not placed on either pan. The answer from the oracle is represented by a binary value $\chi(x; q)$ where $\chi(x; q) = 0$ means the scale is balanced, that is, $q_1 x_1 + \ldots + q_N x_N = 0$ and $\chi(x; q) = 1$ means it is tilted, that is, $q_1 x_1 + \ldots + q_N x_N \neq 0$. In quantum computation, the B-oracle is viewed as a unitary transformation $O_{B,x}$. Namely, $O_{B,x}$ transforms $|q\rangle$ to $(-1)^{\chi(x;q)}|q\rangle$. Throughout this paper, we assume that $k < N/2$ since our B-oracle model is unable to distinguish any N-bit string x from $\bar{x}$ (the bit string obtained by flipping all bits of x).

Notes. An extended version of this paper [12] appears in the arXiv.

2   Upper Bounds

Here is our main result in this paper:

Theorem 1. *The quantum query complexity for finding k false coins among N coins is $O(k^{1/4})$.*

Notice that our algorithm is *exact*, i.e., its output must be correct with probability one to compare our result with the classical case (which has been often studied in the exact setting). Since we use exact amplitude amplification [6] to make our algorithm exact, the assumption that k is known is necessary. But it should be noted that our bounded-error algorithm described in this section works even for unknown k. Also, we note that our algorithm can be easily adapted so that it works when the output of the balance is ternary (while we assume it is binary for simplicity).

Before the proof, we first describe our basic approach, a simulation of the IP oracle by the B-oracle. Recall that the IP oracle (Inner Product oracle) [4] transforms a prequery state $|\widetilde{q}\rangle_{\mathsf{R}}$ to $(-1)^{\widetilde{q} \cdot x}|\widetilde{q}\rangle_{\mathsf{R}}$, where $\widetilde{q} \in \{0, 1\}^N$ in register R is a query string and $x \in \{0, 1\}^N$ is an oracle. Then the Bernstein-Vazirani algorithm (the Hadamard transform) retrieves the string x and we know the k false coins in the case of our problem. Observe that the IP oracle flips the phase of each state if and only if $\widetilde{q} \cdot x$ is odd, in other words, if and only if a multiset $M(\widetilde{q}, x) := \{x_i \mid \widetilde{q}_i = 1\}$ includes an odd number of 1's (or false coins in our case). If $k = 1$, then $M(\widetilde{q}, x)$ includes at most one 1. Hence we can simply replace the IP oracle with the query sequence $\widetilde{q}$ by the B-oracle with a query sequence q such that an arbitrarily one half (the first one half, for instance) of the 1's in $\widetilde{q}$ are changed to -1's, meaning the one half of the coins in $M(\widetilde{q}, x)$ go to the left pan and the remaining one half to the right pan. (As shown in a moment, we can assume without loss of generality that $\widetilde{q}$ includes an even number of 1's.)

Now we consider the general ($k \geq 1$) case. If $M(\widetilde{q}, x)$ includes odd 1's, then the scale is tilted for any such q mentioned above; this is desirable for us. If $M(\widetilde{q}, x)$ includes even 1's, we wish the scale to be balanced. In order for this to happen, however, we must divide the (unknown) false coins in $M(\widetilde{q}, x)$ into the two pans evenly, for which there are no obvious ways other than using randomization. Our idea is to introduce the second register, R′, as follows: On R′, we prepare, with being entangled to each state $\widetilde{q}$ in R, a superposition of all possible states $q_1(\widetilde{q}), q_2(\widetilde{q}), \ldots, q_h(\widetilde{q})$, obtained by flipping one half of 1's in $\widetilde{q}$ into -1's. By using this superposition as a query to the B-oracle, we can achieve a success (being able to detect the scale is balanced) probability of $1/\sqrt{m}$, where m is the number of false coins in $M(\widetilde{q}, x)$. In order to increase this probability, we can use copies of register R′ or, more efficiently, quantum amplitude amplification [6].

As suggested before, we begin with the restriction of the IP oracle without losing its power. The *parity-restricted query* means that the Hamming weights of all superposed queries $\widetilde{q}$, denoted by $wt(\widetilde{q})$, are even.

Lemma 1. *Let* $S_{<N/2} := \{x \in \{0,1\}^N \mid wt(x) < N/2\}$. *Then there is a quantum algorithm to identify an oracle in* $S_{<N/2}$ *by a single parity-restricted query for the IP oracle.*

Proof. For a given oracle $x \in S_{<N/2}$, define $|\psi_x\rangle = \frac{1}{\sqrt{2^{N-1}}} \sum_{\widetilde{q} \in Q_{even}} (-1)^{\widetilde{q} \cdot x} |\widetilde{q}\rangle$. where $Q_{even} = \{\widetilde{q} \in \{0,1\}^N \mid wt(\widetilde{q}) = 0 \bmod 2\}$. Then the Hadamard transform of $|\psi_x\rangle$, $H|\psi_x\rangle$, can be rewritten as follows:

$$H|\psi_x\rangle = \frac{1}{\sqrt{2^{N-1}}} \sum_{\widetilde{q} \in Q_{even}} (-1)^{\widetilde{q} \cdot x} H|\widetilde{q}\rangle = \frac{1}{2^{N-1}\sqrt{2}} \sum_{\widetilde{q} \in Q_{even}} \sum_{z \in \{0,1\}^N} (-1)^{\widetilde{q} \cdot (x \oplus z)} |z\rangle$$

$$= \frac{1}{\sqrt{2}} (|x\rangle + |\bar{x}\rangle) + \frac{1}{2^{N-1}\sqrt{2}} \sum_{\widetilde{q} \in Q_{even}} \sum_{z \neq x, \bar{x}} (-1)^{\widetilde{q} \cdot (x \oplus z)} |z\rangle$$

$$= \frac{1}{\sqrt{2}} (|x\rangle + |\bar{x}\rangle).$$

Note that the last equality in the above equations holds; the second term must vanish because the first term already has a unit length. For any $x \neq y$, $H|\psi_x\rangle = (|x\rangle + |\bar{x}\rangle)/\sqrt{2}$ and $H|\psi_y\rangle = (|y\rangle + |\bar{y}\rangle)/\sqrt{2}$ are orthogonal since $x \neq \bar{y}$ by the restriction of their Hamming weights. This implies that $|\psi_x\rangle$ is orthogonal to $|\psi_y\rangle$ for any $x \neq y$, and hence there is a unitary transformation $W : |x\rangle \mapsto |\psi_x\rangle$. Thus we can design an algorithm similar to Bernstein-Vazirani [4] just replacing the Hadamard transform by W. □

Now we give the proof of our main result.

Proof of Theorem 1. For exposition, we first give a bounded-error algorithm $(Find^*(k))$ and then make it exact $(Find(k))$. In what follows, for a query string $\widetilde{q}$, let $I(\widetilde{q})$ be the set of indices i such that $\widetilde{q}_i = 1$. This set specifies which $wt(\widetilde{q})$ coins of the N coins are placed on the two pans. Let $P_{I(\widetilde{q})}$ be the set of all partitions of the set $I(\widetilde{q})$ of size $wt(\widetilde{q})$ (= even by Lemma 1) into two sets of size $wt(\widetilde{q})/2$. Note that each partition $(Y, \overline{Y})$ in $P_{I(\widetilde{q})}$ specifies how to split the $wt(\widetilde{q})$ coins in half to place them on the left and right pans, and can be identified with the corresponding query q to the B-oracle. Finally, let $\chi(Y, \overline{Y})$ be the answer for the query $(Y, \overline{Y}) \in P_{I(\widetilde{q})}$ to the B-oracle.

Algorithm $Find^*(k)$

1. Prepare N qubits $|0\rangle^{\otimes N}$ in a register R, and apply a unitary transformation W of Lemma 1 to them. Then, we have the state $\frac{1}{\sqrt{2^{N-1}}}\sum_{\widetilde{q}\in Q_{even}}|\widetilde{q}\rangle_R$.

2. For each superposed $\widetilde{q}$, implement Steps 2.1–2.4 on a register R' using $\widetilde{q}$ as a control part.

2.1. Apply a unitary transformation $\mathcal{A}_{\widetilde{q}}$ to the initial state $|0\rangle$ on R' to create a quantum state $\mathcal{A}_{\widetilde{q}}|0\rangle := \frac{1}{\sqrt{|P_{I(\widetilde{q})}|}}\sum_{(Y,\overline{Y})\in P_{I(\widetilde{q})}}|Y,\overline{Y}\rangle_{R'}$, which represents a uniform superposition of all partitions $(Y,\overline{Y})$ in $P_{I(\widetilde{q})}$. Then, the current state is

$$|\xi_{2,1}\rangle = \sum_{\widetilde{q}\in Q_{even}}|\widetilde{q}\rangle_R \sum_{(Y,\overline{Y})\in P_{I(\widetilde{q})}}\gamma\alpha|Y,\overline{Y}\rangle_{R'}$$

$$= \sum_{\widetilde{q}\in Q_{even}\cap Q_e}|\widetilde{q}\rangle_R \sum_{(Y,\overline{Y})\in P_{I(\widetilde{q})}}\gamma\alpha|Y,\overline{Y}\rangle_{R'} + \sum_{\widetilde{q}\in Q_{even}\cap Q_o}|\widetilde{q}\rangle_R \sum_{(Y,\overline{Y})\in P_{I(\widetilde{q})}}\gamma\alpha|Y,\overline{Y}\rangle_{R'}$$

where Q_e (resp. Q_o) denotes the set of all $\widetilde{q}$'s such that $M(\widetilde{q},x)$ includes an even (resp. odd) number of 1's. Also, $\gamma = 1/\sqrt{2^{N-1}}$ and $\alpha = 1/\sqrt{|P_{I(\widetilde{q})}|}$.

2.2. Let $\overline{\chi}$ be the Boolean function defined by $\overline{\chi}(Y,\overline{Y}) = 1$ if and only if $\chi(Y,\overline{Y}) = 0$ (that is, the scale is balanced). Then, under the above $\mathcal{A}_{\widetilde{q}}$ and $\overline{\chi}$, run the amplitude amplification algorithm **QSearch**$(\mathcal{A}_{\widetilde{q}},\overline{\chi})$ when the initial success probability of $\mathcal{A}_{\widetilde{q}}$ is unknown (Theorem 3 in [6]). Here "success" means the scale is balanced and hence we use $\overline{\chi}$, not χ, in **QSearch**. Then we obtain a state in the form of

$$|\xi_{2,2}\rangle = \sum_{\widetilde{q}\in Q_{even}\cap Q_e}|\widetilde{q}\rangle_R \sum_{(Y,\overline{Y})\in P_{I(\widetilde{q})}}\gamma\beta_Y|Y,\overline{Y},g_Y\rangle_{R'} + \sum_{\widetilde{q}\in Q_{even}\cap Q_o}|\widetilde{q}\rangle_R \sum_{(Y,\overline{Y})\in P_{I(\widetilde{q})}}\gamma\alpha|Y,\overline{Y},g_Y\rangle_{R'}$$

where $|g_Y\rangle$ is the garbage state. Note that, in the first term, the amplitudes β_Y such that $\overline{\chi}(Y,\overline{Y}) = 1$ are now large by amplitude amplification while the second term does not change since the scale is always tilted.

2.3. If Step 2.2 finds a "solution," i.e., a partition $(Y,\overline{Y})$ such that $\overline{\chi}(Y,\overline{Y}) = 1$, then do nothing. Otherwise, flip the phase (and then the phase is kickbacked into R). Notice that when $M(\widetilde{q},x)$ includes an odd number of 1's, the phase is always flipped, while when it includes an even number of 1's, the phase is not flipped with high amplitude. Now the current state is

$$|\xi_{2,3}\rangle = \sum_{\widetilde{q}\in Q_{even}\cap Q_e}|\widetilde{q}\rangle_R \sum_{(Y,\overline{Y})\in P_{I(\widetilde{q})}}\gamma\beta_Y(-1)^{\chi(Y,\overline{Y})}|Y,\overline{Y},g_Y\rangle_{R'} - \sum_{\widetilde{q}\in Q_{even}\cap Q_o}|\widetilde{q}\rangle_R \sum_{(Y,\overline{Y})\in P_{I(\widetilde{q})}}\gamma\alpha|Y,\overline{Y},g_Y\rangle_{R'}$$

$$= \sum_{\widetilde{q}\in Q_{even}\cap Q_e}|\widetilde{q}\rangle_R \sum_{(Y,\overline{Y})\in P_{I(\widetilde{q})}}\gamma\beta_Y|Y,\overline{Y},g_Y\rangle_{R'} - \sum_{\widetilde{q}\in Q_{even}\cap Q_o}|\widetilde{q}\rangle_R \sum_{(Y,\overline{Y})\in P_{I(\widetilde{q})}}\gamma\alpha|Y,\overline{Y},g_Y\rangle_{R'}$$

$$- 2\sum_{\widetilde{q}\in Q_{even}\cap Q_e}|\widetilde{q}\rangle_R|err_{\widetilde{q}}\rangle_{R'}$$

where $|err_{\widetilde{q}}\rangle_{R'} = \sum_{(Y,\overline{Y})\in P_{I(\widetilde{q})}:\chi(Y,\overline{Y})=1}\gamma\beta_Y|Y,\overline{Y},g_Y\rangle_{R'}$.

2.4. Reverse the quantum transformation done in Steps 2.1 and 2.2. Notice that the reversible transformation is done on R' in parallel for each $\tilde{q}$ while the contents of R does not change since it is the control part. Therefore, the state becomes

$$|\xi_{2,4}\rangle = \frac{1}{\sqrt{2^{N-1}}} \sum_{\tilde{q} \in Q_{even} \cap Q_e} |\tilde{q}\rangle_R |0\rangle_{R'} - \frac{1}{\sqrt{2^{N-1}}} \sum_{\tilde{q} \in Q_{even} \cap Q_o} |\tilde{q}\rangle_R |0\rangle_{R'} - 2 \sum_{\tilde{q} \in Q_{even} \cap Q_e} |\tilde{q}\rangle_R |err'_{\tilde{q}}\rangle_{R'}$$

$$= \frac{1}{\sqrt{2^{N-1}}} \sum_{\tilde{q} \in Q_{even}} (-1)^{\tilde{q} \cdot x} |\tilde{q}\rangle_R |0\rangle_{R'} - 2 \sum_{\tilde{q} \in Q_{even} \cap Q_e} |\tilde{q}\rangle_R |err'_{\tilde{q}}\rangle_{R'}$$

where $|err'_{\tilde{q}}\rangle_{R'}$ is the transformed state of $|err_{\tilde{q}}\rangle_{R'}$.

3. Apply W^{-1} to the state in R. Then we obtain a final state

$$|\xi_3\rangle = |x\rangle_R |0\rangle_{R'} - 2W^{-1} \left(\sum_{\tilde{q} \in Q_{even} \cap Q_e} |\tilde{q}\rangle_R |err'_{\tilde{q}}\rangle_{R'} \right).$$

Then measure R in the computational basis. (End of Algorithm)

For justifying the correctness of $Find^*(k)$, it suffices to show that the squared magnitude of the second term of $|\xi_3\rangle$ is a small constant, say, $1/400$, since we then measure the desired value x with probability at least $9/10$ (in fact, at least $(1 - \sqrt{1/400})^2 > 9/10$). By the unitarity, its squared magnitude is equal to that of the last term of $|\xi_{2,3}\rangle$, that is, we want to evaluate the following value ϵ.

$$\epsilon := 4 \left\| \sum_{\tilde{q} \in Q_{even} \cap Q_e} |\tilde{q}\rangle_R |err_{\tilde{q}}\rangle_{R'} \right\|^2 = 4 \sum_{\tilde{q} \in Q_{even} \cap Q_e} \| |\tilde{q}\rangle_R \| \| |err_{\tilde{q}}\rangle_{R'} \|^2.$$

Lemma 2. ϵ is at most $1/400$.

Proof. Consider an arbitrary $\tilde{q}$ in $Q_{even} \cap Q_e$. When $M(\tilde{q}, x)$ includes m ($\leq k$) 1's (where m is even), the state $\mathcal{A}_{\tilde{q}}|0\rangle$ includes a partition $(Y, \overline{Y})$ such that $\overline{\chi}(Y, \overline{Y}) = 1$ with probability at least $p = \frac{\binom{m}{m/2}\binom{wt(\tilde{q})-m}{(wt(\tilde{q})-m)/2}}{\binom{wt(\tilde{q})}{wt(\tilde{q})/2}} = \Omega(1/\sqrt{m}) = \Omega(1/\sqrt{k})$. By Theorem 3 in [6], it is guaranteed that, in the algorithm $\mathbf{QSearch}(\mathcal{A}_{\tilde{q}}, \overline{\chi})$, an expected number of applications of the Grover-like subroutine to find a "solution," i.e., a partition $(Y, \overline{Y})$ such that $\overline{\chi}(Y, \overline{Y}) = 1$, is bounded by $O(1/\sqrt{p}) = O(k^{1/4})$. The subroutine consists of (i) $\mathcal{A}_{\tilde{q}}$, (ii) its inverse, (iii) the transformation $O_{\overline{\chi}}$ defined by $O_{\overline{\chi}}|Y, \overline{Y}\rangle = (-1)^{\overline{\chi}(Y, \overline{Y})}|Y, \overline{Y}\rangle$, and (iv) the transformation U_0 defined by $U_0|z\rangle = |z\rangle$ if $z \neq 0$ and $-|z\rangle$ if $z = 0$, where $\mathcal{A}_{\tilde{q}}$ (and hence its inverse) and U_0 can be implemented without any query to the B-oracle, and $O_{\overline{\chi}}$ can be implemented with one query to the B-oracle. Thus the expected number of queries to find a "solution" is $O(k^{1/4})$. By setting the number of applications of the subroutine to $c_0 k^{1/4}$ where c_0 is a large constant, Step 2.2 finds a "solution" with probability at least $1599/1600$. This means that for any $\widetilde{q} \in Q_{even} \cap Q_e$, $\sum_{(Y,\overline{Y}) \in P_{I(\widetilde{q})} : \overline{\chi}(Y,\overline{Y})=0} \beta_Y |Y, \overline{Y}, g_Y\rangle_{\mathsf{R}'}$ has squared magnitude at most $1/1600$. Recalling $\gamma = 1/\sqrt{2^{N-1}}$ we have $\epsilon = 4\gamma^2 \sum_{\widetilde{q} \in Q_{even} \cap Q_e} \left\| \sum_{(Y,\overline{Y}) \in P_{I(\widetilde{q})} : \overline{\chi}(Y,\overline{Y})=0} \beta_Y |Y, \overline{Y}, g_Y\rangle_{\mathsf{R}'} \right\|^2 \leq 1/400$. This completes the proof of Lemma 2.   □

Finally, it is easy to see from the above proof that the query complexity of $Find^*(k)$ is $O(k^{1/4})$ since it makes $O(k^{1/4})$ queries in Step 2 and no queries in Steps 1 and 3.

Now we consider the exact algorithm $Find(k)$. By the symmetric structure of algorithm $Find^*(k)$, the success probability of identifying x correctly is independent of x (recall that the oracle candidates are $\binom{N}{k}$ N-bit strings x with Hamming weight k). Thus we can use the so-called exact amplitude amplification algorithm (Theorem 4 in [6]) to convert it into the exact algorithm.

Here is the brief description of $Find(k)$. First, we implement $Find^*(k)$. As shown above, $Find^*(k)$ produces the correct output (i.e., k false coins) with a constant probability ($\geq 9/10$) larger than $1/4$. Notice that we can make the success probability exactly $1/4$ by an appropriate adjustment. We need an algorithm for checking if the output is correct to amplify the success probability to 1. Namely, an algorithm $Check$ needs to judge whether k coins are indeed all false, which can be implemented classically in $O(\log k)$ weighings. Then we can implement the exact amplitude amplification: Like the $1/4$-Grover's algorithm [5], flip the phase if $Check$ judges that the output is correct, and apply the reflection about the state obtained after $Find^*(k)$. It is not difficult to see that $Find(k)$ always finds k false coins and the total complexity is $O(k^{1/4})$. Therefore, the proof of Theorem 1 is completed.   □

3   Lower Bounds

3.1   Basic Ideas

In this section, we discuss the lower bound of finding k false coins from N coins. We conjecture that the upper bound $O(k^{1/4})$ is tight but, unfortunately, we have not been able to show whether it is true or not. Instead, we show that if there would be an algorithm that improves the upper bound essentially, then it would have a completely different structure from our algorithm.

Before describing our results, we observe two properties of our algorithm $Find(k)$. First, $Find(k)$ essentially uses only "big pans," i.e., it always places at least $\Omega(N)$ coins on the pans, which is called the *big-pan property*. (The algorithm $Find^*(k)$ in Section 2 uses "small pans" but it can be adapted with no essential change so that it works even if the size of pans must be big. Second, the B-oracle is always used in such a way that once the coins placed on the two pans are determined, the partition of them into the two pans is done uniformly at random, which is called the *random-partition property*. What we show in this section is that the current upper bound is best achievable for any algorithm that satisfies at least one of these two properties.

For this purpose, we revisit one version of the (nonnegative) quantum adversary method, called *the strong weighted adversary method* in [19], due to Zhang [21]. Let f be a function from a finite set S to another finite set S'. Recall that in a query complexity model, an input $x \in S$ is given as an oracle. An algorithm $\mathcal{A}$ would like to compute $f(x)$ while it can obtain the information about x by a unitary transformation $O_x|q, a, z\rangle = |q, a \oplus \zeta(x; q), z\rangle$, where $|q\rangle$ is the register for a query string q from a finite set Q, $|a\rangle$ is the register for the binary answer $\zeta(x; q)$, a function from $S \times Q$ to $\{0, 1\}$, and $|z\rangle$ is the work register. Note that the adversary method usually assumes the so-called index oracle, namely q is an integer $1 \leq i \leq N$ and $\zeta(x; q)$ is the ith bit (0 or 1) of $x \in \{0, 1\}^N$. However, one can easily see that the above generalization to $\zeta(x; q)$ requires no essential changes for its proof. Thus Theorem 14 of [21] can be restated as follows:

Lemma 3. *Let w, w' denote a weight scheme as follows:*

1. *Every pair $(x, y) \in S \times S$ is assigned a nonnegative weight $w(x, y) = w(y, x)$ that satisfies $w(x, y) = 0$ whenever $f(x) = f(y)$.*
2. *Every triple $(x, y, q) \in S \times S \times Q$ is assigned a nonnegative weight $w'(x, y, q)$ that satisfies $w'(x, y, q) = 0$ whenever $\zeta(x; q) = \zeta(y; q)$ or $f(x) = f(y)$, and $w'(x, y, q)w'(y, x, q) \geq w^2(x, y)$ for all x, y, q such that $\zeta(x; q) \neq \zeta(y; q)$ and $f(x) \neq f(y)$.*

For all x, q, let $\mu(x) = \sum_y w(x, y)$ and $\nu(x, q) = \sum_y w'(x, y, q)$. Then, the quantum query complexity of f is at least

$$\Omega\left(\max_{w, w'} \min_{\substack{x, y, q: \ w(x, y) > 0, \\ \zeta(x; q) \neq \zeta(y; q)}} \sqrt{\frac{\mu(x)\mu(y)}{\nu(x, q)\nu(y, q)}}\right).$$

3.2   Big Pan Lower Bounds

First, we show that our upper bound is tight under the big-pan property. In what follows, $L \geq l$ denotes the restriction that at least l coins must be placed on the pans whenever the balance is used.

Theorem 2. *If $L \geq l$, we need $\Omega((lk/N)^{1/4})$ weighings to find k false coins. In particular, $\Omega(k^{1/4})$ weighings are necessary if there is some constant c such that $L \geq N/c$.*

Proof. Let $l = N/d$. Then the lower bound we should show is $\Omega((k/d)^{1/4})$. We can assume that $d \leq k/3$ (otherwise, the lower bound becomes trivial). To use Lemma 3, let $S = \{x \in \{0, 1\}^N \mid wt(x) = k\}$, $Q = Q_{\geq N/d} := \bigcup_{l \geq N/d} Q_l$, $\zeta(x; q) = \chi(x; q)$, and $f(x) = x$. Our weight scheme is as follows: Let $w(x, y) = 1$ for any pair $(x, y) \in S \times S$ such that $x \neq y$, and let $w'(x, y, q) = 1$ for all $(x, y, q) \in S \times S \times Q_{\geq N/d}$ such that $\chi(x; q) \neq \chi(y; q)$ and $x \neq y$. It is easy to check that this satisfies the condition of a weight scheme. Then, for any x, we have $\mu(x) = \sum_y w(x, y) = \binom{N}{k} - 1$. We need to evaluate $\nu(x, q)\nu(y, q)$ for pairs

(x, y) such that $\chi(x; q) = 1$ and $\chi(y; q) = 0$ or $\chi(x; q) = 0$ and $\chi(y; q) = 1$. Fix $q \in Q_{\geq N/d}$ arbitrarily and assume that $q \in Q_{N/c}$ where $c \leq d$. When $\chi(x; q) = 1$ (i.e., the scale is tilted for query q when x is the input), notice that $\nu(x, q) = \sum_y w'(x, y, q)$ is the number of all y's such that the scale is balanced when N/c coins are placed on each of the two pans according to q. Therefore, by summing up all the cases such that those N/c coins include m false ones,

$$\nu(x, q) = \gamma(N, k, c) := \sum_{m=0}^{k/2} \binom{N/c}{m}^2 \binom{(1 - 2/c)N}{k - 2m}.$$

Since $\chi(y; q) = 0$, we have $\nu(y, q) = \sum_x w'(x, y, q) = \binom{N}{k} - \gamma(N, k, c)$ by counting all x's such that the scale is tilted. Then the product $\nu(x, q)\nu(y, q)$ is $\gamma(N, k, c)\left(\binom{N}{k} - \gamma(N, k, c)\right)$. Similarly, when $\chi(x; q) = 1$ we can see that the product is also $\gamma(N, k, c)\left(\binom{N}{k} - \gamma(N, k, c)\right)$. By Lemma 3 the quantum query complexity of our problem is at least

$$\Omega\left(\min_{c: c \leq d} \sqrt{\frac{(\binom{N}{k} - 1)^2}{\gamma(N, k, c)(\binom{N}{k} - \gamma(N, k, c))}}\right) = \Omega\left(\min_{c: c \leq d} \sqrt{\frac{\binom{N}{k}}{\gamma(N, k, c)}}\right). \quad (1)$$

Then, by using the fact that $\gamma(N, k, c)/\binom{N}{k}$ means the probability that the scale is balanced when N/c coins (N coins include k false ones) are randomly placed on each of the two pans, we can show the following lemma.

Lemma 4. $\gamma(N, k, c)/\binom{N}{k} = O(\sqrt{c/k})$ for any $2 \leq c \leq d \ (\leq k/3)$.

This implies the desired bound $\Omega((k/d)^{1/4})$ by Eq.(1). $\square$

On the contrary, we can show that any algorithm that uses only "small pans" also needs $\Omega(k^{1/4})$ queries. For instance, we cannot break the current bound $k^{1/4}$ by any algorithm that places $O(N/k)$ coins on the pans. (Notice that the pan includes only a constant number of false coins with high probability in this case and therefore we can achieve a better success probability for the even false-coin case, but at the same time, we cannot use a wide range of superpositions). Moreover, we can obtain another lower bound for the case where "big pans" and "small pans" are both available but "medium pans" are not. Unfortunately one can see that there is still a gap between the sizes of the big pans and small pans even for a weakest nontrivial $(\omega(1))$ lower bound.

3.3   Lower Bounds for the Quasi B-Oracle

Second, we show that our upper bound is tight under the random-partition property. Notice that in this case, if the coins include an odd number of false ones, then the scale is always tilted, and if the coins include an even number $(=m)$ of false ones, the scale will be balanced with probability $1/\sqrt{m}$. Thus in order to show a lower bound, we need to generalize the adversary method that works for such "stochastic" oracles: Now $\zeta(x; q)$ is a random variable and the stochastic version of O_x, denoted by $\widetilde{O}_x$, is defined as (we should be careful not to lose its unitarity):

$$\widetilde{O}_x|q, a, z\rangle = \sqrt{\Pr[\zeta(x; q) = 0]}|q, a, z\rangle + (-1)^a \sqrt{\Pr[\zeta(x; q) = 1]}|q, a \oplus 1, z\rangle.$$

Now Lemma 3 changes to the following:

Lemma 5. *Let w, w' denote a weight scheme as Lemma 3 except replacing Condition 2 to*

2' *Every triple $(x, y, q) \in S \times S \times Q$ is assigned a nonnegative weight $w'(x, y, q)$ that satisfies $w'(x, y, q) = 0$ whenever $\Pr[\zeta(x; q) = \zeta(y; q)] = 1$ or $f(x) = f(y)$, and $w'(x, y, q)w'(y, x, q) \geq w^2(x, y)$ for all x, y, q such that $\Pr[\zeta(x; q) \neq \zeta(y; q)] > 0$ and $f(x) \neq f(y)$.*

Then, the quantum query complexity of f is at least

$$\Omega\left(\max_{w, w'} \min_{\substack{x, y, q: \ w(x, y) > 0, \\ \Pr[\zeta(x; q) \neq \zeta(y; q)] > 0}} \sqrt{\frac{\mu(x)\mu(y)}{\nu(x, q)\nu(y, q)}} \frac{1}{\sqrt{P_{01, q}} + \sqrt{P_{10, q}}}\right),$$

where $P_{ab, q} = \Pr[\zeta(x; q) = a]\Pr[\zeta(y; q) = b]$.

Now we define the stochastic version of our B-oracle by setting

$$\Pr[\zeta(x; q) = 0] = \begin{cases} 0 & \text{(if } wt(x \wedge q) \text{ is odd)} \\ \sqrt{1/wt(x \wedge q)} & \text{(if } wt(x \wedge q) \text{ is positive and even)} \\ 1 & \text{(if } wt(x \wedge q) = 0), \end{cases}$$

where x and q are N-bit strings, and $x \wedge q$ is the N-bit string obtained by the bitwise AND of x and q. We call this oracle the *quasi B-oracle* and one can see that it simulates the B-oracle with the random-partition property. Now we are ready to give the upper and lower bounds for the query complexity of this quasi B-oracle. Assume that $wt(x) = k$. The upper bound is easy by modifying Theorem 1 so that Step 2 in $Find^*(k)$ can be replaced with $O(k^{1/4})$ repetitions of the quasi B-oracle.

Theorem 3. *There is an $O(k^{1/4})$-query quantum algorithm to find x using the quasi B-oracle.*

On the contrary, we can obtain the tight lower bound by using Lemma 5. The weight scheme contrasts with that of Theorem 2; $w(x, y)$ is nonzero only if the Hamming distance between x and y is 2.

Theorem 4. *Any quantum algorithm with the quasi B-oracle needs $\Omega(k^{1/4})$ queries to find x.*

Acknowledgements. We are grateful to Mario Szegedy for directing our interest to the topic of this paper, and an anonymous referee for a helpful idea to improve the earlier upper bounds for general k significantly. We are also grateful to Seiichiro Tani and Shigeru Yamashita for helpful discussions.

References

1. Ambainis, A.: Quantum lower bounds by quantum arguments. J. Comput. Syst. Sci. 64, 750–767 (2002)
2. Ambainis, A.: Polynomial degree vs. quantum query complexity. J. Comput. Syst. Sci. 72, 220–238 (2006)
3. Barnum, H., Saks, M.E., Szegedy, M.: Quantum query complexity and semi-definite programming. In: Proc. 18th CCC, pp. 179–193 (2003)
4. Bernstein, E., Vazirani, U.: Quantum complexity theory. SIAM J. Comput. 26, 1411–1473 (1997)
5. Boyer, M., Brassard, G., Høyer, P., Tapp, A.: Tight bounds on quantum searching. Fortschritte Der Physik 46, 493–505 (1998)
6. Brassard, G., Høyer, P., Mosca, M., Tapp, A.: Quantum amplitude amplification and estimation. In: Quantum Computation and Quantum Information: A Millennium Volume. AMS Contemporary Mathematics Series, vol. 305, pp. 53–74 (2002)
7. van Dam, W., Shparlinski, I.: Classical and quantum algorithms for exponential congruences. In: Kawano, Y., Mosca, M. (eds.) TQC 2008. LNCS, vol. 5106, pp. 1–10. Springer, Heidelberg (2008)
8. Grover, L.K.: A fast quantum mechanical algorithm for database search. In: Proc. 28th STOC, pp. 212–219 (1996)
9. Guy, R.K., Nowakowski, R.J.: Coin-weighing problems. Amer. Math. Monthly 102, 164–167 (1995)
10. Halbeisen, L., Hungerbühler, N.: The general counterfeit coin problem. Discrete Mathematics 147, 139–150 (1995)
11. Høyer, P., Lee, T., Špalek, R.: Negative weights make adversaries stronger. In: Proc. 39th STOC, pp. 526–535 (2007)
12. Iwama, K., Nishimura, H., Raymond, R., Teruyama, J.: Quantum counterfeit coin problems (2010), arXiv:1009.0416
13. Laplante, S., Magniez, F.: Lower bounds for randomized and quantum query complexity using Kolmogorov arguments. SIAM J. Comput. 38, 46–62 (2008)
14. Liu, W.A., Zhang, W.G., Nie, Z.K.: Searching for two counterfeit coins with two-arms balance. Discrete Appl. Math. 152, 187–212 (2005)
15. Magniez, F., Santha, M., Szegedy, M.: Quantum algorithms for the triangle problem. SIAM J. Comput. 37, 413–424 (2007)
16. Manvel, B.: Counterfeit coin problems. Mathematics Magazine 50, 90–92 (1977)
17. Reichardt, B.: Span programs and quantum query complexity: The general adversary bound is nearly tight for every boolean function. In: Proc. 50th FOCS, pp. 544–551 (2009)
18. Reichardt, B.: Reflections for quantum query algorithms (2010), arXiv:1005.1601
19. Špalek, R., Szegedy, M.: All quantum adversary methods are equivalent. Theory of Computing 2, 1–18 (2006)
20. Terhal, B.M., Smolin, J.A.: Single quantum querying of a database. Phys. Rev. A 58, 1822–1826 (1998)
21. Zhang, S.: On the power of Ambainis's lower bounds. Theoret. Comput. Sci. 339, 241–256 (2005)

Priority Range Trees

Michael T. Goodrich and Darren Strash

Department of Computer Science, University of California, Irvine, USA

Abstract. We describe a data structure, called a *priority range tree*, which accommodates fast orthogonal range reporting queries on prioritized points. Let S be a set of n points in the plane, where each point p in S is assigned a weight $w(p)$ that is polynomial in n, and define the rank of p to be $r(p) = \lfloor \log w(p) \rfloor$. Then the priority range tree can be used to report all points in a three- or four-sided query range R with rank at least $\lfloor \log w \rfloor$ in time $O(\log W/w + k)$, and report k highest-rank points in R in time $O(\log \log n + \log W/w' + k)$, where $W = \sum_{p \in S} w(p)$, w' is the smallest weight of any point reported, and k is the output size. All times assume the standard RAM model of computation. If the query range of interest is three sided, then the priority range tree occupies $O(n)$ space, otherwise $O(n \log n)$ space is used to answer four-sided queries. These queries are motivated by the Weber–Fechner Law, which states that humans perceive and interpret data on a logarithmic scale.

1 Introduction

Range searching is a classic problem that has received much attention in the Computational Geometry literature (e.g., see [2,3,7,10,12,15,19,21,22,25]). In what is perhaps the simplest form of range searching, called *orthogonal range reporting*, we are given a rectangular, axis-aligned query range R and our goal is to report the points p contained inside R for a given point set, S.

A recent challenge with respect to the deployment and use of range reporting data structures, however, is that modern data sets can be massive and the responses to typical queries can be overwhelming. For example, at the time of this writing, a Google query for "`range search`" results in approximately 363,000,000 hits! Dealing with this many responses to a query is clearly beyond the capacity of any individual.

Fortunately, there is a way to deal with this type of information overload—*prioritize* the data and return responses in an order that reflects these priorities. Indeed, the success of the Google search engine is largely due to the effectiveness of its PageRank prioritization scheme [9,26]. Motivated by this success, our interest in this paper is on the design of data structures that can use similar types of data priorities to organize the results of range queries.

An obvious solution, of course, is to treat priority as a dimension and use existing higher-dimensional range searching techniques to answer such three-dimensional queries (e.g., see [2,3]). However, this added dimension comes at a cost, in that it either requires a logarithmic slowdown in query time or an increase in the storage costs in order to obtain a logarithmic query time [4]. Thus, we are interested in prioritized range-searching solutions that can take advantage of the nature of prioritized data to avoid viewing priority as yet another dimension.

O. Cheong, K.-Y. Chwa, and K. Park (Eds.): ISAAC 2010, Part I, LNCS 6506, pp. 97–108, 2010.

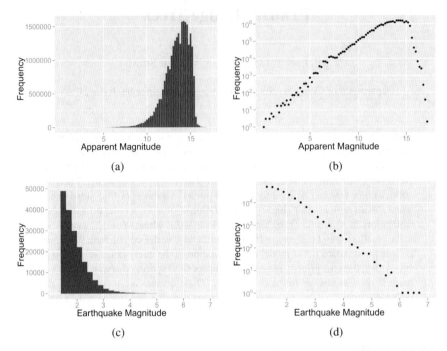

Fig. 1. (a) Frequency of celestial bodies by apparent magnitude for 25 million celestial bodies in the Guide Star Catalog (GSC) 1.2 [24], (b) plotted on a log-linear scale. (c) Frequency of earthquakes by Richter magnitude for 20 years of California earthquakes, (d) plotted on a log-linear scale. Note that because the measurements are made on a logarithmic scale, the straight line in the log-linear plots implies that there is a power law at work.

1.1   The Weber-Fechner Law and Zipf's Law

Since data priority is essentially a perception, it is appropriate to apply perceptual theory to classify it. Observed already in the 19th century, in what has come to be known as the *Weber–Fechner Law* [13,20], Weber and Fechner observed that, in many instances, there is a logarithmic relationship between stimulus and perception. Indeed, this relationship is borne out in several real-world prioritization schemes.

For instance, Hellenistic astronomers used their eyesight to classify stars into six levels of brightness [16]. Unknown to them, light intensity is perceived by the human eye on a logarithmic scale, and therefore their brightness levels differ by a constant factor. Today, star brightness, which is referred to as *apparent magnitude*, is still measured on a logarithmic scale [28]. Furthermore, the distribution of stars according apparent magnitude follows an exponential scale (see Fig. 1(a)–(b)).

Logarithmic scale measurements are not confined to the intensity of celestial bodies, however. Charles Richter's scale [8,27] for measuring earthquake magnitude is also logarithmic, and earthquake magnitude frequency follows an exponential distribution as well (see Fig. 1(c)–(d)). Moreover, as with astrophysical objects, range searching on geographic coordinates for earthquake occurrences is a common scientific query.

Similar in spirit to the Weber-Fechner Law, Zipf's Law is an empirical statement about frequencies in real-world data sets. Zipf's Law (e.g., see [18]) states that the frequency of a data value in real world data sets, such as words in documents, is inversely proportional to its rank. In other words, the relationship between frequency and rank follows a power law. For example, the popularity of web pages on the Internet follows such a distribution [5].

1.2   Problem Statement

Conventional range searching has no notion of priority. All points are considered equal and are dealt with equally. Nevertheless, as demonstrated by the Weber-Fechner Law and Zipf's Law, there are many real-world applications where data points are not created equal—they are prioritized. We therefore aim to develop range query data structures that handle these priorities directly. Specifically, we seek to take advantage of prioritization in two ways: we would like query time to vary according to the priority of items affecting the query, and we want to allow for items beyond a priority threshold not to be involved in a given query.

Because of the above-mentioned laws, we feel we can safely sacrifice some granularity in item weight, focusing instead on their logarithm, to fulfill these goals. The ultimate design goal, of course, is that we desire data structures that provide meaningful prioritized responses but do not suffer the logarithmic slowdown or an increase in space that would come from treating priority as a full-fledged data dimension. To that end, given an item x, let us assume that it is given a priority, $p(x)$, that is positively correlated to x's importance. So as to normalize item importance, if such priorities are already defined on a logarithmic scale (like the Richter scale), then we define x's rank, $r(x)$, as $r(x) = \lfloor p(x) \rfloor$ and we define x's weight, $w(x)$, as $w(x) = 2^{p(x)}$. Otherwise, if priorities are defined on a uniform scale (like hyperlink in-degree on the World-wide web), then we define the $w(x) = p(x)$ and we define $r(x) = \lfloor \log w(x) \rfloor$. We further assume that weight is polynomial in the number of inputs. This assumption implies that there are $O(\log n)$ possible ranks, and that $\log W/w = O(\log n)$, where w is a weight polynomial in n and W is the sum of n such weights. Given these normalized definitions of rank and weight, we desire efficient data structures that can support the following types of prioritized range queries:

– *Threshold query*: Given a query range, R, and a weight, w, report the points in R with rank greater than or equal to $\lfloor \log w \rfloor$.
– *Top-k query*: Given a query range R and an integer, k, report the top k points in R based on rank.

1.3   Prior Work

As mentioned above, range reporting data structures are well-studied in the Computational Geometry literature (e.g., see the excellent surveys by Agarwal [2] and Agarwal and Erickson [3]). In $\mathbf{R}^2$, 2- and 3-sided range queries can be answered optimally using McCreight's priority search tree [22], which uses $O(n)$ space and $O(\log n + k)$ query time. Using the range trees of Bentley [7], and the fractional cascading technique of Chazelle and Guibas [12], 4-sided queries can be answered using $O(n \log n)$ space and

$O(\log n + k)$ time. In the RAM model of computation, 4-sided queries can be answered using $O(n \log^\varepsilon n)$ space and $O(\log n + k)$ time [11]. Alstrup, Brodal, and Reuhe [4] further showed that range reporting in $\mathbf{R}^3$ can be done with $O(n \log^{1+\varepsilon} n)$ space and $O(\log n + k)$ query time in the RAM model.

More recently, Dujmović, Howat, and Morin [15] developed a data structure called a biased range tree which, assuming that 2-sided ranges are drawn from a probability distribution, can perform 2-sided range counting queries efficiently. Afshani, Barbay, and Chan [1] generalized this result, showing the existence of many instance-optimal algorithms. Their methods can be viewed as solving an orthogonal problem to the one studied here, in other words, in that their points have no inherent weights in and of themselves and it is the distribution of ranges that determines their importance.

1.4 Our Results

Given a set S of n points in the plane, where each point p in S is assigned a weight $w(p)$ that is polynomial in n, we provide a data structure, called a *priority range tree*, which accommodates fast three-sided orthogonal range reporting queries. In particular, given a three-sided query range R and a weight w, our data structure can be used to answer a threshold query, reporting all points p in R such that $\lfloor \log w(p) \rfloor \geq \lfloor \log w \rfloor$ in time $O(\log W / w + k)$, where W is the sum of the weights of all points in S. In addition, we can also support top-k queries, reporting k points that have the highest $\lfloor \log w(.) \rfloor$ value in R in time $O(\log \log n + \log W / w + k)$, where w is the smallest weight among the reported points. The priority range tree data structure occupies linear space, and operates under the standard RAM model of computation. Then, with a well-known technique for converting a 3-sided range reporting structure into a 4-sided range reporting structure, we show how to construct a data structure for answering prioritized 4-sided range queries with similar running times to those for our 3-sided query structure. The space for our 4-sided query data structure is larger by a logarithmic factor.

1.5 A Note about Distributions

A key feature of the priority range tree is that it is distribution agnostic. This distinction is crucial, since if the distribution of priorities is fixed to be exponential, then a trivial data structure achieves the same results: for i = 1 to $\lfloor \log w_{\max} \rfloor$, create a priority search tree P_i containing all points with weight 2^i and above. Because the distribution is exponential, the number of elements in P_i is at most $n/2^i \leq W/2^i$, and hence, querying $P_{\lfloor \log w \rfloor}$ correctly answers the query and takes time $O(\log W / w + k)$. Furthermore, the space used for all data structures is $\sum_i n/2^i = O(n)$.

However, such a strategy does not work for other distributions (including power law distributions, which commonly occur in practice) since the storage for each data structure becomes too great to meet the desired query time and maintain linear space. Thus, our data structure provides query times approaching the information theoretic lower bound, in linear space, regardless of the distribution of the priorities.

2 Preliminary Data Structuring Techniques

In this section, we present some techniques that we use to build up our priority range tree data structure.

2.1   Weight-Balanced Priority Search Trees

Consider the following one-dimensional range reporting problem: Given a set S of n points in $\mathbf{R}$, where each point p has weight $w(p)$, we would like to preprocess S into a data structure so that we can report all points in the query interval $[a,b]$ with weight greater or equal to w. Storing the points in a priority search tree [22], affords $O(\log n + k)$ query time using linear space. We can obtain a query time of $O(\log W/w + k)$ by ensuring that the priority search tree is *weight balanced*.

Definition 1. *We say a tree is weight balanced if item i with weight w_i is stored at depth $O(\log W/w_i)$, where W is the sum of the weights of all items stored in the tree.*

To build this search tree, we use a trick similar to Mehlhorn's rule 2 [23]. We first choose the item with the highest weight to be stored at the root. We then divide the remaining points into two sets A and B such that the x-value of every point in A is less than or equal to the x-value of every point in B and $|\sum_{a \in A} w(a) - \sum_{b \in B} w(b)|$ is minimized. Finally, we store the maximum x value from A in the root to facilitate searching, and then recursively build the left and right subtrees on sets A and B. We call this technique *split by weight*. Priority search trees are built much the same way, except that A and B are chosen to have approximately the same cardinality, which we call *split by size*.

The resulting search tree is both weight balanced and heap ordered by weight, and can therefore be used to answer range reporting queries with the same procedure as the priority search tree.

Lemma 1. *The weight-balanced priority search tree consumes $O(n)$ space, and can be used to report all points in a query range $[a,b]$ with weight at least w in time $O(\log(W/w) + k)$ where W is the sum of the weights of all points in the tree.*

2.2   Persistent Heaps

The well-known BuildHeap algorithm can transform any complete binary tree into a heap in linear time [17]. Using the node-copying method for making data structures persistent [14], we can maintain a record of the heap as it exists during each step of the BuildHeap algorithm, allowing us to store a heap on every subtree in linear space. We call this data structure a *persistent heap*.

Lemma 2. *Let T be a tree with n nodes. If the BuildHeap algorithm runs in time $O(n)$ on T, then we can augment every node of T with a heap of the elements in its subtree using extra space $O(n)$.*

Proof. For each swap operation of the BuildHeap algorithm, we do not swap within the tree, but we create two extra nodes, add the swapped elements to these nodes, and add links to the heaps from the previous stages of the algorithm (see Fig. 2).   □

Given n points in $\mathbf{R}^2$, this strategy can be used as a substitute for the priority search tree, by first building complete binary search tree on the x values, and then building a persistent heap using the y values as keys.

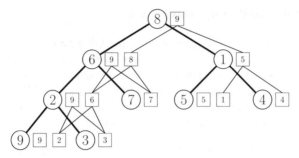

Fig. 2. A persistent heap. Circular nodes represent the original binary tree and square nodes represent heaps at each step of the BuildHeap algorithm.

2.3 Layers of Maxima

We now turn our attention to the following two problems: Given l points in $\mathbf{Z}_m \times \mathbf{R}$, preprocess the points into a data structure, to quickly answer the following queries.

1. *Domination Query*: Given a query point (x,y), report all points (p_x, p_y) such that $x \leq p_x$ and $y \leq p_y$.
2. *Maximization Query*: Given a query value y, report a point with the largest x-value, such that its y-value is greater than y.

This problem can be solved optimally using two techniques: we form the layers of maxima of the points and use fractional cascading to reduce search time.

A point $p \in \mathbf{R}^2$, *dominates* a point $q \in \mathbf{R}^2$ iff each coordinate of p is greater than that of q. A point p is said to be a *maximum* of a set S iff no point in S dominates p. Given a set S, if we find the set of maxima of S, remove the maxima and repeat, then the resulting sets of points are called the *layers of maxima* of S.

We begin by constructing the layers of maxima of the l points. We then form a graph from the layers of maxima by creating a vertex for each point and connecting vertices that are in the same layer in order by x coordinate.

We first fractionally cascade the points from bottom to top, sending up every other point from one layer to the next, including points copied from previous layers (see Fig. 3(a)). We then repeat the same procedure, fractionally cascading the points from left to right (see Fig. 3(b)). For bottom-to-top fractional cascading, we create m entry points into the top layer our data structure stored as an array indexed by x value. Each entry point stores one pointer to the maxima in the top layer that succeeds it in x-value.

To answer domination queries, we enter the catalog at index x, reports all points on the current layer that match the query, then jump down to the next layer and repeat. Each answer can be found with a constant amount of searching. Therefore, the query takes time $O(\max\{k,1\})$ where k is the output size.

To answer maximization queries, we create a catalog of $O(l)$ entry points on the right. Each entry point stores a pointer to a point (copied or not) on the top layer of maxima. Since the domain of y is not constrained to the integers, we perform a search for our query y among the entry points and immediately jump to our answer. This data structure gives us $O(l)$ space and $O(\log l)$ query time.

We now have all the machinery to discuss the priority range tree data structure.

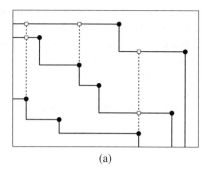 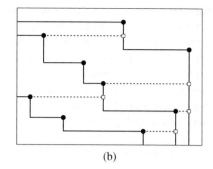

(a)   (b)

Fig. 3. The layers of maxima with fractional cascading (a) from bottom to top and (b) from left to right

3   The Priority Range Tree

In this section, we present a data structure for three-sided range queries on prioritized points. We assume that each point p has a weight $w(p)$, and we define $r(p) = \lfloor \log w(p) \rfloor$ to be the rank of p. Given a range $R = [x_1, x_2] \times [y, \infty)$ our data structure accommodates the following queries.

1. *Threshold Queries*: Given a query weight w, report all points p in R whose weight satisfies $\lfloor \log w(p) \rfloor \geq \lfloor \log w \rfloor$.
2. *Top-k Queries*: Given an integer k, report the k points in R with the highest $\lfloor \log w(\cdot) \rfloor$ value.

We first describe a data structure that has significant storage requirements, to illustrate how to perform each query. We then show how to reduce the space requirements.

For our underlying data structure, we build up a weight-balanced priority search tree on the x-values of our points. On top of this tree, we build one persistent heap for each different rank. That is, given rank r, we build a persistent heap on the y-values of points that have rank r. Of course, points with different rank must be compared in this scheme, therefore, when building a persistent heap for rank r, we treat points with rank not equal to r as dummies with y-value $-\infty$. Once we are done building up these persistent heaps, each node has $O(\log n)$ heaps rooted at it, one for each rank. For each node, we store a catalog, which is an array of roots of each of the $O(\log n)$ heaps, indexed by rank. On top of each catalog, we build the fractionally-cascaded layers-of-maxima data structure, described in the previous section, storing a coordinate (*rank, y-value*) for each root of the $O(\log n)$ heaps.

3.1   Threshold Queries

We first search for x_1 and x_2 down to depth $O(\log W/w)$ in our weight-balanced priority search tree, checking each point for membership as we make our way down the tree.

Each node on the search paths to x_1 and x_2 may have left or right subtree whose x values are entirely in the range $[x_1, x_2]$. For each such subtree, we query the layers-of-maxima data structure to find all points in the catalog that dominate $(\lfloor \log w \rfloor, y)$. If any points are returned, then we perform a layer-by-layer search through the heaps stored for each rank. We return any points that satisfy the query y value.

Each layers-of-maxima search can be charged to the search depth or an answer, and each search within a heap can be charged to an answer. Therefore, we get the desired running time of $O(\log W / w + k)$.

3.2 Top-k Queries

This query type is slightly more involved, so we begin by describing how to find one point of maximum rank in a query range.

Max-reporting. Given a range $R = [x_1, x_2] \times [y, \infty)$, the *max-reporting* problem is to report one point in R with maximum rank.

A first attempt is to search for x_1 and x_2, and run a maximization query for each layers-of-maxima data structure along the search path, maintaining the point in R with maximum rank found so far. Although this is a correct algorithm, there are two issues with this approach, which are brought about because we do not have a query weight:

1. The search may reach depth $\omega(\log W / w')$, where w' is the weight of the answer.
2. Each query to a layers-of-maxima data structure takes time $O(\log \log n)$.

Therefore, we maintain a depth limit, initially ∞, telling our algorithm when to stop searching. If there is no point in R, then our search reaches a leaf at depth $O(\log n)$ and stops. If we find a point p in R, we decrease the depth limit to $c \log W / w(p)$, where c is the constant hidden in the big-oh notation for the weight-balanced priority search tree. If our search reaches the depth limit, then there are no points with greater rank lower in the tree, and we can stop searching. Otherwise, every time we encounter point in p' in R with higher rank $r' = \lfloor \log w' \rfloor$ we decrease the depth limit to $c \log W / w'$.

We reduce the layers-of-maxima query time by fractionally cascading the layers-of-maxima data structure across the entire search tree, allowing us to do one $O(\log \log n)$-time query in the root catalog, and $O(1)$ extra work in the catalogs on the search path.

With these two changes, the search takes time $O(\log \log n + \log W / w')$ total ($O(\log n)$ if no such point exists).

Top-k Reporting. We now extend the max-reporting algorithm to report k points with highest rank in time $O(\log \log n + \log W / w' + k)$ under the standard RAM model.

If the top k points all have the same rank, then we can use our max-reporting algorithm to find the point with highest rank, and use the threshold queries to recover all k points. However, if we have to find points with lower rank, we want to avoid doing an expensive search for each rank. We can accomplish this goal with a priority queue.

Perform an initial max-reporting search. Every point in R encountered during our search is inserted into a priority queue with key equal to its rank. Along with each point we store a link back to the location in the layers-of-maxima data structure where it was found. When we finish the initial max-reporting query, we iterate the following process:

1. Remove the point p with maximum rank r from the priority queue.
2. Enter the layers-of-maxima data structure at point p, and insert both the predecessor of p on the same layer and on the layer below into the priority queue. Each one of these points are candidates for reporting. We then mark points to ensure that duplicates are not added to the priority queue.
3. Search in the heap data structure where point p was found, and report any additional points it contains that are in R (without exceeding k points).
4. If we have reported k points then we are done. Otherwise, look at the point with maximum rank in the priority queue. If its rank $r' = \lfloor \log w' \rfloor$ is less than r, then we increase our search depth to $c \log W / w'$, and continue searching, adding points the priority queue as before.

This priority queue can be efficiently implemented in the standard RAM model. We store our priority queue as an array P, indexed by key. We store in cell $P[r]$ a linked list of elements with key r. Additionally, we maintain two values, r_{max} and r_{min}, which is the maximum (minimum) key, of all elements in the priority queue. We insert an item with key k, by adding it to the linked list $P[k]$ in $O(1)$ time, and updating r_{max} and r_{min}. To remove an item with the maximum key k, we remove it from the linked list $P[k]$. If $P[k]$ becomes empty, then we update r_{max} (and possibly r_{min}).

We now show that our top-k reporting algorithm has running time $O(\log \log n + \log W / w' + k)$. We spend an initial $O(\log \log n)$-time search for our fractional cascading. Our initial search and subsequent extensions of the search path takes time $O(\log W / w')$, by virtue of our depth limit. For each heap in which we perform a layer-by-layer search, we can charge the search time to answers reported. All that remains to be shown is that the priority queue operations do not take too much time.

For each point discovered in a layers-of-maxima query along the search path, we perform at most one insert into the priority queue, thus we do $O(\log W / w')$ of these insertions into the priority queue, each one taking constant time. Furthermore, we can charge all of our priority queue remove operations (excluding pointer updates) to answers. Updating the priority queue pointers does not negatively impact the running time, since the total number of array cells we march through to do the pointer updates is $O(\log W / w')$. For each remove operation, we may perform up to two subsequent insertions, which we can charge to answers. Therefore, we get a total running time of $O(\log \log n + \log W / w' + k)$.

As described, the priority range tree consumes $O(n \log^2 n)$ space, since each persistent heap may consume $O(n \log n)$ space and we store $O(\log n)$ such persistent heaps.[1]

3.3   Reducing the Space Requirements

We can reduce the space to $O(n)$ by making several modifications to the search tree. We build our underlying tree using the split-by-weight strategy down to depth $\frac{1}{2} \log n$, then switch to a split-by-size strategy for the deeper elements, forcing these split-be-size subtrees to be complete. By switching strategies we do not lose the special properties that we desire: the tree is still weight balanced and heap ordered on weights. Finally,

---

[1] Each persistent heap uses space $O(n \log n)$ instead of $O(n)$ because there is no guarantee that BuildHeap will run in linear time on our underlying tree.

we do not store auxiliary data structures for split-by-size subtrees with between $\frac{1}{2}\log n$ and $2\log n$ elements in them, we only store one catalog for each such subtree. We call these subtrees *buckets*.

Lemma 3. *The priority range tree consumes $O(n)$ space.*

Proof. Each layers-of-maxima data structure uses $O(\log n)$ storage. We store one such data structure with each split-by-weight node and one for each split-by-size node excepting those in subtrees with between $1/2\log n$ and $2\log n$ elements. There are $O(\sqrt{n})$ split-by-weight nodes, since they are all above depth $1/2\log n$. Each subtree T_i rooted at depth $1/2\log n$ of size n_i will have at most $2n_i/\log n$ nodes that store the auxiliary data structure. Therefore, the total space for layers-of-maxima data structures is $O(\sqrt{n}\log n) + \sum_i O(n_i/\log n)O(\log n) = O(n)$.

For the persistent heaps, we ensure that each subtree T_i rooted at depth $1/2\log n$ is complete, on which we know BuildHeap will run in linear time. If T_i contains fewer than $1/2\log n$ nodes, then T_i has $O(\log n)$ heap nodes. Otherwise, each subtree T_i stores $O(n_i/\log n)$ heap nodes per rank.

Each node above depth $1/2\log n$ can contribute $O(\log n)$ swaps for each heap. Each heap therefore requires $O(\sqrt{n}\log n + \sum_i n_i/\log n) = O(n/\log n)$ space. Since we have $O(\log n)$ heaps, our data structure requires $O(n)$ space.   □

These structural changes affect our query procedures. In particular, there are three instances where the bucketing affects the query:

1. If the initial search phase hits a bucket, then we exhaustively test the $O(\log n)$ points in the bucket. We only reach a bucket if the search path is of length $\Omega(\log n)$ and therefore we can amortize this exhaustive testing over the search.
2. If we reach a bucket during our layer-by-layer search through a heap, then exhaustively searching through the bucket is not an option, as this would take too much time. Instead, we augment the layers-of-maxima data structure for the bucket so we can walk through lower y values with the same rank as the maxima (possibly producing duplicates).
3. By the very nature of the BuildHeap algorithm, it is possible that points in R were DownHeaped into the buckets, and that information is also lost. Therefore, when looking layer by layer through the heaps we also need to test for membership of the tree nodes in addition to the heap nodes (also possibly producing duplicates).

Each point matching the query will be encountered at most three times, once in each search phase. Therefore, we can avoid returning duplicates with a simple marking scheme without increasing the reporting time by more than a constant factor.

Theorem 1. *The priority range tree consumes $O(n)$ space and can be used to answer three-sided threshold reporting queries with rank above $\lfloor \log w \rfloor$ in time $O(\log W/w + k)$, and top-k reporting queries in time $O(\log\log n + \log W/w' + k)$, where W is the sum of the weights of all points in S, w' is the smallest weight of the reported points, and k is the number of points returned by the query.*

4   Four-Sided Range Reporting

Our techniques can be extended to four-sided range queries with an added logarithmic factor in space using a twist on a well-known transformation. Given a set S of points in $\mathbf{R}^2$, we form a weight-balanced binary search tree on the x-values of the points, such that the points are stored in leaves (e.g., a biased search tree [6]). For each internal node, we store the range of x-values of points contained in its subtree. For each internal node x (except the root), we store a priority range tree P_x on all points in the subtree: if x is a left child then P_x answer queries of the form $[a, \infty) \times [c, d]$, if x is a right child then P_x answers queries of the form $(-\infty, b] \times [c, d]$. We can answer four-sided query given the range $[a, b] \times [c, d]$, by doing the following:

We first search for a and b in in the weight-balanced binary search tree. Let s be the node where the search for a and b diverges, then a must be in s's left subtree left(s) and b must be in s's right subtree right(s). Then we query $P_{\text{left}(s)}$ for points in $[a, \infty) \times [c, d]$ and query $P_{\text{right}(s)}$ for points in $(-\infty, b] \times [c, d]$ and merge the results. In the case of top-k queries, we must carefully coordinate the search and reporting in each priority range tree, otherwise we may search too deeply in one of the trees or report incorrect points.

5   Conclusion

Our priority range tree data structure can be used to report points in a three-sided range with rank greater than or equal to $\lfloor \log w \rfloor$ in time $O(\log W / w + k)$, and report the top k points in time $O(\log \log n + \log W / w' + k)$, where w' is the smallest weight of the reported points, using linear space. These query times extend to four-sided ranges with a logarithmic factor overhead in space. Our results are possible because of our reasonable assumptions that the weights are polynomial in n, and that the magnitude of the weights, rather than the specific weights themselves, are more important to our queries.

Acknowledgments. This work was supported in part by NSF grants 0724806, 0713046, 0830403, and ONR grant N00014-08-1-1015.

References

1. Afshani, P., Barbay, J., Chan, T.M.: Instance-optimal geometric algorithms. In: Proc. 5th IEEE Symposium on Foundations of Computer Science, pp. 129–138 (2009)
2. Agarwal, P.K.: Range searching. In: Goodman, J.E., O'Rourke, J. (eds.) Handbook of Discrete and Computational Geometry, pp. 575–598. CRC Press LLC, Boca Raton (1997)
3. Agarwal, P.K., Erickson, J.: Geometric range searching and its relatives. In: Chazelle, B., Goodman, J.E., Pollack, R. (eds.) Advances in Discrete and Computational Geometry, Contemporary Mathematics, vol. 223, pp. 1–56. AMS, Providence (1999)
4. Alstrup, S., Stolting Brodal, G., Rauhe, T.: New data structures for orthogonal range searching. In: Proc. 41st IEEE Symposium on Foundations of Computer Science, pp. 198–207 (2000)
5. Baeza-Yates, R., Castillo, C., López, V.: Characteristics of the web of spain. International Journal of Scientometrics, Informetrics, and Bibliometrics 9 (2005)

6. Bent, S.W., Sleator, D.D., Tarjan, R.E.: Biased search trees. SIAM J. Comput. 14, 545–568 (1985)

7. Bentley, J.L.: Multidimensional divide-and-conquer. Commun. ACM 23(4), 214–229 (1980)

8. Boore, D.: The Richter scale: its development and use for determining earthquake source parameters. Tectonophysics 166, 1–14 (1989)

9. Brin, S., Page, L.: The anatomy of a large-scale hypertextual web search engine. Comput. Netw. ISDN Syst. 30(1-7), 107–117 (1998)

10. Chazelle, B.: Filtering search: a new approach to query-answering. SIAM J. Comput. 15, 703–724 (1986)

11. Chazelle, B.: A functional approach to data structures and its use in multidimensional searching. SIAM J. Comput. 17(3), 427–462 (1988)

12. Chazelle, B., Guibas, L.J.: Fractional cascading: I. A data structuring technique. Algorithmica 1(3), 133–162 (1986)

13. Dehaene, S.: The neural basis of the Weber-Fechner law: a logarithmic mental number line. Trends in Cognitive Sciences 7(4), 145–147 (2003)

14. Driscoll, J.R., Sarnak, N., Sleator, D.D., Tarjan, R.E.: Making data structures persistent. Journal of Computer and System Sciences 38(1), 86–124 (1989)

15. Dujmović, V., Howat, J., Morin, P.: Biased range trees. In: Proc. 19th ACM-SIAM Symposium on Discrete Algorithms, pp. 486–495. SIAM, Philadelphia (2009)

16. Evans, J.: The history & practice of ancient astronomy. Oxford University Press, Oxford (October 1998)

17. Floyd, R.W.: Algorithm 245: Treesort 3. Commun. ACM 7(12), 701 (1964)

18. Frakes, W.B., Baeza-Yates, R. (eds.): Information retrieval: data structures and algorithms. Prentice-Hall, Inc., Upper Saddle River (1992)

19. Fries, O., Mehlhorn, K., Näher, S., Tsakalidis, A.: A $\log \log n$ data structure for three-sided range queries. Inform. Process. Lett. 25, 269–273 (1986)

20. Hecht, S.: The visual discrimination of intensity and the Weber-Fechner law. Journal of General Physiology 7(2), 235–267 (1924)

21. Lueker, G.S.: A data structure for orthogonal range queries. In: Proc. 19th IEEE Symposium on Foundations of Computer Science, pp. 28–34 (1978)

22. McCreight, E.M.: Priority search trees. SIAM J. Comput. 14(2), 257–276 (1985)

23. Mehlhorn, K.: Nearly optimal binary search trees. Acta Informatica 5(4), 287–295 (1975)

24. Morrison, J., Roser, S., McLean, B., Bucciarelli, B., Lasker, B.: The guide star catalog, version 1.2: An astronomic recalibration and other refinements. The Astronomical Journal 121, 1752–1763 (2001)

25. Overmars, M.H.: Efficient data structures for range searching on a grid. J. Algorithms 9, 254–275 (1988)

26. Page, L., Brin, S., Motwani, R., Winograd, T.: The PageRank citation ranking: Bringing order to the web. Technical Report 1999-66, Stanford InfoLab, previous number = SIDL-WP-1999-0120 (November 1999), http://ilpubs.stanford.edu:8090/422/

27. Richter, C.F.: Elementary Seismology. W.H. Freeman and Co., New York (1958)

28. Satterthwaite, G.E.: Encyclopedia of Astronomy. Hamlyn (1970)

Should Static Search Trees Ever Be Unbalanced?

Prosenjit Bose and Karim Douïeb*

School of Computer Science, Carleton University, Herzberg Building
1125 Colonel By Drive, Ottawa, Ontario, K1S 5B6 Canada
{jit,karim}@cg.scs.carleton.ca
http://cg.scs.carleton.ca

Abstract. In this paper we study the question of whether or not a static search tree should ever be unbalanced. We present several methods to restructure an unbalanced k-ary search tree T into a new tree R that preserves many of the properties of T while having a height of $\log_k n + 1$ which is one unit off of the optimal height. More specifically, we show that it is possible to ensure that the depth of the elements in R is no more than their depth in T plus at most $\log_k \log_k n + 2$. At the same time it is possible to guarantee that the average access time $P(R)$ in tree R is no more than the average access time $P(T)$ in tree T plus $O(\log_k P(T))$. This suggests that for most applications, a balanced tree is always a better option than an unbalanced one since the balanced tree has similar average access time and much better worst case access time.

1 Introduction

The dictionary problem is fundamental in computer science, it asks for a data structure that efficiently stores and retrieves data. Binary search trees are simple, powerful and commonly used dictionaries. The problem of building static search trees has been intensively studied in the past decades. Depending on the performance required one can build a perfectly balanced search tree that guarantees an optimal worst-case search time or one can build a biased search tree matching the entropy bound thereby providing an optimal expected search time. The search tree that minimizes the expected search cost can be unbalanced thereby behaving badly in the worst-case. Thus one may prefer to build a search tree of bounded height, i.e., with a certain guarantee on the worst-case search time that also minimizes the expected search time. In this paper we address the issue of the increase in the expected search cost imposed by restricting the height of the constructed tree.

Since a search tree T minimizing the expected search cost may behave badly in worst-case, one may want to construct another tree R on the same set of keys in such a way that the worst-case search time is improved but the expected search time does not differ too much from the initial tree. One way to achieve this is to guarantee that R has bounded height and that the depth of a key in R is not much more than its depth in T. This is known as the *restructuring search*

---

* Research partially supported by NSERC and MRI.

O. Cheong, K.-Y. Chwa, and K. Park (Eds.): ISAAC 2010, Part I, LNCS 6506, pp. 109–120, 2010.

tree problem. Moreover, the problem of designing such search trees is directly related to the design of good codes. Thus the results obtained in this paper on search trees also has straightforward applications in coding theory.

Preliminaries. Consider the set $x_1, x_2, \ldots, x_n$ of keys contained in a search tree T. We are given $2n + 1$ weights $p_1, p_2, \ldots, p_n$ and $q_0, q_1, \ldots, q_n$ such that $\sum_{i=1}^{n} p_i + \sum_{i=0}^{n} q_i = 1$. Here, p_i is the probability to query the key x_i (successful search) and q_i is the probability to query a key lying between x_i and x_{i+1} (unsuccessful search), q_0 and q_n are the probabilities to query a key that is less or greater, respectively, than any key contained in the tree.

Static *multiway search trees* (or k-ary trees) generalize most of the other static search tree structures. A successful search ends up in an internal node of a k-ary tree that contains the requested key. Each internal node of a k-ary tree contains at most $k - 1$ keys and has between 1 and k children. An unsuccessful search ends up in one of the $n + 1$ leaves of the k-ary tree. A leaf in a k-ary tree does not contain any key. The *weighted path length* of a k-ary tree T (referred to as path length in the remainder of this paper), a measure of the average number of nodes traversed during a search, is defined as

$$P(T) = \sum_{i=1}^{n} p_i(d_T(x_i) + 1) + \sum_{i=0}^{n} q_i d_T(x_{i-1}, x_i), \tag{1}$$

where $d_T(x_i)$ is the depth in terms of number of links from the root node to the internal node containing the key x_i, $d_T(x_{i-1}, x_i)$ is the depth of the leaf reached at the end of the unsuccessful search for a key lying between x_{i-1} and x_i. In the context of binary search trees (when $k = 2$) in the comparisons-based model, the path length corresponds to the average number of comparisons performed during a search. In the external memory model, the path length corresponds to the average number of I/Os performed during a search in the case where each node is stored as one disk block. Note that this is the usual way to store a multiway search tree in external memory.

1.1   Related Work

Optimal Search Trees. Knuth [12] showed that an optimal binary search tree can be built in $O(n^2)$ time using $O(n^2)$ space. Mehlhorn [15] gave an $O(n)$ time algorithm to build a binary search tree that is near-optimal. Concerning the more general case of k-ary trees, Vaishnavi *et al.* [17] showed that an optimal k-ary tree can be built in $O(kn^3)$ time. Becker [2] gave an $O(kn^\alpha)$ time algorithm, with $\alpha = 2 + \log_k 2$, to build an optimal B-tree (subclass of k-ary tree) that satisfies the original constraints fixed by Bayer and McCreight [1]. These constraints require that every leaf in the B-tree have the same depth and that every internal node contains between $k/2$ and k keys except for the root node. In the remainder of this paper, we consider a more general model of k-ary tree. The only constraint is that an internal node contains at most $k - 1$ keys. Recently Bose and Douïeb [3] presented a method to build a k-ary tree in $O(n)$ time (independent of k) that gives the best upper bound on the path length of a k-ary tree and produces a near-optimal k-ary tree for any $k \geq 2$.

The problem of building an optimal search tree when only unsuccessful searches occur, i.e., when $\sum_{i=1}^{n} p_i = 0$, is known as the optimal *alphabetic search tree* problem. Hu and Tucker [8] developed an $O(n^2)$ time and $O(n)$ space algorithm for constructing an optimal alphabetic binary search tree. This was improved by two other algorithms, the first one was by Knuth [11] and the second by Garsia and Wachs [7]. Both algorithms use $O(n \log n)$ time and $O(n)$ space.

Optimal Search Trees with Restricted Height. The problem of building an optimal binary search tree with restricted maximal height has been addressed by Garey [6]. The best algorithms solving this problem have been independently developed by Wessner [18] and Itai [9]. They both produce the optimal binary search tree, with h as the height restriction, in $O(hn^2)$ time. For the problem of building an optimal alphabetic binary search tree with restricted maximal height h, Larmore and Przytycka [14] presented a $O(hn \log n)$ time algorithm.

Restructuring Search Trees. The problem of restructuring a search tree T consists of building another tree R, on the same set of keys, with restricted height such that the path length of R is as close as possible to the path length of T. The *drop* of a node x is defined as $\Delta(x) = d_R(x) - d_T(x)$. This problem was initially posed by Bose. Evans and Kirkpatrick [4] developed a technique to restructure a binary search tree T into a tree R of height $\lceil \log n \rceil + 1$ such that $\Delta(x) \leq \log \log n$ for every node x in T. They also showed that restructuring an alphabetic binary search tree can be done with the guarantee that $\Delta(x) \leq 2$ for every node x. Their work mainly focused on understanding the tradeoff between the height restriction of the restructured tree and the worst-case drop realized by a node. Gagie [5] gave an alternate way to restructure a binary search tree into a tree of height $\log n + 1$ that guarantees a slightly larger worst-case drop but aims at reducing the total drop as opposed to the worst case individual drop. He provided an algorithm where the path length of the restructured tree R satisfies the following $P(R) \leq P(T) + (1 + \epsilon) \log(P(T) + 1) + \log((1/\epsilon) + 1) + 2$ with $1 < \epsilon \leq 2$.

1.2 Our Results

We present several methods to restructure a binary search tree that improves the previous best upper bounds on both the local drop of an individual node as well as the total drop of all nodes. The methods and the proofs are all based on a simple but general technique. We show that our method generalizes and are the first to study how to restructure multiway search trees (previous work only considers binary search trees). Our results are then used to prove new tighter upper bounds on the path length of optimal height-restricted multiway search trees.

In Section 2.2, we present new tree restructuring methods that focus on reducing the worst-case drop of any given key. We first focus our attention on restructuring a given alphabetic k-ary search tree into another one of height $\log_k n + 1$ such that at least a quarter of the leaves do not drop at all, the maximum drop realized by all but one of the leaves is at most 1 and exactly one leaf drops at most 2 levels. Second, we present a restructuring method for the general case of k-ary search trees that builds another k-ary tree on the same keys with a guaranted worst-case drop of at most $\log_k \log_k n$. In fact, this method potentially gives a better bound since it takes into consideration the balance of the initial tree. The more unbalanced the initial tree, the better the guarantee on the drop. For example, if the initial tree is a path, then this method guarantees that the worst-case drop is at most 1.

In Section 2.3, we develop a method focused on the relative drop. By this, we mean that in the worst case, the amount that a node will drop is proportional to its depth in the original tree as opposed to being proportional to the number of nodes in the tree. For a given node x_i, the maximum drop is at most $\log_k(d_T(x_i) + 1) + (1 + \epsilon) \log_k \log(d_T(x_i) + 2) + \log_k \frac{1+\epsilon}{\epsilon} + 1$. As a consequence of this, the path length of the restructured tree is close to the path length of the initial tree but the restructured tree has height at most $\log_k n + 1$. In Section 2.4 we combine the worst-case and relative drop approaches to obtain a hybrid method that guarantees simultaneously the best upper bounds in term of relative and worst-case drop plus a small constant.

Finally we show in Section 3 how the results on relative node drop can be used to obtain tighter upper bounds on the path length of optimal height-restricted multiway search trees.

2   Restructuring Multiway Search Trees

Restructuring a search tree T consists of building a new tree R, on the same set of keys, such that R satisfies a precise constraint on its height. The problem is to determine how the tree R differs from T and how it is efficiently constructed. The main idea of our approach, similar to [5], is to define a weight distribution on the keys based on their depth in the initial tree T. The weights of the keys are defined differently depending on what kind of guarantee on the drop we want to achieve. We distinguish between two types of guarantees on the drop: *local* or *global*. A local guarantee specifies the maximum drop realized by any node. A global guarantee specifies the maximum increase of the path length. Given these newly defined weights, we build a near-optimal search tree using a technique described in the next section.

2.1   Method to Construct Near-Optimal Multiway Search Tees

We describe a technique to build near-optimal multiway search trees, developed by Bose and Douïeb [3] and initially inspired from Mehlhorn's technique [15] when access probabilities are known. This technique guarantees the best theoretical upper bound on the path length of optimal multiway search trees. Note that any other technique to build search trees can be used for the purpose of restructuring trees but we use [3] because it guarantees the best properties.

Let $p_1, p_2, \ldots, p_n$ be the access probabilities of the internal nodes and $q_0, q_1, \ldots, q_n$ be the access probabilities of the leaves. Let T' be the tree built with the method [3]. The following two lemmas characterize the depth of the elements in T', we distinguish the cases where T' has a branching number equal to 2 or when it is greater. We define the value $m = \max\{n - 3P, P\} - 1 \geq \frac{n}{4} - 1$ where P is the number of increasing or decreasing sequences in the access probability distribution on the ordered leaves. The value $q_{rank[i]}$ is the ith smallest access probability among the leaves except for the extremal ones (i.e. we exclude $(-\infty, x_1)$ and (x_n, ∞) from consideration).

Lemma 1. *The depth of the elements in T' satisfy the following*

$$d_{T'}(x_i) \leq \lfloor \log_k \frac{1}{p_i + q_{min}} \rfloor \qquad \text{for} \quad i = 1, \ldots, n \,,$$

$$d_{T'}(x_{i-1}, x_i) \leq \lfloor \log_k \frac{2}{q_i} \rfloor + 1 \qquad \text{for} \quad i = 0, \ldots, n.$$

The following lemma is not explicitly described in [3], additional details will appear in the journal version of this paper.

Lemma 2. *In the case where $k = 2$, the depth of the elements in T' satisfy the following*

$$d_{T'}(x_i) \leq \lfloor \log_2 \frac{1}{p_i + q_{min}} \rfloor \qquad \text{for} \quad i = 1, \ldots, n \,,$$

$$d_{T'}(x_{i-1}, x_i) \leq \lfloor \log_2 \frac{1}{q_i} \rfloor + 2 \qquad \text{for} \quad \text{one leaf } (x_{i-1}, x_i),$$

$$d_{T'}(x_{j-1}, x_j) \leq \lfloor \log_2 \frac{1}{q_j} \rfloor + 1 \qquad \text{for} \quad \text{all leafs } (x_{j-1}, x_j) \neq (x_{i-1}, x_i),$$

$$d_{T'}(x_{j-1}, x_j) \leq \lfloor \log_2 \frac{1}{q_j} \rfloor \qquad \text{for at least } m + 2 \text{ leaves}(x_{j-1}, x_j).$$

Theorem 1. *The path length of the tree T' is at most*

$$UB(k) = \frac{H}{\log_2 k} + 1 + \sum_{i=0}^{n} q_i - q_0 - q_n - \sum_{i=0}^{m} q_{rank[i]},$$

where $H = \sum_{i=1}^{n} p_i \log_2(1/p_i) + \sum_{i=0}^{n} q_i \log_2(1/q_i)$ is the entropy of the probability distribution. In the case $k = 2$ the path length of T' is at most

$$UB(2) = H + 1 - q_0 - q_n + q_{max} - \sum_{i=0}^{m'} pq_{rank[i]},$$

where the value $m' = \max\{2n - 3P, P\} - 1 \geq \frac{n}{2} - 1$, $pq_{rank[i]}$ is the ith smallest access probability among every key and every leaf (except the extremal leaves) and q_{max} is the greatest leaf probability including external leaves.

2.2   Worst Case Drop

In this section we consider the problem of minimizing the maximum drop independently realized by each node.

Alphabetical Tree

An alphabetic search tree is a tree where only unsuccessful searches occur, i.e., when $\sum_{i=1}^{n} p_i = 0$. In order to restructure an alphabetic tree T, we first define a weight for each leaf in T based on its depth in T. Namely the weight of a leaf node (x_{i-1}, x_i) is defined as

$$w(x_{i-1}, x_i) = \max\left(\frac{1}{k^{d_T(x_{i-1}, x_i)}}, \frac{1}{(k-1)n}\right).$$

Let $W = \sum_{i=0}^{n} w(x_{i-1}, x_i)$ which is always strictly smaller than $1 + \frac{n}{(k-1)n} = \frac{k}{(k-1)}$ by Kraft's inequality [13]. These weights are used to define the access probabilities of each leaf. The access probability of a leaf (x_{i-1}, x_i) is defined as $q_i = w(x_{i-1}, x_i)/W$ and the access probability of an internal node x_i as $p_i = 0$. These probabilities are then used as input to the algorithm described in Section 2.1 to build a near-optimal binary search tree giving the restructured tree R on the same keys.

Theorem 2. *An alphabetic multiway tree T can be restructured into a tree R such that the height of R is at most $\log_k n + 1$ and the maximum drop of a leaf is at most 1 if $k > 2$. When $k = 2$ a drop of 2 is realized by only one leaf, the drop of any other is at most 1. In general, at least $m \geq \frac{n}{4} + 2$ leafs do not drop.*

Proof. By Lemma 1, the greatest depth reached by an internal node is $\lfloor \log_k \frac{1}{q_{min}} \rfloor$ $< \log_k \frac{k(k-1)n}{(k-1)} = \log_k n + 1$. As a consequence the greatest depth of a leaf is at most $\log_k n + 1$, which corresponds to the maximum height of the restructured tree.

The depth of a leaf (x_{i-1}, x_i) in the restructured tree R is at most

$$\lfloor \log_k \frac{2}{q_i} \rfloor + 1 < \lfloor \log_k \frac{2k^{d_T(x_{i-1}, x_i)+1}}{k-1} \rfloor + 1 = \lfloor \log_k \frac{2}{k-1} \rfloor + d_T(x_{i-1}, x_i) + 2.$$

Thus for $k > 2$, the depth of a leaf (x_{i-1}, x_i) is at most $d_T(x_{i-1}, x_i) + 1$ which implies a maximum leaf drop of 1. Using Lemma 2, similar arguments verify the theorem in the case where $k = 2$.   □

So this simple method generalizes to k-ary alphabetic search trees the result of Evans and Kirkpatrick [4]. It also gives a more precise guarantee on the maximal drop of a leaf in the binary alphabetic search tree case, since we guarantee that only one leaf drops two levels, all other leaves drop 1 level with a quarter of the leaves not dropping at all. Note that for some binary search trees any restructuring method produces a drop of 2 (see [4]).

General k-ary Search Tree

Here the weight of an internal node x_i is defined as follows

$$w(x_i) = \max\left(\frac{1}{k^{d_T(x_i)}}, \frac{W'}{(k-1)n}\right),$$

where $W' = \sum_{i=1}^{n} \frac{1}{k^{d_T(x_i)}} \leq (k-1)\log_k n$ by the generalization of Kraft's inequality's [16]. Let $W = \sum_{i=1}^{n} w(x_i) < W' + \frac{W'}{(k-1)} = \frac{k}{(k-1)}W' \leq k\log_k n$. These weights are used to construct a probability distribution on the nodes. The access probability of an internal node x_i is $p_i = w(x_i)/W$ whereas the access probability of a leaf is null, i.e., $q_i = 0$ for all leaves. These probabilities are used to build the restructured tree R with the technique described in Section 2.1.

Theorem 3. *A multiway search tree T can be restructured into a tree R such that the height of R is at most $\log_k n + 1$ and the maximum drop of a node is at most $\lfloor \log_k \frac{W'}{k-1} \rfloor \leq \log_k \log_k n$.*

Proof. By Lemma 1, the depth of an internal node x_i is at most $\lfloor \log_k \frac{1}{p_i} \rfloor = \lfloor \log_k \frac{W}{w(x_i)} \rfloor$. The greatest depth reached by an internal node is

$$\max_i \log_k \frac{W}{w(x_i)} < \log_k \frac{\frac{kW'}{(k-1)}}{\frac{W'}{(k-1)n}} = \log_k n + 1.$$

As a consequence the greatest depth of leaf is at most $\log_k n + 1$, which corresponds to the maximum height of the restructured tree. The depth of an internal node x_i in the restructured tree R is at most

$$\lfloor \log_k \frac{1}{p_i} \rfloor < \lfloor \log_k \frac{k}{k-1} W' k^{d_T(x_i)} \rfloor = d_T(x_i) + \lfloor \log_k \frac{W'}{k-1} \rfloor + 1.$$

The maximum drop is $\lfloor \log_k \frac{W'}{k-1} \rfloor \leq \log_k \log_k n$ for both internal nodes and leaves since the drop of a leaf is the same as the drop of its parent (an internal node). $\square$

This method generalizes to k-ary search trees the result of Evans and Kirkpatrick [4]. For the binary search tree case, the worst-case drop guaranteed with this method is similar to the one given by Evans and Kirkpatrick. Indeed there are some instances for which our method produces a drop of $\log_k \log_k n$. But for most instances the guarantee is better since our method takes into consideration the balance of the initial tree. For example if the tree is a list than the worst-case drop is constant. The value W' is the expression of the balance of the initial tree, W' is $O(1)$ for a highly unbalanced tree and $\Omega(\log n)$ when the tree is unbalanced.

2.3 Relative Drop

Generally a static unbalanced search tree is needed when frequently accessed elements have to be accessed much faster than the other elements. In this context, if we want to restructure an unbalanced tree in order to satisfy a precise constraint on its height, it is important that elements located close to the root in the original tree remain close to the root in the restructured tree. To achieve this, we bound the maximum drop of an element with respect to its depth in the original tree. This optimization differs from the previous one as it aims to reduce the global instead of local drop.

First we define the weight of an internal element x_i as

$$w(x_i) = \max \left(\frac{1}{D(x_i)\,(d_T(x_i)+1)\,\log^{1+\epsilon}(d_T(x_i)+2)}, \frac{1+\epsilon}{\epsilon n(k-1)} \right),$$

with $1 < \epsilon \leq 2$ and $D(x_i)$ is the number of elements at depth $d_T(x_i)$ in the tree T, thus $D(x_i) \leq (k-1)k^{d_T(x_i)}$. Let $W = \sum_{i=1}^{n} w(x_i)$ which is strictly smaller than $\sum_{i=1}^{n} \frac{1}{i\log^{1+\epsilon}(i+1)} + \frac{(1+\epsilon)n}{\epsilon n(k-1)} < \frac{k(1+\epsilon)}{(k-1)\epsilon}$. These weights define a probability distribution on the nodes so that the access probability of an internal node x_i is given by $p_i = w(x_i)/W$. We consider the leaves to have an access probability of zero, i.e., $q_i = 0$ for all leaves. These probabilities are used to build the restructured tree R with the technique described in Section 2.1.

Theorem 4. *Define* $f(y) = \log_k y + (1 + \epsilon)\log_k \log(y + 1) + \log_k \frac{1+\epsilon}{\epsilon} + 1$. *A multiway search tree T can be restructured into a tree R of height $\log_k n + 1$ where the drop of an internal node x_i is at most $f(d_T(x_i)+1)$ and the drop of a leaf (x_{i-1}, x_i) is at most $f(d_T(x_{i-1}, x_i)) - 1$.*

Proof. According to Lemma 1, the depth of a internal node is at most $\lfloor \log_k \frac{1}{p_i} \rfloor = \lfloor \log_k \frac{W}{w(x_i)} \rfloor$. The greatest depth that an internal node can reach is

$$\max_i \log_k \frac{W}{w(x_i)} < \log_k \left(\frac{k(1+\epsilon)}{(k-1)\epsilon} \frac{\epsilon\,n(k-1)}{(1+\epsilon)} \right) = \log_k n + 1.$$

As a consequence the greatest depth of a leaf is at most $\log_k n + 1$, which corresponds to the maximum height of the restructured tree.

The depth of an internal node x_i in R is at most

$$\lfloor \log_k \frac{W}{w(x_i)} \rfloor < \lfloor \log_k \frac{k(1+\epsilon)}{(k-1)\epsilon} D(x_i)\,(d_T(x_i)+1)\,\log^{1+\epsilon}(d_T(x_i)+2) \rfloor$$

$$\leq d_T(x_i) + \log_k(d_T(x_i)+1) + (1+\epsilon)\log_k \log(d_T(x_i)+2) +$$

$$\log_k \frac{1+\epsilon}{\epsilon} + 1.$$

The maximum depth of a leaf in R is the same as the maximum depth of its parent node in R. Thus the depth of a leaf (x_{i-1}, x_i) is at most

$$d_T(x_{i-1}, x_i) - 1 + \log_k(d_T(x_{i-1}, x_i)) + (1+\epsilon)\log_k \log(d_T(x_{i-1}, x_i)+1) + \log_k \frac{1+\epsilon}{\epsilon} + 1.$$

$\square$

Theorem 5. *Define* $m = \lfloor \log_k n \rfloor + 1$. *A search multiway tree* T *can be restructured into a tree* R *such that the height of* R *is at most* h *(with* $h \geq m$*) and the depth of an internal node* x_i *satisfies*

$$d_R(x_i) \leq d_T(x_i) + f(d_T(x_i) + 1 - h + m) \quad \text{if} \quad h - m \leq d_T(x_i) < h,$$
$$\leq d_T(x_i) \quad \textbf{otherwise.}$$

For a leaf (x_{i-1}, x_i)*,*

$$d_R(x_{i-1}, x_i) \leq d_T(x_{i-1}, x_i) + f(d_T(x_{i-1}, x_i) - h + m) \text{ if } h - m \leq d_T(x_{i-1}, x_i) < h,$$
$$\leq d_T(x_{i-1}, x_i) \quad \textbf{otherwise.}$$

Proof. Consider the subtrees of T rooted at the elements at depth $h - m$. Apply the restructuring procedure described in the beginning of this section to each of those subtrees seen as independent trees. This restructuring does not affect the depth of elements at depth strictly smaller than $h - m$. According to Theorem 4, the maximal drop of the other internal nodes x_i is proportional to the depth inside the subtree that contains them, i.e., $d_R(x_i) \leq f(d_T(x_i) + 1 - (h - m))$. The maximum drop of a leaf (x_{i-1}, x_i) is at most the maximum drop of its parent node, i.e., $d_R(x_{i-1}, x_i) \leq f(d_T(x_{i-1}, x_i) - (h - m))$. □

We show how to restructure a tree T into a tree R with nearly minimum height such that the increase of the path length is small. This new restructuring tree method slightly improves the result of Gagie [5] and arguably simplifies the proof technique (knowledge about relative entropy is not required). Evans and Kirkpatrick [4] guaranteed a worst-case drop of $\log \log n$. Since this does not take into consideration the original depth of the element in the tree, this could lead to a situation where the depth of the root in the restructured tree is $\log \log n$ times greater then its depth in the initial tree.

2.4 Hybrid Drop

The first method presented in Section 2.2 gives the best upper bound on the worst case drop which is $\log_k \log_k n$. The problem is that the restructured tree produced by this method can have a path length which is $\log_k \log_k n$ times larger than the path length of the original tree. The method introduced in Section 2.3 avoids this problem by guaranteeing a drop that is proportional to the depth of the elements in the original tree, but the guarantee on the worst-case drop is a bit worst than the previous method. Here we present a hybrid method for restructuring a k-ary search tree that guarantees simultaneously the best upper bounds in term of relative and worst-case drop plus a small constant.

Let d' be the value that satisfies $(d' + 1) \log^{1+\epsilon}(d' + 2) = \log_k n$ with $1 < \epsilon \leq 2$. The weight of an internal node x_i is defined as follows

$$w(x_i) = \begin{cases} \max\left(\dfrac{1}{D(x_i)\,(d_T(x_i)+1)\,\log^{1+\epsilon}(d_T(x_i)+2)}, \dfrac{1+2\epsilon}{\epsilon n(k-1)}\right) & \text{for } (d_T(x_i) + 1) \leq d', \\[2ex] \max\left(\dfrac{1}{D(x_i)\log_k n}, \dfrac{1+2\epsilon}{\epsilon n(k-1)}\right) & \text{for } d' < (d_T(x_i) + 1) \leq \log_k n, \\[2ex] \dfrac{1+2\epsilon}{\epsilon n(k-1)} & \text{for } (d_T(x_i) + 1) > \log_k n. \end{cases}$$

The total weight is

$$W = \sum_{i=1}^{n} w(x_i)$$

$$\leq \sum_{j=0}^{d'} \frac{1}{(j+1)\log^{1+\epsilon}(j+2)} + \sum_{j=d'+1}^{\log_k n} \frac{1}{\log_k n} + \frac{n(1+2\epsilon)}{\epsilon n(k-1)}$$

$$< \frac{1}{\epsilon} + 1 + 1 + \frac{(1+2\epsilon)}{\epsilon(k-1)}$$

$$< \frac{k(1+2\epsilon)}{(k-1)\epsilon}.$$

Those weights are used to build the restructured tree R with the technique described in Section 2.1. The access probability of an internal node x_i is given by $p_i = w(x_i)/W$ whereas the access probability of a leaf is null, i.e., $q_i = 0$ for all leaves.

Theorem 6. *A k-aray search tree T can be restructured into a tree R such that the height of R is at most $\log_k n + 1$ and the drop of a key x_i is at most*

$$\min\{\log_k \log_k n, \log_k(d_T(x_i)+1) + (1+\epsilon)\log_k \log(d_T(x_i)+2)\} + \log_k \frac{1+2\epsilon}{\epsilon} + 1.$$

Proof. By Lemma 1, the depth of an internal node x_i is at most $\lfloor \log_k \frac{1}{p_i} \rfloor = \lfloor \log_k \frac{W}{w(x_i)} \rfloor$. The largest depth reached by an internal node is

$$\max_i \log_k \frac{W}{w(x_i)} < \log_k \frac{\frac{k(1+2\epsilon)}{(k-1)\epsilon}}{\frac{1+2\epsilon}{\epsilon n(k-1)}} = \log_k n + 1.$$

As a consequence the largest depth of a leaf is at most $\log_k n + 1$ which corresponds to the maximum height of the restructured tree. Using the same type of argument than in the proof of Theorem 4, an internal node x_i with $(d_T(x_i)+1) \leq d'$ realizes a drop of at most

$$\log_k(d_T(x_i) + 1) + (1+\epsilon)\log_k \log(d_T(x_i) + 2) + \log_k \frac{1+2\epsilon}{\epsilon} + 1$$

which is at most $\log_k \log_k n + \log_k \frac{1+2\epsilon}{\epsilon} + 1$ by the definition of d'. An internal node x_i with $d' < (d_T(x_i) + 1) \leq \log_k n$ realizes a drop of at most

$$\log_k \log_k n + \log_k \frac{1+2\epsilon}{\epsilon} + 1$$

which is at most $\log_k(d_T(x_i) + 1) + (1+\epsilon)\log_k \log(d_T(x_i) + 2) + \log_k \frac{1+2\epsilon}{\epsilon} + 1$ by the definition of d'. $\square$

3   Applications

Nice applications of the results provided in the Section 2.3 about the relative drop occurs in the context of building optimal height-restricted multiway search trees. We are interested in measuring the maximum increase of the path length imposed by a height restriction. We investigate the difference between the path length of the optimal multiway tree and the optimal multiway tree with a height restriction. We give the best upper bound on the path length of an optimal multiway tree with a height restriction. Note that to prove the bound we assume that the access probabilities to the nodes and leaves are given.

Theorem 7. *Consider T^* the optimal multiway tree built over the set of keys $x_1, \ldots, x_n$ and let T_h^* define the optimal multiway tree build on the same set of keys and such that its height is no more than $h \geq \lfloor \log_k n \rfloor + 1$. The following is always satisfied*

$$P(T_h^*) \leq P(T^*) + f(\max\{1, P(T^*) - h + m\}),$$

where $f(y) = \log_k y + (1+\epsilon)\log_k \log(y+1) + \log_k \frac{1+\epsilon}{\epsilon} + 1$ and $m = \lfloor \log_k n \rfloor + 1$.

Proof. Using the method described in Section 2.3 we can restructure T^* into the tree R_h which has a maximum height h. By definition we have $P(T_h^*) \leq P(R_h)$. Using Theorem 5 and by Jensen's inequality [10] we show

$$P(R_h) = \sum_{i=1}^{n} p_i(d_{R_h}(x_i) + 1) + \sum_{i=0}^{n} q_i d_{R_h}(x_{i-1}, x_i)$$

$$\leq \sum_{i=1}^{n} p_i(d_{T^*}(x_i) + 1 + f(\max\{1, d_{T^*}(x_i) + 1 - h + m\}))$$

$$+ \sum_{i=0}^{n} q_i(d_{T^*}(x_{i-1}, x_i) + f(\max\{1, d_{T^*}(x_{i-1}, x_i) - h + m\}))$$

$$\leq P(T^*) + f(\max\{1, P(T^*) - (h - m)\}).$$

□

Among other things this theorem states that a height restricted optimal multiway tree has a path length that differs from the optimal path length $P(T^*)$ without the height restriction by roughly $2 \log_k P(T^*)$ (even if the height restriction is nearly maximum, i.e., $\log n + 1$). This casts doubt on the necessity of using unbalanced search trees.

Theorem 8. *There exists a linear running time algorithm which builds a multiway search tree R_h with a height smaller than $h \geq \lfloor \log_k n \rfloor + 1$ and such that*

$$P(R_h) \leq UB(k) + f(\max\{1, UB(k) - h + m\})$$

where $UB(k)$ is defined in Theorem 1.

Proof. We use the technique described in Section 2.1 to build a near-optimal multiway search tree T in $O(n)$ time. This guarantees that $P(T) \leq UB(k)$. Then we restructure T into R_h in $O(n)$ time using the technique developed in Section 2.3. We can deduce from Theorem 5 the correctness of the theorem.   □

Acknowledgment

The authors wish to thank Travis Gagie, Pat Morin and Michiel Smid for fruitful discussions. The first author wishes to especially thank Luc Devroye for his many key insights on this problem and for encouraging him to never give up.

References

1. Bayer, R., McCreight, E.M.: Organization and maintenance of large ordered indexes. Acta Informatica 1(3), 173–189 (1972)
2. Becker, P.: A new algorithm for the construction of optimal B-trees. Nordic Journal of Computing 1, 389–401 (1994)
3. Bose, P., Douïeb, K.: Efficient construction of near-optimal binary and multiway search trees. In: Proceedings of the 11th Workshop on Algorithms and Data Structures (WADS), pp. 230–241 (2009)
4. Evans, W., Kirkpatrick, D.: Restructuring ordered binary trees. Journal of Algorithms 50, 168–193 (2004)
5. Gagie, T.: Restructuring binary search trees revisited. Inf. Process. Lett. 95(3), 418–421 (2005)
6. Garey, M.: Optimal binary search trees with restricted maximal depth. SIAM J. Comput. 3, 101–110 (1974)
7. Garsia, A.M., Wachs, M.L.: A new algorithm for minimum cost binary trees. SIAM Journal on Computing 6, 622–642 (1977)
8. Hu, T.C., Tucker, A.C.: Optimal computer search trees and variable-length alphabetical codes. SIAM Journal on Applied Mathematics 21(4), 514–532 (1971)
9. Itai, A.: Optimal alphabetic trees. SIAM J. Comput. 5, 9–18 (1976)
10. Jensen, J.L.W.V.: Sur les fonctions convexes et les inégalités entre les valeurs moyennes. Acta Mathematica 30(1), 175–193 (1906)
11. Knuth, D.: The Art of Computer Programming. Sorting and Searching, vol. 3. Addison-Wesley, Reading (1973)
12. Knuth, D.: Optimum binary search trees. Acta Informatica 1, 79–110 (1971)
13. Kraft, L.G.: A device for quantizing grouping and coding amplitude modulated pulses. Master's thesis, Electrical Eng. Dept., MIT, Cambridge (1949)
14. Larmore, L.L., Przytycka, T.M.: A fast algorithm for optimum height-limited alphabetic binary trees. SIAM J. Comput. 23(6), 1283–1312 (1994)
15. Mehlhorn, K.: A best possible bound for the weighted path length of binary search trees. SIAM Journal on Computing 6, 235–239 (1977)
16. Prisco, R.D., Santis, A.D.: New lower bounds on the cost of binary search trees. Theor. Comput. Sci. 156(1&2), 315–325 (1996)
17. Vaishnavi, V.K., Kriegel, H.P., Wood, D.: Optimum multiway search trees. Acta Informatica 14(2), 119–133 (1980)
18. Wessner, R.: Optimal alphabetic search trees with restricted maximal height. Information Processing Letters 4(4), 90–94 (1976)

Levelwise Mesh Sparsification for
Shortest Path Queries

Yuichiro Miyamoto[1], Takeaki Uno[2], and Mikio Kubo[3]

[1] Sophia University, Kioicho 7-1, Chiyoda-ku, Tokyo, Japan
miyamoto@sophia.ac.jp
[2] National Institute of Informatics, Hitotsubashi 2-1-2, Chiyoda-ku, Tokyo, Japan
uno@nii.jp
[3] Tokyo University of Marine Science and Technology, Etchujima 2-1-6, Koto-ku, Tokyo, Japan
kubo@kaiyodai.ac.jp

Abstract. In this paper, we address the shortest path query problem, i.e., constructing a data structure of a given network to answer the shortest path length of two given points in a short time. We present a method named Levelwise Mesh Sparsification for the problem. The key idea is to divide the network into meshes and to sparsify the network in each mesh by removing unnecessary edges and vertices that are never used when the shortest path passes through the mesh. In large real-world road networks in the United States, our method is about 1,500 times faster than Dijkstra's algorithm, which is competitive with existing methods. The time taken to construct the data structure is a few hours on a typical PC. Unlike previous methods, our geometric partition method succeeded in reducing the data for connecting the sparsified network. As a result, our method uses additional data that is only about 10% of the original data size, while existing methods use more than 2000%. Our method has considerable extensibility because it is independent of search algorithms. Thus, it can be used with Dijkstra's algorithm and A*-search among others, and with several models such as negative costs, time-dependent costs, and so on. These are rarely handled by previous methods.

1 Introduction

The shortest path problem is one of the most fundamental problems in computer science. It has been studied well, especially in the areas of optimization and algorithms, and it has many applications in theory and practice. For example, dynamic programming is basically solving the shortest path on a table, and computation of the edit distance between two strings is reduced to the shortest path problem. In the real world, car navigation systems utilize shortest path algorithms, and Internet packet routing needs a short path to the packet destination. In artificial intelligence and model checking, shortest path problems sometimes need to be solved to determine the feasibility of the problem. Applications of the shortest path continue to increase.

The shortest path problem can be solved with Dijkstra's algorithm [5] in $O(m + n \log n)$ time, where n and m are respectively the number of vertices and edges in the network to be solved. Recently, sophisticated quasi-linear time algorithms have been proposed [15]. However, some recent applications require the problem to be solved in a very short time. For example, an on-line car navigation system must solve the shortest

O. Cheong, K.-Y. Chwa, and K. Park (Eds.): ISAAC 2010, Part I, LNCS 6506, pp. 121–132, 2010.

path problem in a network having a huge number of edges, e.g., the US road network, which has 58 million edges, in a sufficiently short time. In such applications, linear time is too long. In these applications, the input network is usually fixed, or it does not change frequently or drastically. Thus, this is a natural motivation to construct an index (data structure) that reduces the computational time. Such an approach can be considered as a database query, which we call the *shortest path query*. In this paper, we address the problem and present *Levelwise Mesh Sparsification* (LMS), a method based on geometrical partition to reduce the search space.

In the last decade, three representative methods are presented: bit vector [10], highway hierarchy [13,14], and transit node routing [3,2]. These methods preserve the optimality of the solution, while most previous techniques (that are popular in modern car navigation systems) have no assurance of optimality.

The bit vector method [10] adds additional data for each edge of the given network. The additional data indicate that the corresponding edge is on a shortest path for a given target. The data reduce the search space in query processing, especially when the edge is far from the target. Though the bit vector method is simple and fast in query processing, it takes a long time in preprocessing (and updating the preprocessed data). The highway hierarchy method constructs a highway network composed of highway edges. A highway edge is defined as an intermediate edge of a shortest path between two sufficiently distant points. The recursive extraction of highway edges helps to reduce the search space in query processing. Though highway hierarchy achieves a short preprocessing time and query time, it only allows bi-directional search in query processing. Transit node routing [3,2] selects the transit nodes in preprocessing such that, for any pair of distant vertices, at least one transit node is included in their shortest path. Then, we compute and store the all-pairs shortest paths of the transit nodes. In the shortest path search, we can directly move from a transit vertex near the origin to a transit vertex near the destination. Though transit node routing achieves a very quick response for most queries, the preprocessing (and updating the preprocessed data) takes a long time.

Our LMS is a simple sparsification method for a given network. The simplicity ensures that several search methods can be used in preprocessing and query processing. Furthermore, computational results show that the geometric implementation of LMS achieves competitive performance, i.e., quick response for queries, using less memory. In Table 1, the advantages of each method are summarized.

Other techniques preserving optimality are also known: A^* search with landmarks [7], reach [6,8], arc-flags [11], SPAH based on HiTi [9], etc. Moreover, the shortest path query has many extensions, such as finding k nearest neighbors [12], time dependent costs [4], etc, but we omit the details here.

Table 1. Advantages of each method

	Response time	Preprocessing time	Update cost	Memory usage	Negative cost/ Time dependent cost
Bit Vector	fair	poor	poor	poor	good
Highway Hierarchy	fair	good	fair	fair	poor
Transit Node Routing	good	poor	poor	poor	good
LMS (ours)	fair	fair	good	good	good

2   Preliminaries

Let $\mathbb{R}^+$ and $\mathbb{Z}$ be a set of positive real numbers and a set of integers, respectively. Let $G = (V, A, d)$ be a simple directed network: V is a vertex set, A is a directed edge set, and $d : A \to \mathbb{R}^+$ is a positive distance function on the edges. An ordered sequence of edges $((v_1, v_2), (v_2, v_3), \ldots, (v_{k-1}, v_k))$ is called a v_1-v_k path of G. The vertices v_1 and v_k are called the end vertices of the path. The length of a path is the sum of the distances of all edges of the path. The shortest s-t path is one whose length is shortest among all s-t paths. In general, the shortest s-t path is not unique. For a given network and a query specified by an origin s and a destination t, the shortest s-t path problem is to determine one of the shortest s-t paths. In certain contexts, the shortest s-t path problem requires only the length of the shortest s-t path. In this paper, we assume that the shortest s-t path is unique. There is no loss of generality since we can use a symbolic perturbation for the edge distance function or use a shortest path algorithm to determine the canonical shortest path.

For a vertex v, let $N(v)$ be the set of neighbors of $v \in V$ defined by $N(v) = \{w \in V \mid (w, v) \in A$ or $(v, w) \in A\}$. Let $N(S)$ be the set of neighbors of $S \subseteq V$ defined by $N(S) = \cup_{v \in S} N(v) \setminus S$. The subnetwork of G induced by $U \subseteq V$ is denoted by $G[U]$. The union of two graphs G_1 and G_2 is defined by a graph whose vertex set and edge set are the union of the vertex sets, and edge sets, respectively, of the G_1 and G_2 graphs.

3   Levelwise Mesh Sparsification

Here, we describe levelwise mesh sparsification (LMS) when a network $G = (V, A, d)$ is given. The purpose of the method is to extract edges included in some shortest paths between distant vertices.

3.1   Geometric Implementation

Hereafter we assume that all vertices are embedded on a 2-dimensional plane; thus, each vertex has 2-dimensional coordinates. This assumption is reasonable in the context of geometric networks such as road networks and railroad networks. Let $x(v)$ and $y(v)$ be the x- and y-coordinates of $v \in V$, respectively. A square region on the 2-dimensional plane is specified by the region's height (width) and the point with the smallest coordinates. For a given real constant c and given integers i, j, let $R(c, i, j)$ be a square region such that $R(c, i, j) = \{(x, y) \in \mathbb{R}^2 \mid i \cdot c \le x < (i + 1)c, \ j \cdot c \le y < (j + 1)c\}$. Clearly, the set of $R(c, i, j)$, $\forall i, j \in \mathbb{Z}$ partitions the 2-dimensional plane. We define an outer region $R_{outer}(c, i, j)$ of $R(c, i, j)$ as $R_{outer}(c, i, j) = \{(x, y) \in \mathbb{R}^2 \mid (i - 1)c \le x < (i + 2)c, \ (j - 1)c \le y < (j + 2)c\}$. Let $V(c, i, j)$ and $V_{outer}(c, i, j)$ be subsets of V defined by $V(c, i, j) = \{v \in V \mid (x(v), y(v)) \in R(c, i, j)\}$ and $V_{outer}(c, i, j) = \{v \in V \mid (x(v), y(v)) \in R_{outer}(c, i, j)\}$, respectively. Clearly $V_{outer}(c, i, j)$ includes $V(c, i, j)$. Without loss of generality, we assume that $0 \le x(v), y(v)$ and that $\exists L \in \mathbb{R}^+$, $x(v), y(v) < L$, $\forall v \in V$. Under this assumption, it is sufficient to consider a finite number of subsets $V(c, i, j)$ and $V_{outer}(c, i, j)$ for a given constant c.

Our idea to find edges that can be used in some shortest paths is based on the observation that, in real-world networks, only a few edges in an area are used in the middles of the shortest paths connecting distant vertex pairs. This motivates us to remove all edges that will never be included in the shortest paths connecting a pair of vertices outside $R_{\text{outer}}(c, i, j)$. When we find a shortest path connecting vertices outside $R_{\text{outer}}(c, i, j)$, we only have to refer to the remaining edges in the middle of the search. Thus, we may drastically reduce the number of edges to be examined. The reason for using $R_{\text{outer}}(c, i, j)$ instead of directly using $R(c, i, j)$ is that almost all edges near the boundary of $R(c, i, j)$ might be used by a shortest path, and we would not gain much reduction. Now let us define the *sparsified network* that is obtained through implementation of the above idea.

Definition 1. *For given constants $c \in \mathbb{R}^+$ and $i, j \in \mathbb{Z}$, a sparsified network of G derived by $V(c, i, j)$ is the subnetwork of $G[V(c, i, j) \cup N(V(c, i, j))]$ composed of edges such that each edge is included in the shortest path of G connecting a pair of vertices in $N(V_{\text{outer}}(c, i, j))$. We denote the sparsified network of G derived by $V(c, i, j)$ as $G'(c, i, j)$.*

Note that a shortest path connecting two vertices outside $R_{\text{outer}}(c, i, j)$ always passes through some vertices in $N(V_{\text{outer}}(c, i, j))$ if it includes a vertex in $V(c, i, j)$. Thus, this definition characterizes the edges that may be included in such shortest paths.

The operation "*replacement of $G'(c, i, j)$*" constructs the network obtained from G by removing edges in $G[V(c, i, j)]$ and adding edges in $G'(c, i, j)$.

Lemma 1. *For any pair of vertices s and t outside $R_{\text{outer}}(c, i, j)$, a shortest s-t path is included in the network obtained from G by the $G'(c, i, j)$ replacement operation.*

Suppose that regions $R(c_1, x_1, y_1), \cdots, R(c_h, x_h, y_h)$ are pairwise disjoint, and G' is the graph obtained by replacing all $R(c_1, x_1, y_1), \cdots, R(c_h, x_h, y_h)$.

Lemma 2. *For any pair of vertices s and t outside $R_{\text{outer}}(c_i, x_i, y_i)$ for any i, a shortest s-t path is included in G'.*

We say an edge is *inside* $G'(c, i, j)$ if both end vertices are in $V(c, i, j)$. Next, let us define a *sparsified mesh*, which is an edge-contracted version of the sparsified network. *Edge contractions* of a network G involve the following procedures.

- One way: If only two inside edges (u, v) and (v, w) are incident to a vertex $v \in V(G)$, remove them and add (u, w) whose distance is $d(u, v) + d(v, w)$.
- Bi-direction: If only four inside edges (u, v), (v, u), (v, w), and (w, v) are incident to a vertex $v \in V(G)$, remove them and add edges (u, w) and (w, u) whose distances are $d(u, v) + d(v, w)$ and $d(w, v) + d(v, u)$, respectively.

Intuitively, edge contraction is the replacement of an "inner path" of the sparsified network by an edge whose length is the same as the inner path.

Definition 2. *A sparsified mesh of G derived by $V(c, i, j)$ is a minimal network obtained by edge contractions.*

We denote a sparsified mesh of G derived by $V(c, i, j)$ as $G(c, i, j)$. The replacement of a sparsified mesh is defined as that of the sparsified network.

Lemma 3. *The length of the shortest s-t path of G is equal to that of G replaced by $G(c, i, j)$, if both s and t are not in $V_{outer}(c, i, j)$.*

We say that a sparsified mesh $G(c, i, j)$ is *valid* for a pair of vertices (query) $\{s, t\}$ if both s and t are not in $V_{outer}(c, i, j)$. This means that we can replace the network G with valid meshes to find the shortest s-t path. From the definition of the valid sparsified mesh and from Lemma 3, the following lemma is immediate.

Lemma 4. *The length of the shortest s-t path of G is equal to that of G replaced by all valid meshes for $\{s, t\}$.*

The replacement by the sparsified mesh preserves the length of the shortest path between distant vertices. However, the shortest s-t path of G replaced by valid sparsified meshes is not the shortest path of G in general because some paths of sparsified networks are replaced by edges. If the shortest path query requires the path itself (not only the length of the path), we must restore the original path corresponding to the edges of the sparsified meshes. For the restoration, we make pointers from each edge of the sparsified meshes to a path (sequence of edges) of G when we contract the edges. Because each edge of G belongs to at most one edge of sparsified meshes, the number of pointers is at most the number of edges of G.

If the size of a square region c is small, the number of square regions (i.e., the number of sparsified meshes) will be large; thus, the network to be processed will not be sparse. In contrast, we cannot use sparsified meshes in a large area around the origin and destination if c is large. To offset these disadvantages, we introduce a *levelwise sparsified mesh* that uses a hierarchic structure of sparsified meshes.

Definition 3. *For a given constant C, a levelwise sparsified mesh of G is a collection of sparsified meshes $G(2^k C, i, j)$ for all i, j, $k \in \mathbb{Z}$ over a sufficiently large range, where $k > 0$.*

By use of $R(2^k C, i, j)$ and $R_{outer}(2^k C, i, j)$, the sparsified level-k mesh is defined in the same way. A sparsified mesh in $G(2^k C, i, j)$ is called a *level-k mesh*. In particular, we define $G[V(C, i, j) \cup N(V(C, i, j))]$ as a level-0 mesh. The collection of all level-0 meshes is a given original network.

When a query (the origin s and destination t) is given, we use a *query network* that is constructed from a combination of valid meshes of many levels, which minimizes the number of meshes. The construction details of the query network are described in the next subsection.

3.2 Shortest Path Query

For a given query s and t, a valid level-k mesh is called *maximal* if its region is not covered by the region of any valid level-$(k + 1)$ mesh. Since any region of a level-k mesh is partitioned by disjoint regions of level-$(k - 1)$ meshes, any two maximal valid meshes are disjoint. The definition of a *query network* is as follows.

Definition 4. *For a given query* $\{s, t\}$, *a query network is the network obtained by replacing all the maximal valid meshes for* $\{s, t\}$.

From Lemmas 2 and 4, the shortest path length is the same as that in G and the query network. Thus, we have the following theorem.

Theorem 1. *For any embedded network G and query* $\{s, t\}$, *the length of the shortest s-t path is invariant in G and in the query network for* $\{s, t\}$ *of G.*

Note that for a different query $\{s', t'\}$, the shortest s'-t' distance may differ between G and the query network for s and t, so the query network is basically valid only for s and t.

To obtain the edges of G in the shortest path in the query network, we have to extract them from the contracted edges. To extract from an edge in a level-$(k + 1)$ mesh, we similarly put pointers from the edge to the corresponding edges in the level-k meshes. We apply this extraction recursively to the extracted edges. The existence of the edges is ensured by the following lemma.

Lemma 5. *For each edge e of level-$(k+1)$ mesh $G(2^{k+1}C, i, j)$, there is a path on the union of level-k meshes $G(2^kC, 2i, 2j) \cup G(2^kC, 2i + 1, 2j) \cup G(2^kC, 2i, 2j + 1) \cup G(2^kC, 2i + 1, 2j + 1)$ such that the edge contraction of the path is the same as e.*

3.3   Construction of Levelwise Sparsified Mesh

LMS requires preprocessing to construct a levelwise sparsified mesh. A straightforward approach is to solve the all-pairs shortest path problem on the boundary of each outer region. Because this involves heavy computation, we use a method to reduce the computational time.

The construction is done in a bottom-up way. We first compute all the level-1 meshes. This is done in a straightforward way, and it does not take long because the outer regions are quite small. Next, we compute the level-2 meshes by finding the shortest paths connecting all pairs of vertices on the boundary of each outer region. Note that for any pair of vertices on the boundary, the sparsified meshes of some internal areas of the outer region are valid. Thus, in this process, we can replace induced subgraphs of G with level-1 meshes. In general, to compute level-k mesh $G(2^kC, i, j)$, we replace induced subgraph $G[V_{\text{outer}}(2^kC, i, j) \cup N(V_{\text{outer}}(2^kC, i, j))]$ with maximal meshes of a lower level that are valid for any pair of boundary vertices $N(V_{\text{outer}}(2^kC, i, j))$. The replacement by lower level meshes does not decrease the accuracy of the constructed mesh from Lemma 5. This recursive use of sparsified meshes not only reduces the computational time in preprocessing but also makes it easier to find hierarchical pointers between the edges of levelwise sparsified meshes for the restoration of the shortest path. The construction of the levelwise sparsified mesh allows simple parallel computation, since $G(2^kC, i, j)$ and $G(2^kC, i', j')$ can be computed independently.

In real-world data, the precision of the network differs between areas. Thus, we should prepare low-level meshes for areas with a dense network and should not prepare low-level meshes for sparse areas having only a few edges. Because we want to automatically determine a suitable size, we use *granularity* as a tuning parameter. Intuitively, granularity g is the upper bound of the number of vertices included in the lowest level (smallest) meshes. For a given granularity g, we calculate maximum size C such that each region $R(C, i, j)$ includes at most g vertices. As a result, the highest level of meshes is automatically determined by the granularity, e.g., if $R(2^k C, 0, 0)$ includes all vertices, the level-k (and higher level) mesh has no edge. We store the levelwise meshes in a quad-tree and delete all empty meshes. This guarantees that the number of stored meshes is at most the number of vertices for any input data. Furthermore, when the number of vertices included in $R(2^k C, i, j)$ is at most g, we directly copy the original network to the mesh $G(2^k C, i, j)$. This reduces the memory usage in practice.

3.4   Query Computation

When we are given a query $\{s, t\}$, we first construct a query network by combining the maximal valid meshes. To collect all the valid meshes, we use quad-trees representing the inclusion relations of all level meshes. The root of a quad-tree corresponds to a top-level mesh, and the four children of the root correspond to the four lower-level meshes included in the top-level mesh. Similarly, each vertex of the quad-tree corresponds to a level-k mesh, and its (at most) four children correspond to level-$(k-1)$ meshes included in the level-k mesh.

Observe that any pair of maximal valid meshes are disjoint, so they can never be ancestors or descendants in the quad-trees each other. To find all maximal valid meshes, we search each quad-tree starting from its root vertex. If the mesh corresponding to the vertex (on the tree) is valid, we take it and terminate the search on this quad-subtree since any descendant will not correspond to a maximal valid mesh. If the corresponding mesh is not valid, we recursively search all four children. We recursively search the tree until we reach the bottom of the tree or find a valid mesh. In this way, we can collect all maximal valid meshes in time linear to the number of maximal valid meshes.

A simple way to execute a shortest path algorithm on the query network is to construct the query network explicitly by combining all maximal valid meshes. This requires time linear to the query network size. Instead of this, we can either dynamically or implicitly construct the network. In the dynamic version, we first construct the working network composed of a level-0 mesh that includes the origin. When the front line of the search arrives at the boundary, we attach the neighboring maximal valid mesh to the working network. In this way, we can omit this attachment operation for the meshes that the search does not access.

The other way is to implicitly combine the maximal valid meshes. In this way, a vertex is represented by a pair of the vertex and the mesh (or level of the mesh) we are currently searching at the vertex. We also start from the level-0 mesh, and when the search is going to cross the boundary of the mesh, we find the neighboring maximal valid mesh M. Note that a vertex is included in many meshes of different levels, and thus the difference in the meshes is their included edges. Then, the end vertex v of the edge crossing the boundary is given by the pair of v and M, so a further search will be done in the mesh M.

These techniques are efficient when the number of edges accessed by the shortest path algorithm is much less than that in the query network. However, in our computational experiments, the number of edges accessed by a normal Dijkstra algorithm is up to 1/3 of all edges. The improvements by these techniques would be limited, so we use the simple way in our computational experiments.

4   General Framework

The idea of LMS has many possibilities for extensions and generalizations. This section discusses some of these possibilities.

We defined the outer region $R_{\text{outer}}(c, i, j)$ around $R(c, i, j)$ as $R_{\text{outer}}(c, i, j) = \{(x, y) \in \mathbb{R}^2 \mid (i - 1)c \leq x < (i + 2)c, (j - 1)c \leq y < (j + 2)c\}$. In general, by using non-negative real r, we could define $R_{\text{outer}}(r, c, i, j)$ as $R_{\text{outer}}(r, c, i, j) = \{(x, y) \in \mathbb{R}^2 \mid (i - r)c \leq x < (i + r + 1)c, (j - r)c \leq y < (j + r + 1)c\}$. As the radius r increases, the sparsified networks and meshes become sparser. However, the cost of construction increases, and we can not use sparse networks near s and t. Thus, there would be an optimal value for r. This involves another optimization problem.

Another extension is the use of variable shapes for regions and outer regions. We can use non-square areas for the partition, such as non-square rectangles, hexagons, and circles. Another idea is to use four outer regions shifted in the vertical and horizontal directions for each region. This enables us to use much larger sparsified networks in areas close to s or t. When t is north of s, we could use outer regions shifted to the north. Accordingly, the southern parts of the outer regions will be small; thus, we could use sparsified networks of much larger outer regions even if they are close to s. In contrast, we could use outer regions shifted to the south near t. In a sense, LMS considers only stable geometrical partitions, but this shifting introduces "directions" to the geometrical partition.

Direct application of LMS is infeasible when the given network is not embedded on a 2-dimensional plane. In this case, we can use a generalization of LMS. Essentially, our sparsification method depends on the definitions of the (inner) region and outer region. The key to the sparseness is that any vertex outside the outer region is far from the vertices in the (inner) region. Thus, if we can define (inner) regions and outer regions by using vertex sets satisfying such a property, we can obtain the sparsified networks in the same way. We can partition the vertex set into many groups so that each group consists of many edges and try to minimize the number of edges between two groups. We can define region R as a group of vertices and the outer region of R as vertices near some vertices in R. Another way is to recursively bi-partition the vertex set by using a minimal cut to obtain the regions, and to define the outer region as a larger region including the (inner) region. In such a way, we can partition non-geometric networks and use an LMS-like method on them.

5   Approximation for Further Speed-Up

In this section, we present an approximation technique for speeding up LMS in pre-processing and shortest path queries. In preprocessing, LMS constructs each sparsified mesh in the bottom-up manner. For this construction, LMS finds shortest path trees iteratively. Because the procedure may be the heaviest calculation, we omit the number of iterations of finding shortest path trees.

Our approximation technique is a minor change in the preprocessing. For the construction of each sparsified mesh, we restrict the number of iterations of finding shortest path trees. Precisely, we uniformly choose each origin of $N(V_{\text{outer}}(c, i, j))$ in clockwise order.

Because the resulting sparsified meshes may lose edges that are necessary for the optimal shortest path search, the outputs in queries of LMS using the meshes are approximate shortest paths. The approximation technique might reduce preprocessing costs and query costs, where cost means the time and the memory usage, but it also loses accuracy.

6 Computational Experiments

We evaluated LMS by using a large real-world road network of the United States (US). The network was obtained from the TIGER/Line Files [16]. Since the network is undirected, we transformed it into a bi-directional network. There are 23,947,347 vertices and 58,333,344 edges in the network. The network data contains the longitude and latitude of each vertex, so we used longitude and latitude as the x-coordinate and y-coordinate, respectively, in LMS. The network data were composed of edges and vertices. An edge is identified by its end vertices and its length. The length of each edge represents the travel time between the end vertices in the network data.

We implemented our algorithms in C and ran all our experiments on a Mac Pro with a 3GHz Intel Xeon CPU and 2-GByte RAM, running Mac OS 10.5.7. In the preprocessing and shortest path queries, we used a simple binary heap implementation of Dijkstra's algorithm. We evaluated the performance of LMS in comparison with that of the normal Dijkstra's algorithm. However, the performances of the algorithms are detailed in the original reports [7,10,14], so we can compare the results.

We evaluated the performance of LMS in terms of the preprocessing CPU time, the query CPU time, and the size of the additional data. The additional data means the constructed levelwise sparsified meshes whose levels are at least 1; this is because the union of level-0 meshes is input network data.

Preprocessing CPU time is amount of the computational time for constructing levelwise sparsified meshes. Additional size is the number of edges of levelwise meshes whose levels are at least 1 that are constructed and used in LMS. Additional ratio is the ratio of the additional number of edges to the number of original edges. The number of

Table 2. Preprocessing costs and query speed of LMS in US

Granularity		10	20	40	80	160
Highest level of meshes		19	17	17	16	16
Preproc. CPU time		119 m 52 s	128 m 04 s	140 m 07 s	170 m 31 s	215 m 41 s
	Additional size	17,965,824	11,192,879	7,155,933	4,206,795	2,212,340
	Additional ratio	62.26 %	38.79 %	24.80 %	14.58 %	8.75 %
Query	# settled vertices (avg.)	3,609.564	3,751.591	4,090.936	4,856.587	6,474.404
	# settled vertices (worst)	5,866	6,007	6,580	7,985	10,617
	Speed-up (avg.)	3,180.235	3,059.591	2,806.024	2,363.648	1,773.022

undirected edges of the original is 29,166,672. We used the number of settled vertices in Dijkstra's algorithm as the measure of the speed of the shortest path search (response time). This measure has been used in previous studies as a fairly natural measure of query difficulty because it does not depend on computational environments. The average/worst number of settled vertices in LMS was calculated over 1,000 random queries. Query speedup is the average of the number of settled vertices on the original network divided by the number of settled vertices in LMS calculated over 1,000 random queries. From Table 2, we can see that if 60% of the original data and a few hours is used, LMS achieves over 3,000-times speed-up compared to the simple Dijkstra's algorithm. According to the results in the previous papers, the acceleration by bit vector is up to 600 times [10], by A^* landmark is 100 [7], by highway hierarchy is 10,000 [14], and by the transit node routine is more than 100,000 [2,3], for large-scale data such as the road network of the US. LMS is not the fastest, but it is close to the fastest.

In query processing, there is a trade-off between the number of levels we use and the query speed-up. Table 3 shows the trade-offs between the size of additional data and the query speed of LMS in the road network of the US. If we use only the top 11 levels (i.e., we do not use the bottom 8 levels except for the lowest level), LMS uses only 10% of the original data size and achieves about 1,500-times speed-up. Moreover, if we use only the top 9 levels, LMS uses only 1% of the original data size and still achieves 200-times speed-up. This is useful if the computation is performed on portable systems such as a PDA. Basically, the bit vector method uses considerable memory, such as 50 bytes for each vertex, to obtain high performance [10]. The highway hierarchy method also uses additional data of about 50 bytes for each vertex [14]. A^* landmark and transit node routing use much more memory to achieve high performance. The size of the additional data used in LMS is about 2 or 3 bytes for each vertex, so the additional data size is about 1/20 of that used by the other methods. Though the previous methods allow parameter tuning, none can reduce the size of the additional data while still maintaining speeding-up. We implemented only sparsification; use of a sophisticated search method in conjunction with it would reduce the number of settled vertices even further.

Next we show approximation results. Though LMS takes only a few hours in preprocessing, there may be a practical situation that cannot afford such a long preprocessing time. For this purpose, we also developed an approximation technique that reduces preprocessing time. Table 4 shows the trade-offs between the size of additional data and query speed of LMS in the road network of the US.

The approximation ratio is [the length of shortest paths found by LMS with the approximation technique] divided by [the length of the optimal shortest path]. The speed-up and the approximation ratio are averages over 1,000 random queries. UB is the upper bound of the number of iterations of finding shortest path trees for obtaining each sparsified mesh. If the upper bound is infinity (Inf.), the preprocessing retains optimality and the approximation ratio is exactly 1. In the exact preprocessing, the number of iterations increases when the mesh level is higher in general. As a result, in the approximated preprocessing, higher-level meshes are sometimes rough, meaning they are not optimal. From Table 4, the approximation technique achieves 5–10 times speed-up in preprocessing while still having high accuracy.

Table 3. Additional data and query speed of LMS in US

Levels	Add. ratio	Speed-up
9	0.86%	230
10	3.03%	626
11	9.34%	1,437
12	21.91%	2,390
13	35.05%	2,958
14	48.78%	3,150
15	61.51%	3,180
16	62.24%	3,180
17	62.26%	3,180
18	62.26%	3,180
19	62.26%	3,180

Table 4. Approximation results of LMS

Gra.	UB	PreProc.	Add. ratio	Speed-up	Approx. ratio
10	4	12m37s	27.0%	4,749	1.7182
	8	16m03s	29.1%	3,655	1.0475
	16	21m18s	29.9%	3,368	1.0137
	32	30m20s	30.2%	3,305	1.0014
	Inf.	119m52s	62.2%	3,180	1.0000
20	4	12m07s	16.2%	4,456	1.3937
	8	15m22s	17.7%	3,546	1.0230
	16	20m44s	18.4%	3,270	1.0047
	32	29m59s	18.6%	3,175	1.0016
	Inf.	128m52s	38.8%	3,060	1.0000
40	4	10m59s	10.0%	3,938	1.6281
	8	14m12s	11.0%	3,232	1.0103
	16	19m26s	11.5%	3,000	1.0019
	32	28m46s	11.7%	2,919	1.0002
	Inf.	140m07s	24.8%	2,806	1.0000

Our LMS finds "important" edges that are frequently used in shortest paths connecting distant vertices. When the length of each edge is the geometric length instead of the travel time length, the importance of edges should be less distinct since the difference between fast and slow roads decreases. The geometric length of each edge is also presented in the US road network data. We also tested LMS on this length. When the granularity was 20, the highest level was 17, the preprocessing time was about 173 minutes, the ratio of additional to original data size was 58%, the average number of settled vertices was 20,304, the worst number of settled vertices was 40,045, and the average speed-up ratio was about 584 over 1,000 random queries. Although the preprocessing costs increased and query speed decreased in this case, the results show that our sparsification method is still useful.

7   Concluding Remarks

In this paper, we presented a LMS, a method for the shortest path query problem. LMS uses the sparsified meshes obtained by removing unnecessary edges to compute optimal shortest paths for speed-up. Using meshes of several sizes, we can efficiently reduce the number of edges to be searched, with a much smaller additional size of memory used. Unlike other methods, our sparsification method can use any shortest path algorithm, so it can handle negative costs and time-dependent costs. The sparsification is based only on the local information, so minor change of the network will occur in only a few meshes. From these advantages, LMS would be a suitable method for real-world small computer systems such as car navigation systems and mobile devices with limited memory resources.

Our computational experiments used a simple implementation of Dijkstra's algorithm. Using more sophisticated algorithms would further improve performance, and could be implemented for dealing negative costs and time-dependent costs. Improving preprocessing is also future work. The computational experiments did not examine any of the generalizations and extensions mentioned in this paper. These will be investigated in the future. Theoretical evaluation [1] of our method is also future work.

References

1. Abraham, I., Fiat, A., Goldberg, A.V., Werneck, R.F.: Highway dimension, shortest paths, and provably efficient algorithms. In: Proceedings of the 21st Annual ACM-SIAM Symposium on Discrete Algorithms, pp. 782–793 (2010)
2. Bast, H., Funke, S., Sanders, P., Schultes, D.: Fast routing in road networks with transit nodes. Science 316(5824), 566 (2007)
3. Bast, H., Funke, S., Matijevic, D., Sanders, P., Schultes, D.: In transit to constant shortest-path queries in road networks. In: Proceedings of the Workshop on Algorithm Engineering and Experiments (ALENEX), pp. 46–59 (2007)
4. Delling, D., Nannicini, G.: Bidirectional core-based routing in dynamic time-dependent road networks. In: Hong, S.-H., Nagamochi, H., Fukunaga, T. (eds.) ISAAC 2008. LNCS, vol. 5369, pp. 812–823. Springer, Heidelberg (2008)
5. Dijkstra, E.W.: A note on two problems in connexion with graphs. Numerische Mathematik 1, 269–271 (1959)
6. Goldberg, A.V., Kaplan, H., Werneck, R.: Reach for A^*: efficient point-to-point shortest path algorithms. In: Proceedings of the Workshop on Algorithm Engineering and Experiments (ALNEX), pp. 129–143 (2006)
7. Goldberg, A.V., Harrelson, C.: Computing the shortest path: A search meets graph theory. In: Proceedings of the 16th Annual ACM-SIAM Symposium on Discrete Algorithms, pp. 156–165. SIAM, Philadelphia (2005)
8. Gutman, R.J.: Reach-based routing: A new approach to shortest path algorithms optimized for road networks. In: Proceedings of the 6th Workshop on Algorithm Engineering and Experiments and the 1st Workshop on Analytic Algorithmics and Combinatorics (ALENEX/ANALC), pp. 100–111. SIAM, Philadelphia (2004)
9. Jung, S., Pramanik, S.: An efficient path computation model for hierarchically structured topographical road maps. IEEE Transactions on Knowledge and Data Engineering 14(5), 1029–1046 (2002)
10. Köhler, E., Möhring, R.H., Schilling, H.: Acceleration of shortest path and constrained shortest path computation. In: Nikoletseas, S.E. (ed.) WEA 2005. LNCS, vol. 3503, pp. 126–138. Springer, Heidelberg (2005)
11. Möhring, R.H., Schilling, H., Schütz, B., Wagner, D., Willhalm, T.: Partitioning graphs to speed up Dijkstra's algorithm. In: Nikoletseas, S.E. (ed.) WEA 2005. LNCS, vol. 3503, pp. 189–202. Springer, Heidelberg (2005)
12. Samet, H., Sankaranarayanan, J., Alborzi, H.: Scalable network distance browsing in spatial databases. In: Proceedings of the ACM SIGMOD International Conference on Management of Data 2008, pp. 43–54 (2008)
13. Sanders, P., Schultes, D.: Highway hierarchies hasten exact shortest path queries. In: Brodal, G.S., Leonardi, S. (eds.) ESA 2005. LNCS, vol. 3669, pp. 568–579. Springer, Heidelberg (2005)
14. Sanders, P., Schultes, D.: Engineering highway hierarchies. In: Azar, Y., Erlebach, T. (eds.) ESA 2006. LNCS, vol. 4168, pp. 804–816. Springer, Heidelberg (2006)
15. Thorup, M.: Undirected single-source shortest paths with positive integer weights in linear time. Journal of the Association for Computing Machinery 46(3), 362–394 (1999)
16. U.S. Census Bureau, Washington, DC. UA Census 2000 TIGER/Line Files (2000), http://www.census.gov/geo/www/tiger/tigerua/ua_tgr2k.html

Unit-Time Predecessor Queries
on Massive Data Sets

Andrej Brodnik[1,2] and John Iacono[3,4,*]

[1] University of Primorska, Slovenia
[2] University of Ljubljana, Slovenia
[3] Polytechnic Institute of New York University, Brooklyn, New York, USA
[4] University of Aarhus, Aarhus, Denmark

Abstract. New data structures are presented for very fast predecessor queries on integer data sets stored on multiple disks. A structure is presented that supports predecessor queries in one disk seek performed in parallel over multiple disks, no matter how large the data set. For truly massive data sets, the space requirement of this structure approaches twice the space needed to simply store the data on disk.

A second structure is presented that supports predecessor queries in the time it takes to perform two disk seeks, but has more moderate space requirements. Its space usage approaches the space needed to store the data on disk, and has manageable space requirements for smaller massive data sets.

1   Introduction

Massive data sets do not fit one one disk. Once you have multiple disks, there is the ability to make requests to them in parallel. In this paper we describe data structures that use this ability to reduce the number of disk accesses performed in parallel to perform predecessor queries to a constant, no matter how much data is stored. This constant is very low; we present one structure with a constant of one and another with a constant of two. The main difference between these structures is the space usage. Both require a certain minimum number of disks, plus space to store the data. The unit-time structure has a substantially larger minimum-disk requirement, and once this is met uses roughly double the amount of disks than would would be needed to simply store the data. The cost-two structure has a substantially lower minimum-disk requirement, and once this is met uses roughly the same number of disks as would be needed to simply store the data. Our structures are dynamic and support expected amortized unit-cost insertions and delations. (This is not optimal as amortized sub-constant-time insertions and deletions are possible on disk. See [5]).

---
[*] Research partially supported by NSF grants CCF-0430849, OISE-0334653, CCF-1018370, and an Alfred P. Sloan fellowship, and by MADALGO—Center for Massive Data Algorithmics, a Center of the Danish National Research Foundation, Aarhus University.

O. Cheong, K.-Y. Chwa, and K. Park (Eds.): ISAAC 2010, Part I, LNCS 6506, pp. 133–144, 2010.

Table 1. Table of results. Predecessor queries are performed on n k-bit integers, using disks of size D bits and block size B bits. The constants (α, β) are determined by a tradeoff depending on the choice of underlying hash table; typical values are $(2,1)$ and $(3, 0.19)$. The results of the third structure hold for any integer $e \geq 1$.

Structure	Space, measured in disks	Disk accesses
First	$\dfrac{(2n-1)(k+1)(1+\beta)}{D} + (k+1)\alpha$	1
Second	$\dfrac{(\frac{nk}{B}-1)(k+1)(1+\beta)}{D} + \dfrac{nk(1+\beta)}{D} + (k+1)\alpha$	2
Third	$\left(1 + \dfrac{1}{2^e}\right)\cdot \dfrac{nk}{D}\cdot(1+\beta) + \dfrac{k\alpha}{\log_2 B - e}$	2

The unit-cost structure, due to its space needs, is best suited to when the difference between answering predecessor queries in one and two disk accesses is crucial. The cost-two structure would be a reasonable choice in any application with massive data where predecessor queries were the dominant operation.

The formulae for space usage space usage are presented in table 1. To show what space our structures would require for some typical combinations of parameters, section 4 presents numerical results.

Our structure is closely related to the structure for constant-time priority queue operations on a specialized memory known as the Yggdrasil [6]. This memory is a hierarchically created memory where different memory addresses may share bits (as opposed to traditional memory where each memory word has disjoint bits). The ideas behind this kind of specialized memory were first proposed as the *RAM with byte overlap* RAMBO in [9], where they introduce it citing [14]. While novel, the data structure for priority queue operations proposed in [6] suffers from large space requirements; storing k-bit data requires $O(2^k)$ bits of memory. Prototypes of this specialized memory were built [10].

Our results here reduce this memory requirement, and in section 5 we discuss how our structures for disks can be adapted to internal memory with a custom memory controller, and use only $O(nk)$ bits of memory.

The model. In the DAM model [1], a computer is modeled as having a memory of size M and a disk. Consecutive blocks of size B bits can be moved between the memory and the disk at unit cost. Because disks are so much slower than CPU and internal memory (roughly a factor of one million), the cost of any operations solely in internal memory are ignored and counted as zero for the purposes of the model.

In [16], the *Parallel Disk Model* PDM was introduced. In the DAM model, conceptually there is only one disk. However, disks do not have unlimited size, and massive data sets do not fit on one disk. When there are multiple disks attached to one computer, block transfers can be done in parallel from all disks simultaneously. In the PDM model, we use d to denote the number of disks, and for simplicity we assume all disks are of the same size D. For an overview of these models see [15].

We are of course mindful that both of these models have limitations. Neither model is good at modeling the fact that scans are much faster on disk than random searches. In our application, if we were to assume the predecessor searches are random, then at least one random seek on a disk is unavoidable. In the DAM, we make the assumption that the algorithms being run in internal memory are relatively simple so that they do not dominate the runtime. In the PDM, the number of disks and the amount of data transferred in parallel should not overload the CPU or the various buses connecting the disks to the CPU. As a very rough calculation of what is possible, with a 5ms seek time, and a 1k block size, 1.6 megabits per second can be transferred by as single disk reading random blocks continuously. If all drives shared the same Fibre Channel 4Gb connection, then it would take approximately 2500 dives performing random accesses to saturate the interface. Currently, quad-4Gb fibre channel cards are readily available.

The problem. The problem addressed here is predecessor queries on a set S of n k-bit integers. Given a query value x, the result of the query is the largest value in S no larger than x. The predecessor query problem is one of the most fundamental problems in data structures. The standard data structure for predecessor queries in external memory is the B-Tree [3], which is widely used in practice, executing predecessor queries in $O(\log_B n)$ in the DAM model. Through the use of striping predecessor queries are supported in $O(\log_{dB} n)$ in the PDM model (e.g. [13]). Our main contribution is a method to support predecessor queries in a single disk access.

2   The Static Data Structures

2.1   Preliminaries

In this section we describe a sequence of several increasingly sophisticated data structures that can support predecessor queries in the PDM model. These structures maintain a static set S of n k-bit integers using disks of size D and block size B, and an internal memory of size M. All sizes are measured in bits. For each structure we describe the time needed in the PDM model to perform a predecessor query, and the space usage of the structure, measured in terms of the number of disks needed. Both of these values are given in the worst-case, in terms of n, k, D, and B. All of our structures require that memory is large enough to simultaneously store one block from each disk. (With 1000 disks with 1K blocks, this would only be 1MB of memory).

Disk-Independent Arrays. Given a set of p arrays, where the size of the ith array is γ_i, we call a placement of these arrays on disks disk-independent if no disk stores elements from more than one array. The number of disks needed to store these arrays disk-independently is

$$d_{\min} = \sum_{i=1}^{p} \left\lceil \frac{\gamma_i}{D} \right\rceil < \frac{\sum_{i=1}^{p} \gamma_i}{D} + p \qquad (1)$$

All p arrays can have up to B consecutive bits simultaneously retrieved from each in unit time on a PDM.

Hashing

Lemma 1. *For any fixed $\kappa \leq B$, and for any integer $\alpha > 1$ there is a data structure that supports dictionary queries (exact matches) on a set of n items of κ-bit data in unit time on an PDM with α disks and uses $n\kappa(1 + \beta)$ space, and supports insertions and deletions in $O(1)$ expected amortized time, for some β. We call such a structure an (α, β) dictionary.*

Proof. Such a structure is the cuckoo hash table [12]. In cuckoo hashing, there are $\alpha \geq 2$ simple hash tables, each of which has its own random hash function. A dictionary query consists of looking for the queried item in all α hash tables, as if it is currently in the set being stored then it is guaranteed to be in one of these locations. These queries into each hash table can be done in unit time on a PDM, provided the arrays storing the individual hash tables are stored disk-independently.

The larger the value of α, the smaller the extra needed space β, and an tradeoff was proven in [8]. For example, cuckoo hashing has proven tradeoffs of $\alpha = 2$ and $\beta \leq 1$, or $\alpha = 3$ and $\beta \leq 0.19$, though experimentally when $\alpha = 3$ β appears to be 0.10 [8].

Insertion and deletion are supported in expected amortized constant time.

Cuckoo hashing is a form of dynamic perfect hashing. The traditional structure for dynamic perfect hashing [7] also executes queries with two table lookups, but the second query depends on the first query, and thus they can not be parallelized.

2.2   The Reference Tree

All notation in the description of our structure is paramatrized implicitly by k, S, and B.

Let T be the trie of all k-bit long binary strings. Level i of T contains all binary strings of length i. Each node corresponds to a binary string of length at most k. Every element of S corresponds to a leaf of T.

For all $t \in T$, we color it black or white, as a function of S. The color is determined recursively as follows: for a leaf, t, which is k bits long, we color it black if $k \in S$ and white otherwise. For all internal nodes t, we color t black if there is at least one black node in both its left and its right subtree. For every black node t we define its *middle-predecessor*, $mp(t)$, to be the name of the largest (rightmost) black leaf in its left subtree.

Let $m(x, i)$ be the binary string formed by the i most significant bits of x written as a k-digit binary number. The node in T with label x has the node with label $m(x, i)$ as its ancestor at level i of the tree.

Lemma 2. *For any k-bit long integer x, let $P(x)$ be the set of the middle predecessors of all the black ancestors of the leaf labeled x in T, along with the maximum element of S and $-\infty$. Formally,*

$$P(x) = \{t | \exists_{i \in [0..k]} t = mp(m(x, i)) \wedge m(x, i) \text{ is black}\} \cup \{\max S\} \cup \{-\infty\}$$
The predecessor of x in S is the predecessor of x in $P(x)$.

Proof. If x is larger than $\max S$ then its predecessor is $\max S$, and if it is smaller than $\min S$ then its predecessor is $-\infty$. These values are in $P(x)$, by construction. If neither of these cases hold, then x has both a predecessor and successor in S.

Let w be the predecessor of x in S, and let y be the successor of x in S. Let l be internal node of T that is the least common ancestor (LCA) of the leaves labeled w and y in T. Note that since w and y are in S, and are thus black, l must be black as well. Since l is the LCA of w and y, and $w < y$, w is in the left subtree of l, and furthermore, must be the largest such item. Thus the predecessor of x is $w = mp(l)$, which is in $P(x)$, since l is a black ancestor of x.

2.3   First Static Structure

Description of structure. See figure 1. The first static structure consists of $k + 1$ (α, β)-dictionaries $h_0, h_2, \ldots h_k$. Each of these dictionaries k_i stores the labels of the black nodes of level i in the tree T described in the previous section. The labels of these nodes t are augmented with $mp(t)$. Each (α, β)-dictionary table, by definition, is implemented with α arrays, for a total of $(k + 1)\alpha$ arrays. For speed, each of these arrays is stored disk-independently. Thus, trivially, $(k+1)\alpha$ disks are required.

Additionally, a single variable containing $\max S$ is stored.

Space usage. The tree T has $2n - 1$ black nodes. For each black node, we wish to store the name of the node and the mp value. However, noting that the name of a node is a prefix of its mp value, only the k-bit-long mp value needs to be stored. Thus the total amount of actual data stored in the (α, β) dictionary is $(2n - 1)(k + 1)$, which takes total space $(2n - 1)(k + 1)(1 + \beta)$ bits by definition. If these tables were to be stored without regard to the requirement that they be stored on separate disks, $\left\lceil \frac{(2n-1)(k+1)(1+\beta)}{D} \right\rceil$ disks would suffice. However, the disk-independence requirement raises this number, using equation (1) to at most

$$\frac{(2n - 1)(k + 1)(1 + \beta)}{D} + (k + 1)\alpha \tag{2}$$

Queries. A predecessor query can be answered by querying each (α, β) dictionary table h_i for $m(x, i)$, for all $i \in [0..k]$. This can be done in unit time on a PDM, since this is just a single block access to $(k + 1)\alpha$ disk-independently stored arrays. and the results fit into memory. The return values of these queries are exactly the needed values for the definition of the set $P_k(x)$ (along with $\max S$ and $-\infty$). Thus, by Lemma 2, the answer to the predecessor query can be answered by looking for the the predecessor of x in $P_k(x)$, which now resides in internal memory. According to the PDM model, this takes zero time. Thus, the predecessor query can be answered by our data structure in unit time in the PDM model.

First Structure

$$S = \{0010, 0011, 0110, 1100, 1110, 1111\}$$

$$k = 4$$

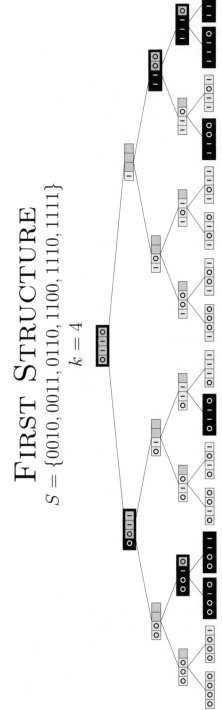

Contents of the tables h_i

Contains all black nodes at level i

Example: $\left(\overbrace{11}^{mp(x)}, \underbrace{00}_{x} \right)$

h_0 $\{(\lambda, 0101)\}$
h_1 $\{(0, 011)\}$
h_2 $\{(11, 00)\}$
h_3 $\{(001, 1), (111, 0)\}$
h_4 $\{(0010, \lambda), (0011, \lambda), (0110, \lambda), (1100, \lambda), (1110, \lambda), (1111, \lambda)\}$

Computing the predecessor of 0101

i	$m(0101, i)$	$h_i(m(0101, i))$	$mp(m(0101, i))$
0	λ	$(\lambda, 0101)$	0101
1	0	$(0, 011)$	0011
2	01	not found	-
3	010	not found	-
4	0101	not found	-

The mp values in the right column are loaded into memory and the predecessor of 0101 is the predecessor of 0101 among these values (0101 and 0011). This is computed in internal memory, and the result, 0011 is returned.

Fig. 1. Illustration of the first structure. Black nodes are colored black, and white nodes are shaded yellow. Blue bits, when combined with the back bits, store the mp values of the black nodes.

2.4   Second Static Structure

The second static structure is a simple refinement to use far less space, by re-ducing the constant in the first term of equation (2). The idea is simple. Let the set S' consist of every B/kth element of S', starting with the minimum element. Store S' using the first structure, except for the (α, β) hash table h_k. This last table, h_k, which stores the black leaves leaves of T, will have every node aug-mented with the elements of S located between the value of its label and its successor in S'.

Space Usage. Each entry of h_k will be of size B bits rather than the k bits for the elements of the other hash tables. The space is usage for the h_i, $i < k$ is computed using the same analysis as for the first structure, where we substitute $\frac{kn}{B}$ for n, and change the number of black nodes from $2n - 1$ to $n - 1$ since only internal nodes are being stored. We then add the $\frac{nk(1+\beta)}{D}$ space needed to store the leaves in h_k in a (α, β) dictionary to give the total space usage in disks of:

$$\frac{(\frac{nk}{B} - 1)(k + 1)(1 + \beta)}{D} + \frac{nk(1 + \beta)}{D} + (k + 1)\alpha \qquad (3)$$

For large values of n, and when $B = \omega(k^2)$, the second term dominates and this approaches the naive space requirement of $\left\lceil \frac{nk(1+\beta)}{D} \right\rceil$ to simply store the data in a hash table (but not support predecessor queries).

Queries. Using the same analysis as in the first structure, the predecessor of x in S' can be found in unit time; call this value y. Looking up y in h_k gives B bits containing the value y and all elements of S from the predecessor of and up to but not including the successor of x in S. The predecessor of x in S is thus located here. The steps of computing y using the first structure, and loading $h_k(y)$ must be done sequentially. Thus predecessor queries using the second structure take time two in the PDM model.

2.5   Third Static Structure

This structure is, once again, a refinement of the space usage. This structure is only of use when the $+(k + 1)\alpha$ third term of the space usage of the second structure, equation (3), dominates the space usage, as it is likely to for all non–very-massive data sets. In this case, almost all of this disk space is empty and wasted due to the disk-independence requirement being placed on (relatively) small sets h_i, so a tradeoff is done whereby we reduce the number of disks by increasing the total amount of data stored.

As a starting point, we begin with the second data structure. The idea is simple: instead of having a separate disk for each of the k non-leaf levels of the tree T, we combine levels. The last level, which contains the leaves, is unchanged. Let c be the number of levels combined into one, and for simplicity we assume k is a multiple of c. We partition T into subtrees of height c whose roots are at levels divisible by c. Each subtree has $2^c - 1$ nodes. These subtrees form k/c levels. We refer to each subtree as a *supernode* of T. We call a supernode black if any of its nodes is black and white if all of its nodes is white. This technique is has appeared before and is known as a *level-compressed trie*.

The data structure consists of k/c (α, β) hash tables, h'_j. Each of these hash tables stores all of the back supernodes in level j of the condensed tree, indexed by the name of the root of the subtree in each supernode.

Space usage. Since each of these supernodes can contain up to $2^c - 1$ nodes, and we previously observed that storing the name of a node, along with its mp value takes space k, the space requirement per supernode is $k(2^c - 1)$. Since each black supernode contains at least one black node, the number of internal back supernodes could be equal to the number of black internal nodes, $n - 1$. Thus a total of at most $n - 1$ supernodes of size $k(2^c - 1)$ are stored in the k/c (α, β) disk-independently stored hash tables for a disk usage of at most:

$$\frac{(\frac{kn}{B} - 1) \cdot k(2^c - 1) \cdot (1 + \beta)}{D} + \frac{k\alpha}{c}$$

Now, we must add in the space needed by the hash table h_k, which is stored unchanged from the second structure, to give a total space usage of at most:

$$\frac{(\frac{kn}{B} - 1) \cdot k(2^c - 1) \cdot (1 + \beta)}{D} + \frac{nk(1 + \beta)}{D} + \frac{k\alpha}{c}$$

Setting $c = \log_2 B - e$ for some integer e, $1 \le e \le \frac{1}{2} \log_2 B$ gives a space usage of at most:

$$\frac{\left(\frac{n}{B} - 1\right) \cdot k(\frac{B}{\cdot 2^e} - 1) \cdot (1 + \beta)}{D} + \frac{nk(1 + \beta)}{D} + \frac{k\alpha}{\log_2 B - e}$$

$$\le \left(1 + \frac{1}{2^e}\right) \cdot \frac{nk}{D} \cdot (1 + \beta) + \frac{k\alpha}{\log_2 B - e}$$

Setting e to an appropriate constant can make the first term arbitrarily close to the best disk usage to store a (α, β) hash table, $\frac{nk}{D}(1 + \beta)$, while the second term reduces the additive term by a factor of $\Theta(\log B)$ over the previous structures. This has the effect, for small values of n to reduce the number of disks needed by a $\Theta(\log B)$ factor.

Queries. Queries are executed as in the second structure, except that instead of looking for the node labeled $m(x, i)$ in $h(i)$, this information can be found in the contents of the supernode with root $m\left(x, \lceil \frac{i}{c} \rceil\right)$ in $h'_{\lceil \frac{i}{c} \rceil}$. Thus all of the information needed to compute the predecessor of x in S' can be done with one query to each h', which can be done in unit time in a PDM, since these are stored disk-independently. The second query, to h_k remains unchanged from the second structure. Thus the cost for a predecessor query in the third structure is two in the PDM model.

3   Making the Structures Dynamic

First Structure. Inserting a new value x in S requires several changes to the structure to maintain the invariants. A new black leaf and a new internal node must be added. The mp value of the new internal node must be set, and the node with a mp value equal to the predecessor of x is S might need to be updated to x instead. All of these changes can be done in $O(1)$ parallel operations on the hash table, and deletions are done in the same manner. Only three of the hash tables are changed, and thus the total runtime of $O(1)$ expected time on the PDM. (This argument is needed because if all k hash tables were to be updated in parallel, he expected time on the PDM would be $\omega(1)$, since the expected maximum of k expected $O(1)$-time cuckoo hashing insertion or deletion operations grows with k. Fortunately, the total number of changes to all the hash tables is only a constant).

Second Structure. The key difference with the first structure is that each of the elements of h_k are full with $\frac{B}{k}$ consecutive elements of S. This does not leave any room for additions. The easy solution is to only fill each of these elements of h_k initially with $\frac{B}{2k}$ elements of S and to split them in half when they get full. This would require doubling the space required to store h_k. If this is not acceptable, then the elements of h_k can be kept within a constant factor of full, for any small constant, with the additional complication that when they become full they must be combined with a constant number number of predecessors using list management teachniques (that depends on how full they are) before splitting. Thus the cost of insertion and deletion is $O(1)$, but the space usage increases slightly to, for any $\epsilon > 0$: $\frac{(\frac{nk}{B}-1)(k+1)(1+\beta)}{D} + \frac{nk(1+\beta)(1+\epsilon)}{D} + (k+1)\alpha$.

Third Structure. The grouping of levels has no significant implications beyond what was described for the first two structures. Insertions and deletions can be supported in time $O(1)$ in the PDM model. The space usage, as in the second structure, increases slighly due to the need to keep extra space in elements of h_k as in the second structure to at most, for any $\epsilon > 0$ and any integer $1 \le e \le \frac{1}{2}\log_2 B$: $\left(1 + \frac{1}{2^e}\right) \cdot \frac{nk}{D} \cdot (1+\beta)(1+\epsilon) + \frac{k\alpha}{\log_2 B - e}$.

4   Numerical Interpretation of Results

To demonstrate the typical disk usage required of our structures, presented in table 2, we have picked typical values for the block size ($B =$4KB, 2^{15} bits), disk size ($D =$1TB, 2^{43} bits), size of the integers stored ($k = 64$ bits). The values of n considered are 2^{37}, 2^{42} and 2^{47}, which gives 1TB, 32TB, and 1PB of data since 64-bit integers are being stored. Also listed is $n = 1$, representing the minimum number of disks needed by our structures no matter how small the data set. For each possibility we choose either $\alpha = 2, \beta = 1$ or $\alpha = 3, \beta = 0.1$ based on whichever yields the smallest space requirement (these are the experimentally observed values). The third structure was also optimized over e. Optimal values of each parameter are listed in the table.

Table 2. Numerical evaluation of the space requirements of our structures. Disks are of size 1TB with block size 4KB.

	Search Time	Space Usage in 1TB disks			
		$n = 1$	$n = 2^{37}$ 1 TB data	$n = 2^{42}$ 32 TB data	$n = 2^{47}$ 1 PB data
Min Disks Needed		1	1	32	1,024
First Structure	1	130 $\alpha = 2$ $\beta = 1$	134 $\alpha = 2$ $\beta = 1$	260 $\alpha = 2$ $\beta = 1$	2,483 $\alpha = 3$ $\beta = 0.1$
Second Structure	2	130 $\alpha = 2$ $\beta = 1$	132 $\alpha = 2$ $\beta = 1$	195 $\alpha = 2$ $\beta = 1$	1,323 $\alpha = 3$ $\beta = 0.1$
Third Structure	2	10 $\alpha = 2$ $\beta = 1$ $e = 2$	13 $\alpha = 2$ $\beta = 1$ $e = 2$	57 $\alpha = 3$ $\beta = 0.1$ $e = 5$	1,163 $\alpha = 3$ $\beta = 0.1$ $e = 8$

Interpretation. The initial required number of disks is high for the first structure (130 disks), but is more reasonable for the third (10 disks). As the amount of data grows, the space differential between the space the raw data takes up and the third structure shrinks; when 1PB of data is being stored the difference is only 13%. The first structure has very high storage costs for small data sizes, but for larger data sizes it approaches having twice the storage requirement of what is needed for the raw data. So, for large amounts of data, the tradeoff is clear, the first structure compared to the third structure is twice as fast and uses twice the space. As data sets get larger, our structures' search times will not degrade.

5   Notes

- Sucessor. Our data structures were presented for predecessor search only. There are a number of alternate ways to support successor search. One would be to symmetrically store *middle successor* pointers. This would have the impact of increasing the space requirement for the first structure from to $\frac{(2n-1)(k+1)(1+\beta)}{D} + (k+1)\alpha$ to $\frac{2(2n-1)(k+1)(1+\beta)}{D} + (k+1)\alpha$, which would be a constant factor for larger data sets. For the third structure the space would change very slightly from $\left(1 + \frac{1}{2^e}\right) \cdot \frac{nk}{D} \cdot (1+\beta) + \frac{k\alpha}{\log_2 B - e}$ to $\left(1 + \frac{1}{2^{e-1}}\right) \cdot \frac{nk}{D} \cdot (1+\beta) + \frac{k\alpha}{\log_2 B - e}$.
- Alignment. We have not paid any attention to the alignment of our data vis-à-vis the alignment of blocks on disk. This care could be taken, but in reality the costs are virtually identical to retrieve any B consecutive bits from any random location on disk. Also, today the "blocks" are reported by the disk controller do not represent any physical unit inherent to the disk.
- Throughput. We note that our structures are optimized for one thing only, the time to execute a predecessor query. Undoubtedly, in many applications a worse predecessor query time can be tolerated in exchange for a better space usage, throughput or insertion/deletion time than our structure.
- Augmented data. Often one wishes to store more than just the key values, but to also store some affiliated data. While this can be done by using our

structure as an index into some other structure that has this data, this would require one more disk access. In the first structure, the mp values can be augmented to include the auxiliary data of the node it refers to. This, in essence will require that all such data be stored twice, but will maintain unit-cost predecessor queries. In the second and third structure, the augmented data may simply be placed in the bottom hash table, h_k.

- Top levels of the structure. The top levels of the structure are very small, and in fact the first level of the first structure stores but a single item. These should be stored in internal memory, to the extent that they fit. For a 1GB internal memory, this optimization would reduce the space usage of the first structure by 25 disks when $k = 64$. The third structure, as a result of compacting the levels, would typically only save one or two disks.
- Scans and range queries. Our first structure is not optimized at all for range queries, finger queries, or scans of the data in sorted order; they are certainly supported by walking through the predecessors and successors. The second and third structure supports scans of f consecutive items in time $2\left(1 + \frac{f}{B}\right)$.

Significance for internal memory. The methods we have described for our first data structure can be adapted to internal memory, by using banks of memory working in parallel, rather than disks. However, this can not be done with conventional hardware, as specialized memory controllers would need to be developed to implement the parts of our algorithm that are currently being executed in internal memory. Given a word size w, these controllers would need to be able to compute αw hash functions to generate αw memory requests to αw separate banks of memory. The results of these memory requests, the set $P(x)$, could then be combined using standard logic gates to give the w-bit predecessor back to the CPU. The total size of the control circuitry would be a reasonable $O(w^c)$, for some c. This would result in something more efficient and practical than *content-addressable memory (CAM)*, since fast conventional memory can be used with all of the customization being in the controller. Also, our solution would only look at one word in each of the αw banks of memory per operation. This puts it at an advantage with regards to power and heat dissipation compared to a CAM.

The bottleneck in such a solution is the multiplication needed to evaluate the hash functions. Although they can all be done in parallel, it still requires a circuit depth of $\Theta\left(\frac{\log w}{\log \log w}\right)$. However, this circuit depth is needed as it has been shown that the depth must be $\Omega\left(\frac{\log w}{\log \log w}\right)$ when the memory size is $2^{\mathrm{polylog}(n)}$ [2] in the circuit-RAM model (Note that there is a technical reason why this lower bound does not directly apply to our structure: the definition of the circuit RAM does not allow circuitry between the RAM and the CPU, which our structure needs. However, the result of [2] can be easily extended to allow this circuitry).

Theoretically, by supporting constant-time predecessor queries this structure has circumvented the $\Theta\left(\sqrt{\frac{\log n}{\log \log n}}\right)$ lower (and matching upper) bound [4] in the cell-probe model [11]. In our structure, the communication both to and from memory is w bits. This is in contrast to the assumption in the cell-probe lower bound that the communication to memory is size $\log n$ (which is no larger then w), and the communication from memory is size w.

References

1. Aggarwal, A., Vitter, J.S.: The input/output complexity of sorting and related problems. Communications of the ACM 31(9), 1116–1127 (1988)
2. Andersson, A., Miltersen, P.B., Riis, S., Thorup, M.: Static dictionaries on AC^0 rams: Query time $\Theta(\sqrt{\log n/\log \log n})$ is necessary and sufficient. In: FOCS, pp. 441–450 (1996)
3. Bayer, R., McCreight, E.M.: Organization and maintenance of large ordered indexes. Acta Informatica 1(3), 173–189 (1972)
4. Beame, P., Fich, F.E.: Optimal bounds for the predecessor problem and related problems. J. Comput. Syst. Sci. 65(1), 38–72 (2002)
5. Brodal, G.S., Demaine, E.D., Fineman, J.T., Iacono, J., Langerman, S., Munro, J.I.: Cache-oblivious dynamic dictionaries with optimal update/query tradeoff. In: Proceedings of the 21st Annual ACM-SIAM Symposium on Discrete Algorithms (SODA 2010), Austin, Texas, January 17-19, pp. 1448–1456 (2010)
6. Brodnik, A., Carlsson, S., Fredman, M.L., Karlsson, J., Munro, J.I.: Worst case constant time priority queue. Journal of Systems and Software, 249–259 (2005)
7. Dietzfelbinger, M., Karlin, A., Mehlhorn, K., Meyer a uf der Heide, F., Rohnert, H., Tarjan, R.E.: Dynamic perfect hashing: upper and lower bounds. SIAM J. Comput. 23, 738–761 (1994)
8. Fotakis, D., Pagh, R., Sanders, P., Spirakis, P.G.: Space efficient hash tables with worst case constant access time. Theory Comput. Syst. 38(2), 229–248 (2005)
9. Fredman, M.L., Saks, M.E.: The cell probe complexity of dynamic data structures. In: STOC, pp. 345–354 (1989)
10. Leben, R., Miletić, M., Špegel, M., Trost, A., Brodnik, A., Karlsson, J.: Design of a high performance memory module on pc100. In: Proceedings Electrotechnical and Computer Science Conference, Portorož, Slovenia, vol. A, pp. 75–78 (1999)
11. Miltersen, P.B.: Cell probe complexity — a survey. In: Advances in Data Structures (1999)
12. Pagh, R., Rodler, F.F.: Cuckoo hashing. J. Algorithms 51(2), 122–144 (2004)
13. Seeger, B., Larson, P.-Å.: Multi-disk b-trees. In: SIGMOD Conference, pp. 436–445 (1991)
14. Stallone, S.: First blood. United Artists (1982)
15. Vitter, J.S.: Algorithms and data structures for external memory. Foundations and Trends in Theoretical Computer Science 2(4), 305–474 (2006)
16. Vitter, J.S., Shriver, E.A.M.: Algorithms for parallel memory i: Two-level memories. Algorithmica 12(2/3), 110–147 (1994)

Popularity at Minimum Cost[*]

Telikepalli Kavitha[1], Meghana Nasre[2], and Prajakta Nimbhorkar[3]

[1] Tata Institute of Fundamental Research, India
kavitha@tcs.tifr.res.in
[2] Indian Institute of Science, India
meghana@csa.iisc.ernet.in
[3] The Institute of Mathematical Sciences, India
prajakta@imsc.res.in

Abstract. We consider an extension of the *popular matching* problem in this paper. The input to the popular matching problem is a bipartite graph $G = (\mathcal{A} \cup \mathcal{B}, E)$, where $\mathcal{A}$ is a set of people, $\mathcal{B}$ is a set of items, and each person $a \in \mathcal{A}$ ranks a subset of items in an order of preference, with ties allowed. The popular matching problem seeks to compute a matching M^* between people and items such that there is no matching M where more people are happier with M than with M^*. Such a matching M^* is called a popular matching. However, there are simple instances where no popular matching exists.

Here we consider the following natural extension to the above problem: associated with each item $b \in \mathcal{B}$ is a non-negative price $\mathsf{cost}(b)$, that is, for any item b, new copies of b can be added to the input graph by paying an amount of $\mathsf{cost}(b)$ per copy. When G does not admit a popular matching, the problem is to "augment" G at minimum cost such that the new graph admits a popular matching. We show that this problem is NP-hard; in fact, it is NP-hard to approximate it within a factor of $\sqrt{n_1}/2$, where n_1 is the number of people. This problem has a simple polynomial time algorithm when each person has a preference list of length at most 2. However, if we consider the problem of *constructing* a graph at minimum cost that admits a popular matching that matches all people, then even with preference lists of length 2, the problem becomes NP-hard. However, when the number of copies of each item is *fixed*, the problem of computing a minimum cost popular matching or deciding that no popular matching exists can be solved in $O(mn_1)$ time.

1 Introduction

The *popular matching* problem deals with matching people to items, where each person ranks a subset of items in an order of preference, with ties allowed. The input is a bipartite graph $G = (\mathcal{A} \cup \mathcal{B}, E)$ where $\mathcal{A}$ is the set of people, $\mathcal{B}$ is the set of items and the edge set $E = E_1 \cup \cdots \cup E_r$ (E_i is the set of edges of rank i). For any $a \in \mathcal{A}$, we say a prefers item b to item b' if the rank of edge (a, b) is smaller than the rank of edge (a, b'). If the ranks of (a, b) and (a, b') are the same, then a is indifferent between b and b'. The goal is to match people with items

---

[*] Work done as part of the DST-MPG partner group "Efficient Graph Algorithms".

O. Cheong, K.-Y. Chwa, and K. Park (Eds.): ISAAC 2010, Part I, LNCS 6506, pp. 145–156, 2010.

in an *optimal* manner, where the definition of optimality will be a function of the preferences expressed by the elements of $\mathcal{A}$. The problem of computing such an optimal matching is a well studied problem and several notions of optimality have been considered so far; for instance, pareto-optimality [1], rank-maximality [4], and fairness. One criterion that does not use the absolute values of the ranks is the notion of *popularity*. Let $M(a)$ denote the item to which a person a is matched in a matching M. We say that a person a *prefers* matching M to M' if (i) a is matched in M and unmatched in M', or (ii) a is matched in both M and M', and a prefers $M(a)$ to $M'(a)$. M is *more popular than* M', denoted by $M \succ M'$, if the number of people who prefer M to M' is higher than those that prefer M' to M. A matching M^* is *popular* if there is no matching that is more popular than M^*.

Popular matchings were first introduced by Gärdenfors [3] in the context of stable matchings. Popular matchings can be considered to be stable as no majority vote of people can force a migration to another matching. On the flip side, popularity does not provide a complete answer since there exist simple instances that do not admit any popular matching. Abraham et al. [2] designed efficient algorithms for determining if a given instance admits a popular matching and computing one, if it exists.

Our problem. Here we consider a natural generalization to the popular matching problem: *augment* the input graph G such that the new graph admits a popular matching. Our input consists of $G = (\mathcal{A} \cup \mathcal{B}, E)$ and a function $\mathsf{cost} : \mathcal{B} \to \mathbb{R}^+$, where $\mathsf{cost}(b)$ for any $b \in \mathcal{B}$ is the cost of making a new copy of item b. The set $\mathcal{B}$ is a set of items, say books or DVDs, and new copies of any $b \in \mathcal{B}$ can be obtained by paying $\mathsf{cost}(b)$ for each new copy of b. We assume that the initial copy of every item comes for free. There is no restriction on the number of copies of any item that can be made. The only criterion that we seek to optimize is the total cost of augmenting G. We call this the *min-cost augmentation* problem.

A related problem. In a related problem, we do not have a starting graph G. We are given a set $\mathcal{A}$ of people and their preference lists over a universe U of items where each $b \in U$ has a price $\mathsf{cost}(b) \geq 0$ associated with it. The problem is to "construct" an input graph $G = (\mathcal{A} \cup \mathcal{B}, E)$ where $\mathcal{B}$ is a multiset of some elements in U such that G admits a popular matching and the cost of constructing G, that is, $\sum_{b \in \mathcal{B}} \mathsf{cost}(b)$, is minimized. Here we also require that the popular matching should leave no person unmatched, otherwise we have a trivial solution of $\mathcal{B} = \emptyset$. We call this the *min-cost popular instance* problem.

Our Results. We show the following results in this paper:

1. The min-cost popular instance problem is NP-hard, even when each preference list has length at most 2.
2. The min-cost augmentation problem has a polynomial time algorithm when each preference list has length at most 2. However, for general lists the problem is NP-hard. In fact, it is NP-hard to approximate to within a factor of $\sqrt{n_1}/2$, where $n_1 = |\mathcal{A}|$.

Our NP-hardness results hold even when preference lists are derived from a *master list*. A master list is a total ordering of the items according to some global objective criterion. Thus if b_1 precedes b_2 in the master list and if a person a has both b_1 and b_2 in her list, then b_1 has to precede b_2 in a's list as well.

The NP-hardness results for the min-cost augmentation/min-cost popular instance problems are because the number of copies of each of the items are not fixed. We now define the *min-cost popular matching* problem where each item b has a *fixed* number of copies along with a cost associated with it.

∗ The *min-cost popular matching* problem is to determine if $G = (\mathcal{A} \cup \mathcal{B}, E)$ admits a popular matching or not and if so, to compute the one with minimum cost. Every $b \in \mathcal{B}$ has a *fixed* number of copies and non-negative price $\mathsf{cost}(b)$. The cost of a matching M be the sum of costs of items that are matched in M. We show that this problem can be solved in $O(mn_1)$ time where $m = |E|$ and $n_1 = |\mathcal{A}|$. Manlove and Sng [10] considered the above problem without costs in the context of house allocation where items were called houses and houses had capacities. They called it Capacitated House Allocation with Ties (CHAT) and gave an $O((\sqrt{C} + n_1)m)$ algorithm for the CHAT problem, where C is the sum of capacities of all the houses. Thus, our algorithm improves on the algorithm in [10].

1.1 Background

Abraham et al. [2] considered the problem of determining if a given graph $G = (\mathcal{A} \cup \mathcal{B}, E)$ admits a popular matching or not, and if so, computing one. They also gave a structural characterization of graphs that admit popular matchings. Section 1.2 outlines this characterization and the algorithm that follows from it.

Subsequent to this work, there have been several variants of the popular matchings problem considered. One line of research has been on generalizations of the popular matchings problem while the other direction has been to deal with instances that do not admit any popular matchings. The generalizations include the capacitated version studied by Manlove and Sng [10], the weighted version studied by Mestre [13] and random popular matchings considered by Mahdian [9]. Kavitha and Nasre [6] as well as McDermind and Irving [12] independently studied the problem of computing an *optimal* popular matching for strict instances where the notion of optimality is specified as a part of the input. They also consider the min-cost popular matchings but in that case the costs are associated with edges whereas in our case costs are associated with items. The line of research pursued for instances that do not admit popular matchings includes the NP-hardness of least unpopular matchings [11], the existence and algorithms for popular mixed matchings [5] and NP-hardness of the popular matchings problem with variable job capacities [7].

1.2 Preliminaries

We review the characterization of popular matchings given in [2]. Let $G_1 = (\mathcal{A} \cup \mathcal{B}, E_1)$ be the graph containing only rank-1 edges. Then [2, Lemma 3.1] shows that a matching M is popular in G only if $M \cap E_1$ is a maximum matching of G_1. Maximum matchings have the following important properties, which we use throughout the rest of the paper.

$M \cap E_1$ defines a partition of $\mathcal{A} \cup \mathcal{B}$ into three disjoint sets: a vertex $u \in \mathcal{A} \cup \mathcal{B}$ is *even* (resp. *odd*) if there is an even (resp. odd) length alternating path in G_1 (w.r.t. $M \cap E_1$) from an unmatched vertex to u. Similarly, a vertex u is *unreachable* if there is no alternating path from an unmatched vertex to u. Denote by $\mathcal{E}$, $\mathcal{O}$ and $\mathcal{U}$ the sets of even, odd, and unreachable vertices, respectively.

Lemma 1 (Gallai-Edmonds Decomposition). *Let $\mathcal{E}$, $\mathcal{O}$ and $\mathcal{U}$ be the sets of vertices defined by G_1 and $M \cap E_1$ above. Then,*
(a) $\mathcal{E}$, $\mathcal{O}$ and $\mathcal{U}$ are pairwise disjoint, and independent of the maximum matching $M \cap E_1$.
(b) In any maximum matching of G_1, every vertex in $\mathcal{O}$ is matched with a vertex in $\mathcal{E}$, and every vertex in $\mathcal{U}$ is matched with another vertex in $\mathcal{U}$. The size of a maximum matching is $|\mathcal{O}| + |\mathcal{U}|/2$.
(c) No maximum matching of G_1 contains an edge between a vertex in $\mathcal{O}$ and a vertex in $\mathcal{O} \cup \mathcal{U}$. Also, G_1 contains no edge between a vertex in $\mathcal{E}$ and a vertex in $\mathcal{E} \cup \mathcal{U}$.

As every maximum cardinality matching in G_1 matches all vertices $u \in \mathcal{O} \cup \mathcal{U}$, these vertices are called *critical* as opposed to vertices $u \in \mathcal{E}$ which are called *non-critical*. Using this partition of vertices, we define the following.

Definition 1. *For each $a \in \mathcal{A}$, define $f(a)$ to be the set of top choice items for a. Define $s(a)$ to be the set of a's most-preferred non-critical items in G_1.*

Theorem 1 ([2]). *A matching M is popular in G iff (i) $M \cap E_1$ is a maximum matching of $G_1 = (\mathcal{A} \cup \mathcal{B}, E_1)$, and (ii) for each person a, $M(a) \in f(a) \cup s(a)$.*

The algorithm for solving the popular matching problem is now straightforward: each $a \in \mathcal{A}$ determines the sets $f(a)$ and $s(a)$. A matching that is maximum in G_1 and that matches each a to an item in $f(a) \cup s(a)$ needs to be determined. If no such matching exists, then G does not admit a popular matching.

2   Min-cost Popular Instance

In this section we consider the min-cost popular instance problem. Our input is a set $\mathcal{A}$ of people where each $a \in \mathcal{A}$ has a preference list over items in a universe U, where each item $b \in U$ has a price $\mathsf{cost}(b) \geq 0$. The problem is to "construct" a graph G or equivalently, set suitable values of $\mathsf{copies}(b)$ for each $b \in U$, in order to ensure that the resulting graph G admits a popular matching that matches all $a \in \mathcal{A}$, at the least possible cost.

We show that the above problem is NP-hard by showing a reduction from the monotone 1-in-3 SAT problem to this problem. The monotone 1-in-3 SAT problem is a variant of the 3SAT problem where each clause contains exactly 3 literals and no literal appears in negated form. The monotone 1-in-3 SAT

a_1^i	u_{j_1}	u_{j_2}
a_2^i	u_{j_2}	u_{j_3}
a_3^i	u_{j_1}	u_{j_3}

a_4^i	u_{j_1}	p_1^i
a_5^i	u_{j_2}	p_2^i
a_6^i	u_{j_3}	p_3^i

a_7^i	p_1^i	q^i
a_8^i	p_2^i	q^i
a_9^i	p_3^i	q^i

Fig. 1. The preference lists of people corresponding to the i-th clause in $\mathcal{I}$

problem asks if there exists a satisfying assignment to the variables such that each clause has exactly 1 literal set to true. This problem is NP-hard [14].

Let $\mathcal{I}$ be an instance of the monotone 1-in-3 SAT problem. Let $C_1, \ldots, C_m$ be the clauses in $\mathcal{I}$ and let $X_1, \ldots, X_n$ be the variables in $\mathcal{I}$. We construct from $\mathcal{I}$ an instance of the min-cost popular instance problem as follows:

Corresponding to each clause $C_i = (X_{j_1} \vee X_{j_2} \vee X_{j_3})$, we have 9 people $A_i = \{a_1^i \ldots, a_9^i\}$. Their preference lists are shown in Fig. 1. In this case, every person has a preference list of length 2, for instance, a_1^i treats item u_{j_1} as its rank-1 item and item u_{j_2} as its rank-2 item.

The items $u_{j_1}, u_{j_2}, u_{j_3}$ are called *public* items and the items p_1^i, p_2^i, p_3^i, and q^i are called *internal* items. The internal items appear only on the preference lists of the people of A_i, while the public items appear on preference lists of people in A_i as well as some other people. The public item u_j corresponds to the variable X_j. In every clause C_i that X_j belongs to, the item u_j appears in the preference lists of some of the people in the set A_i as shown in Fig. 1.

The set $\mathcal{A}$ of people in our instance is $\cup_i A_i$. The universe U of all items is union of $\{u_1, \ldots, u_n\}$ (the n public items) and the set $\cup_i \{p_1^i, p_2^i, p_3^i, q^i\}$ of all internal items. It remains to describe the costs of the items. For each i, the cost of each p_t^i for $t = 1, 2, 3$, is 1 unit, while the cost of q^i is zero units. The cost of each u_j, for $j = 1, \ldots, n$, is 3 units. Recall that our problem is to determine a set $\mathcal{B}$ of items with suitable copies so that the graph $(\mathcal{A} \cup \mathcal{B}, E)$ admits a popular matching that matches all $a \in \mathcal{A}$ and we want to do this at the least possible cost. We first show the following lemma.

Lemma 2. *Any instance $(\mathcal{A} \cup \mathcal{B}, E)$ that admits a popular matching that matches all $a \in \mathcal{A}$ has cost at least $14m$.*

Proof. Let us focus on the set A_i of people corresponding to clause C_i. Since we seek an instance where all the people get matched, for every person a we are forced to spend an amount that is at least as much as the cost of the cheapest item on a's preference list. This implies that we spend at least $9(= 3 \times 3)$ units for a_1^i, a_2^i, a_3^i, at least $3(= 3 \times 1)$ units for a_4^i, a_5^i, a_6^i and possibly 0 units for a_7^i, a_8^i, a_9^i, totalling to an amount of 12 units for people in A_i.

However, it is not possible to spend just 12 units for the people in A_i. This is because, in the first place, we are forced to have non-zero copies for at least 2 of the 3 items in $\{u_{j_1}, u_{j_2}, u_{j_3}\}$. Suppose u_{j_1} has non-zero copies, then if we seek to match a_4^i to p_1^i while u_{j_1} is around, then p_1^i is a_4^i's second choice item. Since p_1^i is a_7^i's top choice item, we also have to match a_7^i to p_1^i since a popular matching has to be a maximum cardinality matching on rank-1 edges (see Theorem 1). Thus, it is not possible to match a_7^i to q^i in a popular matching while p_1^i gets matched to a_4^i who regards this item as a second choice item because u_{j_1} is around.

It is now easy to see that although we have several options, the cheapest one is to match a_4^i, a_5^i, a_6^i to p_1^i, p_2^i, p_3^i respectively, and at least 2 of these 3 people are getting matched to their second choice items. Hence, at least 2 out of the 3 people among a_7^i, a_8^i, a_9^i will also have to be matched to their top choice items in order to ensure that the resulting matching is popular. This implies a cost of at least 14 for A_i. This holds for people corresponding to each clause and since there are m clauses, it amounts to at least $14m$ in total for all the clauses.   □

The following lemma 3 (proof omitted, refer to [8]) establishes the correctness of our reduction.

Lemma 3. *There exists an instance $(\mathcal{A} \cup \mathcal{B}, E)$ with cost $14m$ that admits a popular matching that matches all $a \in \mathcal{A}$ iff there exists a 1-in-3 satisfying assignment for $\mathcal{I}$.*

Note that the preference lists of all the people in our instance G are strict and of length at most 2. Also, the preference lists are drawn from a *master list*. We have thus shown the following theorem.

Theorem 2. *The min-cost popular instance problem is NP-hard, even when each preference list has length at most 2 which are derived from a master list.*

3   Min-cost Augmentation

In this section we show various results for the min-cost augmentation problem. Recall that the input here is a graph $G = (\mathcal{A} \cup \mathcal{B}, E)$ where each item $b \in \mathcal{B}$ has a non-negative $\mathsf{cost}(b)$ associated with it. The problem is to determine how to make extra copies of items in $\mathcal{B}$ so that the resulting graph admits a popular matching and the cost of the extra copies is minimized. We first show a polynomial time algorithm for the problem when the preference lists are strict and restricted to length at most 2. We assume that we add at the end of each a's preference list a dummy item called the *last item* ℓ_a, where a being matched to ℓ_a amounts to a being left unmatched.

3.1   Preference Lists of Length 2

For any $a \in \mathcal{A}$, a's preference list consists of a top choice item (call it f_a), and possibly a second choice item (call it z_a) and then of course, the last item ℓ_a that we added for convenience. Let G_1 be the graph G restricted to rank-1 edges. Let the graph $G' = (\mathcal{A} \cup \mathcal{B}, E')$, where E' consists of : (i) all the top ranked edges $(a, f(a))$: one such edge for each $a \in \mathcal{A}$, and (ii) the edges $(a, s(a))$, where a is *even* in G_1 and $s(a)$ is a's most preferred item that is *even* in G_1. Thus, $s(a) = z_a$ when z_a is nobody's rank-1 item, else $s(a) = \ell_a$.

It follows from Theorem 1 that G admits a popular matching if and only if G' admits an $\mathcal{A}$-complete matching. Since we assume that G does not admit a popular matching, there exists a set S of people such that the neighborhood $N(S)$ of S in G' satisfies $|N(S)| < |S|$. Let S denote a minimal such set of people. Every $a \in S$ is *even* in G_1 and must satisfy $s(a) = z_a$. Otherwise $s(a) = \ell_a$ and since no vertex in $\mathcal{A}$ other than a has an edge to ℓ_a, such an a will be always matched in any maximum cardinality matching in G'. Hence, such an a cannot belong to S due to its minimality. Further note that for any such minimal set S, the set $N(S)$ is a set of items that are all *odd* in the graph G' with respect to a maximum cardinality matching in G'.

Since $s(a) = z_a$ for every $a \in S$, and the preference lists are of length at most 2, there are no items sandwiched between $f(a)$ and $s(a)$ in a's preference list for every $a \in S$. Thus, in order to ensure that these people get matched in any popular matching, we need to make extra copies of items in $N(S)$ or equivalently items that are *odd* in the graph G'. Our algorithm precisely does this and in order to get a min-cost augmentation, it chooses the item which has least cost. Our algorithm to add extra copies is the following:

- Let M be a maximum cardinality matching in G'.
- While M is not $\mathcal{A}$-complete do
  - Identify all the items that are *odd* with respect to M; call this set $\mathcal{O}'$.
  - Let b be a cheapest item in $\mathcal{O}'$. Set copies(b) = copies(b) + 1.
  - Augment M along the new augmenting path now available.

Our algorithm maintains the invariant that no person a changes her s-item due to the increase in copies. Note that the set $\mathcal{O}'$ of items is identified by constructing alternating paths from an unmatched vertex which in this case is an *even* person. No item b that was unreachable in G_1 ever occurs on any such alternating path because b cannot be $f(a)$ or $s(a)$ for any *even* person a. An item b that is *odd* in G_1 does occur on these alternating paths however, copies(b) is always bounded by degree of b in G' which is the same as degree of b in G_1. Thus, even with the extra copies, an item that was critical in G_1 remains critical in the augmented graph restricted to rank-1 edges. This implies that for every person a its most preferred *even* item or its $s(a)$ remains unchanged.

Let $\tilde{G}$ be the graph with extra copies returned by our algorithm. It is clear that the graph $\tilde{G}'$ admits an $\mathcal{A}$-complete matching or $\tilde{G}$ admits a popular matching. To see that the instance returned is a min-cost instance, observe that there is no alternating path between an item b which got duplicated in our algorithm and an item b' whose cost is strictly smaller than the cost(b). Otherwise $b' \in \mathcal{O}'$ and it would have been picked up by our algorithm. It is easy to see that the running time of the algorithm is $O(n_1^2)$, where n_1 is the size of $\mathcal{A}$. Hence we have shown the following theorem.

Theorem 3. *The min-cost augmentation problem with strict preference lists of length at most 2 can be solved in $O(n_1^2)$ time.*

3.2   Hardness for the General Case

We now show that the min-cost augmentation problem in the general case is NP-hard. The reduction is again from the monotone 1-in-3 SAT problem (refer

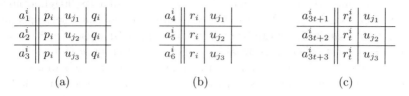

Fig. 2. (a) & (b) Preference lists of the 6 people in A_i, (c) Preference list of people in the t-th triplet

to Section 2). Let $\mathcal{I}$ be an instance of the monotone 1-in-3 SAT problem with $X_1, \ldots, X_n$ as the variables and $C_1, \ldots, C_m$ being the clauses. We construct from $\mathcal{I}$ an instance of the min-cost augmentation problem as follows.

Let C_i be $(X_{j_1} \vee X_{j_2} \vee X_{j_3})$. Corresponding to this clause we have 6 people $A_i = \{a_1^i, a_2^i, a_3^i, a_4^i, a_5^i, a_6^i\}$ and 3 internal items $D_i = \{p_i, q_i, r_i\}$. We also have public items $u_{j_1}, u_{j_2}, u_{j_3}$ which belong to the preference lists of people in A_i and whenever X_j occurs in a clause C_i, the item u_j will belong to the preference lists of some people in A_i. The public items have unit cost whereas each internal item $b \in D_i$ has cost 2. Fig. 2(a) and (b) show the preference lists of people in A_i.

The set $\mathcal{B}$ of items is the union of $\cup_{i=1}^m D_i$ (the set of all the internal items) and $\{u_1, \ldots, u_n\}$ (consisting of all the public items, where vertex u_j corresponds to the j-th variable X_j). The set $\mathcal{A}$ of people is the union of $\cup_{i=1}^m A_i$ and $\{x_1, \ldots, x_n\}$, where the vertex x_j corresponds to the variable X_j. The preference list of each x_j is of length 1, it consists of the item u_j. To see that the graph G does not admit a popular matching note that, with a single copy of every item, all the public items are critical in G_1. Now consider the people in A_i: for each $a_t^i \in \{a_1^i, a_2^i, a_3^i\}$, we have $f(a_t^i) = \{p_i\}$ and $s(a_t^i) = \{q_i\}$. Since there are only 2 items p_i, q_i for the 3 people a_1^i, a_2^i, a_3^i to be matched to in any popular matching, G does not admit a popular matching.

Let $\tilde{G}$ be an augmentation of G such that $\tilde{G}$ admits a popular matching. We note that for the people in each A_i we need to spend at least 1 unit in order to obtain a popular matching. Since there are m such sets, we need to spend at least m units. The following lemma (proof omitted, refer to [8]) establishes the correspondence between the instance $\mathcal{I}$ and the instance G that we constructed.

Lemma 4. $\tilde{G}$ has cost at most m iff there exists a 1-in-3 satisfying assignment for the instance $\mathcal{I}$.

We can now conclude the following theorem.

Theorem 4. The min-cost augmentation problem is NP-hard, even for strict lists of length at most 3. Further, the lists can be derived from a master list.

3.3   Inapproximability of Min-cost Augmentation

We extend the above reduction from $\mathcal{I}$ to show that this problem is NP-hard to approximate to within a factor of $\sqrt{n_1}/2$, where n_1 is the size of $\mathcal{A}$. We construct a graph H on at most $4m^4$ people that satisfies the following property:

(∗) If $\mathcal{I}$ is a *yes* instance for 1-in-3 SAT, then H can be augmented at a cost of
m to admit a popular matching. If $\mathcal{I}$ is a *no* instance for 1-in-3 SAT, then
H needs a cost strictly greater than m^3 to admit a popular matching.

The construction of H is as follows. Let us call the group of 3 people (a_4^i, a_5^i, a_6^i)
in Fig. 2(b) as a *triplet*. Instead of having just one triplet in A_i, as was the case
in the previous section, here we have $m^3 + 1$ such triplets. The preference list
for one particular triplet $(a_{3t+1}^i, a_{3t+2}^i, a_{3t+3}^i)$ is shown in Fig. 2(c).

The set $\mathcal{A} = \cup_i A_i \cup \{x_1, \dots, x_n\}$. We can bound the size of $\mathcal{A}$ by $3m^4 + 9m \le$
$4m^4$, because each A_i contains $3 + 3(m^3 + 1)$ people and $n \le 3m$.

Recall that for each j, the preference list of x_j is of length 1, which consists
of only u_j. The costs of the items are as follows: the cost of each of the *internal*
items, i.e., p^i, q^i, and r_k^i, for $k = 1, \dots, m^3 + 1$ is m^3, and the cost of each u_j
for $j = 1, \dots, n$ is 1. We refer the reader to [8] for the proof that the instance
constructed as above satisfies the property (∗).

Now suppose that the min-cost augmentation problem admits a $\sqrt{n_1}/2$ approximation algorithm (call it Algo1). If $\mathcal{I}$ is a yes instance, then Algo1 has to
return an augmentation of cost $\le 1/2.\sqrt{4m^4}.m = m^3$, otherwise there is no
augmentation of cost at most m^3, so Algo1 returns an answer of cost $> m^3$.
Thus using Algo1 it is possible to determine whether $\mathcal{I}$ has a 1-in-3 satisfying
assignment or not, a contradiction. Hence we conclude the following theorem.

Theorem 5. *It is NP-hard to approximate the min-cost augmentation problem*
on $G = (\mathcal{A} \cup \mathcal{B}, E)$ *within* $\sqrt{|\mathcal{A}|}/2$.

4   Min-cost Popular Matchings

In this section we present an $O(mn_1)$ time algorithm for the min-cost popular
matchings problem. Our input is an instance $G = (\mathcal{A} \cup \mathcal{B}, E)$ where each item $b \in$
$\mathcal{B}$ has associated with it the number copies(b) (denoting the maximum number
of people that can be matched to b) and a price cost$(b) \ge 0$. Whenever a person
gets matched to b, an amount of cost(b) has to be paid. Thus, if $k \le$ copies(b)
copies of b get used in a matching M, then a cost of $k \cdot$ cost(b) has to be paid by
M with respect to the item b. As done in the earlier sections, we will add a last
item ℓ_a at the end of a's preference list for each person $a \in \mathcal{A}$. The cost of ℓ_a is
0, since using the edge (a, ℓ_a) amounts to leaving a unmatched.

Our problem here is to decide whether G admits a popular matching or not
and if so, to compute the one with minimum cost. In order to solve the min-cost
popular matchings problem, for each $b \in \mathcal{B}$, we could make copies(b) duplicates of
each vertex b, call them $b_1, \dots, b_{\text{copies}(b)}$, where each b_i has the same neighborhood
as the original vertex b. However, such a graph has too many vertices and edges,
hence we will stick to the original graph $G = (\mathcal{A} \cup \mathcal{B}, E)$ and simulate the
larger graph in G itself. Note that a *matching* in G can contain up to copies(b)
pairs (a_i, b). It is easy to see that the structural characterization for popular
matchings from [2] holds for our problem as well. That is, any popular matching
in our graph G has to be a maximum cardinality matching on rank-1 edges and

every person a has to be matched to an item in $f(a) \cup s(a)$. This is because by making copies(b) many duplicates of every item b in G our problem becomes equivalent to the original popular matchings problem.

4.1   Our Algorithm

Our algorithm to compute a min-cost popular matching can be broadly partitioned into two stages. In the first stage we build the graph G', i.e. the graph where every person adds edges to their f and s-items. The second stage then computes a min-cost popular matching in the graph G' if one exists.

The first stage. We first construct the graph G_1 on rank-1 edges. In order to find a maximum cardinality matching in the graph G_1, we use Ford-Fulkerson max-flow algorithm. The following transformation from G_1 into a flow network is based on the standard transformation from the bipartite matching problem to the maximum flow problem:

(a) add a vertex s and an edge directed from s to each person $a \in \mathcal{A}$ with an edge capacity of 1 on this edge.

(b) add a vertex t and an edge directed from each item $b \in \mathcal{B}$ to t with an edge capacity of copies(b) on this edge.

(c) direct every edge (a, b) of G from a to b and set an edge capacity of 1 for each such edge.

Let $F(G_1)$ denote the above graph. It is easy to see that a valid flow from s to t in the graph $F(G_1)$ can be translated to a *matching* in G_1 in which every person is matched to at most 1 item and every item b is matched up to copies(b) people. We compute a maximum cardinality matching M_0 of G_1 by computing a max-flow from s to t in $F(G_1)$. Using the matching M_0, our goal is to obtain a partition of $\mathcal{A} \cup \mathcal{B}$ into $\mathcal{O}$ (*odd*), $\mathcal{E}$ (*even*) and $\mathcal{U}$ (*unreachable*). This can be done in O(number of edges) provided we create copies(b) many duplicates of each item b and duplicate the neighborhood of b for each copy of b. However, this is too expensive. The main point to note is that all the copies(b) many duplicates of b, for each item b, have the same *odd/even/unreachable* status. We note that it is possible to remain in the graph G_1 and determine the *odd/even/unreachable* status of all the vertices in linear time. The idea is to build Hungarian trees rooted at unmatched people in M_0 and at items that are matched to fewer than copies(b) many people in M_0. We mark every vertex the first time we see it, so that every vertex appears in only one of the Hungarian trees. We omit the details of the procedure for lack of space, refer to [8].

Assuming that we have obtained the partition, it is now possible to define $s(a)$ for every person a as the set of most preferred *even* items of a. Let the graph G' be the graph G_1 along with the edges (a, b) where $a \in \mathcal{E}$ and $b \in s(a)$.

Since a popular matching is a maximum cardinality matching on rank-1 edges, all items that are *critical* in G_1, that is, all items in $\mathcal{O} \cup \mathcal{U}$ have to be fully matched in every popular matching M^* of G. The only choice we have is in choosing which items of $\mathcal{E}$ should participate with how many copies in the min-cost popular matching. We make this choice in the second stage.

The second stage. Our goal in the second part of the algorithm is to augment the matching M_0 to find a min-cost popular matching. However, we start with the matching M_1, where $M_1 = M_0 \setminus \{$all edges (a, b) where $a \in \mathcal{O}\}$. Thus, M_1 consists only of edges (a, b) where $b \in \mathcal{O} \cup \mathcal{U}$. We take M_1 to be our starting matching rather than M_0 because it may be possible to match people $\mathcal{O} \cap \mathcal{A}$ to cheaper rank-1 neighbors. Recall that while computing the max-flow M_0, the costs of items played no role.

Let M be the current matching (M is initialized to M_1). Let ρ be an augmenting path with respect to M, i.e., one end of ρ is an unmatched person and the other end of ρ is item b that is not fully matched. The cost of augmenting M along ρ is the cost of b. By augmenting M along ρ, every item other than b that is currently matched stays matched to the same number of people and the item b gets matched to one more person. Thus, the cost of the new matching is the cost of the old matching $+$ $\mathsf{cost}(b)$. In order to match an unmatched person a, our algorithm always chooses the cheapest augmenting path starting from the person a.

To find the cheapest augmenting path we build a Hungarian tree T_a rooted at every person that is unmatched in M. Initially, all vertices are unmarked and while building T_a every visited vertex gets marked so that each vertex occurs at most once in T_a. If T_a has no augmenting path then we quit and declare "G does not admit a popular matching". If an augmenting path is found in T_a, then we do not terminate the construction of T_a as soon as we find such a path, but we build T_a completely in order to find a min-cost item b which is not fully matched to which there is an augmenting path from a. The matching M is augmented along this cheapest augmenting path. The algorithm terminates when all vertices in $\mathcal{A}$ are matched.

Correctness. We first note that if there is no augmenting path in T_a, where a is an unmatched person in M_i, then there is no popular matching in G. This is because every popular matching is a maximum cardinality matching on rank-1 edges and has to match every $a \in \mathcal{A}$ to a item in $f(a) \cup s(a)$. To prove that the matching returned is a min-cost popular matching, we work with the *cloned* graph where each item b has $\mathsf{copies}(b)$ many duplicates. If M is not a min-cost matching then let OPT be such a matching. Consider $\mathsf{OPT} \oplus M$ which is a collection of even cycles and even paths. Since the cycles do not contribute to any change of cost, it suffices to consider only paths. Let ρ be a path in $\mathsf{OPT} \oplus M$. Let β_0 and β_M be the endpoints of this path, where OPT leaves β_M unmatched while M leaves β_0 unmatched. It is possible to prove that $\mathsf{cost}(\beta_M) \leq \mathsf{cost}(\beta_0)$ because in the second stage of our algorithm we always selected the cheapest augmenting path. We refer to [8] for the details of the proof.

Time complexity. The difference between our algorithm and that of Manlove and Sng for the CHAT problem in the first stage is that they use Gabow's algorithm to find a matching on rank-1 edges whereas we use Ford-Fulkerson max-flow algorithm. Gabow's algorithm runs in time $O(\sqrt{C}m)$ where $C = \sum_{i=1}^{|\mathcal{B}|} \mathsf{copies}(b_i)$ whereas since the value of max-flow in the graph $F(G_1)$ is upper bounded by $|\mathcal{A}| = n_1$, Ford-Fulkerson algorithm takes $O(mn_1)$ time. Also, the total time taken by our algorithm to partition vertices into $\mathcal{O}, \mathcal{E}$, and $\mathcal{U}$ is $O(m+n)$ where $n = |\mathcal{A}| + |\mathcal{B}|$. It is easy to see that the time spent by our algorithm in the second stage is also $O(mn_1)$ since it takes $O(m)$ time to build the tree T_a and there are at most n_1 such trees that we build. We can now conclude the following theorem.

Theorem 6. *There exists an $O(mn_1)$ time algorithm to decide whether a given instance G of the min-cost popular matchings problem admits a popular matching and if so, to compute one with min-cost.*

References

1. Abraham, D.J., Cechlárová, K., Manlove, D.F., Mehlhorn, K.: Pareto-optimality in house allocation problems. In: Proceedings of 15th Annual International Symposium on Algorithms and Computation, pp. 3–15 (2004)
2. Abraham, D.J., Irving, R.W., Kavitha, T., Mehlhorn, K.: Popular matchings. SIAM Journal on Computing 37(4), 1030–1045 (2007)
3. Gärdenfors, P.: Match making: assignments based on bilateral preferences. Behavioural Sciences 20, 166–173 (1975)
4. Irving, R.W., Kavitha, T., Mehlhorn, K., Michail, D., Paluch, K.: Rank-maximal matchings. ACM Transactions on Algorithms 2(4), 602–610 (2006)
5. Kavitha, T., Mestre, J., Nasre, M.: Popular mixed matchings. In: Proceedings of the 36th International Colloquium on Automata, Languages and Programming, pp. 574–584 (2009)
6. Kavitha, T., Nasre, M.: Note: Optimal popular matchings. Discrete Applied Mathematics 157(14), 3181–3186 (2009)
7. Kavitha, T., Nasre, M.: Popular matchings with variable job capacities. In: Proceedings of 20th Annual International Symposium on Algorithms and Computation, pp. 423–433 (2009)
8. Kavitha, T., Nasre, M., Nimbhorkar, P.: Popularity at minimum cost. In: CoRR (2010) abs/1009.2591
9. Mahdian, M.: Random popular matchings. In: Proceedings of the 8th ACM Conference on Electronic Commerce, pp. 238–242 (2006)
10. Manlove, D., Sng, C.: Popular matchings in the capacitated house allocation problem. In: Proceedings of the 14th Annual European Symposium on Algorithms, pp. 492–503 (2006)
11. McCutchen, R.M.: The least-unpopularity-factor and least-unpopularity-margin criteria for matching problems with one-sided preferences. In: Proceedings of the 15th Latin American Symposium on Theoretical Informatics, pp. 593–604 (2008)
12. McDermid, E., Irving, R.W.: Popular matchings: Structure and algorithms. In: Proceedings of 15th Annual International Computing and Combinatorics Conference, pp. 506–515 (2009)
13. Mestre, J.: Weighted popular matchings. In: Proceedings of the 33rd International Colloquium on Automata, Languages and Programming, pp. 715–726 (2006)
14. Schaefer, T.: The complexity of satisfiability problems. In: Proceedings of the 10th Annual ACM Symposium on Theory of Computing, pp. 216–226 (1978)

Structural and Complexity Aspects of Line Systems of Graphs

Jozef Jirásek and Pavel Klavík

Department of Applied Mathematics, Faculty of Mathematics and Physics,
Charles University, Malostranské náměstí 25, 118 00 Prague, Czech Republic
jirasekjozef@gmail.com, pavel@klavik.cz

Abstract. We study line systems in metric spaces induced by graphs.
A line is a subset of vertices defined by a relation of betweenness.

We show that the class of all graphs having exactly k different lines
is infinite if and only if it contains a graph with a bridge. We also study
lines in random graphs—a random graph almost surely has $\binom{n}{2}$ different
lines and no line containing all the vertices.

We call a pair of graphs isolinear if their line systems are isomorphic.
We prove that deciding isolinearity of graphs is polynomially equivalent
to the Graph Isomorphism Problem.

Similarly to the Graph Reconstruction Problem, we question the reconstructability of graphs from their line systems. We present a polynomial-
time algorithm which constructs a tree from a given line system. We give
an application of line systems: This algorithm can be extended to decide
the existence of an embedding of a metric space into a tree metric and to
construct this embedding if it exists.

1 Introduction

In this paper, we present several results motivated by an open problem of Chen
and Chvátal [9]. We study systems of lines in metric spaces induced by graphs.
Lines considered in this paper are sets of vertices defined by a relation of betweenness, as introduced by Menger [17]. A line containing all the vertices is
called a *universal line*. Similar properties, concerning distances in graphs, are
studied in metric graph theory, see a survey by Bandelt and Chepoi [4].

The problem of Chen and Chvátal asks whether the following generalization
of the de Bruijn-Erdős Theorem [7] holds:

> *True or false? Every graph with n vertices defines at least n different
> lines or it contains a universal line.*

We note that this question was originally stated about general discrete metric
spaces, but we consider only graphs. This problem is still open. The best known
lower bound is $\Omega(n^{2/7})$, proved by Chiniforooshan and Chvátal [10].

In this paper, we study properties of line systems. First, we present several
results concerning their structure. We study the structure of graphs with exactly

O. Cheong, K.-Y. Chwa, and K. Park (Eds.): ISAAC 2010, Part I, LNCS 6506, pp. 157–168, 2010.

k different lines. We show that subdividing a bridge does not change the number of different lines. Surprisingly, this is the only operation to create an infinite class of graphs with exactly k different lines.

We also show that lines in random graphs behave in a similar way to lines between random points in the plane. A random graph almost surely contains no universal line and it almost surely contains $\binom{n}{2}$ different lines. This implies that almost every graph satisfies the question of Chen and Chvátal.

In the second part of the paper, we introduce the following question: How much can a line system tell us about the structure of a graph? It is easy to show that graphs are not uniquely determined by their line systems. We call two graphs *isolinear* if their line systems are isomorphic. We prove that deciding the isolinearity of two given graphs is equally hard as the Graph Isomorphism Problem.

We also introduce the problem of line reconstruction: For a given system of sets, find a graph with this system as its line system. In general, the complexity of this problem remains open. We present a polynomial-time algorithm for the reconstruction of trees. Surprisingly, this algorithm also gives an application of line systems: It allows us to find an embedding of a given metric into a tree metric in time $\mathcal{O}(n^4)$.

Finally, we study graphs with line systems satisfying the Helly property. A system of sets has the Helly property if every subset of pairwise intersecting sets has a common intersection. We call these graphs *Helly graphs*. The structure of Helly graphs is very restricted. Surprisingly, every Helly graph has a universal line.

This paper has the following structure. The next section contains definitions of lines and line systems. In Section 3, we present structural properties of line systems. In Section 4, we study the problem of line reconstruction. In the last section, we conclude the article with related questions and open problems. All the proofs omitted in this paper will be contained in a journal version.

2   Definitions

Graph Metrics. In this paper, we consider only simple, connected and undirected graphs. For a graph G, we denote its vertices by $V(G)$ and its edges by $E(G)$. A graph G induces a *graph metric* $(V(G), d)$, where $d(u, v)$ is defined as the number of edges on a shortest path between u and v.

A subgraph G' of G is called an *isometric subgraph* if the distances of G' are preserved in G—for every $u, v \in G'$ is $d_{G'}(u, v) = d_G(u, v)$. For example, the shortest cycle is an isometric subgraph. Observe that every isometric subgraph is also an induced subgraph.

Lines. First, we describe the betweenness relation. A vertex b lies *between* vertices a and c, denoted by $[abc]$, if

$$d(a, b) + d(b, c) = d(a, c).$$

In other words, this equality holds if and only if b lies on a shortest path between a and c.

For two distinct vertices a and b, the *line* $\overleftrightarrow{ab}$ is the following subset of vertices:

$$\overleftrightarrow{ab} = \{x \mid [xab] \vee [axb] \vee [abx]\}.$$

A line containing all the vertices of the graph is called a *universal line*.

Line Systems. For a graph G, the *line system*, denoted by $\mathcal{L}(G)$, is the set of all the lines of G:

$$\mathcal{L}(G) = \{\overleftrightarrow{ab} \mid a, b \in V, a \neq b\}.$$

For an example, see Figure 1.

Fig. 1. Examine the example graph W_4. First, consider the line $\overleftrightarrow{ab}$. The vertices c and d lie on $\overleftrightarrow{ab}$, because $[abc]$ and $[dab]$. On the other hand, the vertex e does not lie on $\overleftrightarrow{ab}$, because the distances $d(a, b)$, $d(a, e)$ and $d(b, e)$ are equal. Next, $\overleftrightarrow{ae}$ contains the vertex c but it does not contain the vertices b and d. The line $\overleftrightarrow{ac}$ is universal, it contains all the vertices of the graph. Finally, the line system $\mathcal{L}(G)$ is the following:

$$\mathcal{L}(G) = \{\{a, c, e\}, \{b, d, e\}, \{a, b, c, d\}, \{a, b, c, d, e\}\}.$$

3   Structural Properties of Line Systems

We study structural properties of line systems of graphs, especially the number of different lines. For a graph G, we denote the number of different lines of G by $|\mathcal{L}(G)|$. In this section, we present two main results: The structure of graphs with exactly k different lines and properties of lines in random graphs.

3.1   Basic Properties

We present some basic properties of lines and line systems. These provide insight into the definitions and are also useful in the following sections.

Lemma 1. *Every line induces a connected subgraph.*

Edge Lines. We call a line $\overleftrightarrow{ab}$ an *edge line* if ab is an edge. Observe that a vertex c lies on the edge line $\overleftrightarrow{ab}$ if and only if $d(a, c) \neq d(b, c)$.

Bipartite graphs can be characterized by their edge lines in the following way:

Proposition 1. *A graph is bipartite if and only if all of its edge lines are universal.*

(a)   (b)

Fig. 2. (a) The structure of shortest paths of a cycle. (b) A line $\overleftrightarrow{uv}$ in an odd/even cycle contains everything except the vertices opposite to the uv.

To prove this proposition we use the following well-known characterization of bipartite graphs: A graph is bipartite if and only if, for every edge uv and every vertex w, $d(u,w) \neq d(v,w)$.

The next lemma gives a lower bound on the number of different lines using an isometric subgraph.

Lemma 2. *Let G be a graph and H be its isometric subgraph. Then $|\mathcal{L}(G)| \geq |\mathcal{L}(H)|$.*

Lemma 3. *Let $\overleftrightarrow{uv}$ be a line of C_n. A vertex $x \in \overleftrightarrow{uv}$ if and only if $[uxv]$ or distance of x from u or v is at most $\lfloor \frac{n}{2} \rfloor - d(u,v)$.*

Proof. Figure 2a shows the structure of shortest paths. The cycle is split into two half-cycles where the shortest paths to u go along different sides of the cycle. Using the relation between lines and shortest paths, observe that $\overleftrightarrow{uv}$ is a union of two such half-cycles, see Figure 2b.   □

Finally, we estimate the number of lines in selected graph classes.

Lemma 4. *The number of different lines of basic graph classes is the following:*

$$|\mathcal{L}(K_n)| = \binom{n}{2}, \qquad |\mathcal{L}(S_n)| = \binom{n}{2} + 1 \qquad and \qquad |\mathcal{L}(C_n)| = \Theta(n^2),$$

where S_n is a star with n leaves.

3.2   Graphs with Exactly k Different Lines

We present several results concerning the number of different lines in a graph. A graph G with $|\mathcal{L}(G)| = k$ is called k-*linear*. We denote the class of all k-linear graphs by $\mathfrak{L}_k$. We study structural properties of these classes.

First, consider a bridge. Its subdivision does not change the structure of shortest paths. Therefore, the number of different lines of the graph is not changed.

Lemma 5. *Let G be a graph with a bridge uv and G' be a graph obtained by subdividing uv with a vertex w. Then $|\mathcal{L}(G)| = |\mathcal{L}(G')|$.*

Therefore, if a k-linear graph G has a bridge, we can subdivide it over and over to create an infinite sequence of k-linear graphs. The existence of a k-linear graph with a bridge implies that $\mathfrak{L}_k$ is infinite. Surprisingly, this condition is not only sufficient, but also necessary.

Theorem 1. *The class $\mathfrak{L}_k$ is infinite if and only if it contains a graph with a bridge.*

To prove the converse, we first show several properties of k-linear graphs.

Lemma 6. *For every k, there exists a constant Δ_k such that every graph in $\mathfrak{L}_k$ has the maximum degree at most Δ_k.*

Blocks of a graph are its maximal biconnected subgraphs. Bridges are blocks isomorphic to K_2, we call them *trivial blocks*. The blocks are connected by a tree structure called the *block tree*.

Lemma 7. *For every k, there exists a constant β_k such that every graph in $\mathfrak{L}_k$ has at most β_k non-trivial blocks.*

Lemma 8. *For every k, there exists a constant d_k such that every biconnected graph in $\mathfrak{L}_k$ has a diameter of at most d_k.*

We are now ready to prove Theorem 1:

Proof (Theorem 1). Let $\mathfrak{L}_k$ be an infinite class. Therefore, it contains a graph G with a sufficiently large number of vertices. We want to show that G contains a bridge. By Lemma 7, the number of non-trivial blocks of G is bounded by β_k. Each block is a graph with at most k different lines. By lemmas 6 and 8, its degree is bounded by Δ_k and its diameter is bounded by d_k. This implies a bounded number of vertices in every block. If the number of vertices of G is sufficiently large, there are some trivial blocks. Therefore, G has a bridge.   □

3.3   Lines in Random Graphs

In this section, we use the probabilistic method to show that almost all graphs satisfy the question of Chen and Chvátal. If you are not familiar with the probabilistic method, check a wonderful book by Alon and Spencer [2].

A random graph $G_{n,p}$, $0 < p < 1$, is a graph on n vertices with probability p of having an edge between a pair of vertices.

Theorem 2. *Let $G_{n,p}$ be a random graph. For $n \to \infty$, this graph has almost surely no universal line and the maximum possible number of different lines, i.e* $|\mathcal{L}(G_{n,p})| = \binom{n}{2}$.

A sketch of the proof: To prove the first part, we show that for every line there is almost surely a vertex not contained in this line. For every two lines there almost surely exists a vertex that distinguishes then which implies the second part of the theorem.

Surprisingly, points and lines in the plane have similar properties. A random set of points in the plane is almost surely non-collinear and in general position, i.e. they define $\binom{n}{2}$ different lines.

4 Reconstructing Graphs from Line Systems

In this section we ask how well are graphs determined by their line systems. We propose several questions: Are there non-isomorphic graphs having isomorphic line systems? For a given system of sets, how hard is it to show whether there exists a graph having this system as its line system? In this section, we try to give answers to these questions.

Similar ideas are well-studied in graph theory. We remind of the famous Reconstruction Conjecture, posed by Kelly and Ulam in 1941. It states that every graph with at least three vertices is uniquely determined by the sequence of its vertex-deleted subgraphs, which is called the deck. This conjecture is still open for general graphs. Complexity results concerning the reconstruction are studied in [14,13]. For more references, see a survey [6].

The famous Graph Isomorphism Problem (GI) asks whether two given graphs are isomorphic. This problem is known to be in NP. On the other hand, it is not known to be either in P or NP-complete. Moreover, it is not likely to be NP-complete. Problems polynomially equivalent to GI are called GI-complete. For example, the isomorphism of hypergraphs is GI-complete. For more details, see [18].

4.1 Definitions

Graph Isomorphism. Graphs G and H are *isomorphic*, denoted by $G \cong H$, if there exists a bijection $f : V(G) \to V(H)$, such that $uv \in E(G)$ if and only if $f(u)f(v) \in E(H)$. Such a mapping f is called an isomorphism.

Observe that we can represent a line system $\mathcal{L}(G)$ by a hypergraph, whose vertices are $V(G)$ and hyperedges are lines of G. The graph isomorphism is naturally extendable to hypergraphs as a mapping preserving hyperedges.

Graph Isolinearity. We call graphs G and H *isolinear*, denoted by $G \stackrel{\backsim}{=} H$, if $\mathcal{L}(G)$ and $\mathcal{L}(H)$ are isomorphic (as hypergraphs). The graph isolinearity is an equivalence relation like the graph isomorphism. The isomorphism of two graphs implies their isolinearity. The converse does not hold: For example, consider P_3 (a path of length three) and C_4. It is easy to see that they are isolinear, but not isomorphic. However, from the complexity point of view, we show that the graph isomorphism and the graph isolinearity are polynomially equivalent.

Line Reconstruction. We use definitions similar to the Graph Reconstruction Problem, see [6]. A *reconstruction* of a line system $\mathcal{L}$ is a graph G, such that $\mathcal{L}(G) \cong \mathcal{L}$. A graph G is *reconstructible* if every reconstruction of $\mathcal{L}(G)$ is isomorphic to G. A class of graphs $\mathcal{C}$ is *reconstructible* if every graph in $\mathcal{C}$ is reconstructible. We call a class $\mathcal{C}$ *recognizable*, if for every graph G in $\mathcal{C}$, all the reconstructions of $\mathcal{L}(G)$ belong to $\mathcal{C}$. Finally, a class $\mathcal{C}$ is *weakly reconstructible*, if for every graph G, all its reconstructions in $\mathcal{C}$ are isomorphic to G. This means that graphs in $\mathcal{C}$ are uniquely distinguished from all the other graphs in $\mathcal{C}$ by their lines. Observe that $\mathcal{C}$ is reconstructible if it is recognizable and weakly reconstructible.

There exists an infinite number of graphs that are not reconstructible. The reader can verify that $K_{m,n}$ and $K_{m,n}$ without an edge are isolinear. Therefore, a conjecture similar to the Graph Reconstruction Problem does not hold. The question of line reconstructability becomes a question of which graphs are reconstructible.

4.2 Graph Isolinearity Problem

We prove that determining the Graph Isolinearity is GI-complete.

Isolinearity Problem. The Graph Isolinearity is the following problem, similar to Graph Isomorphism:

Problem:	Graph Isolinearity
Input:	Graphs G and H.
Output:	Yes if $G \cong H$, no otherwise.

Theorem 3. *The Graph Isolinearity Problem is GI-complete.*

To prove Theorem 3, we use the following *split construction*, introduced by Booth and Lueker [15] in the proof that the isomorphism of chordal graphs is GI-complete.

Let G be a graph. The split construction of G is a graph G', such that

$$V(G') = V(G) \cup E(G),$$
$$E(G') = \{uv \mid u,v \in V(G)\} \cup \{eu, ev \mid e = uv \in E(G)\}.$$

In other words, $V(G)$ induces a complete subgraph in G' and, for every edge in G, we add a triangle to G'. See Figure 3.

For a graph G, we denote its split construction by G'. We denote the minimum degree of the graph G by $\delta(G)$. We show two properties of the split construction, the first proven in [15].

Lemma 9 ([15]). *Let G and H be graphs with $\delta(G), \delta(H) \geq 3$. Then $G \cong H$ if and only if $G' \cong H'$.* $\square$

Lemma 10. *Let G and H be graphs with $\delta(G), \delta(H) \geq 3$. Then $G' \cong H'$ if and only if $G' \cong H'$.*

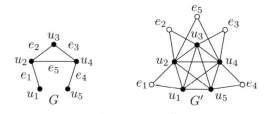

Fig. 3. An example of the split construction

Proof (Theorem 3). Graph Isolinearity is solvable using hypergraph isomorphism which is polynomially equivalent to GI, see [18].

To show the other reduction, we use the split construction. GI is equally hard when restricted to graphs with a minimum degree of three. For input graphs G and H of GI, construct G' and H'. Output whether $G' \stackrel{\sim}{=} H'$. By Lemma 9, we know that $G \cong H$ if and only if $G' \cong H'$. Lemma 10 shows that $G' \cong H'$ if and only if $G' \stackrel{\sim}{=} H'$. The reduction is clearly polynomial.   □

Line Check. Another problem we consider is checking whether a given system of sets is a line system of a given graph.

Problem:	Line Check
Input:	A graph G and a system of sets $\mathcal{L}$.
Output:	Yes if $\mathcal{L}(G) \cong \mathcal{L}$, no otherwise.

Proposition 2. *Line Check is GI-complete.*

Proof. Using the isomorphism of hypergraphs $\mathcal{L}(G)$ and $\mathcal{L}$, we can solve Line Check. Conversely, Graph Isolinearity can be solved using Line Check. Let G and H be input graphs of Graph Isolinearity. Output the result of Line Check for G and $\mathcal{L}(H)$.   □

4.3   Reconstruction of Trees

Reconstruction Problems. We present the following problem:

Problem:	Line Reconstruction
Input:	A system of sets $\mathcal{L}$.
Output:	Yes if there exists a graph G, such that $\mathcal{L}(G) \cong \mathcal{L}$, no otherwise.

In general, the complexity of this problem remains open.

Reconstruction of Trees. We solve this problem for trees. We show a polynomial-time algorithm which identifies line systems of trees. Moreover, if the given system of sets is a line system of a tree, the algorithm reconstructs a tree with this line system.

Theorem 4. *Trees are reconstructible in time $\mathcal{O}(n^4)$ where n is the number of vertices.*

First, we describe the structure of lines in a tree. A line $\overleftrightarrow{ab}$ contains the path ab and subtrees connected to a and b, see Figure 4.

We call a vertex of a tree *large* if its degree is at least three. Otherwise, we call the vertex *small*. A path with all the inner vertices small is called a *non-branching path*. We consider maximal non-branching paths, which cannot be enlarged. Their end vertices are either leaves or large vertices.

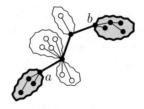

Fig. 4. A line $\overleftrightarrow{ab}$ in a tree

Let T be a tree. Consider the relation $R \subseteq V(T) \times V(T)$, such that $(a, b) \in R$ if and only if every line of T containing the vertex a also contains b. Observe that $(u, v) \in R$ for exactly all the vertices v contained in the intersection of all the lines containing u. We derive the structure of the tree from this relation.

Lemma 11. *Let u be a large vertex. For every vertex v, $(u, v) \in R$ implies $u = v$.*

Lemma 12. *Let u be a small vertex. For every vertex v, $(u, v) \in R$ implies that v lies on the maximal non-branching path P containing u.*

Lemma 13. *Let u and v be large vertices. The intersection of all the lines containing both u and v is the path P between u and v.*

A sketch of the algorithm: Compute the relation R. Using these lemmas, identify small and large vertices, and edges between them.

Corollary 1. *Trees are weakly reconstructible.*

4.4 Embedding Metrics into Weighted Trees

Metric embeddings are well studied, motivated by many practical applications. An embedding of a metric space into another metric space preserves distances. A natural question is whether there exists an embedding into a simpler metric space, for example a tree metric. Often an approximation of a complicated metric by a simple metric is sufficient.

Matoušek [16] describes approximations of metric spaces by points in d-dimensional Euclidean spaces. Buneman [8] studies the structure of tree metrics and describes the Four Point Condition. Bandelt [3] describes a connection between tree metrics and ultrametrics. Moreover, Bandelt presents an algorithm for testing whether a given metric is a tree metric in time $\mathcal{O}(n^2 \log n)$. Approximations of metrics by trees are consider in Alon et. al. [1] and Bartal [5].

Metric Embedding. An embedding of a metric X to a metric Y is a bijective mapping $f : X \to Y$, such that $d_X(a, b) = d_Y(f(a), f(b))$ for every $a, b \in X$.

Tree Metric. A weighted tree (V, E, w) defines a tree metric (V, d), such that $d(u, v)$ is equal to the sum of the weights of edges on the path between u and v. Lines in a weighted tree are defined similarly by the betweenness relation.

Weights of the edges of a tree do not affect the structure of shortest paths. Therefore, lines in a weighted tree are exactly the same as in a non-weighted tree—the relation of betweenness is independent on the weights.

The algorithm described in Theorem 4 can be slightly modified for the following useful application.

Theorem 5. *For a given finite n-point metric X, we can find its embedding into a weighted tree, or show there is no such embedding, in time $\mathcal{O}(n^4)$.*

A sketch of the proof: First, we calculate lines of the metric X. If the metric X can be embedded into a tree, the algorithm for tree reconstruction gives a tree T. Finally, we assign weights to the edges of T, according to distances in X.

4.5   Line Systems with Helly Property

One of the well-studied properties of set systems is the *Helly property*. A system of sets $\mathcal{S}$ has the Helly property if for every set $\mathcal{S}' \subseteq \mathcal{S}$ of pairwise intersecting sets their common intersection is non-empty. The Helly property appears naturally in combinatorics and geometry. For instance, a set of intervals on the real line has the Helly property. For more references, see a survey [11].

We study graphs that have line systems with the Helly property. We call such graphs *Helly graphs* and we denote them by

$$HELLY = \{\text{a graph } G \mid \mathcal{L}(G) \text{ has the Helly property}\}.$$

The Helly property is a property of a line system, not a graph. Therefore the class $HELLY$ is recognizable. We describe the structure of Helly graphs.

Lemma 14. *Let G be a Helly graph. Then it contains no cycle of length three, five or more as an isometric subgraph.*

We denote $TREE$ the class of all trees, BIP the class of all bipartite graphs and $UNIV$ the class of all graphs having a universal line.

Proposition 3. *Classes $TREE$, $HELLY$, BIP and $UNIV$ are in the following inclusions:*
$$TREE \subsetneq HELLY \subsetneq BIP \subsetneq UNIV.$$

Fig. 5. Inclusion of the classes $TREE$, $HELLY$, BIP and $UNIV$

Proof. TREE $\subseteq$ *HELLY*: By Lemma 1, all lines of a tree are its subtrees. Subtrees of a tree have the Helly property, see [12].

HELLY $\subseteq$ *BIP*: If the graph contains an odd cycle, consider the shortest odd cycle. Observe that this cycle is an isometric subgraph. By Lemma 14, a Helly graph contains no odd isometric cycles. Therefore, the graph is not Helly.

BIP $\subseteq$ *UNIV*: By Proposition 1, all edge lines of a bipartite graph are universal.

All the inclusions are proper: See examples in Figure 5.   □

Surprisingly, Proposition 3 implies that all Helly graphs have a universal line. In other words, line systems of graphs are special in a way that having the Helly property implies the existence of a set containing all the elements.

5   Open Problems and Related Questions

We conclude this paper by stating several open problems.

Theorem 1 shows that a graph without a bridge has at least $f(n)$ different lines, where $f(n)$ is non-decreasing and unbounded.

Problem 1. Find a closer estimation of $f(n)$. Similarly to the question of Chen and Chvátal, is $f(n)$ a linear function?

For example, the graph in Figure 1 has no bridge, five vertices and only four different lines. This is the only bridgeless graph having $|\mathcal{L}(G)| < |V(G)|$ we know.

Problem 2. What is the complexity of Line Reconstruction?

We believe that Line Reconstruction is NP-complete. We note that Kratsch and Hemaspaandra [14] proved that Deck Legitimity, a similar problem concerning Graph Reconstruction, is GI-hard. Unfortunately, we have not found an analogous relation between Graph Isomorphism and Line Reconstruction. Also, it would be interesting to prove that these problems are polynomially solvable for some restricted classes of graphs, for example k-trees or chordal graphs.

We show that trees are weakly reconstructible. Proposition 3 implies that there is no tree isolinear with a non-bipartite graph—we can distinguish them by the Helly property. We conjecture that trees are "almost" reconstructible:

Conjecture 1. The only non-tree isolinear with a tree is C_4.

Proposition 1 gives us a reason to believe that the following conjecture holds:

Conjecture 2. Bipartite graphs are recognizable.

We know that they are not reconstructible, as for example $K_{m,n}$ and $K_{m,n}$ without an edge are isolinear.

Acknowledgments

We would like to thank Jan Kratochvíl for introducing us to the problem and guiding our research. We would also like to thank Martin Loebl for pointing out possible relations with the Graph Reconstruction Conjecture. We are grateful to Ondřej Bílka, Martin Doucha, Pavel Valtr and Jan Volec, for fruitful discussions, and also to Jiří Matoušek for references to metric embedddings. Also, we would like to thank to anonymous reviewers for several remarks.

References

1. Alon, N., Karp, R.M., Peleg, D., West, D.: A Graph-Theoretic Game and its Application to the k-Server Problem. SIAM J. Comput. 24(1), 78–100 (1995)
2. Alon, N., Spencer, J.: The Probabilistic Method. John Wiley, New York (1991)
3. Bandelt, H.J.: Recognition of tree metrics. SIAM J. Discret. Math. 3(1), 1–6 (1990)
4. Bandelt, H.J., Chepoi, V.: Metric graph theory and geometry: a survey. Contemporary Mathematics 453, 49–86 (2008)
5. Bartal, Y.: On approximating arbitrary metrices by tree metrics. In: STOC 1998: Proceedings of the Thirtieth Annual ACM Symposium on Theory of Computing, pp. 161–168. ACM, New York (1998)
6. Bondy, J.A., Hemminger, R.L.: Graph reconstruction—a survey. Journal of Graph Theory 1(3), 227–268 (1977)
7. de Bruijn, N.G., Erdős, P.: On a combinatorial problem. Indagationes Mathematicae 10, 421–423 (1948)
8. Buneman, P.: A note on the metric properties of trees. Journal of Combinatorial Theory, Series B 17(1), 48–50 (1974)
9. Chen, X., Chvátal, V.: Problems related to a de Bruijn-Erdős theorem. Discrete Appl. Math. 156(11), 2101–2108 (2008)
10. Chiniforooshan, E., Chvátal, V.: A de Bruijn-Erdős theorem and metric spaces. arXiv:0906.0123v1 [math.CO] (2009)
11. Danzer, L., Grünbaum, B., Klee, V.: Helly's theorem and its relatives. In: Proc. Symp. Pure Math., vol. 7, pp. 101–179 (1963)
12. Golumbic, M.: Algorithmic Graph Theory and Perfect Graphs (2004)
13. Hemaspaandra, E., Hemaspaandra, L.A., Radziszowski, S.P., Tripathi, R.: Complexity results in graph reconstruction. Discrete Appl. Math. 155(2), 103–118 (2007)
14. Kratsch, D., Hemaspaandra, L.A.: On the complexity of graph reconstruction. Math. Syst. Theory 27(3), 257–273 (1994)
15. Lueker, G.S., Booth, K.S.: A linear time algorithm for deciding interval graph isomorphism. J. ACM 26(2), 183–195 (1979)
16. Matoušek, J.: Lectures on Discrete Geometry. Springer, New York (2002)
17. Menger, K.: Untersuchungen über allgemeine Metrik. Mathematische Annalen 100(1), 75–163 (1928)
18. Zemlyachenko, V.N., Korneenko, N.M., Tyshkevich, R.I.: Graph isomorphism problem. Journal of Mathematical Sciences 29(4), 1426–1481 (1985)

Neighbor Systems, Jump Systems, and Bisubmodular Polyhedra

Akiyoshi Shioura

Graduate School of Information Sciences, Tohoku University, Sendai 980-8579, Japan
shioura@dais.is.tohoku.ac.jp

Abstract. The concept of neighbor system, introduced by Hartvigsen (2009), is a set of integral vectors satisfying a certain combinatorial property. In this paper, we reveal the relationship of neighbor systems with jump systems and with bisubmodular polyhedra. We firstly prove that for every neighbor system, there exists a jump system which has the same neighborhood structure as the original neighbor system. This shows that the concept of neighbor system is essentially equivalent to that of jump system. We next show that the convex closure of a neighbor system is an integral bisubmodular polyhedron. In addition, we give a characterization of neighbor systems using bisubmodular polyhedra. Finally, we consider the problem of minimizing a separable convex function on a neighbor system. By using the relationship between neighbor systems and jump systems shown in this paper, we prove that the problem can be solved in weakly-polynomial time for a class of neighbor systems.

1 Introduction

The concept of neighbor system, introduced by Hartvigsen [14], is a set of integral vectors satisfying a certain combinatorial property. The definition of neighbor system is as follows. Throughout this paper, let n be a positive integer and $E = \{1, 2, \ldots, n\}$. For a vector $x \in \mathbf{R}^n$, we define $\mathrm{supp}(x) = \{i \in E \mid x(i) \neq 0\}$. Let $\mathcal{F}$ be a set of integral vectors in $\mathbf{Z}^n$. For $x, y \in \mathcal{F}$, we say that y is a *neighbor* of x if there exist some vector $d \in \{0, +1, -1\}^n$ with $|\mathrm{supp}(d)| \leq 2$ and a positive integer α such that $y = x + \alpha d$ and $x + \alpha' d \notin \mathcal{F}$ if $0 < \alpha' < \alpha$. For vectors $x, y, z \in \mathbf{Z}^n$, z is said to be *between* x *and* y if $\min\{x(i), y(i)\} \leq z(i) \leq \max\{x(i), y(i)\}$ holds for all $i \in E$. The set $\mathcal{F}$ is called an *(all-)neighbor system* if it satisfies the following (note that $i \in \mathrm{supp}(x - y)$ if and only if $x(i) \neq y(i)$):

> for every $x, y \in \mathcal{F}$ and $i \in \mathrm{supp}(x - y)$, there exists a neighbor $z \in \mathcal{F}$ of x such that $i \in \mathrm{supp}(z - x)$ and z is between x and y.

See Figure 1 for an example of a 2-dimensional neighbor system. Given a positive integer k, a neighbor system $\mathcal{F}$ is said to be an N_k-*neighbor system* if we can always choose a neighbor z in the axiom above such that $\|z - x\|_1 \leq k$. For example, the neighbor system in Figure 1 is an N_k-neighbor system for every $k \geq 3$, but not for $k = 1, 2$ since for $x = (0, 2)$ and $y = (3, 5)$ we do not have such a neighbor z with $\|z - x\|_1 \leq 2$.

O. Cheong, K.-Y. Chwa, and K. Park (Eds.): ISAAC 2010, Part I, LNCS 6506, pp. 169–181, 2010.

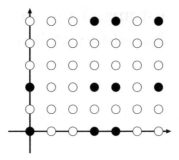

Fig. 1. An example of 2-dimensional neighbor system, where the black dots represents integral vectors in the neighbor system

Neighbor system is a common generalization of various concepts such as matroid, integral polymatroid, delta-matroid, integral bisubmodular polyhedron, and jump system. Below we review these concepts; see [12] for more accounts.

Matroids. The concept of matroid is introduced by Whitney [21]. One of the important results on matroids, from the viewpoint of combinatorial optimization, is the validity of a greedy algorithm for linear optimization (see, e.g., [11]).

Integral Polymatroids. The concept of polymatroid is introduced by Edmonds [10] as a generalization of matroids. A polymatroid is a polyhedron defined by a monotone submodular function, and a greedy algorithm for matroids naturally extends to polymatroids. The minimization of separable-convex function can be also done in a greedy way, and efficient algorithms have been proposed (see, e.g., [13,15]). An integral polymatroid is a polymatroid which is an integral polyhedron, i.e., all extreme points are given by integer vectors.

Delta-Matroids. The concept of delta-matroid (or pseudomatroid) is introduced by Bouchet [5] and Chandrasekaran and Kabadi [7]. A delta-matroid can be seen as a family of subsets of a ground set with a nice combinatorial structure, and generalizes the concept of matroid. A more general greedy algorithm works for the linear optimization on a delta-matroid.

Integral Bisubmodular Polyhedron. The concept of bisubmodular polyhedron (or polypseudomatroid), introduced by Chandrasekaran and Kabadi [7] (see also [6,12]), is a common generalization of polymatroid and delta-matroid. For the following discussion, we give a precise definition. We denote $3^E = \{(X,Y) \mid X, Y \subseteq E, \ X \cap Y = \emptyset\}$. A function $\rho : 3^E \to \mathbf{R} \cup \{+\infty\}$ is said to be *bisubmodular* if it satisfies the bisubmodular inequality:

$$\rho(X_1, Y_1) + \rho(X_2, Y_2) \geq \rho((X_1 \cup X_2) \setminus (Y_1 \cup Y_2), (Y_1 \cup Y_2) \setminus (X_1 \cup X_2))$$
$$+ \rho(X_1 \cap X_2, Y_1 \cap Y_2) \quad (\forall (X_1, Y_1), (X_2, Y_2) \in 3^E).$$

For a function $\rho : 3^E \to \mathbf{R} \cup \{+\infty\}$ with $\rho(\emptyset, \emptyset) = 0$, we define a polyhedron $\mathrm{P}_*(\rho) \subseteq \mathbf{R}^n$ by

$$\mathrm{P}_*(\rho) = \{x \in \mathbf{R}^n \mid \textstyle\sum_{i \in X} x(i) - \sum_{i \in Y} x(i) \leq \rho(X, Y) \ ((X, Y) \in 3^E)\}, \quad (1)$$

which is called a *bisubmodular polyhedron* if ρ is bisubmodular. Bisubmodular polyhedra constitute an important class of polyhedra on which a simple greedy algorithm works for linear optimization. In addition, separable convex function minimization can be done in a greedy manner [2]. In this paper, we are mainly interested in integral bisubmodular polyhedra.

Jump Systems. The concept of jump system is introduced by Bouchet and Cunningham [6] as a common generalization of delta-matroid and the set of integral vectors in an integral bisubmodular polyhedron. Interesting examples of jump systems can be found in the set of degree sequences of subgraphs in undirected and directed graphs; for example, jump systems arise from matchings and b-matchings in undirected graphs [6,8,16] and even-factors in directed graphs [17]. Validity of certain greedy algorithms is shown for the linear optimization [6] and for separable-convex function minimization [3]. Moreover, a polynomial-time algorithm for separable-convex function minimization is given [20].

We give a precise definition of jump systems. For $i \in E$, the characteristic vector $\chi_i \in \{0,1\}^n$ is given by $\chi_i(i) = 1$ and $\chi_i(j) = 0$ for $j \in E \setminus \{i\}$. Denote by U the set of vectors $+\chi_i, -\chi_i$ $(i \in E)$. For vectors $x, y \in \mathbf{Z}^n$, define

$$\mathrm{inc}(x,y) = \{p \in U \mid x + p \text{ is between } x \text{ and } y\}. \tag{2}$$

A set $\mathcal{J} \subseteq \mathbf{Z}^n$ is a *jump system* if it satisfies the following axiom:

> **(J)** for every $x, y \in \mathcal{J}$ and every $p \in \mathrm{inc}(x,y)$, if $x + p \notin \mathcal{J}$ then there exists $q \in \mathrm{inc}(x+p, y)$ such that $x + p + q \in \mathcal{J}$.

It is shown that a jump system is equivalent to an N_2-neighbor system [14].

We give two additional examples of neighbor systems which are not jump systems; such an example is also given in Figure 1.

Example 1.1 (Expansion of jump systems). For a jump system $\mathcal{J} \subseteq \mathbf{Z}^n$ and a positive integer k, the set $\{kx \in \mathbf{Z}^n \mid x \in \mathcal{J}\}$ is an N_{2k}-neighbor system [14].

Example 1.2 (Rectilinear grid). Let $u \in \mathbf{Z}_+^n$ be a nonnegative vector, and for $i \in E$, let $\pi_i : [0, u(i)] \to \mathbf{Z}$ be a strictly increasing function. Then, the set of $(\pi_i(x(i)) \mid i \in E)$ for vectors $x \in \mathbf{Z}^n$ with $\mathbf{0} \leq x \leq u$ is an all-neighbor system.

These examples, in particular, show that a neighbor system may have a "hole," as in the case of jump system, and it can be arbitrarily large.

Neighbor systems provide a systematic and simple way to characterize matroids and its generalizations for which linear optimization can be solved by greedy algorithms. Indeed, it is shown that linear optimization on a neighbor system can be solved by a greedy algorithm, and that the greedy algorithm runs in polynomial time for N_k-neighbor systems for every fixed k [14].

The main aim of this paper is to reveal the relationship of neighbor systems with jump systems and with bisubmodular polyhedra. We firstly prove that for every neighbor system $\mathcal{F} \subseteq \mathbf{Z}^n$, there exists a jump system $\mathcal{J} \subseteq \mathbf{Z}^n$ which has the same neighborhood structure as the neighbor system $\mathcal{F}$ (see Theorem 3.1

for the precise statement). This means that the concept of neighbor system is essentially equivalent to that of jump system, although the class of neighbor systems properly contains that of jump systems. Our result implies that every property of jump systems can be restated in terms of neighbor systems by using the equivalence. Indeed, we show in Section 5 that several useful properties of jump systems naturally extend to neighbor systems.

We then discuss the relationship between neighbor systems and bisubmodular polyhedra. It is known that the convex closure of a jump system, which is a special case of neighbor systems, is an integral bisubmodular polyhedron [6]. We show that the convex closure of a neighbor system is also an integral bisubmodular polyhedron (see Theorem 4.1). In addition, we give a characterization of neighbor systems using bisubmodular polyhedra, stating that a set of integral vectors is a neighbor system if and only if the convex closure of its restriction with an interval is an integral bisubmodular polyhedron (see Theorem 4.2). This result implies, in particular, that a simple greedy algorithm for the linear optimization on a bisubmodular polyhedron described below can be also used for neighbor systems (see [9], [12, §3.5 (b)] for the greedy algorithm of this type).

Greedy algorithm for the minimization of a linear function $\sum_{i=1}^{n} w(i)x(i)$.
Step 0: Let x_0 be any vector in $\mathcal{F}$ and put $x := x_0$. Order the elements in $E = \{e_1, e_2, \ldots, e_n\}$ and compute an integer k so that

$$|w(e_1)| \geq \cdots \geq |w(e_k)| > |w(e_{k+1})| = \cdots = |w(e_n)| = 0.$$

Step 1: For $i = 1, 2, \ldots, k$, do the following: if $w(e_i) \geq 0$ (resp., $w(e_i) < 0$), then fix the components $x(e_1), x(e_2), \ldots, x(e_{i-1})$ and decrease (resp., increase) $x(e_i)$ as much as possible under the condition $x \in \mathcal{F}$.

As an application of the results shown in this paper, we consider the separable convex optimization problem on neighbor systems. Given a family of univariate convex functions $f_i : \mathbf{Z} \to \mathbf{R}$ $(i \in E)$ and a finite neighbor system $\mathcal{F} \subseteq \mathbf{Z}^n$, we consider the following problem:

(SC) Minimize $f(x) \equiv \sum_{i=1}^{n} f_i(x(i))$ subject to $x \in \mathcal{F}$.

For a special case where $\mathcal{F}$ is a jump system, it is shown that the problem (SC) can be solved in pseudo-polynomial time by a greedy-type algorithm [3], and in weakly-polynomial time by an algorithm called the domain reduction algorithm [20]. We extend these algorithms to neighbor systems.

To do this, we show that the problem (SC) on a neighbor system can be reduced to the problem (SC') of minimizing a separable convex function on a jump system by using the relationship between neighbor systems and jump systems shown in this paper. Note that this reduction does not yield efficient algorithms for neighbor systems since it requires exponential time. Instead, we extend the properties used in the algorithms for jump systems to neighbor systems, which enables us to develop efficient algorithms for neighbor systems.

The organization of this paper is as follows. Section 2 is devoted to preliminaries on the fundamental concepts discussed in this paper. We discuss the relationship of neighbor systems with jump systems and with bisubmodular polyhedra in Sections 3 and 4, respectively. In Section 5, we propose efficient algorithms for (SC). Due to the page limit, most of the proofs are omitted.

2   Preliminaries

We denote by $\mathbf{Z}, \mathbf{Z}_+, \mathbf{Z}_{++}$ the sets of integers, nonnegative integers, and positive integers, respectively. We denote by $\mathbf{R}$ the set of real numbers. For vectors $\ell \in (\mathbf{Z} \cup \{-\infty\})^n$ and $u \in (\mathbf{Z} \cup \{+\infty\})^n$ with $\ell \leq u$, we define the integer interval $[\ell, u]$ as the set of integral vectors $x \in \mathbf{Z}^n$ with $\ell(i) \leq x(i) \leq u(i)$ $(\forall i \in E)$.

We review the original definition of neighbor systems in [14], which uses the concept of neighbor function. A *neighbor function*, denoted by N, is a function that takes as input any set $\mathcal{F} \subseteq \mathbf{Z}^n$ with any $x \in \mathcal{F}$ and outputs a subset of the neighbors of x in $\mathcal{F}$, denoted by $N(\mathcal{F}, x)$. In particular, $N^a(\mathcal{F}, x)$ (resp., $N_k(\mathcal{F}, x)$) denotes the set of all neighbors of x in $\mathcal{F}$ (resp., the set of all neighbors y of x in $\mathcal{F}$ with $\|y - x\|_1 \leq k$). For vectors $x, y, z \in \mathbf{Z}^n$, z is said to be *between* x *and* y if

$$\min\{x(i), y(i)\} \leq z(i) \leq \max\{x(i), y(i)\}$$

for all $i \in E$. Given a set $\mathcal{F} \subseteq \mathbf{Z}^n$ and a neighbor function N, we say that $\mathcal{F}$ is an N-*neighbor system* if the following condition holds:

> **(NS)** for every $x, y \in \mathcal{F}$ and every $i \in \mathrm{supp}(x - y)$, there exists $z \in N(\mathcal{F}, x)$ such that $i \in \mathrm{supp}(z - x)$ and z is between x and y.

An N-neighbor system is said to be an all-neighbor system if $N = N^a$, and an N_k-neighbor system if $N = N_k$.

In the following discussion, we also use the following equivalent axiom of neighbor systems. Recall the definition of $\mathrm{inc}(x, y)$ $(x, y \in \mathbf{Z}^n)$ in (2).

> **(NS')** for every $x, y \in \mathcal{F}$ and every $p \in \mathrm{inc}(x, y)$, there exist $q \in \mathrm{inc}(x, y) \cup \{\mathbf{0}\} \setminus \{p\}$ and $\alpha \in \mathbf{Z}_{++}$ such that $x' \equiv x + \alpha(p + q) \in N(\mathcal{F}, x)$ and x' is between x and y.

We note that the axiom (NS') is similar to the axiom (J) for jump systems.

The class of neighbor systems is closed under the following operations.

Proposition 2.1 (cf. [14]). *Let $\mathcal{F} \subseteq \mathbf{Z}^n$ be an N-neighbor system.*
(i) For a positive integer $m > 0$, define a set $\mathcal{F}' = \{mx \mid x \in \mathcal{F}\}$ and a neighbor function N' by $N'(\mathcal{F}', mx) = \{my \mid y \in N(\mathcal{F}, x)\}$. Then, $\mathcal{F}'$ is an N'-neighbor system.
(ii) For a vector $s \in \{+1, -1\}^n$, we define a set $\mathcal{F}_s = \{(s(e)x(e) \mid e \in E) \mid x \in \mathcal{F}\}$ and a neighbor function N_s by $N_s(\mathcal{F}_s, y) = \{(s(e)x'(e) \mid e \in E) \mid x' \in N(\mathcal{F}, x)\}$ for $y = (s(e)x(e) \mid e \in E) \in \mathcal{F}_s$. Then, $\mathcal{F}_s$ is an N_s-neighbor system.
(iii) For vectors $\ell, u \in \mathbf{Z}^n$ with $\ell \leq u$ and $\mathcal{F} \cap [\ell, u] \neq \emptyset$, the set $\mathcal{F} \cap [\ell, u]$ is an N-neighbor system.
(iv) For a vector $a \in \mathbf{Z}^n$, the set $\mathcal{F} + a = \{x + a \mid x \in \mathcal{F}\}$ is an N'-neighbor system, where $N'(\mathcal{F} + a, x) = \{y + a \mid y \in N(\mathcal{F}, x)\}$.

We introduce a concept of *proper neighbor* for neighbor systems. Recall that U is the set of vectors $+\chi_i, -\chi_i$ $(i \in E)$. For a neighbor system $\mathcal{F}$ and vectors $x, y \in \mathcal{F}$, we say that y is a *proper neighbor of x in $\mathcal{F}$* if y is a neighbor of x satisfying the condition (i) or (ii), where

> (i) there exist some $\alpha \in \mathbf{Z}_{++}$ and $p \in U$ such that $y - x = \alpha p$,
> (ii) there exist some $\alpha \in \mathbf{Z}_{++}$ and $p, q \in U$ with $\mathrm{supp}(p) \neq \mathrm{supp}(q)$ such that $y - x = \alpha(p + q)$ and $x + \alpha'p \notin \mathcal{F}$ for all α' with $0 < \alpha' \leq \alpha$.

To illustrate the concept of proper neighbor, consider the neighbor system in Figure 1. The vector $y = (6, 2)$ is a proper neighbor of $x = (4, 0)$ since $(5, 0)$ and $(6, 0)$ are not in $\mathcal{F}$ and therefore the condition (ii) holds with $p = (1, 0)$, $q = (0, 1)$, and $\alpha = 2$. The vector $(6, 5)$ is a neighbor of $(3, 2)$, but not a proper neighbor of $(3, 2)$ since $(4, 2), (3, 5) \in \mathcal{F}$.

For jump system, which is a special case of neighbor system, the definition of proper neighbor can be simplified as follows; for a jump system $\mathcal{J}$ and vectors $x, y \in \mathcal{J}$, the vector y is said to be a *proper neighbor of x in $\mathcal{J}$* if y is a neighbor of x satisfying the condition (i) or (ii), where

> (i) $\exists p \in U : y - x = p$,   (ii) $\exists p, q \in U : y - x = p + q$ and $x + p \notin \mathcal{J}$.

3   Relationship between Neighbor Systems and Jump Systems

We discuss the relationship between neighbor systems and jump systems. It is shown that for every neighbor system, there exists a jump system which has the same neighborhood structure as the original neighbor system.

Theorem 3.1. *Let $\mathcal{F} \subseteq \mathbf{Z}^n$ be an all-neighbor system. Then, there exist a jump system $\mathcal{J} \subseteq \mathbf{Z}^n$ and a bijective function $\pi : \mathcal{J} \to \mathcal{F}$ satisfying the following:*

> *(∗) for every $x, y \in \mathcal{F}$, the vector x is a proper neighbor of y in $\mathcal{F}$ if and only if $\pi^{-1}(x)$ is a proper neighbor of $\pi^{-1}(y)$ in $\mathcal{J}$, where $\pi^{-1} : \mathcal{F} \to \mathcal{J}$ is the inverse function of π.*

Below we give an outline of the proof. Let $\mathcal{F} \subseteq \mathbf{Z}^n$ be an all-neighbor system. By Proposition 2.1 (iv), we may assume, without loss of generality, that $\mathcal{F}$ contains the zero vector $\mathbf{0}$. For $i \in E$, we define a set $\mathcal{F}_i = \{\alpha \mid \alpha \in \mathbf{Z}, \exists x \in \mathcal{F} \text{ s.t. } x(i) = \alpha\}$. We also define the numbers $u(i) \in \mathbf{Z} \cup \{+\infty\}$ and $\ell(i) \in \mathbf{Z} \cup \{-\infty\}$ by

$$u(i) = |\{\alpha \mid \alpha \in \mathcal{F}_i, \ \alpha > 0\}|, \quad l(i) = -|\{\alpha \mid \alpha \in \mathcal{F}_i, \ \alpha < 0\}|.$$

We also define a function $\pi_i : [\ell(i), u(i)] \to \mathbf{Z}$ by $\pi_i(0) = 0$ and

$$\pi_i(k) = \text{ the } k\text{-th smallest positive integer in } \mathcal{F}_i \quad (\text{if } 0 < k \leq u(i)),$$
$$\pi_i(-k) = \text{ the } k\text{-th largest negative integer in } \mathcal{F}_i \quad (\text{if } \ell(i) \leq -k < 0).$$

Then, each π_i is a strictly increasing function in the interval $[\ell(i), u(i)]$. We define a set $\mathcal{J} \subseteq \mathbf{Z}^n$ and a function $\pi : \mathcal{J} \to \mathcal{F}$ by

$$\mathcal{J} = \{z \in \mathbf{Z}^n \mid (\pi_i(z(i)) \mid i \in E) \in \mathcal{F}\}, \qquad \pi(z) = (\pi_i(z(i)) \mid i \in E) \quad (z \in \mathcal{J}).$$

By the definitions of π_i and $\mathcal{J}$, the function π is bijective. It can be shown that the jump system $\mathcal{J}$ and the bijective function π satisfy the desired condition $(*)$ in Theorem 3.1, where the details are omitted.

4 Polyhedral Structure of Neighbor Systems

We investigate the polyhedral structure of the convex closure of neighbor systems. For a set $\mathcal{F} \subseteq \mathbf{Z}^n$, we denote by $\mathrm{conv}(\mathcal{F})$ $(\subseteq \mathbf{R}^n)$ the convex closure (closed convex hull) of $\mathcal{F}$.

Theorem 4.1. *For every all-neighbor system $\mathcal{F} \subseteq \mathbf{Z}^n$, its convex closure $\mathrm{conv}(\mathcal{F})$ is an integral bisubmodular polyhedron.*

This is an extension of the known result that the convex closure of a jump system is an integral bisubmodular polyhedron [6]. It should be noted that Theorem 4.1 does not follow immediately from this fact and Theorem 3.1.

We also provide a characterization of neighbor systems by the property that the convex closure is a bisubmodular polyhedron.

Theorem 4.2. *A nonempty set $\mathcal{F} \subseteq \mathbf{Z}^n$ is an all-neighbor system if and only if for every vectors $\ell, u \in \mathbf{Z}^n$ satisfying $\ell \le u$ and $\mathcal{F} \cap [\ell, u] \ne \emptyset$, the convex closure $\mathrm{conv}(\mathcal{F} \cap [\ell, u])$ is an integral bisubmodular polyhedron.*

We firstly give an outline of the proof of Theorems 4.1. Let $\mathcal{F} \subseteq \mathbf{Z}^n$ be a neighbor system, and $\rho : 3^E \to \mathbf{R} \cup \{+\infty\}$ is a function defined by

$$\rho(X, Y) = \sup\{\textstyle\sum_{i \in X} x(i) - \sum_{i \in Y} x(i) \mid x \in \mathcal{F}\} \qquad ((X, Y) \in 3^E).$$

Note that $\rho(\emptyset, \emptyset) = 0$ and the value $\rho(X, Y)$ is integer if $\rho(X, Y) < +\infty$. To prove Theorem 4.1, it suffices to show that ρ is a bisubmodular function satisfying $\mathrm{conv}(\mathcal{F}) = \mathrm{P}_*(\rho)$; recall the definition of $\mathrm{P}_*(\rho)$ in (1).

Due to the page limit, we consider only the case where $\mathcal{F}$ is a finite set. Then, $\rho(X, Y) < +\infty$ holds for all $(X, Y) \in 3^E$. We give a key property to show Theorem 4.1, where the proof is omitted.

Lemma 4.1. *For every $(A, B) \in 3^E$ with $A \cup B = E$ and $k \ (\ge 1)$ subsets $V_1, V_2, \ldots, V_k$ of E with $V_1 \subset V_2 \subset \cdots \subset V_k$, there exists some $x \in \mathcal{F}$ such that $x(V_t \cap A, V_t \cap B) = \rho(V_t \cap A, V_t \cap B)$ for $t = 1, 2, \ldots, k$.*

To show the bisubmodularity of ρ, we use the following characterization.

Lemma 4.2 ([4, Theorem 2]). *A function* $\rho : 3^E \to \mathbf{R}$ *is bisubmodular if and only if* ρ *satisfies the following conditions:*

$$\rho(X \cap A, X \cap B) + \rho(Y \cap A, Y \cap B)$$
$$\geq \rho((X \cup Y) \cap A, (X \cup Y) \cap B) + \rho((X \cap Y) \cap A, (X \cap Y) \cap B)$$
$$(\forall (A, B) \in 3^E \text{ with } A \cup B = E, \forall X, Y \in 2^E), \quad (3)$$
$$\rho(X \cup \{i\}, Y) + \rho(X, Y \cup \{i\}) \geq 2\rho(X, Y)$$
$$(\forall (X, Y) \in 3^E, \ \forall i \in E \setminus (X \cup Y)). \quad (4)$$

Note that the condition (3) is equivalent to the submodularity of the function $\rho_{A,B} : 2^E \to \mathbf{R}$ defined by $\rho_{A,B}(X) = \rho(X \cap A, X \cap B)$ $(X \in 2^E)$. By using this characterization, we can prove that the function ρ is bisubmodular, where the details are omitted.

To show $\mathrm{conv}(\mathcal{F}) = \mathrm{P}_*(\rho)$, we use the following characterization of extreme points in a bounded bisubmodular polyhedron.

Lemma 4.3 ([12, Corollary 3.59]). *Let* $\rho : 3^E \to \mathbf{R}$ *be a bisubmodular function. A vector* $x \in \mathbf{R}^n$ *is an extreme point of* $\mathrm{P}_*(\rho)$ *if and only if there exist* $(A, B) \in 3^E$ *with* $A \cup B = E$ *and subsets* $V_0, V_1, \ldots, V_n$ *of* E *with* $\emptyset = V_0 \subset V_1 \subset \cdots \subset V_{n-1} \subset V_n = E$ *such that* $x(V_t \cap A, V_t \cap B) = \rho(V_t \cap A, V_t \cap B)$ *for* $t = 1, 2, \ldots, n$.

By the definition of $\mathrm{P}_*(\rho)$, we have $\mathcal{F} \subseteq \mathrm{P}_*(\rho)$, and therefore $\mathrm{conv}(\mathcal{F}) \subseteq \mathrm{P}_*(\rho)$ follows. To show the reverse inclusion, it suffices to show that every extreme point of $\mathrm{P}_*(\rho)$ is contained in $\mathcal{F}$, which follows from Lemmas 4.1 and 4.3. Hence, we have $\mathrm{conv}(\mathcal{F}) = \mathrm{P}_*(\rho)$.

We then prove Theorem 4.2. The "only if" part is immediate from Theorem 4.1 and Proposition 2.1 (iii). In the following, we prove the "if" part. Let $x, y \in \mathcal{F}$ and $i \in \mathrm{supp}(x - y)$. Assume, without loss of generality, that $x(i) < y(i)$.

We define the vectors $\ell, u \in \mathbf{Z}^E$ by $\ell(e) = \min\{x(e), y(e)\}$ and $u(e) = \max\{x(e), y(e)\}$ for $e \in E$. Since $x \in \mathcal{F} \cap [\ell, u]$, the set $\mathcal{F} \cap [\ell, u]$ is nonempty, and therefore its convex closure $S = \mathrm{conv}(\mathcal{F} \cap [\ell, u])$ is an integral bisubmodular polyhedron. By the definitions of ℓ and u, the vector x is an extreme point of S. We consider the tangent cone $\mathrm{TC}(x)$ of S at x, which is given by

$$\mathrm{TC}(x) = \{\alpha z \mid x \in \mathbf{R}^E, \ x + z \in S, \ \alpha \in \mathbf{R}, \ \alpha \geq 0\}.$$

Lemma 4.4. *There exists an extreme ray* $d \in \mathbf{R}^E$ *of* $\mathrm{TC}(x)$ *that is a positive multiple of either* $+\chi_i$, $+\chi_i + \chi_k$, *or* $+\chi_i - \chi_k$ *for some* $k \in E \setminus \{i\}$.

By Lemma 4.4, there exists an extreme ray $d \in \mathbf{R}^E$ of $\mathrm{TC}(x)$ that is a positive multiple of either $+\chi_i$, $+\chi_i + \chi_k$, or $+\chi_i - \chi_k$ for some $k \in E \setminus \{i\}$. Since d is an extreme ray of $\mathrm{TC}(x)$, there exists some vector $z_0 \in S$ such that z_0 is an extreme point of S and $z_0 - x$ is a positive multiple of d. Since $d(i) > 0$, we have $z_0(i) > x(i)$. The vector z_0 is contained in $\mathcal{F} \cap [\ell, u]$ since it is an extreme point of $S = \mathrm{conv}(\mathcal{F} \cap [\ell, u])$. This implies, in particular, $z_0 \in \mathcal{F}$ and z_0 is between x and y. Let α be the minimum positive number such that $\alpha z_0 + (1 - \alpha)x \in \mathcal{F}$, and put $z = \alpha z_0 + (1 - \alpha)x$. Then, z is a neighbor of x between x and y and satisfies $z(i) > x(i)$. This concludes the proof of Theorem 4.2.

5   Separable Convex Optimization on Neighbor Systems

We consider the problem (SC) of minimizing a separable convex function on a finite neighbor system $\mathcal{F}$. Define the *size* of $\mathcal{F}$ by $\Phi(\mathcal{F}) = \max_{e \in E}[\max_{x \in \mathcal{F}} x(e) - \min_{x \in \mathcal{F}} x(e)]$; $\Phi(\mathcal{F})$ is just the length of the longest edge in the bounding box of $\mathcal{F}$. We propose a greedy algorithm for the problem (SC) and show that it runs in pseudo-polynomial time, i.e., time polynomial in n and in $\Phi(\mathcal{F})$. We then show that if $\mathcal{F}$ is an N_k-neighbor system with a fixed k, then the problem (SC) can be solved in weakly polynomial time, i.e., time polynomial in n and in $\log \Phi(\mathcal{F})$.

It is assumed that we are given a membership oracle for $\mathcal{F}$, which enables us to check whether a given vector is contained in $\mathcal{F}$ or not in constant time. For simplicity, we mainly assume in this section that $\mathcal{F}$ is an N_k-neighbor system for some k; note that a finite all-neighbor system can be seen as an N_k-neighbor system with $k = \Phi(\mathcal{F})$.

5.1   Theorems

We show some useful properties in developing efficient algorithms for (SC). The next theorem shows that the optimality of a vector can be characterized by a local optimality.

Theorem 5.1. *A vector $x \in \mathcal{F}$ is an optimal solution of (SC) if and only if $f(x) \leq f(y)$ for every proper neighbor y of x.*

The next property shows that a given nonoptimal vector in $\mathcal{F}$ can be easily separated from an optimal solution.

Theorem 5.2. *Let $x \in \mathcal{F}$ be a vector which is not an optimal solution of (SC). Let $x' \equiv x + \alpha_*(p_* + q_*)$ be a proper neighbor of x in $\mathcal{F}$ such that $p_* \in U$, $q_* \in U \cup \{0\} \setminus \{+p_*, -p_*\}$, $\alpha_* \in \mathbf{Z}_{++}$, and $f(x') < f(x)$. Suppose that x' minimizes the value $\{f(x + \alpha_* p_*) - f(x)\}/\alpha_*$ among all such vectors. Then, there exists an optimal solution x_* of (SC) satisfying*

$$\begin{cases} x_*(i) \leq x(i) - \alpha_- & (\text{if } p_* = -\chi_i), \\ x_*(i) \geq x(i) + \alpha_+ & (\text{if } p_* = +\chi_i), \end{cases}$$

where $\alpha_- = \min\{x(i) - y(i) \mid y \in \mathcal{F}, \ y(i) < x(i)\}$ and $\alpha_+ = \min\{y(i) - x(i) \mid y \in \mathcal{F}, \ y(i) > x(i)\}$.

To prove Theorems 5.1 and 5.2, we show that the problem (SC) can be reduced to the problem (SC′) of minimizing a separable convex function on a jump system by using the relationship between neighbor systems and jump systems shown in Section 3.

We define vectors $\ell, u \in \mathbf{Z}^n$, a jump system $\mathcal{J} \subseteq \mathbf{Z}^n$, and a family of strictly increasing functions $\pi_i : [\ell(i), u(i)] \to \mathbf{Z}$ $(i \in E)$ as in Section 3. We define functions $g_i : [\ell(i), u(i)] \to \mathbf{R}$ $(i \in E)$ by

$$g_i(\alpha) = f_i(\pi_i(\alpha)) \qquad (\alpha \in [\ell(i), u(i)]).$$

Note that g_i is a convex function since f_i is a convex function. The problem (SC) for a neighbor system $\mathcal{F}$ can be reduced to the following problem:

$$\textbf{(SC')} \quad \text{Minimize } g(x) \equiv \sum_{i=1}^{n} g_i(x(i)) \text{ subject to } x \in \mathcal{J},$$

which is the minimization of a separable convex function g on a jump system $\mathcal{J}$.

For the problem (SC'), the following properties are known.

Theorem 5.3 (cf. [3, Cor. 4.2]). *A vector $x \in \mathcal{J}$ is an optimal solution of (SC') if and only if $g(x) \leq g(y)$ for all proper neighbors y of x in $\mathcal{J}$.*

Theorem 5.4 (cf. [20, Theorem 4.2]). *Let $x \in \mathcal{J}$ be a vector that is not an optimal solution of (SC'). Let $x' \equiv x + p_* + q_*$ be a proper neighbor of x in $\mathcal{J}$ such that $p_* \in U$, $q \in U \cup \{0\}$, and $g(x') < g(x)$, and suppose that x' minimizes the value $g(x+p_*)$ among all such vectors. Then, there exists an optimal solution $x_* \in \mathcal{J}$ of (SC') satisfying $x_*(i) \leq x(i) - 1$ if $p_* = -\chi_i$ and $x_*(i) \geq x(i) + 1$ if $p_* = +\chi_i$.*

Theorems 5.1 and 5.2 are just the restatement of Theorems 5.3 and 5.4 by using Theorem 3.1 and the equivalence between (SC) and (SC').

We then show that the check of local optimality in the sense of Theorem 5.1 and the computation of a proper neighbor x' in Theorem 5.2 can be done efficiently. The next theorem implies that the both operations can be done in $O(n^2 k)$ time.

Theorem 5.5. *Let $\mathcal{F} \subseteq \mathbf{Z}^n$ be an N_k-neighbor system. For $x \in \mathcal{F}$, all proper neighbors of x can be computed in $O(n^2 k)$ time.*

5.2 Pseudopolynomial-Time Algorithm

Based on Theorems 5.1 and 5.2, we propose a greedy algorithm for solving the problem (SC). The greedy algorithm maintains an interval $[a, b]$ containing an optimal solution of (SC). Note that $\mathcal{F} \cap [a, b]$ is a neighbor system by Proposition 2.1 (iii). The vectors a and b are updated by using Theorem 5.2 so that the value $\|b - a\|_1$ reduces in each iteration. Recall that $\mathcal{F}$ is assumed to be a finite N_k-neighbor system. We assume that an initial vector $x_0 \in \mathcal{F}$ is given.

Algorithm GREEDY
Step 0: Let $x := x_0 \in \mathcal{F}$. Set $a(e) := a_{\mathcal{F}}(e)$ and $b(e) := b_{\mathcal{F}}(e)$, where

$$a_{\mathcal{F}}(e) := \min\{x(e) \mid x \in \mathcal{F}\}, \quad b_{\mathcal{F}}(e) := \max\{x(e) \mid x \in \mathcal{F}\} \quad (e \in E). \quad (5)$$

Step 1: If $f(x) \leq f(y)$ for all proper neighbors y of x in $\mathcal{F} \cap [a, b]$, then stop (x is optimal).
Step 2: Let $x' \equiv x + \alpha_*(p_* + q_*)$ be a proper neighbor of x in $\mathcal{F} \cap [a, b]$ such that $p_* \in U$, $q_* \in U \cup \{0\} \setminus \{+p_*, -p_*\}$, $\alpha_* \in \mathbf{Z}_{++}$, and $f(x') < f(x)$, and suppose that x' minimizes the value $\{f(x + \alpha_* p_*) - f(x)\}/\alpha_*$ among all such vectors.
Step 3: Modify a or b as follows:

$$\begin{cases} b(i) := x(i) - \alpha_- \text{ (if } p_* = -\chi_i), \\ a(i) := x(i) + \alpha_+ \text{ (if } p_* = +\chi_i), \end{cases} \quad (6)$$

where α_-, α_+ are defined by

$$\begin{aligned}\alpha_- &= \min\{x(i) - y(i) \mid y \in \mathcal{F} \cap [a, b],\ y(i) < x(i)\},\\ \alpha_+ &= \min\{y(i) - x(i) \mid y \in \mathcal{F} \cap [a, b],\ y(i) > x(i)\}.\end{aligned} \qquad (7)$$

Set $x := x'$. Go to Step 1. □

We show the validity of the algorithm. By Theorem 5.1, the output x of the algorithm is a minimizer of the function f in the set $\mathcal{F} \cap [a, b]$. We see from Theorem 5.2 that the set $\mathcal{F} \cap [a, b]$ always contains an optimal solution of (SC). Hence, the output x of the algorithm is an optimal solution of (SC).

The time complextiy of the algorithm is given as follows.

Theorem 5.6. *The algorithm* GREEDY *finds an optimal solution of the problem* (SC). *The running time is* $O(n^3\,\Phi(\mathcal{F})^2)$ *if* $\mathcal{F}$ *is an all-neighbor system, and* $O(n^3\,k\,\Phi(\mathcal{F}))$ *if* $\mathcal{F}$ *is an* N_k-*neighbor system and the value* k *is given.*

5.3   Polynomial-Time Algorithm

We propose a faster algorithm for (SC) based on the domain reduction approach. The domain reduction approach is used in [19,20] to develop polynomial-time algorithms for discrete convex function minimization problems. We show that the proposed algorithm runs in weakly polynomial time if $\mathcal{F}$ is an N_k-neighbor system with a fixed k and the value k is known a priori.

Given an N_k-neighbor system $\mathcal{F} \subseteq \mathbf{Z}^n$, we define a set $\mathcal{F}^\bullet \subseteq \mathbf{Z}^n$ by $\mathcal{F}^\bullet = \mathcal{F} \cap [a_\mathcal{F}^\bullet, b_\mathcal{F}^\bullet]$, where $a_\mathcal{F}, b_\mathcal{F} \in \mathbf{Z}^n$ are defined by (5) and

$$a_\mathcal{F}^\bullet(e) = a_\mathcal{F}(e) + \left\lfloor \frac{b_\mathcal{F}(e) - a_\mathcal{F}(e)}{nk} \right\rfloor, \quad b_\mathcal{F}^\bullet(e) = b_\mathcal{F}(e) - \left\lfloor \frac{b_\mathcal{F}(e) - a_\mathcal{F}(e)}{nk} \right\rfloor \quad (e \in E).$$

The set $\mathcal{F}^\bullet$ has the following properties.

Theorem 5.7. *Let* $\mathcal{F} \subseteq \mathbf{Z}^n$ *be an* N_k-*neighbor system.*
(i) *The set* $\mathcal{F}^\bullet$ *is nonempty and hence an* N_k-*neighbor system.*
(ii) *A vector in* $\mathcal{F}^\bullet$ *can be found in* $O(n^3 k \log \Phi(\mathcal{F}))$ *time, provided a vector in* $\mathcal{F}$ *is given.*

The algorithm is as follows. Assume that an initial vector $x_0 \in \mathcal{F}$ is given.

Algorithm DOMAIN_REDUCTION
Step 0: Set $a := a_\mathcal{F}$ and $b := b_\mathcal{F}$.
Step 1: Find a vector $x \in (\mathcal{F} \cap [a, b])^\bullet$.
Step 2: If $f(x) \leq f(y)$ for all proper neighbors y of x in $\mathcal{F} \cap [a, b]$, then stop.
Step 3: Let $x' \equiv x + \alpha_*(p_* + q_*)$ be a proper neighbor of x in $\mathcal{F} \cap [a, b]$ satisfying the same condition as in Step 2 of GREEDY.
Step 4: Modify a or b by (6). Go to Step 1. □

The validity of this algorithm can be shown in a similar way as the algorithm GREEDY. In the analysis of the time complexity, the following property is the key to obtain a polynomial bound.

Lemma 5.1. *Let p_* be the vector chosen in Step 3 of the m-th iteration, and i is the unique element in* $\mathrm{supp}(p_*)$. *Then, we have* $b_{m+1}(i) - a_{m+1}(i) < (1 - 1/nk)(b_m(i) - a_m(i))$, *where a_t and b_t are the vectors a and b in the t-th iteration.*

Theorem 5.8. *The algorithm* DOMAIN_REDUCTION *finds an optimal solution of the problem* (SC) *in* $\mathrm{O}(n^5 k^2 (\log \Phi(\mathcal{F}))^2)$ *time if $\mathcal{F}$ is an N_k-neighbor system.*

References

1. Ando, K., Fujishige, S.: On structure of bisubmodular polyhedra. Math. Programming 74, 293–317 (1996)
2. Ando, K., Fujishige, S., Naitoh, T.: A greedy algorithm for minimizing a separable convex function over an integral bisubmodular polyhedron. J. Oper. Res. Soc. Japan 37, 188–196 (1994)
3. Ando, K., Fujishige, S., Naitoh, T.: A greedy algorithm for minimizing a separable convex function over a finite jump system. J. Oper. Res. Soc. Japan 38, 362–375 (1995)
4. Ando, K., Fujishige, S., Naitoh, T.: A characterization of bisubmodular functions. Discrete Math. 148, 299–303 (1996)
5. Bouchet, A.: Greedy algorithm and symmetric matroids. Math. Programming 38, 147–159 (1987)
6. Bouchet, A., Cunningham, W.H.: Delta-matroids, jump systems and bisubmodular polyhedra. SIAM J. Discrete Math. 8, 17–32 (1995)
7. Chandrasekaran, R., Kabadi, S.N.: Pseudomatroids. Discrete Math. 71, 205–217 (1988)
8. Cunningham, W.H.: Matching, matroids, and extensions. Math. Program. 91, 515–542 (2002)
9. Dunstan, F.D.J., Welsh, D.J.A.: A greedy algorithm for solving a certain class of linear programmes. Math. Programming 5, 338–353 (1973)
10. Edmonds, J.: Submodular functions, matroids, and certain polyhedra. Combinatorial Structures and their Applications, pp. 69–87. Gordon and Breach, New York (1970)
11. Edmonds, J.: Matroids and the greedy algorithm. Math. Programming 1, 127–136 (1971)
12. Fujishige, S.: Submodular Functions and Optimization, 2nd edn. Elsevier, Amsterdam (2005)
13. Groenevelt, H.: Two algorithms for maximizing a separable concave function over a polymatroid feasible region. European J. Operational Research 54, 227–236 (1991)
14. Hartvigsen, D.: Neighbor system and the greedy algorithm (extended abstract), RIMS Kôkyûroku Bessatsu (to appear, 2010)
15. Hochbaum, D.S.: Lower and upper bounds for the allocation problem and other nonlinear optimization problems. Math. Oper. Res. 19, 390–409 (1994)
16. Kobayashi, Y., Szabo, J., Takazawa, K.: A proof to Cunningham's conjecture on restricted subgraphs and jump systems. TR-2010-04. Egervary Research Group, Budapest (2010)
17. Kobayashi, Y., Takazawa, K.: Even factors, jump systems, and discrete convexity. J. Combin. Theory, Ser. B 99, 139–161 (2009)

18. Lovász, L.: The member ship problem in jump systems. J. Combin. Theory, Ser. B 70, 45–66 (1997)
19. Shioura, A.: Minimization of an M-convex function. Discrete Appl. Math. 84, 215–220 (1998)
20. Shioura, A., Tanaka, K.: Polynomial-time algorithms for linear and convex optimization on jump systems. SIAM J. Discrete Math. 21, 504–522 (2007)
21. Whitney, H.: On the abstract properties of linear dependence. Amer. J. Math. 57, 509–533 (1935)

Generating Trees on Multisets

Bingbing Zhuang and Hiroshi Nagamochi

Graduate School of Informatics, Kyoto University
{zbb,nag}@amp.i.kyoto-u.ac.jp

Abstract. Given a multiset $M = V_1 \cup V_2 \cup \cdots \cup V_C$ of n elements and a capacity function $\Delta : [1, C] \rightarrow [2, n-1]$, we consider the problem of enumerating all unrooted trees T on M such that the degree of each vertex $v \in V_i$ is bounded from above by $\Delta(i)$. The problem has a direct application of enumerating isomers of tree-like chemical graphs. We give an algorithm that generates all such trees without duplication in $O(1)$-time delay per output in the worst case using $O(n)$ space, with $O(n)$ initial preprocessing time.

1 Introduction

The problem of enumerating (i.e., listing) all graphs with bounded size is one of the most fundamental and important issues in graph theory. Many algorithms for particular classes of graphs have been studied [3, 7, 8, 13].

One of the common ideas behind efficient enumeration algorithms (e.g., [9, 10, 12]) is to define a unique representation for each graph in a graph class as its "parent," which induces a rooted tree that connects all graphs in the class, called the *family tree* $\mathcal{F}$, where each node in $\mathcal{F}$ corresponds to a graph in the class. Then all graphs in the class will be enumerated one by one according to the depth-first traversal of the family tree $\mathcal{F}$.

Our research group has been developing algorithms for enumerating chemical graphs that satisfy given various constraints [4,5,6]. The pioneering work for enumerating chemical graphs performed by Caley [1] was dedicated to enumerating structural isomers of alkanes, and a century later several studies based on computational methods followed [2]. We have designed efficient branch-and-bound algorithms for enumerating tree-like chemical graphs [4, 6], which are based on the tree enumeration algorithm [10], and implementations of these algorithms are available on our web server[1].

Several algorithms to generate all trees with n vertices without repetition have been already known. One of the best algorithms runs in time proportional to the number of trees, i.e., the time delay is $O(1)$ on average [13]. Nakano and Uno [11] gave an $O(1)$-time delay algorithm to generate all rooted unordered trees with exactly n vertices and a given diameter d without repetition, where they use "left-heavy trees" as canonical forms of trees. In our companion paper [16], we show that, given a number n of vertices and a capacity function $\Delta \geq 2$, all

---

[1] http://sunflower.kuicr.kyoto-u.ac.jp/tools/enumol/

O. Cheong, K.-Y. Chwa, and K. Park (Eds.): ISAAC 2010, Part I, LNCS 6506, pp. 182–193, 2010.

unrooted trees T with exactly n vertices such that the degree of each vertex is at most Δ can be generated in $O(1)$-time delay per output in the worst case using $O(n)$ space. The main difficulty for generating unrooted trees under a capacity constraint is to keep a center of the trees as the root of rooted trees, and we have developed a novel technique to maintain the center of unrooted trees to define a parent-child relationship among all the trees.

However, for applications to chemical graph enumerations, tree-like chemical graphs wherein each kind of atoms in a chemical graph may have a different valence are modeled as colored trees such that each vertex receives a color and the degree of the vertex is bounded from above by a capacity that depends on the color. For example, alkane isomers C_nH_{2n+2} can be regarded as unrooted trees with exactly n carbon atoms (neglecting hydrogen atoms) such that the degree of each vertex is at most four. However, the known algorithm for generating colored trees [12] enumerates *rooted trees* with *at most n* vertices, where neither the number of vertices with a specific color nor the degree of vertices can be controlled by the algorithm. One of the difficulties in extending the result in [12] to a problem of generating unrooted trees on a multiset, where the number of vertices with each color is fixed, is that the representation (encoded label sequences) of colored trees in the previous result cannot uniquely determine "center" of such a colored tree. To overcome the difficulty, we restrict construction of colored trees only based on the tree structure ignoring assigned colors before we make use of information of colors to determine "canonical form" of unrooted trees. This new idea enables us to solve the problem just by applying the idea of "left-heavy" representation twice, yielding an $O(1)$-time delay enumeration algorithm for unrooted colored trees on a given multiset under a prescribed degree constraint. The output will be the character sequences of the trees. This means that tree-like chemical graphs on a specified set of atoms can be generated in $O(1)$ time per each. The proofs omitted due to the space limitation can be found in a full version of the paper [17], and also the details of the main algorithm ENUMERATE.

2   Preliminaries

For two sequences A and B over a set of elements for which a total order is defined, let $A > B$ mean that A is lexicographically larger then B, and let $A \geq B$ mean that $A > B$ or $A = B$. Let $A \sqsupset B$ mean that B is a prefix of A and $A \neq B$, and let $A \gg B$ mean that $A > B$ but B is not a prefix of A. Let $A \sqsupseteq B$ mean that $A \sqsupset B$ or $A = B$, i.e., B is a prefix of A.

A graph stands for a simple undirected graph, which is denoted by a pair $G = (V, E)$ of a vertex set V and an edge set E. The set of vertices and the set of edges of a given graph G are denoted by $V(G)$ and $E(G)$, respectively. The *degree deg(v; G)* of a vertex v in a graph G is the number of neighbours of v in G. A path is a sequence of distinct vertices $(v_0, v_1, \ldots, v_k)$ such that (v_{i-1}, v_i) is an edge for $i = 1, 2, \ldots, k$. The *length* of a path is the number of edges in the path. The distance between a pair of vertices u and v is the minimum length of a path between u and v. The *diameter* of G is the maximum distance between two vertices in G.

Unrooted Trees. A tree (unrooted tree) is a connected graph without cycles. For two vertices u and v in a tree, let $P_T(u, v)$ be the unique path that connects u and v in T. In an unrooted tree, there are at most two vertices the maximum distance from which to other vertices is minimized. If such a vertex v is unique (i.e., the diameter of T is even), then we call v the *center* of T, and define the depth of a vertex u to be the distance from u to the center. On the other hand, if there are two such vertices v and v' (i.e., the diameter of T is odd), then we call the (v, v') the *center* of T, and define the *depth* $dep(u; T)$ of a vertex u to be the distance from u to the endvertices of the center, i.e., the length of the path from u to the center (v, v') including the edge (v, v').

A multiset M of n elements is denoted by a disjoint union $V_1 \cup V_2 \cup \cdots \cup V_C$ of sets of colored vertices, where V_i is the set of vertices of color $i \in [1, C]$. Let $col(v) \in [1, C]$ denote the color assigned to a vertex v. Let $\mathcal{T}(M)$ denote the set of all unrooted trees with n vertices such that exactly $|V_i|$ vertices are colored with color i. Let $\Delta : [1, C] \rightarrow [2, n-1]$ be a capacity function, where $\Delta(i)$ is the maximum degree of any vertex of color i. An unrooted tree $T \in \mathcal{T}(M)$ is called Δ-*bounded* if

$$deg(v; T) \leq \Delta(col(v)) \text{ for all } v \in V(T). \tag{1}$$

Let $\mathcal{T}(M, \Delta)$ denote the set of all Δ-bounded unrooted trees on a multiset M.

In this paper, we show the following result.

Theorem 1. *For a given multiset* $M = V_1 \cup V_2 \cup \cdots \cup V_C$ *of* $n \geq 3$ *elements and a capacity function* $\Delta : [1, C] \rightarrow [2, n-1]$, *all unrooted trees in* $\mathcal{T}(M, \Delta)$ *can be generated in* $O(1)$-*time delay in the worst case using* $O(n)$ *space after an* $O(n)$-*time initialization.*

Let $\mathcal{T}_{odd}(M, \Delta)$ (resp., $\mathcal{T}_{even}(M, \Delta)$) denote the set of all unrooted trees in $\mathcal{T}(M, \Delta)$ with an odd (resp., even) diameter. We treat an unrooted tree T with an odd diameter as a tree T' centered at a dummy vertex r^*, which is inserted in the center edge of T, where $col(r^*) = C + 1$ and $\Delta(C + 1) = 2$. In the rest of the paper, we focus on the problem of generating unrooted trees in $\mathcal{T}_{even}(M, \Delta)$. The argument for generating trees in $\mathcal{T}_{even}(M, \Delta)$ can be modified easily so that a specified vertex r^* is always used as the center, implying that we can also generate trees in $\mathcal{T}_{odd}(M, \Delta)$ in the same time complexity.

Unordered Rooted Trees. We represent unrooted trees as "rooted trees." A *rooted* tree is a tree with one vertex r designated as its root. If $P_T(r, v)$ has exactly k edges then we say that the *depth* $dep(v; T)$ of v is k. The *parent* of $v \neq r$ is its neighbour on $P_T(r, v)$, and the ancestors of $v \neq r$ are the vertices on $P_T(r, v)$. The parent of the root r and the ancestors of r are not defined. We say that if v is the parent of u then u is a child of v, and if v is an ancestor of u then u is a descendant of v. A *leaf* is a vertex that has no child. Note that $P_T(r, v)$ denotes the set of all ancestors of a vertex v in a rooted tree T, where $v \in P_T(r, v)$.

Now we show how to convert the problem of generating unrooted trees in $\mathcal{T}_{even}(M, \Delta)$ to a problem of generating rooted trees in some classes. Given a capacity function Δ, let us call a rooted tree T Δ-*bounded* if it satisfies (1).

We call a rooted tree T *centered* if T has an even diameter and r is the center of T, i.e., there are two children c_1 and c_2 of the root r such that the subtrees T_i at c_i, $i = 1, 2$ attain $dep(T_1) = dep(T_2) = dep(T) - 1$. Let $\mathcal{RT}(M, \Delta)$ denote the set of all Δ-bounded centered trees. In what follows, we consider how to generate rooted trees in $\mathcal{RT}(M, \Delta)$.

Ordered Trees. Rooted trees are then represented as "ordered trees." An *ordered tree* (o-tree, for short) is a rooted tree with a left-to-right ordering specified for the children of each vertex. For an o-tree T and a vertex in T, let $T(v)$ denote the ordered subtree induced from T by the set of v and descendants of v, preserving the left-to-right ordering for the children of each vertex. Fig. 1 shows three ordered trees T_1, T_2 and T_3 of the same rooted tree.

For an o-tree T', a leaf v in T' is called the *leftmost* (resp., *rightmost*) leaf if v is a descendant of the leftmost (resp., rightmost) child of any ancestor of v in T'. Let $lml(T')$ (resp., $rml(T')$) denote the leftmost (resp., rightmost) leaf in an o-tree T'. See Fig. 1(c).

Let T be an o-tree with n vertices, and $(v_1, v_2, \ldots, v_n)$ be the list of the vertices of T in preorder, i.e., vertices are indexed in the order of DFS. For two vertices $u = v_i$ and $v = v_j$, we write $u >_T v$ if $i < j$. Consider two vertices $u = v_i$ and $v = v_j$, $i \leq j$ in T. Let $[u, v]$ denote the set of all vertices $v_{i'}$ with $i \leq i' \leq j$, and let $T[u, v]$ denote the graph induced from T by the vertex set $[u, v]$. Also let $lca(u, v)$ denote the least common ancestor of u and v in T. Let $lca_L(u, v)$ denote the child w of $lca(u, v)$ such that $w \in P_T(r, u)$, where we let $lca_L(u, v) = u$ if $lca(u, v) = u$. Similarly, $lca_R(u, v)$ denotes the child w' of $lca(u, v)$ such that $w' \in P_T(r, v)$, where we let $lca_R(u, v) = v$ if $lca(u, v) = v$. We denote the children of the root r in an o-tree by $c_1, c_2, \ldots, c_p$ from left to right. Let $\mathcal{OT}(T)$ denote the set of all o-trees obtained from a rooted tree T.

3   Representation for Trees on a Multiset

This section defines a "canonical representation" of trees on a multiset based on "left-heavy trees," a special type of unordered trees, and treats "semi-paths," a path-like canonical tree separately from "non-semi-paths."

Left-heavy Trees. Since all o-trees in $\mathcal{OT}(T)$ of the same tree T are isomorphic, we choose a particular o-tree as the representative of T. For this, we use "left-heavy trees" [11]. For an o-tree T, we define the *depth sequence* $L^d(T)$ to be

$$L^d(T) = [dep(v_1; T), dep(v_2; T), \ldots, dep(v_n; T)].$$

If $L^d(T_1) > L^d(T_2)$ for two ordered trees T_1 and T_2, then we say that $L^d(T_1)$ is *heavier* than $L^d(T_2)$. An o-tree $T_1 \in \mathcal{OT}(T)$ of a rooted tree T is called *left-heavy* tree if $L^d(T_1) \geq L^d(T_2)$ holds for all o-trees $T_2 \in \mathcal{OT}(T)$. For two vertices v_i and v_j, $i \leq j$, let $L_{i,j}^d(T) = [dep(v_i; T), dep(v_{i+1}; T), \ldots, dep(v_j; T)]$. It is known that left-heavy trees can be characterized as follows.

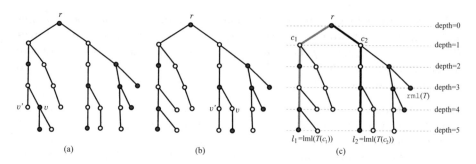

Fig. 1. (a) An ordered tree T_1 on a multiset $M = V_1 \cup V_2$ which is not left-heavy; (b) a left-heavy tree T_2 on M which is not canonical, where the white vertices belongs to V_1 and the black vertices to V_2; (c) a canonical tree T_2 on M

Lemma 1. [11] *An o-tree $T \in \mathcal{OT}(T')$ is the left-heavy tree of a rooted tree T' if and only if, for a non-root vertex v and its immediate right sibling v' of v (if any) in T, it holds $L^{\mathsf{d}}(T(v)) \geq L^{\mathsf{d}}(T(v'))$.*

See Fig. 1, where T_1 is not left-heavy since $L^{\mathsf{d}}(T(v)) < L^{\mathsf{d}}(T(v'))$ holds in T_1 whereas T_2 and T_3 are left-heavy among the three ordered trees.

   For each rooted tree T, a left-heavy tree in $\mathcal{OT}(T)$ is unique up to the iso-morphism with respect to the root. In what follows, we assume that unordered rooted trees are represented by left-heavy trees.

   By definition of left-heavy trees, we can easily observe that the following inequality on depth also holds.

Lemma 2. *For a non-root vertex v and its immediate right sibling v' of v (if any) in a left-heavy tree T, it holds $dep(T(v)) \geq dep(T(v'))$.*

In particular, $dep(T(c_1)) = dep(T(c_2)) \geq \cdots \geq dep(T(c_p))$ holds for the children $c_1, c_2, \ldots, c_p$ of the root r in a left-heavy and centered tree T. We call a left-heavy and centered T *distinguished* if, for each $i = 1, 2$, the number of leaves with the maximum depth in $T(c_i)$ is 1 (i.e., no other leaf than $\mathrm{lml}(T(c_i))$ attains $dep(T(c_i))$).

   For the leftmost and second leftmost children c_1 and c_2 of the root r in a left-heavy tree T, let ℓ_i, $i = 1, 2$ denote the leftmost leaf of the subtree $T(c_i)$ rooted at c_i. We call each vertex in $P_T(r, \ell_1) \cup P_T(r, \ell_2)$ a *core vertex*. A left-heavy tree is called a *semi-path* if it has at most two non-core vertices.

   We consider how to add a new leaf along the rightmost path $P_T(r, \mathrm{rml}(T))$ of a left-heavy tree T so that the resulting o-tree remains left-heavy. This problem has been solved by Uno and Nakano [10]. We here use another solution "competi-tors" proposed in our companion paper [14, 15], since "competitors" are easier to handle the case where some left part of a left-heavy tree may change.

   A vertex u in a left-heavy tree T is called *valid* if the o-tree obtained from T by appending a new vertex v at u as the rightmost child of u remains left-heavy.

Let v be a vertex in an o-tree T. For a descendant v_i of v in T, we define the *pre-sequence* $\mathrm{ps}^{\mathrm{d}}(v, v_i)$ of v_i to v to be

$$\mathrm{ps}^{\mathrm{d}}(v, v_i) = [dep(v_k; T), dep(v_{k+1}; T), \ldots, dep(v_{i-1}; T)]$$

for the child v_k of v such that v_k is an ancestor of v_i. For a vertex v_i and a vertex v_j with $j < i$ incomparable v_i, we call v_j *pre-identical* to v_i if $\mathrm{ps}^{\mathrm{d}}(v, v_j) = \mathrm{ps}^{\mathrm{d}}(v, v_i)$ holds, and $\mathrm{lca}_{\mathrm{L}}(v_j, v_i)$ is the immediate left sibling of $\mathrm{lca}_{\mathrm{R}}(v_j, v_i)$ for $v = \mathrm{lca}(v_j, v_i)$ [14,15]. We define the *competitor* of a vertex v_i to be the vertex v_j pre-identical to v_i which has the smallest index j $(< i)$ among all vertices pre-identical to v_i. A vertex v_i has no competitor if no vertex v_j, $j < i$ is pre-identical to v_i.

Lemma 3. [14, 15] *Let T be a left-heavy tree, let $u_0, u_1, \ldots, u_q$ $(= \mathrm{rml}(T))$ denote the rightmost path of T. Then there is an index h^* such that a vertex u_i is valid if and only if $0 \leq i \leq h^*$. Moreover such an index h^* is determined as follows.*
(i) *u_q has no competitor: Then $h^* = q$.*
(ii) *u_q has a competitor v_j: Let v_h be the parent of the vertex v_{j+1} next to v_j in T. Then $h^* = dep(v_h; T)$.*

Let us call such a vertex v_{h^*} the *lowest valid ancestor* of u_q in T. By maintaining vertices $\{v_1, v_2, \ldots, v_n\}$ in an array and the current tree T in a linked data structure, we can compute v_{h^*} from u_q in $O(1)$ time.

We can also compute competitors efficiently.

Lemma 4. [14,15] *we can determine the competitor of a new vertex v in $O(1)$ time per operation of appending a new leaf.*

Canonical Trees. The main new idea for attaining an $O(1)$-time delay enumeration algorithm for colored unrooted trees is how to represent these trees so that the root retains as the center of an unrooted tree while the "canonical representation" is preserved in a parent-child relationship among all trees.

For a rooted colored tree T, let $\mathcal{LH}(T)$ denote the set of ordered trees T' of T such that the tree T' ignoring the colors is a left-heavy tree. For an o-tree T, we define the *depth-color sequence* $L^{\mathrm{c}}(T)$ to be

$$L^{\mathrm{c}}(T) = [dep(v_1; T), col(v_1), dep(v_2; T), col(v_2), \ldots, dep(v_n; T), col(v_n)].$$

A left-heavy tree $T' \in \mathcal{LH}(T)$ is called *canonical* if $L^{\mathrm{c}}(T')$ is lexicographically maximal among all left-heavy trees in $\mathcal{LH}(T)$. We easily observe the next characterization of canonical trees.

Lemma 5. *An o-tree $T \in \mathcal{OT}(T')$ is the canonical tree of a rooted tree T' if and only if, for a non-root vertex v and its immediate right sibling v' of v (if any) in T, it holds either "$L^{\mathrm{d}}(T(v)) > L^{\mathrm{d}}(T(v'))$" or "$L^{\mathrm{d}}(T(v)) = L^{\mathrm{d}}(T(v'))$ and $L^{\mathrm{c}}(T(v)) \geq L^{\mathrm{c}}(T(v'))$."*

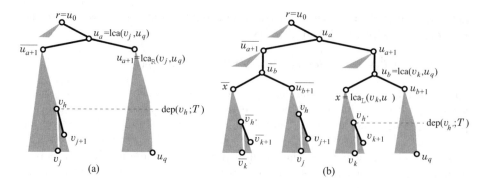

Fig. 2. (a) Competitor v_j of u_q; (b) c-competitor v_k of u_q

See Fig. 1, where the left-heavy tree T_2 is not canonical since $L^d(T(v)) = L^d(T(v'))$ and $L^c(T(v)) < L^c(T(v'))$ hold in T_2 whereas T_3 is canonical among the three ordered trees.

Given an o-tree T and color a, let $T + (v, a)$ denote the o-tree obtained from T by appending a new vertex x with $col(x) = a$ to a vertex v in T as the rightmost child of v. A pair (u, a) of a vertex u in a canonical tree T and a color a is called *c-valid* if $T + (u, a)$ remains canonical. We also call a vertex u in T *c-valid* if it admits a c-valid pair (u, a) for some color a.

We consider how to add a new leaf along the rightmost path $P_T(r, \text{rml}(T))$ of a canonical tree T so that the resulting o-tree remains canonical. We again use the idea of competitors for characterizing c-valid vertices.

Let v be a vertex in an o-tree T. For a descendant v_i of v in T, we define the *pre-c-sequence* $\text{ps}^c(v, v_i)$ of v_i to v to be $L^c_{k,i-1}(T)$ for the child v_k of v such that v_k is an ancestor of v_i, i.e.,

$$[dep(v_k; T), col(v_k), dep(v_{k+1}; T), col(v_{k+1}), \ldots, dep(v_{i-1}; T), col(v_{i-1})].$$

For a vertex v_i and a vertex v_j with $j < i$ incomparable v_i, we call v_j *c-pre-identical* to v_i if $\text{ps}^c(v, v_j) = \text{ps}^c(v, v_i)$ holds, and $\text{lca}_L(v_j, v_i)$ is the immediate left sibling of $\text{lca}_R(v_j, v_i)$ for $v = \text{lca}(v_j, v_i)$. We define the *c-competitor* of a vertex v_i to be the vertex v_j c-pre-identical to v_i which has the smallest index $j \, (< i)$ among all vertices c-pre-identical to v_i. A vertex v_i has no c-competitor if no vertex v_j, $j < i$ is c-pre-identical to v_i. All c-valid pairs can be determined as follows.

Lemma 6. *Let T be a canonical tree, and let $u_0, u_1, \ldots, u_q \, (= \text{rml}(T))$ denote the rightmost path of T. Then there is a pair of an index $\tilde{h}$ and a color $\tilde{a}$ such that a pair of a vertex u_i and color $a \in [1, C]$ is c-valid if and only if "$0 \le i < \tilde{h}$" or "$i = \tilde{h}$ and $a \le \tilde{a}$". Moreover such an index $\tilde{h}$ and a color $\tilde{a}$ are determined as follows.*

(i) u_q has none of a competitor and a c-competitor: Then $\tilde{h} = q$ and $\tilde{a} = C$.

(ii) u_q has a competitor v_j but no c-competitor: Let v_h be the parent of the vertex v_{j+1} next to v_j in T. Then $\tilde{h} = dep(v_h; T)$ and $\tilde{a} = C$.

(iii) u_q *has a competitor* v_j *and a c-competitor* v_k*: Let* v_h *be the parent of the vertex* v_{j+1} *next to* v_j *in* T*, and let* $v_{h'}$ *be the parent of the vertex* v_{k+1} *next to* v_k *in* T*, where* $dep(v_h; T) \leq dep(v_{h'}; T)$ *holds (see Fig. 2). Then* $\tilde{h} = dep(v_h; T)$ *holds. It holds* $\tilde{a} = col(v_{h'})$ *if* $dep(v_h; T) = dep(v_{h'}; T)$ *and the vertex* v_{j+2} *next to* v_{j+1} *belongs to subtree* $T(\mathrm{lca}_{\mathrm{R}}(v_j, u_q))$*; it holds* $\tilde{a} = C$ *otherwise.*

We call the c-valid pair $(u_{\tilde{h}}, \tilde{a})$ in the lemma the *critical pair* of T. Analogously with the case of computation of valid vertices, we can compute the critical pair $(u_{\tilde{h}}, \tilde{a})$ from u_q in $O(1)$ time.

Given a multiset M and a capacity function Δ, we generate each Δ-bounded centered tree in $\mathcal{RT}(M, \Delta)$ in the form of a canonical tree. Let $\mathcal{RT}^1(M, \Delta)$ (resp., $\mathcal{RT}^2(M, \Delta)$) denote the set of canonical trees $T \in \mathcal{RT}(M, \Delta)$ such that T is a semi-path (resp., a non-semi-path). For generating semi-paths in $O(1)$ time delay, we design a procedure for generating all permutations of M with an additional property. We omit describing our algorithm for $\mathcal{RT}^1(M, \Delta)$ due to space limitation. In the next section, we show how to generate all canonical trees in $\mathcal{RT}^2(M, \Delta)$.

4   Generating Non-semi-paths

Parent-trees of non-semi-paths. In this section, we define the "parent-tree" of each non-semi-path T in the class $\mathcal{RT}^2(M, \Delta)$. For ease of applications of the properties on left-heavy trees, we also introduce the class $\mathcal{RT}'(M, \Delta)$ (resp., $\mathcal{RT}''(M, \Delta)$) of canonical trees on subsets $M' \subseteq M$ with $|M'| = |M| - 1$ (resp., $M'' \subseteq M$ with $|M''| = |M| - 1$) and define the "parent-tree" of each tree in $\mathcal{RT}'(M, \Delta) \cup \mathcal{RT}''(M, \Delta)$ so that the parent-child relationship over these classes forms a family tree $\mathcal{F}$. We will design an algorithm that visits all nodes in family tree $\mathcal{F}$ each in $O(1)$-time. However, we output only trees in $\mathcal{RT}(M, \Delta)$ during the traversal of $\mathcal{F}$.

We define the parent-tree of a canonical tree $T \in \mathcal{RT}^2(M, \Delta) \cup \mathcal{RT}'(M, \Delta) \cup \mathcal{RT}''(M, \Delta)$ as follows. Let $v_{\mathtt{last}}$ denote the non-core vertex with the largest preorder index in T.

(1) $T \in \mathcal{RT}^2(M, \Delta) \cup \mathcal{RT}'(M, \Delta)$: The *parent-tree* $\mathcal{P}(T)$ of T is defined to be the o-tree $T - v_{\mathtt{last}}$ obtained from T by removing $v_{\mathtt{last}}$. For example, the parent-tree of T_1 with n vertices (reps., T_2 with $n - 1$ vertices) is T_2 (resp., T_3) in Fig. 3. The inequalities in Lemma 1 still hold in $T - v_{\mathtt{last}}$, and hence $\mathcal{P}(T) = T - v_{\mathtt{last}}$ remains canonical. Clearly $\mathcal{P}(T) = T - v_{\mathtt{last}}$ remains Δ-bounded.

(2) $T \in \mathcal{RT}''(M, \Delta)$: T is obtained from a tree $T' \in \mathcal{RT}^2(M, \Delta)$ by removing two vertices v and v' by definition. Then the *parent-tree* $\mathcal{P}(T)$ of T is defined to be the o-tree obtained from T by appending leaves y and x with $col(y) = \max\{col(v), col(v')\}$ and $col(x) = \min\{col(v), col(v')\}$ to ℓ_1 and ℓ_2. For example, the parent-tree of T_3 with $n - 2$ vertices is T_4 in Fig. 3. The inequalities in Lemma 1 still hold in $\mathcal{P}(T) = (T + (\ell_1, v)) + (\ell_2, v')$, since only the leftmost paths in $T(c_1)$ and $T(c_2)$ extend in the resulting tree. Hence $T - v_{\mathtt{last}}$ remains canonical. Also $\mathcal{P}(T) = (T + (\ell_1, col(y))) + (\ell_2, col(x))$ remains Δ-bounded since $\Delta(i) \geq 2$ for any color i.

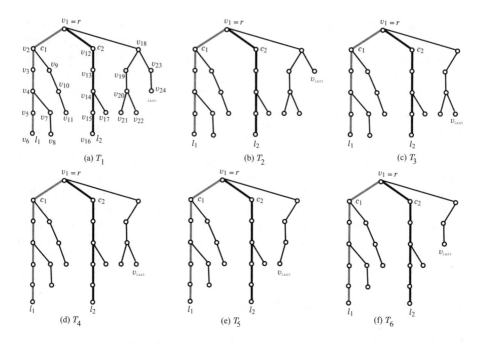

Fig. 3. Examples of canonical trees for $n = 24$, where $T_1, T_4 \in \mathcal{RT}(M, \Delta)$, $T_2, T_5 \in \mathcal{RT}'(M, \Delta)$, and $T_3, T_6 \in \mathcal{RT}''(M, \Delta)$, and T_{i+1} is the parent-tree of T_i, $i = 1, 2, \ldots, 5$

Lemma 7. *For each canonical tree $T \in \mathcal{RT}^2(M, \Delta) \cup \mathcal{RT}'(M, \Delta) \cup \mathcal{RT}''(M, \Delta)$, the parent-tree $\mathcal{P}(T)$ is a Δ-bounded canonical tree which belongs to $\mathcal{RT}^2(M, \Delta) \cup \mathcal{RT}'(M, \Delta) \cup \mathcal{RT}''(M, \Delta)$.*

Child-trees of non-semi-paths. A canonical tree T' is called a *child-tree* of a canonical tree $T \in \mathcal{RT}^2(M, \Delta) \cup \mathcal{RT}'(M, \Delta) \cup \mathcal{RT}''(M, \Delta)$ if T is the parent-tree of T', where T' may not be Δ-bounded. A vertex v in T is called *unsaturated* if $deg(v; T) < \Delta(dep(v; T))$. In what follows, we first characterize the set of all child-tree of a canonical tree $T \in \mathcal{RT}^2(M, \Delta) \cup \mathcal{RT}'(M, \Delta) \cup \mathcal{RT}''(M, \Delta)$. Next we describe an entire algorithm ENUMERATE for enumerating all canonical trees $T \in \mathcal{RT}^2(M, \Delta) \cup \mathcal{RT}'(M, \Delta) \cup \mathcal{RT}''(M, \Delta)$ by a recursive procedure GEN of generating all child-trees of a given canonical tree $T \in \mathcal{RT}^2(M, \Delta) \cup \mathcal{RT}'(M, \Delta) \cup \mathcal{RT}''(M, \Delta)$.

Appending a Leaf to $T \in \mathcal{RT}'(M, \Delta) \cup \mathcal{RT}''(M, \Delta)$. Let $T \in \mathcal{RT}'(M, \Delta) \cup \mathcal{RT}''(M, \Delta)$. By definition of parent-trees, any child-tree T' of T has n or $n - 1$ vertices, and T is obtained from T' by removing the non-core vertex $v_{\text{last}}(T')$ with the largest index in T'. Recall that $T \in \mathcal{RT}'(M, \Delta)$ (resp., $T \in \mathcal{RT}''(M, \Delta)$) uses one vertex (resp., two vertices) less than a tree in $\mathcal{RT}'(M, \Delta)$ uses all vertices in M. Let col_T be the color i of a vertex $v \in V_i$ not used in $T \in \mathcal{RT}'(M, \Delta)$, and $col_T^1 \geq col_T^2$ be the colors of vertices $v \in M$ not used in $T \in \mathcal{RT}'(M, \Delta)$. Thus, T' is obtained from T by appending a new vertex u with $col(u) = a$ to a

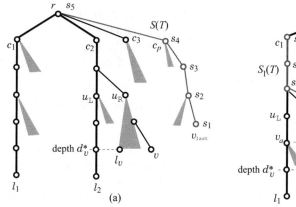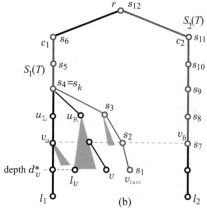

Fig. 4. (a) Spine $S(T) = \{s_1, s_2, \ldots, s_5\}$ in a tree T with $v_{\text{last}} \notin V(T(c_1))$; (b) spine $S(T) = S_1(T) \cup S_2(T) = \{s_1, s_2, \ldots, s_{11}\}$ in a tree T with $v_{\text{last}} \in V(T(c_1))$

vertex v in T so that $T + (u, a)$ is canonical and Δ-bounded, where $a = col_T$ if $T \in \mathcal{RT}'(M, \Delta)$, and $i \in \{col_T^1, col_T^2\}$ if $T \in \mathcal{RT}''(M, \Delta)$. The condition when $T + (u, a)$ is canonical is given as (i) u is a valid vertex in T; (ii) (u, a) is a c-valid pair in T; and (iii) the new vertex v is the non-core vertex $v_{\text{last}}(T + (u, a))$ with the largest index in $T + (u, a)$.

We define the *spine* $S(T)$ of a canonical tree $T \in \mathcal{RT}'(M, \Delta) \cup \mathcal{RT}''(M, \Delta)$ as the set of vertices u at which appending a new leaf (u, v) results in an o-tree T' such that v is the non-core vertex $v_{\text{last}}(T')$ with the largest index in T'. Let $spn(v)$ denote the parent $v' = s_{i+1} \in S(T)$ of a vertex $v = s_i \in S(T)$, where $spn(r) = \emptyset$.

By definition of $S(T)$, we observe the next.

Lemma 8. [16] *For a canonical tree $T \in \mathcal{RT}'(M, \Delta)$, let T' be the o-tree obtained by appending a new leaf (u, v). Then vertex v is the non-core vertex $v_{\text{last}}(T')$ with the largest index in T' if and only if u belongs to $S(T)$.*

We can find all valid vertices in $S(T)$ by identifying the critical pair $(\tilde{h} = s_i, \tilde{a})$ by Lemma 6.

Shortening Depth of $T \in \mathcal{RT}(M, \Delta)$. Let $T \in \mathcal{RT}^2(M, \Delta)$ be a canonical tree. By definition of parent-trees, T with exactly n vertices has at most one child-tree, which is given by the o-tree $T - \{\ell_1, \ell_2\}$ obtained by removing the leaves ℓ_1 and ℓ_2. Recall that the parent-tree of any tree $T \in \mathcal{RT}''(M, \Delta)$ with respect to n is distinguished. In fact, for a distinguished canonical tree $T \in \mathcal{RT}(M, \Delta)$, $T - \{\ell_1, \ell_2\}$ is a child-tree of T if and only if $T - \{\ell_1, \ell_2\}$ is canonical. We show how to examine the canonicality of $T - \{\ell_1, \ell_2\}$ in $O(1)$ time. Assume that $T \in \mathcal{RT}(M, \Delta)$ is distinguished, since otherwise $T - \{\ell_1, \ell_2\}$ cannot be a child-tree of T.

If $T' = T - \{\ell_1, \ell_2\}$ is not canonical, then $L^{\mathrm{d}}(T'(u_{\mathrm{L}})) < L^{\mathrm{d}}(T'(u_{\mathrm{R}}))$ holds for a core vertex u_{L} in $T(c_i)$ ($i = 1$ or 2) and its right sibling u_{R}. Note that $L^{\mathrm{d}}(T'(u_{\mathrm{L}}))$ is obtained from the current sequence $L^{\mathrm{d}}(T(u_{\mathrm{L}}))$ by eliminating the last entry $dep(\ell_i; T)$. If $T' = T - \{\ell_1, \ell_2\}$ remains distinguished, then $L^{\mathrm{d}}(T'(u_{\mathrm{L}})) > L^{\mathrm{d}}(T'(u_{\mathrm{R}}))$ holds for each core vertex u_{L} in $T(c_i)$, $i = 1, 2$ and its right sibling u_{R}, implying that $T' = T - \{\ell_1, \ell_2\}$ remains canonical too.

Lemma 9. [16] *Let $T \in \mathcal{RT}(M, \Delta)$ be a distinguished canonical tree. If $T' = T - \{\ell_1, \ell_2\}$ remains distinguished, then T' is canonical.*

We next show how to check whether T can have a non-distinguished child-tree, i.e., whether the o-tree $T - \{\ell_1, \ell_2\}$ remains canonical or not. We here give only a sketch of our method due to space limitation. If $T - \{\ell_1, \ell_2\}$ is not canonical, then there is a pair of a core vertex u and its immediate right sibling u' such that "$L^{\mathrm{d}}(T(u)) < L^{\mathrm{d}}(T(u'))$" or "$L^{\mathrm{d}}(T(u)) = L^{\mathrm{d}}(T(u'))$ and $L^{\mathrm{c}}(T(u)) < L^{\mathrm{c}}(T(u'))$" by Lemma 5. Consider the case of $u \in V(T(c_1))$. For each vertex v in $T(u')$, we compute the comparison of $L^{\mathrm{d}}(T(u))$ and $L^{\mathrm{d}}(T(u'))$, one of the states, $L^{\mathrm{d}}(T(u)) \sqsupset L^{\mathrm{d}}(T(u'))$, $L^{\mathrm{d}}(T(u)) = L^{\mathrm{d}}(T(u'))$, $L^{\mathrm{d}}(T(u)) \gg L^{\mathrm{d}}(T(u'))$, and $L^{\mathrm{d}}(T(u)) < L^{\mathrm{d}}(T(u'))$. We can compute such states of all vertices in $O(1)$ time per each if we compute the state of v based on the state of the previous vertex in the preorder. We remark that the above subtree $T(u)$ is not the subtree $T'(u)$ of the current tree T' when a vertex v is newly appended, i.e., the above subtree $T(u)$ is obtained from $T'(u)$ by deleting the core vertices which depth is larger than the maximum depth of a vertex in $T(u')$. Based on such states state$_1(v)$, $v \in V(T) - \{r\}$, we can determine whether $L^{\mathrm{d}}(T(u)) \geq L^{\mathrm{d}}(T(u'))$ still holds for all core vertices in $T(c_1)$ of $T - \{\ell_1, \ell_2\}$ by defining state$_2(v)$, $v \in V(T) - \{r\} - V(T(c_1))$ analogously. To detect whether there is a core vertex u in $T(c_1)$ (resp., $T(c_2)$) such that "$L^{\mathrm{d}}(T(u)) = L^{\mathrm{d}}(T(u'))$ and $L^{\mathrm{c}}(T(u)) < L^{\mathrm{c}}(T(u'))$," we define state $c - $state$_1(v)$ (resp., $c - $state$_2(v)$) of a vertex v in a similar manner with state based depth-color sequences L^{c} instead of L^{d}.

The Entire Algorithm. We are ready to obtain an entire algorithm ENU-MERATE for generating all canonical trees in $\mathcal{RT}(M, \Delta)$. Algorithm ENUMERATE constructs semi-paths in $\mathcal{RT}(M, \Delta)$ before it invokes a procedure GEN(T) that recursively generates all descendants of T.

5   Conclusion

To generate tree-like chemical graphs, we consider trees induced by non-hydrogen atoms, which are represented as vertices. For those vertices with degree less than the valence of the atoms in such a tree, we only need to attach hydrogen atoms to them. Hence all tree-like chemical graphs on a certain set of atoms can be generated in $O(1)$ time per output.

References

1. Cayley, A.: On the analytic forms called trees with applications to the theory of chemical combinations. Rep. Brit. Assoc. Adv. Sci. 45, 257–305 (1875)
2. Balaban, T.S., Filip, P.A., Ivanciuc, O.: Computer generation of acyclic graphs based on local vertex invariants and topological indices, derived canonical labeling and coding of trees and alkanes. J. Math. Chem. 11, 79–105 (1992)
3. Beyer, T., Hedetniemi, S.M.: Constant time generation of rooted trees. SIAM J. Computing 9, 706–712 (1980)
4. Fujiwara, H., Wang, J., Zhao, L., Nagamochi, H., Akutsu, T.: Enumerating tree-like chemical graphs with given path frequency. J. Chem. Inf. Mod. 48, 1345–1357 (2008)
5. Imada, T., Ota, S., Nagamochi, H., Akutsu, T.: Enumerating stereoisomers of tree structured molecules using dynamic programming. In: Dong, Y., Du, D.-Z., Ibarra, O. (eds.) ISAAC 2009. LNCS, vol. 5878, pp. 14–23. Springer, Heidelberg (2009)
6. Ishida, Y., Zhao, L., Nagamochi, H., Akutsu, T.: Improved algorithm for enumerating tree-like chemical graphs. In: Ishida, Y., Zhao, L., Nagamochi, H., Akutsu, T. (eds.) GIW 2008. Genome Informatics, vol. 21, pp. 53–64 (2008)
7. Li, G., Ruskey, F.: The advantage of forward thinking in generating rooted and free trees. In: SODA 1999, pp. 939–940 (1999)
8. McKay, B.D.: Isomorph-free exhaustive generation. J. of Algorithms 26, 306–324 (1998)
9. Nakano, S.: Efficient generation of triconnected plane triangulations. Computational Geometry Theory and Applications 27(2), 109–122 (2004)
10. Nakano, S., Uno, T.: Efficient generation of rooted trees, NII Technical Report, NII-2003-005 (2003)
11. Nakano, S., Uno, T.: Constant time generation of trees with specified diameter. In: Hromkovič, J., Nagl, M., Westfechtel, B. (eds.) WG 2004. LNCS, vol. 3353, pp. 33–45. Springer, Heidelberg (2004)
12. Nakano, S., Uno, T.: Generating colored trees. In: Kratsch, D. (ed.) WG 2005. LNCS, vol. 3787, pp. 249–260. Springer, Heidelberg (2005)
13. Wright, R.A., Richmond, B., Odlyzko, A., McKay, B.D.: Constant time generation of free trees. SIAM J. Comput. 15, 540–548 (1986)
14. Zhuang, B., Nagamochi, H.: Enumerating rooted graphs with reflectional block structures, Dept. of Applied Mathematics and Physics, Graduate School of Informatics, Kyoto University, Technical Report 2009-019 (2009), http://www-or.amp.i.kyoto-u.ac.jp/members/nag/Technical_report/TR2009-019.pdf
15. Zhuang, B., Nagamochi, H.: Enumerating rooted graphs with reflectional block structures. In: Calamoneri, T., Diaz, J. (eds.) CIAC 2010. LNCS, vol. 6078, pp. 49–60. Springer, Heidelberg (2010)
16. Zhuang, B., Nagamochi, H.: Constant time generation of trees with degree bounds, Dept. of Applied Mathematics and Physics, Kyoto University, Technical Report 2010-006 (2010), http://www-or.amp.i.kyoto-u.ac.jp/members/nag/Technical-report/TR2010-006.pdf
17. Zhuang, B., Nagamochi, H.: Generating Trees on Multisets, Dept. of Applied Mathematics and Physics, Kyoto University, Technical Report 2010-009 (2010), http://www-or.amp.i.kyoto-u.ac.jp/members/nag/Technical_report/TR2010-009.pdf

Seidel Minor, Permutation Graphs and Combinatorial Properties*

Vincent Limouzy

LIMOS - Univ. Blaise Pascal, Clermont-Ferrand, France
limouzy@isima.fr

Abstract. A permutation graph is an intersection graph of segments lying between two parallel lines. A Seidel complementation of a finite graph at a vertex v consists in complementing the edges between the neighborhood and the non-neighborhood of v. Two graphs are *Seidel complement equivalent* if one can be obtained from the other by a sequence of *Seidel complementations*.

In this paper we introduce the new concept of *Seidel complementation* and *Seidel minor*. We show that this operation preserves cographs and the structure of modular decomposition.

The main contribution of this paper is to provide a new and succinct characterization of permutation graphs namely, a graph is a permutation graph if and only if it does not contain any of the following graphs: C_5, C_7, XF_6^2, XF_5^{2n+3}, $C_{2n}, n \geqslant 6$ and their complements as a Seidel minor. This characterization is in a sense similar to Kuratowski's characterization [Kur30] of planar graphs by forbidden topological minors.

Keywords: Permutation graph; Seidel complementation; Modular decomposition.

1 Introduction

Graph classes are frequently characterized by a list of forbidden induced subgraphs. For instance such a characterizations are known for cographs, interval graphs, chordal graphs and many others. However, it is not always convenient to deal with this kind of characterizations and the list of forbidden subgraphs can be quite large. Some characterizations rely on the use of local operators such as minors, local complementation or Seidel switch. Certainly, Kuratowksi's characterization of planar graphs by forbidden topological minors is one of the most famous [Kur30].

A nice characterization of circle graphs, *i.e.* the intersection graphs of chords in a circle, was given by Bouchet [Bou94], using an operation called *local complementation*. This operation consists in complementing the graph induced by

---

* This work was carried out in part when the author was a postdoctoral fellow at Dept. of Computer Science at the University of Toronto and at the Caesarea Rothchild Institute at the University of Haifa, Israel.

O. Cheong, K.-Y. Chwa, and K. Park (Eds.): ISAAC 2010, Part I, LNCS 6506, pp. 194–205, 2010.

the neighborhood of a vertex. His characterization states that a graph is a circle graph if and only if it does not contain W_5, W_7 and BW_3[1] as vertex minor. This operation has strong connections with a graph decomposition called *rank-width*, this relationship is presented in the work of Oum [Oum05a, Oum05b]. Another example of a local operator is the Seidel switch. The Seidel switch is a graph operator introduced by Seidel in his seminal paper [Sei76]. A Seidel switch in a graph consists in complementing the edges between a subset of vertices S and its complement $V \setminus S$.

Seidel switch has been intensively studied since its introduction; Colbourn *et al.* [CC80] proved that deciding whether two graphs are Seidel switch equivalent is ISO-Complete. The Seidel switch has also applications in graph coloring [Kra03]. Other interesting applications of Seidel switch concerns structural graph properties [Hay96, Her99]. It has also been used by Rotem and Urrutia [RU82] to show that the recognition of circular permutation graphs (CPG for short) can be polynomially reduced to the recognition of permutation graphs. Years later, Sritharan [Sri96] presented a nice and efficient algorithm to recognize CPGs in linear time. Once again it is a reduction to permutation graph recognition, and it relies on the use of a Seidel switch. Montgolfier *et al.* [MR05a, MR05b] used it to characterize graphs completely decomposable *w.r.t.* Bi-join decomposition. Seidel switch is not only relevant to the study of graphs. Ehrenfeucht *et al.* [EHR99] showed the interest of this operation for the study of 2-structures and recently, Bui-Xuan *et al.* extended these results to broader structures called Homogeneous relations [BXHLM07, BXHLM08].

We present in this paper a novel characterization of the well known class of permutation graphs, *i.e.*, the intersection graphs of segments lying between two parallel lines. Permutation graphs were introduced by Even, Lempel and Pnueli [PLE71, EPL72]. They established that a graph is a permutation graph if and only if the graph and its complement are *transitively orientable*. They also gave a polynomial time procedure to find a transitive orientation when it is possible. A linear time algorithm recognition algorithm is presented in [MS99].

For that we introduce a new local operator called Seidel complementation. In few words, the Seidel complementation on an undirected graph at a vertex v consists in complementing the edges between the neighborhood and the non-neighborhood of v. A schema of Seidel complementation is depicted in Figure 1. Our notion of Seidel complementation is a combination of local complementation and Seidel switch. The use of a vertex as *pivot* comes from local complementation, and the transformation from Seidel switch.

This results constitutes, in a sense, an improvement compared to Gallai's characterization of permutation graphs by forbidden induced subgraphs which counts no less than 18 finite graphs, and 14 infinite families [Gal67].

Thanks to this operator and the corresponding minor, we obtain a compact list of Seidel minor obstructions for permutation graphs.

---

[1] W_5 (*resp.* W_7) is the wheel on five (*resp.* seven) vertices, *i.e.* a chordless cycle vertices plus a dominating vertex, and BW_3 is a wheel on three vertices where the cycle is subdivided.

The main result of this paper is a new characterization of permutation graphs. We show that a graph is a permutation graph if and only if it does not contain any of the following graphs C_5, C_7, XF_6^2, XF_5^{2n+3}, $C_{2n}, n \geqslant 6$ or their complements as Seidel minors.

The proof is based on a study of the relationships between Seidel complementation and modular decomposition. We show that any Seidel complementation of a prime graph *w.r.t.* modular decomposition is a prime graph. As a consequence we get that cographs are stable under Seidel complementation. We also present a complete characterization of equivalent cographs, which leads to a linear time algorithm for verifying Seidel complement equivalence of cographs.

The paper is organized as follows. In section 2 we present the definitions of Seidel complementation and Seidel minor. Then we show some structural properties of Seidel complementation and we introduce the definitions and notations used in the sequel of the paper. In section 3 we show the relationships between Seidel complementation and modular decomposition, namely we prove that Seidel complementation preserves the structure of modular decomposition of a graph. Finally we show that cographs are closed under this relation. Section 4 is devoted to prove the main theorem, namely a graph is a permutation graph if and only if it does not contain any of the forbidden Seidel minors.

Due to lack of space several proofs and figures are omitted, however, interested readers can find more details in [Lim09].

2   Definitions and Notations

In this paper only undirected, finite, loop-less and simple graphs are considered. We present here some notations used in the paper. The graph induced by a subset of vertices X is noted $G[X]$. For a vertex v, $N(v)$ denotes the neighborhood of v, and $\overline{N}(v)$ represents the non-neighborhood. Sometimes we need to use a refinement of the neighborhood on a subset of vertices X, noted $N_X(v) = N(v) \cap X$. Let A and B be two disjoint subsets of V, and let $E[A, B] = \{ab \in E : a \in A \text{ and } b \in B\}$ be the set of edges between A and B. For two sets A and B, let $A \Delta B = (A \setminus B) \cup (B \setminus A)$.

Definition 1 (Seidel complement). *Let $G = (V, E)$ be a graph, and let v be a vertex of V. The Seidel complement at v on G (see Figure 1), denoted $G * v$ is defined as follows:*
Swap the edges and the non-edges between $G[N(v)]$ and $G[\overline{N}(v)]$, namely

$$G * v = (V, E \Delta \{xy : vx \in E, vy \notin E\})$$

From the previous definition it is straightforward to notice that $G * v * v = G$.

Proposition 1. *Let G be a graph. If vw is an edge of G, then $G * v * w * v = G * w * v * w$. This operation is denoted $G \star vw$. This proposition remains true even if vw is not an edge.*

Remark 1. *One can remark that the ⋆ operation merely exchanges the vertices v and w without modifying the graph $G[V \setminus \{v\} \cup \{w\}]$.*

Definition 2 (Seidel Minor). *Let $G = (V, E)$ and $H = (V', E')$ be two graphs. H is a Seidel minor of G (noted $H \leqslant_S G$) if H can be obtained from G by a sequence of the following operations:*

- *Perform a Seidel complementation at a vertex v of G,*
- *Delete a vertex of G.*

Definition 3 (Seidel Equivalent Graphs). *Let $G = (V, E)$ and $H = (V, F)$ be two graphs. G and H are said to be Seidel equivalent if and only if there exists a word ω defined on V^* such that $G * \omega \cong H$.*

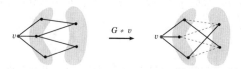

Fig. 1. An illustration of the Seidel complement concept

Lemma 1. *Let G and H be two graphs, G and H are Seidel complement equivalent if and only if they are at distance at most 1.*

At first glance Seidel complementation seems to be just a particular case of Seidel switch, but after a careful examination, one can see that they are not comparable.

3   Modular Decomposition and Cographs

In this section we investigate the relationships between Seidel complementation and modular decomposition. This study is relevant in order to prove the main result. Actually a permutation graph is uniquely representable if and only if it is prime *w.r.t.* to modular decomposition. And one of the results of this section is to prove that if a graph is prime *w.r.t.* modular decomposition this property is preserved by Seidel complementation. As a consequence for permutation graphs, it means that if the graph is uniquely representable so are their Seidel complement equivalent graphs.

Let us now briefly recall the definition of module. A module in a graph is subset of vertices M such that any vertex outside M is either completely connected to M or is completely disjoint from M. Modular decomposition is a decomposition of graph introduced by Gallai [Gal67]. The modular decomposition of a graph G is the decomposition of G into its modules. Without going too deeply into the details, there exists for each graph a unique modular decomposition tree, and it is possible to compute it in linear time (*cf.* [TCHP08]).

In the sequel of this section we show that if G is prime, *i.e.* not decomposable, *w.r.t.* modular decomposition, then applying a Seidel complementation at any vertex of the graph preserves this property. Then we prove that the family of cographs is closed under Seidel minor. And finally show how the modular decomposition tree of a graph is modified by a Seidel complementation.

3.1   Modular Decomposition

Theorem 2. *Let $G = (V, E)$ be graph, and let v be an arbitrary vertex of G. G is prime w.r.t. modular decomposition if and only if $G * v$ is prime w.r.t. to modular decomposition.*

Corollary 1. *Let $G = (V, E)$ and let v be a vertex. And let M be a module of G such that v does not belong to M, then M is also a module in $G * v$.*

3.2   Cographs

Cographs are the graphs which are completely decomposable *w.r.t.* modular decomposition. There exist several characterizations of cographs (see [CPS85]), one of them is given by a forbidden induced subgraph, *i.e.* cographs are the graphs without P_4 –a chordless path on four vertices– as induced subgraph. Another fundamental property of cograph is the fact that its modular decomposition tree –called its co-tree– has only series (1) and parallel (0) nodes as internal nodes. An example of a cograph and its associated co-tree is given in Figure 2(a). A co-tree is a rooted tree, where the leaves represent the vertices of the graph, and the internal nodes of the co-tree encode the adjacency of the vertices of the graph. Two vertices are adjacent iff their Least Common Ancestor[2] (LCA) is a series node (1). Conversely two vertices are disconnected iff their LCA is a parallel node (0). The following theorem shows that the class of cographs is closed under Seidel complementation.

Theorem 3. *Let $G = (V, E)$ be a cograph, and v a vertex of G, then $G * v$ is also a cograph.*

Proof. Let T be the co-tree of G. The Seidel complementation at a vertex v is obtained as follows: Let T' be the tree obtained by $T * v$. $P(v)$, the former parent node of v, becomes the new root of T', and now the parent of v in T' is the former root, namely $R(T)$. In other words by performing a Seidel complementation we have reversed the path from $P(v)$ to $R(T)$.

It is easy to see that $G[N(v)]$ and $G[\overline{N}(v)]$ are not modified. Now to see that the adjacency between $G[N(v)]$ and $G[\overline{N}(v)]$ is reversed, it is sufficient to remark that for two vertices u and w, u belonging to the neighborhood of v and w belonging to the non-neighborhood of v. If u and w are adjacent in G it means that their LCA is a series node. We note that this node lies on the path from v

---

[2] The LCA of two leaves x and y is first node in common on the paths from the leaves to the root.

to $R(T)$. After proceeding to a Seidel complementation their LCA is modified and it is now a parallel node, consequently reversing the adjacency between the neighborhood and the non-neighborhood.

An example of the Seidel complement of the co-tree is given in Figure 2(b).

Remark 4 (Exchange property). *Actually a Seidel complementation on a cograph, or more precisely on its co-tree is equivalent of exchanging the root of the co-tree with the vertex v used to proceed to the Seidel complement,i¿½i.e. the vertex v is attached to the former root of the co-tree and the new root is the former parent of the vertex v.*

Except for this transformation, the other parts of the co-tree remain unchanged, i.e. the number and the types of internal nodes are preserved, and no internal nodes are merged.

(a)   (b)

Fig. 2. (a) An example of a cograph on 5 vertices and its respective co-tree. (b) A schema of a Seidel complement at a vertex v on a co-tree.

Proposition 2. *The Seidel complementation of a cograph on its co-tree can be performed in $O(1)$-time.*

3.3   Modular Decomposition Tree

In this section we will show how the modular decomposition tree of a graphs is modified. Using Theorems 2, 3 and 1.

Let $G = (V, E)$ be a graph, and let $T(G)$ (T for short) be its modular decomposition tree. Modular decomposition tree is a generalization of the cotree for cographs. The only difference with cotree is that the modular decomposition tree can contain prime nodes. Prime nodes corresponds to graphs that are not decomposable *w.r.t.* modular decomposition.

We generalize the operation on the co-tree, described in Theorem 3, to arbitrary modular decomposition tree.

Theorem 5. *Let $G = (V, E)$ be a graph, and let T be its modular decomposition tree. Let v be a vertex of G. By applying a Seidel complement at v the modular decomposition tree of $T * v$ of $G * v$ is obtained by:*

- *performing a Seidel complement in every prime node lying on the path from v to $R(T)$.*
- *making $P(v)$ the root of $T * v$.*
- *Reverse the path from $P(v)$ to $R(T)$: if α and β are prime node in T with $\beta = P(\alpha)$ then $\alpha = P(\beta)$ and β is connected in place of the subtree coming from v.*

4   Permutation Graphs

In this section we show that the class of permutation graphs is closed under Seidel minor, and we prove the main theorem that states that a graph is a permutation graph if and only if it does not contain any of the following graphs: C_5, C_7, XF_6^2, XF_5^{2n+3}, $C_{2n}, n \geqslant 6$ or their complements as Seidel minor.

Definition 4 (Permutation graph). *A graph $G = (V, E)$ is a permutation graph if there exist two permutations σ_1, σ_2 on $V = \{1, \ldots, n\}$, such that two vertices u, v of V are adjacent iff $\sigma_1(u) < \sigma_1(v)$ and $\sigma_2(v) < \sigma_2(u)$. $R = \{\sigma_1, \sigma_2\}$ is called a representation of G, and $G(R)$ is the permutation graph represented by R.*

More properties of permutation can be found in [Gol04]. An example of a permutation graph is presented in Figure 3(b).

Theorem 6 (Gallai'67 [Gal67]). *A permutation graph is uniquely representable iff it is prime w.r.t. modular decomposition.*

Theorem 7 ([Gal67][3]). *A graph is a permutation graph if and only if it does not contain one of the finite graphs as induced subgraphs T_2, X_2, X_3, X_{30}, X_{31}, X_{32}, X_{33}, X_{34}, X_{36} nor their complements and does not contain the graphs given by the infinite families: XF_1^{2n+3}, XF_5^{2n+3}, XF_6^{2n+2}, XF_2^{n+1}, XF_3^n, XF_4^n, the Holes, and their complements.*

Operation S: Let $\sigma = A . v . B$ be a permutations on $[n]$. Let v be an element of $[n]$. The operation S at an element v of $[n]$ noted $\sigma * v$ is done Let $\sigma * v = B . v . A$.

Remark 8. *Let $R = \{\sigma_1, \sigma_2\}$ be a permutation representation of a permutation graph G then $R * v = \{\sigma_1 * v, \sigma_2 * v\}$ is a permutation representation of a graph H.*

Theorem 9. *Let $G = (V, E)$ be a permutation graph, and let v be a vertex of G, and let $R = \{\sigma_1, \sigma_2\}$ be the permutation representation of G. We have $G(R * v) = G * v$.*

Corollary 2. *The Seidel complementation at a vertex v of a permutation graph can be achieved in $O(1)$-time.*

An arbitrary remark. To perform a Seidel complementation at a vertex on a graph can require in the worst case $O(n^2)$-time. It suffices to consider the graph consisting of a star $K_{1,n}$ and a stable S_n, whose size is $2n+1$ with $n+1$ connected components. Applying a Seidel complementation on the vertex of degree n results in a connected graph with $O(n^2)$ edges.

---

[3] http://wwwteo.informatik.uni-rostock.de/isgci/classes/AUTO_3080.html

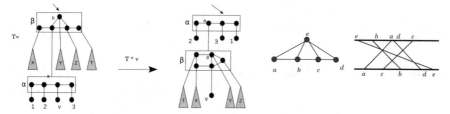

(a) Seidel complementation on a modular decomposition tree.

(b) A permutation graph and its representation

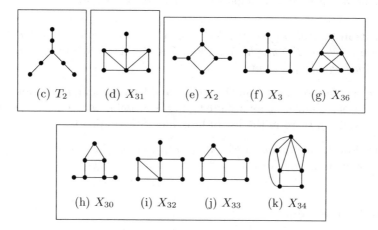

(c) T_2 (d) X_{31} (e) X_2 (f) X_3 (g) X_{36}

(h) X_{30} (i) X_{32} (j) X_{33} (k) X_{34}

Fig. 3. Finite forbidden induced subgraphs for permutation graphs. Those in the same box are Seidel complement equivalent.

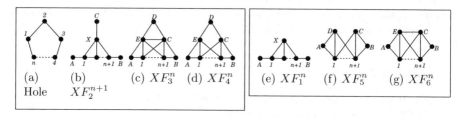

(a) Hole   (b) XF_2^{n+1} (c) XF_3^n (d) XF_4^n (e) XF_1^n (f) XF_5^n (g) XF_6^n

Fig. 4. Forbidden infinite families for permutation graphs. The families in the left box (a)-(d) contains asteroidal triples. The families in the right box (e)-(g) do not contain asteroidal triple, the key point is the parity of the dashed path.

4.1   Finite Families

In this section we show that it is possible to reduce the list of forbidden induced subgraphs by using Seidel Complementation. Actually a lot of forbidden subgraphs are Seidel equivalent. The graphs that are Seidel complement equivalent are in the same box in Figure 3. Thus, the list of finite forbidden graphs is reduced from 18 induced subgraphs to only 6 finite Seidel minors. The forbidden Seidel minors are C_5, C_7, XF_6^2 and their complements.

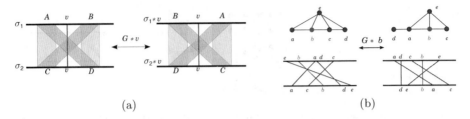

Fig. 5. (a) Schematic view of Seidel complementation on a permutation diagram. (b) An example of a permutation graph and a Seidel complementation at a vertex b.

Proposition 3. *The graphs X_3, X_2, X_{36} (cf. Figure 3(e)-(g)) are Seidel complement equivalent. The graphs X_{30}, X_{32}, X_{33} and X_{34} (cf. Figure 3(h)-(k)) are Seidel complement equivalent.*

The following propositions show that two forbidden finite graphs contain actually an instance of a member of an infinite family as Seidel minor. Thus it is no longer necessary to keep them in the list of forbidden Seidel minors.

Proposition 4. *The graph XF_4^0 is a Seidel minor of T_2 and X_{31}. The graph C_6 is a Seidel minor of XF_4^0.*

4.2   Infinite Families

We show in this section that actually forbidden infinite families under the relation on induced subgraphs are redundant when the Seidel minor operation is considered. Consequently the following propositions allows us to reduce from 14 infinite families with the induced subgraph relation to only 4 infinite families under Seidel minor relation. The forbidden families are XF_5^{2n+3} and $C_{2n}, n \geqslant 6$ and their complements.

Proposition 5

- The Hole is a Seidel minor of XF_3^n, XF_4^n and XF_2^{n+1}.
- XF_5^{2n+1} is a Seidel minor of XF_6^{2n+2}.
- XF_5^{2n+1} is a Seidel minor of XF_1^{2n+3}.
- XF_5^{2n+1} is a Seidel minor of C_{2n+3}.

4.3   Main Theorem

Definition 5 (Seidel Complement Stable). *A graph $G = (V, E)$ is said to be Seidel complement stable if: $\forall v \in V : G \cong G * v$*

Few small graphs are Seidel complement stable, for instance, P_4, C_5, and more trivially K_n the clique on n vertices and S_n the stable on n vertices.

Lemma 2. *The graph XF_5^n is Seidel complement stable.*

Lemma 3. *The Seidel stable class of the hole C_n consists of C_n, XF_4^{n-6}.*

Due to lack of space the proof is omitted, but in a few words, it relies on the "regular" structure of XF_5^n and Lemma 2.

Theorem 10 (Main Theorem). *A graph is a permutation graph if and only if it does not contain as finite graphs C_5, C_7 and XF_6^2 and their complements and as infinite families XF_5^{2n+3} and $C_{2n}, n \geqslant 6$ and their complements as Seidel minor.*

Proof (Sketch of Proof). This theorem relies on Gallai's result (*cf.* Theorem 7). If G is not a permutation graph then it contains one of the graphs listed in Theorem 7 as an induced subgraph. Thanks to previous propositions 3-4 concerning the finite families, and proposition 5concerning the infinite families. We are able to reduce these induced subgraphs into a smaller set of graphs which are now forbidden Seidel minor. It remains to prove that this list is minimal. Concerning the infinite families, Lemma 2 proves that it is not possible to get rid of this families since it is Seidel stable. Concerning even holes (since odd holes are dismissed because they contain XF_5^{2n-1} as Seidel minors) Lemma 3 says that it is not possible to get rid of them. The same kind of argument holds for the finite graphs.

Corollary 3. *The class of permutation graphs is not well quasi ordered under Seidel minor relation.*

5   Conclusion and Perspectives

We have shown that the new paradigm of Seidel minor provides a nice and compact characterization of permutation graphs. A lot of questions remain open. A natural question lies in the fact that Theorem 10 is obtained using Gallai's result on forbidden induced subgraphs. Is it possible to give a direct proof of Theorem 10 without using Gallai's result? Another direction concerns graph decomposition. Oum [Oum05a] has shown that local complementation preserves rank-width. Is there a graph decomposition that is preserved by Seidel complementation? Finally, it could be interesting to generalize the Seidel complement operator to directed graphs.

We hope that this Seidel minor will be relevant in the future as a tool to study graph decomposition and to provide similar characterizations, as the one presented for permutation graphs, to other graph classes.

Acknowledgment

The author wish to thank Derek G. Corneil and Martin C. Golumbic for their help on the paper. The author also thank the anonymous referee for helping to improve the paper.

References

[Bou94]   Bouchet, A.: Circle graph obstructions. Journal of Combinatorial The-
               ory, Series B 60(1), 107–144 (1994)
[BXHLM07]   Bui-Xuan, B.-M., Habib, M., Limouzy, V., de Montgolfier, F.: Unifying
               two graph decompositions with modular decomposition. In: Tokuyama,
               T. (ed.) ISAAC 2007. LNCS, vol. 4835, pp. 52–64. Springer, Heidelberg
               (2007)
[BXHLM08]   Bui-Xuan, B.-M., Habib, M., Limouzy, V., de Montgolfier, F.: A new
               tractable combinatorial decomposition (2008) (submitted)
[CC80]   Colbourn, C.J., Corneil, D.G.: On deciding switching equivalence of
               graphs. Discrete Applied Mathematics 2(3), 181–184 (1980)
[CPS85]   Corneil, D.G., Perl, Y., Stewart, L.: A linear recognition algorithm for
               cographs. SIAM Journal on Computing 14(4), 926–934 (1985)
[EHR99]   Ehrenfeucht, A., Harju, T., Rozenberg, G.: Theory of 2-Structures: A
               Framework for Decomposition and Transformation of Graphs. World
               Scientific, Singapore (1999)
[EPL72]   Even, S., Pnueli, A., Lempel, A.: Permutation graphs and transitive
               graphs. Journal of ACM 19(3), 400–410 (1972)
[Gal67]   Gallai, T.: Transitiv orientierbare Graphen. Acta Mathematica
               Academiae Scientiarum Hungaricae 18, 25–66 (1967)
[Gol04]   Golumbic, M.C.: Algorithmic graph theory and perfect graphs. In:
               Annals of Discrete Mathematics, 2nd edn., vol. 57, p. 314. Elsevier, Am-
               sterdam (2004)
[Hay96]   Hayward, R.B.: Recognizing P_3-structure: A switching approach. Jour-
               nal of Combinatorial Theory, Series B 66(2), 247–262 (1996)
[Her99]   Hertz, A.: On perfect switching classes. Discrete Applied Mathemat-
               ics 94(1-3), 3–7 (1999)
[Kra03]   Kratochvíl, J.: Complexity of Hypergraph Coloring and Seidel's Switch-
               ing. In: Bodlaender, H.L. (ed.) WG 2003. LNCS, vol. 2880, pp. 297–308.
               Springer, Heidelberg (2003)
[Kur30]   Kuratowski, C.: Sur le problème des courbes gauches en topologie. Fun-
               damenta Mathematicae 15, 271–283 (1930)
[Lim09]   Limouzy, V.: Seidel complementation, combinatorial properties (2009)
               (in preparation), http://arxiv.org/abs/0904.1923
[MR05a]   de Montgolfier, F., Rao, M.: The bi-join decomposition. Electronic
               Notes in Discrete Mathematics 22, 173–177 (2005)
[MR05b]   de Montgolfier, F., Rao, M.: Bipartitive families and the bi-join decom-
               position (submitted 2005),
               http://hal.archives-ouvertes.fr/hal-00132862
[MS99]   McConnell, R.M., Spinrad, J.P.: Modular decomposition and transitive
               orientation. Discrete Mathematics 201(1-3), 189–241 (1999)
[Oum05a]   Oum, S.-I.: Graphs Of Bounded Rank Width. PhD thesis. Princeton
               University (2005)
[Oum05b]   Oum, S.-I.: Rank-width and vertex-minors. Journal of Combinatorial
               Theory, Series B 95(1), 79–100 (2005)
[PLE71]   Pnueli, A., Lempel, A., Even, S.: Transitive orientation of graphs and
               identification of permutation graphs. Canadian Journal of Mathemat-
               ics 23(1), 160–175 (1971)

[RU82]   Rotem, D., Urrutia, J.: Circular permutation graphs. Networks 12, 429–437 (1982)
[Sei76]   Seidel, J.J.: A survey of two-graphs. In: Colloquio Internazionale sulle Teorie Combinatorie (Rome, 1973), Tomo I. Atti dei Convegni Lincei, vol. 17, pp. 481–511. Accad. Naz. Lincei, Rome (1976)
[Sri96]   Sritharan, R.: A linear time algorithm to recognize circular permutation graphs. Networks 27(3), 171–174 (1996)
[TCHP08]   Tedder, M., Corneil, D.G., Habib, M., Paul, C.: Simpler Linear-Time Modular Decomposition Via Recursive Factorizing Permutations. In: Aceto, L., Damgård, I., Goldberg, L.A., Halldórsson, M.M., Ingólfsdóttir, A., Walukiewicz, I. (eds.) ICALP 2008, Part I. LNCS, vol. 5125, pp. 634–645. Springer, Heidelberg (2008)

Simultaneous Interval Graphs

Krishnam Raju Jampani and Anna Lubiw

David R. Cheriton School of Computer Science,
University of Waterloo, Ontario, Canada
{krjampan,alubiw}@uwaterloo.ca

Abstract. In a recent paper, we introduced the simultaneous representation problem (defined for any graph class $\mathcal{C}$) and studied the problem for chordal, comparability and permutation graphs. For interval graphs, the problem is defined as follows. Two interval graphs G_1 and G_2, sharing some vertices I (and the corresponding induced edges), are said to be "simultaneous interval graphs" if there exist interval representations R_1 and R_2 of G_1 and G_2, such that any vertex of I is mapped to the same interval in both R_1 and R_2. Equivalently, G_1 and G_2 are simultaneous interval graphs if there exist edges E' between $G_1 - I$ and $G_2 - I$ such that $G_1 \cup G_2 \cup E'$ is an interval graph.

Simultaneous representation problems are related to simultaneous planar embeddings, and have applications in any situation where it is desirable to consistently represent two related graphs, for example: interval graphs capturing overlaps of DNA fragments of two similar organisms; or graphs connected in time, where one is an updated version of the other.

In this paper we give an $O(n^2 \log n)$ time algorithm for recognizing simultaneous interval graphs, where $n = |G_1 \cup G_2|$. This result complements the polynomial time algorithms for recognizing probe interval graphs and provides an efficient algorithm for the interval graph sandwich problem for the special case where the set of optional edges induce a complete bipartite graph.

Keywords: Simultaneous Graphs, Interval Graphs, Graph Sandwich Problem, Probe Graphs, PQ-trees.

1 Introduction

Let $\mathcal{C}$ be any intersection graph class (such as interval graphs or chordal graphs) and let G_1 and G_2 be two graphs in $\mathcal{C}$, sharing some vertices I and the edges induced by I. G_1 and G_2 are said to be *simultaneously representable $\mathcal{C}$ graphs* or *simultaneous $\mathcal{C}$ graphs* if there exist intersection representations R_1 and R_2 of G_1 and G_2 such that any vertex of I is represented by the same object in both R_1 and R_2. The *simultaneous representation problem* for class $\mathcal{C}$ asks whether G_1 and G_2 are simultaneous $\mathcal{C}$ graphs. For example, Figures 1(a) and 1(b) show two simultaneous interval graphs and their interval representations with the property that vertices common to both graphs are assigned to the same interval. Figure 1(c) shows two interval graphs that are not simultaneous interval graphs.

O. Cheong, K.-Y. Chwa, and K. Park (Eds.): ISAAC 2010, Part I, LNCS 6506, pp. 206–217, 2010.

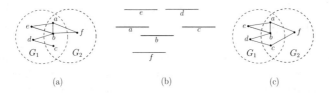

Fig. 1. Graphs in (a) are simultaneous interval graphs as shown by the representations in (b). Graphs in (c) are not simultaneous interval graphs.

Simultaneous representation problems were introduced by us in a recent paper [9] and have application in any situation where two related graphs should be represented consistently. A main instance is for temporal relationships, where an old graph and a new graph share some common parts. Pairs of related graphs also arise in many other situations, e.g: two social networks that share some members; overlap graphs of DNA fragments of two similar organisms, etc.

Simultaneous representations are related to simultaneous planar embeddings: two graphs that share some vertices and edges (not necessarily induced) have a *simultaneous geometric embedding* [3] if they have planar straight-line drawings in which the common vertices are represented by common points. Thus edges may cross, but only if they are in different graphs. Deciding if two graphs have a simultaneous geometric embedding is NP-Hard [4].

In [9], we showed that the simultaneous representation problem can be solved efficiently for chordal, comparability and permutation graphs. We also showed that for any intersection class $\mathcal{C}$, the simultaneous representation problem for G_1 and G_2 is equivalent to the following problem: Do there exist edges E' between $G_1 - I$ and $G_2 - I$ so that the augmented graph $G_1 \cup G_2 \cup E'$ belongs to class $\mathcal{C}$.

The *graph sandwich problem* [7] is a more general augmentation problem defined for any graph class $\mathcal{C}$: given graphs $H_1 = (V, E_1)$ and $H_2 = (V, E_2)$, is there a set E of edges with $E_1 \subseteq E \subseteq E_2$ so that the graph $G = (V, E)$ belongs to class $\mathcal{C}$. This problem has a wealth of applications but is NP-complete for interval graphs, comparability graphs, and permutation graphs [7].

The simultaneous representation problem (for class $\mathcal{C}$) is the special case of the graph sandwich problem (for $\mathcal{C}$) where $E_2 - E_1$ forms a complete bipartite subgraph. A related special case where $E_2 - E_1$ forms a clique is the problem of recognizing *probe graphs*: a graph G with a specified independent set N is a *probe graph* for class $\mathcal{C}$ if there exist edges $E' \subseteq N \times N$ so that the augmented graph $G \cup E'$ belongs to class $\mathcal{C}$.

Probe graphs have several applications [14,8] and have received much attention recently. The first polynomial-time algorithm for recognizing probe interval graphs was due to Johnson and Spinrad [11]. They used a variant of PQ-trees and achieved a run-time of $O(n^2)$. Techniques from modular decomposition provided more speed up [13], but the most recent algorithm by McConnell and Nussbaum [12] reverts to PQ-trees and achieves linear time.

We note that there has also been work [15] on a concept of simultaneous intersection called "polysemy" where two graphs are represented as intersections of sets and their complements.

In this paper, we give an $O(n^2 \log n)$ algorithm for solving the simultaneous representation problem for interval graphs. We use PQ-trees, which were developed by Booth and Lueker for the original linear time interval graph recognition algorithm. They used a PQ-tree to capture the orderings of the maximal cliques of the graph (see [6] for an introduction to interval graphs and PQ-trees).

In the probe interval recognition problem, there is a single PQ-tree (of the graph induced by the probes) and a set of constraints imposed by the non-probes. However in our situation we have two PQ-trees, one for each graph, that we want to re-order to "match" on the common vertex set I. We begin by "reducing" each PQ-tree to contain only vertices from I. This results in PQ-trees that store non-maximal cliques, and our task is to modify each PQ-tree by inserting non-maximal cliques from the other tree while re-ordering the trees to make them the same.

2   Reduction to PQ-Trees

In this section we transform the simultaneous interval graph problem to a problem about "compatibility" of two PQ-trees arising from the two graphs.

Recall that an interval graph is defined to be the intersection graph of intervals on the real line. For any point on the line, the intervals containing that point form a clique in the graph. This leads to the fundamental one-to-one correspondence between the interval representations of an interval graph and its *clique orderings*, defined as follows: A *clique ordering* of G is a sequence of (possibly empty) cliques $\mathcal{S} = Q_1, Q_2, \cdots, Q_l$ that contains all the maximal cliques of G and has the property that for each vertex v, the cliques in $\mathcal{S}$ that contain v appear consecutively. Note that we allow cliques to be empty.

The standard interval graph recognition algorithm attempts to find a clique order of the maximal cliques of a graph by making the maximal cliques into leaves of a PQ-tree, and imposing PQ-tree constraints to ensure that the cliques containing each vertex v appear consecutively. This structure is called *the* PQ-tree of the graph. Note that the children of a P-node may be reordered arbitrarily and the children of a Q-node may only be reversed. We consider a node with 2 children to be a Q-node. In the figures, we use a circle to denote a P-node and a rectangle to denote a Q-node. A *leaf-order* of a PQ-tree is the order in which its leaves are visited in an in-order traversal of the tree, after children of P and Q-nodes are re-ordered as just described.

Note that ignoring non-maximal cliques is fine for recognizing interval graphs; for our purposes, however, we want to consider clique orders and PQ-trees that may include non-maximal cliques. We say that a PQ-tree whose leaves correspond to cliques of a graph is *valid* if for each of its leaf orderings and for each vertex v, the cliques containing v appear consecutively.

Let $\mathcal{S} = Q_1, Q_2, \cdots, Q_l$ be a clique ordering of interval graph G and let the maximal cliques of G be $Q_{i_1}, Q_{i_2}, \cdots, Q_{i_m}$ (appearing in positions $i_1 < i_2 <$

$\cdots < i_m$ respectively). Note that all the cliques in $\mathcal{S}$ between Q_{i_j} and $Q_{i_{j+1}}$ contain $B = Q_{i_j} \cap Q_{i_{j+1}}$. We say that B is the *boundary clique* or *boundary* between Q_{i_j} and $Q_{i_{j+1}}$. Note that B may not necessarily be present in $\mathcal{S}$. The sequence of cliques between Q_{i_j} and $Q_{i_{j+1}}$ that are subsets of Q_{i_j} is said to be the *right tail* of Q_{i_j}. The *left tail* of $Q_{i_{j+1}}$ is defined analogously. Observe that the left tail of a clique forms an increasing sequence and the right tail forms a decreasing sequence (w.r.t set inclusion). Also note that all the cliques that precede Q_{i_1} are subsets of Q_{i_1} and this sequence is called the left tail of Q_{i_1} and all the cliques that succeed Q_{i_m} are subsets of Q_{i_m} and this sequence is called the right tail of Q_{i_m}. Thus any clique ordering of G consists of a sequence of maximal cliques, with each maximal clique containing a (possibly empty) left and right tail of subcliques.

Let Q_0 and Q_{l+1} be defined to be empty sets. An insertion of clique Q' between Q_i and Q_{i+1} (for some $i \in \{0, \cdots, l\}$) is said to be a *subclique insertion* if $Q' \supseteq Q_i \cap Q_{i+1}$ and either $Q' \subseteq Q_i$ or $Q' \subseteq Q_{i+1}$. It is clear that after a subclique insertion the resulting sequence is still a clique ordering of G. A clique ordering $\mathcal{S}'$ is an *extension* of $\mathcal{S}$ if $\mathcal{S}'$ can be obtained from $\mathcal{S}$ by subclique insertions. We also say that $\mathcal{S}$ extends to $\mathcal{S}'$. Furthermore, we say that a clique ordering is *generated* by a PQ-tree, if it can be obtained from a leaf order of the PQ-tree with subclique insertions. The above definitions yield the following Lemma.

Lemma 1. *A sequence of cliques $\mathcal{S}$ is a clique ordering of G if and only if $\mathcal{S}$ can be generated from the PQ-tree of G.*

Let G_1 and G_2 be two interval graphs sharing a vertex set I (i.e. $I = V(G_1) \cap V(G_2)$) and its induced edges. Note that $G_1[I]$ is isomorphic to $G_2[I]$. A clique ordering of $G_1[I]$ is said to be an *I-ordering*.

The *I-restricted* PQ-tree of G_j is defined to be the tree obtained from the PQ-tree of G_j by replacing each clique Q (a leaf of the PQ-tree) with the clique $Q \cap I$. Thus there is a one-to-one correspondence between the two PQ-trees, and the leaves of the I-restricted PQ-tree are cliques of $G_1[I]$.

Let $\mathcal{I} = X_1, X_2, \cdots, X_l$ be an I-ordering. $\mathcal{I}$ is said to be G_j-*expandable* if there exists a clique ordering $\mathcal{O} = Q_1, Q_2, \cdots, Q_l$ of G_j such that $X_i \subseteq Q_i$ for $i \in \{1, \cdots, l\}$. Further, we say that $\mathcal{I}$ expands to $\mathcal{O}$. By the definition of clique-ordering it follows that, if $\mathcal{I}$ is G_j-expandable then it remains G_j-expandable after a subclique insertion. (i.e. any extension of $\mathcal{I}$ is also G_j-expandable). We first observe the following (see the full version [10] of the paper for all missing proofs).

Lemma 2. *The set of G_j-expandable I-orderings is same as the set of orderings that can be generated from the I-restricted PQ-tree of G_j.*

Two I-orderings $\mathcal{I}_1$ and $\mathcal{I}_2$ are said to be *compatible* if both $\mathcal{I}_1$ and $\mathcal{I}_2$ (separately) extend to a common I-ordering $\mathcal{I}$. For e.g. the ordering $\{1\}, \{1,2\}, \{1,2,3,4\}$ is compatible with the ordering $\{1\}, \{1,2,3\}, \{1,2,3,4\}$, as they both extend to the common ordering: $\{1\}, \{1,2\}, \{1,2,3\}, \{1,2,3,4\}$. Note that the compatibility relation is not transitive. Two PQ-trees T_1 and T_2 are said to be *compatible* if there exist orderings $\mathcal{O}_1$ and $\mathcal{O}_2$ generated from T_1 and T_2 (respectively) such that $\mathcal{O}_1$ is compatible with $\mathcal{O}_2$. The following Lemma is our main tool.

Lemma 3. G_1 and G_2 are simultaneous interval graphs if and only if the I-restricted PQ-tree of G_1 is compatible with the I-restricted PQ-tree of G_2.

Our algorithm will decide if the I-restricted PQ-tree of G_1 is compatible with the I-restricted PQ-tree of G_2. We first show how the I-restricted PQ-trees can be simplified in several ways. Two I-orderings $\mathcal{I}_1$ and $\mathcal{I}_2$ are said to be *equivalent* if for any I-ordering $\mathcal{I}'$, $\mathcal{I}_1$ and $\mathcal{I}'$ are compatible if and only if $\mathcal{I}_2$ and $\mathcal{I}'$ are compatible. Note that this is an equivalence relation. The Lemma below follows directly from the definitions of equivalent orderings and subclique insertions.

Lemma 4. Let $\mathcal{I} = X_1, X_2, \cdots, X_l$ be an I-ordering in which $X_i = X_{i+1}$ for some $i \in 1, \cdots, l-1$. Let $\mathcal{I}'$ be the I-ordering obtained from $\mathcal{I}$ by deleting X_{i+1}. Then $\mathcal{I}$ is equivalent to $\mathcal{I}'$.

Further, because equivalence is transitive, Lemma 4 implies that an I-ordering $\mathcal{I}$ is equivalent to the I-ordering $\mathcal{I}'$ in which all consecutive duplicates are eliminated. This allows us to simplify the I-restricted PQ-tree of G_j. Let T be the I-restricted PQ-tree of G_j. We obtain a PQ-tree T' from T as follows.

1. Initialize $T' = T$.
2. As long as there is a non-leaf node n in T' such that all the descendants of n are the same, i.e. they are all duplicates of a single clique X, replace n and the subtree rooted at n by a leaf node representing X.
3. As long as there is a (non-leaf) Q-node n in T' with two consecutive child nodes n_a and n_b (among others) such that all the descendants of n_a and n_b are the same, i.e. they are all duplicates of a single clique X, replace n_a, n_b and the subtrees rooted at these vertices by a single leaf node representing the clique X.

Note that the resulting T' is unique. We call T' the I-reduced PQ-tree of G_j. The following Lemma is easy to prove.

Lemma 5. G_1 and G_2 are simultaneous interval graphs if and only if the I-reduced PQ-tree of G_1 is compatible with the I-reduced PQ-tree of G_2.

3   Labeling and Further Simplification

In section 2, we transformed the simultaneous interval graph problem to a problem of testing compatibility of two I-reduced PQ-trees where I is the common vertex set of the two graphs. These PQ-trees may have nodes that correspond to non-maximal cliques in I. In this section we prove some basic properties of such I-reduced PQ-trees, and use them to further simplify each tree.

   Let $\mathcal{T}$ be the I-reduced PQ-tree of G_j. Recall that each leaf l of $\mathcal{T}$ corresponds to a clique X in $G_j[I]$. If X is maximal in I, then X is said to be a *max-clique* and l is said to be a *max-clique node*, otherwise X is said to be a *subclique* and l is said to be a *subclique node*. When the association is clear from the context, we will sometimes refer to a leaf l and its corresponding clique X interchangeably, or interchange the terms "max-clique" and "max-clique node" [resp. subclique and subclique node]. A node of $\mathcal{T}$ is said to be an *essential node* if it is a non-leaf node or if it is a leaf node representing a max-clique.

Given a node n of $\mathcal{T}$, the *descendant cliques* of n are the set of cliques that correspond to the leaf-descendants of n. Because our algorithm operates by inserting subcliques from one tree into the other, we must take care to preserve the validity of a PQ-tree. For this we need to re-structure the tree when we do subclique insertions. The required restructuring will be determined based on the label $U(n)$ that we assign to each node n as follows.

$U(n)$ or the *Universal set* of n is defined as the set of vertices v such that v appears in all descendant cliques of n.

Note that for a leaf node l representing a clique X, $U(l) = X$ by definition. Also note that along any path up the tree, the universal sets decrease. The following Lemma gives some useful properties of the I-reduced PQ-tree.

Lemma 6. *Let $\mathcal{T}$ be the I-reduced PQ-tree of G_j. Let n be a non-leaf node of $\mathcal{T}$ (n is used in properties 2–6). Then we have:*
0. *Let l_1 and l_2 be two distinct leaf nodes of $\mathcal{T}$, containing a vertex $t \in I$. Let y be the least common ancestor of l_1 and l_2. Then: (a) If y is a P-node then all of its descendant cliques contain t. (b) If y is a Q-node then t is contained in all the descendant cliques of all children of y between (and including) the child of y that is the ancestor of l_1 and the child that is the ancestor of l_2.*
1. *Each max-clique is represented by a unique node of $\mathcal{T}$.*
2. *A vertex u is in $U(n)$ if and only if for every child n_1 of n, $u \in U(n_1)$.*
3. *n contains a max-clique as a descendant.*
4. *If n is a P-node, then for any two child nodes n_1 and n_2 of n, we have $U(n) = U(n_1) \cap U(n_2)$.*
5. *If n is a P-node, then any child of n that is a subclique node represents the clique $U(n)$.*
6. *If n is a Q-node and n_1 and n_2 are the first and last child nodes of n then $U(n) = U(n_1) \cap U(n_2)$.*

Let $\mathcal{T}$ be the I-reduced PQ-tree of G_j. Recall that an essential node is a non-leaf node or a leaf node representing a maximal clique. Equivalently (by Lemma 6.3), an essential node is a node which contains a max-clique as a descendant. The following Lemma shows that in some situations we can obtain an equivalent tree by deleting subclique child nodes of a P-node n. Recall that by Lemma 6.5, such subclique nodes represent the clique $U(n)$.

Lemma 7. *Let $\mathcal{T}$ be the I-reduced PQ-tree of G_j and n be a P-node in $\mathcal{T}$. Then*
1. *If n has at least two essential child nodes, then $\mathcal{T}$ is equivalent to the tree $\mathcal{T}'$, obtained from $\mathcal{T}$ by deleting all the subclique children of n.*

2. *If n has at least two subclique child nodes, then $\mathcal{T}$ is equivalent to the tree $\mathcal{T}'$, obtained from $\mathcal{T}$ by deleting all except one of the subclique children of n.*

We will simplify $\mathcal{T}$ as much as possible by applying Lemma 7 and by converting nodes with two children into Q-nodes. We call the end result a *simplified I-reduced PQ-tree*, but continue to use the term "*I*-reduced PQ-tree" to refer to it. Note that the simplification process does not change the universal sets and preserves the validity of the PQ-tree so Lemma 5 and all the properties given in Lemma 6 still hold. Because we consider nodes with 2 children as Q-nodes Lemma 7 implies:

Corollory 1. *In a [simplified] I-reduced PQ-tree, any P-node has at least 3 children, and all the children are essential nodes.*

4   Algorithm

For $k \in \{1, 2\}$, let $\mathcal{T}_k$ be the [simplified] *I*-reduced PQ-tree of G_k. By Lemma 5, testing whether G_1 and G_2 are simultaneous interval graphs is equivalent to testing whether $\mathcal{T}_1$ and $\mathcal{T}_2$ are compatible. We test this by modifying $\mathcal{T}_1$ and $\mathcal{T}_2$ (e.g. inserting the sub-clique nodes from one tree into the other) so as to make them identical, without losing their compatibility. The following is a high level overview of our approach for checking whether $\mathcal{T}_1$ and $\mathcal{T}_2$ are compatible.

Our algorithm is iterative and tries to *match* essential nodes of $\mathcal{T}_1$ with essential nodes of $\mathcal{T}_2$ in a bottom-up fashion. An essential node n_1 of $\mathcal{T}_1$ is matched with an essential node n_2 of $\mathcal{T}_2$ if and only if the subtrees rooted at n_1 and n_2 are the same, i.e. their essential children are matched, their subclique children are the same and furthermore (in the case of Q-nodes) their child nodes appear in the same order. If n_1 is matched with n_2 then we consider n_1 and n_2 to be identical and use the same name (say n_1) to refer to either of them. Initially, we match each max-clique node of $\mathcal{T}_1$ with the corresponding max-clique node of $\mathcal{T}_2$. Note that every max-clique node appears uniquely in each tree by Lemma 6.1. A sub-clique node may appear in only one tree in which case we must first insert it into the other tree. This is done when we consider the parent of the subclique node.

In each iteration, we either match an unmatched node u of $\mathcal{T}_1$ to an unmatched node v of $\mathcal{T}_2$ (which may involve inserting subclique child nodes of v as child nodes of u and vice versa) or we *reduce* either $\mathcal{T}_1$ or $\mathcal{T}_2$ without losing their compatibility relationship. *Reducing* a PQ-tree means restricting it to reduce the number of leaf orderings. Finally, at the end of the algorithm either we have modified $\mathcal{T}_1$ and $\mathcal{T}_2$ to a "common" tree $\mathcal{T}_I$ that establishes their compatibility or we conclude that $\mathcal{T}_1$ is not compatible with $\mathcal{T}_2$. The common tree $\mathcal{T}_I$ is said to be an *intersection tree* (of $\mathcal{T}_1$ and $\mathcal{T}_2$) and has the property that any ordering generated by $\mathcal{T}_I$ can also be generated by $\mathcal{T}_1$ and $\mathcal{T}_2$. If $\mathcal{T}_1$ and $\mathcal{T}_2$ are compatible, there may be several intersection trees of $\mathcal{T}_1$ and $\mathcal{T}_2$, but our algorithm finds only one of them.

We need the following additional notation for the rest of this paper. A sequence of subcliques $\mathcal{S} = X_1, X_2, \cdots, X_l$ is said to satisfy the *subset property* if $X_i \subseteq X_{i+1}$ for $i \in \{1, \cdots, l-1\}$. $\mathcal{S}$ is said to satisfy the *superset property* if $X_i \supseteq X_{i+1}$ for each i. Note that S satisfies the subset property if and only if $\bar{\mathcal{S}} = X_l, \cdots, X_2, X_1$ satisfies the superset property.

Let d be an essential child node of a Q-node in $\mathcal{T}_k$. We will overload the term "tail" (previously defined for a max clique in a clique ordering) and define the *tails* of d as follows. The left tail (resp. right tail) of d is defined as the sequence of subcliques that appear as siblings of d, to the immediate left (resp. right) of d, such that each subclique is a subset of $U(d)$. Note that the left tail of d should satisfy the subset property and the right tail of d should satisfy the superset property (otherwise $\mathcal{T}_k$ will not be valid). Also note that since the children of a Q-node can be reversed in order, "left" and "right" are relative to the child ordering of the Q-node. We will be careful to use "left tail" and "right tail" in such a way that this ambiguity does not matter. Now suppose d is a matched node. Then in order to match the parent of d in $\mathcal{T}_1$ with the parent of d in $\mathcal{T}_2$, our algorithm has to "merge" the tails of d.

Let $\mathcal{L}_1$ and $\mathcal{L}_2$ be two subclique sequences that satisfy the subset property. Then $\mathcal{L}_1$ is said to be *mergable* with $\mathcal{L}_2$ if the union of subcliques in $\mathcal{L}_1$ and $\mathcal{L}_2$ can be arranged into an ordering $\mathcal{L}'$ that satisfies the subset property. Analogously, if $\mathcal{L}_1$ and $\mathcal{L}_2$ satisfy the superset property, then they are said to be mergable if the union of their subcliques can be arranged into an ordering $\mathcal{L}'$ that satisfies the superset property. In both cases, $\mathcal{L}'$ is said to be the *merge* of $\mathcal{L}_1$ and $\mathcal{L}_2$ and is denoted by $\mathcal{L}_1 + \mathcal{L}_2$.

A *maximal matched node* is a node that is matched but whose parent is not matched. For an unmatched essential node x, the *MM-descendants* of x, denoted by $MMD(x)$ are its descendants that are maximal matched nodes. If x is matched then we define $MMD(x)$ to be the singleton set containing x. Note that the MM-descendants of an essential node is non-empty (since every essential node has a max-clique descendant).

Our algorithm matches nodes from the leaves up, and starts by matching the leaves that are max-cliques. As the next node n_1 that we try to match, we want an unmatched node whose essential children are already matched. To help us choose between $\mathcal{T}_1$ and $\mathcal{T}_2$, and also to break ties, we prefer a node with larger U set. Then, as a candidate to match n_1 to, we want an unmatched node in the other tree that has some matched children in common with n_1. With this intuition in mind, our specific rule is as follows.

Among all the unmatched essential nodes of $\mathcal{T}_1$ union $\mathcal{T}_2$ choose n_1 with maximal $U(n_1)$, minimal $MMD(n_1)$, and maximal depth, in that preference order. Assume without loss of generality that $n_1 \in \mathcal{T}_1$. Select an unmatched node n_2 from $\mathcal{T}_2$ with maximal $U(n_2)$, minimal $MMD(n_2)$ and maximal depth (in that order) satisfying the property that $MMD(n_1) \cap MMD(n_2) \neq \emptyset$. The following Lemma captures certain properties of n_1 and n_2, including why these rules match our intuitive justification.

Lemma 8. *For n_1 and n_2 chosen as described above, let $M_1 = MMD(n_1)$, $M_2 = MMD(n_2)$ and $X = M_1 \cap M_2$. Also let C_1 and C_2 be the essential child nodes of n_1 and n_2 respectively. Then we have:*
1. $M_1 = C_1$ and $X \subseteq C_2$.
Further when T_1 is compatible with T_2, we have:
2. *For every (matched) node l in $M_1 - X$ of T_1, its corresponding matched node l' in T_2 is present outside the subtree rooted at n_2. Analogously, for every (matched) node r' in $M_2 - X$ of T_2, its corresponding matched node r in T_1 is present outside the subtree rooted at n_1.*
3. *If n_1 [resp. n_2] is a Q-node, then in its child ordering, no node of $C_1 - X$ [resp. $C_2 - X$] can be present between two nodes of X.*
4. *If n_1 and n_2 are Q-nodes, then in the child ordering of n_1 and n_2, nodes of X appear in the same relative order i.e. for any three nodes $x_1, x_2, x_3 \in X$, x_1 appears between x_2 and x_3 in the child ordering of n_1 if and only if x_1 also appears between x_2 and x_3 in the child ordering of n_2.*
5. *If $C_1 - X$ (resp.$C_2 - X$) is non-empty then $U(n_1) \subseteq U(n_2)$ (resp. $U(n_2) \subseteq U(n_1)$). Further, if $C_1 - X$ is non-empty then so is $C_2 - X$ and hence $U(n_1) = U(n_2)$.*
6. *Let $C_1 - X$ be non-empty. If n_1 [resp. n_2] is a Q-node, then in its child-ordering either all nodes of $C_1 - X$ [resp. $C_2 - X$] appear before the nodes of X or they all appear after the nodes of X.*

We now describe the main step of the algorithm. Let $n_1, n_2, M_1, M_2, C_1, C_2$ and X be as defined in the above Lemma. We have four cases depending on whether n_1 and n_2 are P or Q-nodes. In each of these cases, we make progress by either matching two previously unmatched essential nodes of T_1 and T_2 or by reducing T_1 and/or T_2 at n_1 or n_2 while preserving their compatibility. We can show (see full version for details) that our algorithm requires at most $O(n \log n)$ iterations and each iteration takes $O(n)$ time. Thus our algorithm runs in $O(n^2 \log n)$ time.

During the course of the algorithm we may also insert subcliques into a Q-node when we are trying to match it to another Q-node. This is potentially dangerous as it may destroy the validity of the PQ-tree. When the Q-nodes have the same universal set, this trouble does not arise. However, in case the two Q-nodes have different universal sets, we need to re-structure the trees. Case 4, when n_1 and n_2 are both Q-nodes, has sub cases to deal with these complications.

Case 1: n_1 and n_2 are both P-nodes

By Corollary 1, the children of n_1 and n_2 are essential nodes, so C_1 and C_2 are precisely the children of n_1 and n_2 respectively. Let X consist of nodes $\{x_1, \cdots, x_{k_0}\}$. If $C_2 - X$ is empty, then by Lemma 8.5, $C_1 - X$ is also empty and hence n_1 and n_2 are the same. So we match n_1 with n_2 and go to the next iteration. Suppose now that $C_2 - X$ is non empty. Let $C_2 - X = \{r_1, \cdots, r_{k_2}\}$. If $C_1 - X$ is empty, then we use the reduction template of Figure 2(a) to modify T_2, matching the new parent of X in T_2 to n_1. It is easy to see that T_1 is compatible with T_2 if and only if T_1 is compatible with the modified T_2.

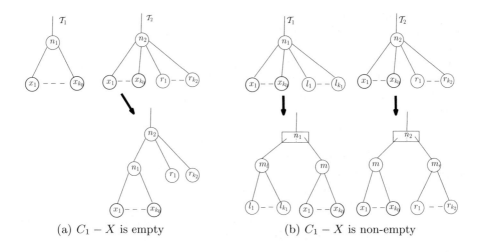

(a) $C_1 - X$ is empty   (b) $C_1 - X$ is non-empty

Fig. 2. Reduction templates for Case 1

Now let $C_1 - X = \{l_1, \cdots, l_{k_1}\}$ be non-empty. In this case we use the reduction template of Figure 2(b) to modify $\mathcal{T}_1$ and $\mathcal{T}_2$ to $\mathcal{T}_1'$ and $\mathcal{T}_2'$ respectively. Note that it is possible to have $k_i = 1$ for some i's, in which case the template is slightly different because we do not make a node with one child, however, the reduction always makes progress as each n_i has at least 3 children.

We now claim that $\mathcal{T}_1$ is compatible with $\mathcal{T}_2$ if and only if $\mathcal{T}_1'$ is compatible with $\mathcal{T}_2'$. The reverse direction is trivial. For the forward direction, let $\mathcal{O}_1$ and $\mathcal{O}_2$ be two compatible leaf orderings of $\mathcal{T}_1$ and $\mathcal{T}_2$ respectively. Recall that by Lemma 8.2, for every [matched] node of $C_1 - X$ in $\mathcal{T}_1$, the corresponding matched node in $\mathcal{T}_2$ appears outside the subtree rooted at n_2. This implies that the descendant nodes of $\{x_1, x_2, \cdots, x_{k_0}\}$ all appear consecutively in $\mathcal{O}_1$. Hence the descendant nodes of $\{x_1, x_2, \cdots, x_{k_0}\}$ also appear consecutively in $\mathcal{O}_2$. Thus we conclude that $\mathcal{T}_1$ and $\mathcal{T}_2$ are compatible if and only if the reduced trees $\mathcal{T}_1'$ and $\mathcal{T}_2'$ are also compatible. Note that both the template reductions take at most $O(n)$ time.

Case 2: n_1 is a P-node and n_2 is a Q-node

If $C_1 - X = \emptyset$, we reduce $\mathcal{T}_1$ by ordering the children of n_1 as they appear in the child ordering of n_2, and changing n_1 into a Q-node (and leading to Case 4). This reduction preserves the compatibility of the two trees.

Now suppose $C_1 - X \neq \emptyset$. Lemma 8.5 implies that, $C_2 - X \neq \emptyset$ and $U(n_1) = U(n_2)$. By Lemma 8.6, we can assume that the nodes in X appear before the nodes in $C_2 - X$ in the child ordering of n_2. Now let $X = x_1, \cdots, x_{k_0}$, $C_1 - X = l_1, \cdots, l_{k_1}$ and $C_2 - X = r_1, \cdots, r_{k_2}$. For $i \in 2, \cdots, k_0$, let $\mathcal{S}_i$ be the sequence of subcliques that appear between x_{i-1} and x_i in the child ordering of n_2. Note that $\mathcal{S}_i$ consists of the right tail of x_{i-1} followed by the left tail of x_i. We let $\mathcal{S}_1$ and $\mathcal{S}_{k_0+1}$ denote the left and right tails of x_1 and x_{k_0} respectively. We now reduce the subtree rooted at n_1 as shown in Figure 3, changing it into a Q-node. Clearly $U(n_1)$ is preserved in this operation. The correctness of this operation follows by Lemma 8.2. It is easy to see that both the template reductions run in $O(n)$ time.

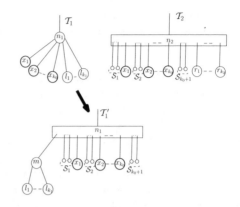

Case 3: n_1 is a Q-node and n_2 is a P-node

This case is similar to Case 2 (see full version for details).

Fig. 3. Reduction template for Case 2, when $C_1 - X \neq \emptyset$

Case 4: n_1 and n_2 are both Q-nodes

Let $X = \{x_1, \cdots, x_{k_0}\}$ appear in that order in the child ordering of n_1 and n_2. (They appear in the same order because of Lemma 8.4.) Let p_1 and p_2 be the parents of n_1 and n_2 respectively.

If n_1 and n_2 have no other children than X, we match n_1 with n_2 and proceed to the next iteration. More typically, they have other children. These may be essential nodes to one side or the other of X (by Lemma 8.6) or subclique nodes interspersed in X as tails of the nodes of X. We give a high-level outline of Case 4, beginning with a discussion of subclique nodes.

For $i \in \{1, \cdots, k_0\}$, let $\mathcal{L}_i$ and $\mathcal{R}_i$ be the left and right tails of x_i in $\mathcal{T}_1$ and, $\mathcal{L}_i'$ and $\mathcal{R}_i'$ be the left and right tails of x_i in $\mathcal{T}_2$. The only way to deal with the subclique nodes is to do subclique insertions in both trees to merge the tails. This is because in any intersection tree $\mathcal{T}_I$ obtained from $\mathcal{T}_1$ and $\mathcal{T}_2$, the tails of x_i in $\mathcal{T}_I$ must contain the merge of the tails of x_i in $\mathcal{T}_1$ and $\mathcal{T}_2$. So long as $|X| \geq 2$, the ordering $x_1, \cdots, x_{k_0}$ completely determines which pairs of tails must merge: $\mathcal{L}_i$ must merge with $\mathcal{L}_i'$ and $\mathcal{R}_i$ must merge with $\mathcal{R}_i'$.

The case $|X| = 1$ is more complicated because the intersection tree may merge $\mathcal{L}_1$ with $\mathcal{L}_1'$ and $\mathcal{R}_1$ with $\mathcal{R}_1'$ or merge $\mathcal{L}_1$ with $\bar{\mathcal{R}}_1$ and $\mathcal{R}_1$ with $\bar{\mathcal{L}}_1$. This decision problem is referred to as the *alignment problem*. We can prove that in case both choices give mergable pairs, then either choice yields an intersection tree, if an intersection tree exists (see the full paper).

Case 4 has subcases depending on whether $U(n_1) = U(n_2)$ and whether n_1 and n_2 have the same essential children. If both these conditions hold, then we merge the tails of the nodes of X and match n_1 with n_2. (In other words we replace $\mathcal{L}_i$ and $\mathcal{L}_i'$ with $\mathcal{L}_i + \mathcal{L}_i'$, and replace $\mathcal{R}_i$ and $\mathcal{R}_i'$ with $\mathcal{R}_i + \mathcal{R}_i'$). When $U(n_1) \neq U(n_2)$ or when n_1 and n_2 do not have the same essential children then we have three sub cases depending on whether $U(n_1) \supseteq U(n_2)$ or not and whether $C_1 - X$ is empty or not. Details of these sub cases are available in the full version of the paper.

5   Open Problem

Simultaneous graphs can be generalized in a natural way to more than two graphs: when $G_1 = (V_1, E_1), G_2 = (V_2, E_2), \cdots, G_k = (V_k, E_k)$ are k graphs in class $\mathcal{C}$, sharing a vertex set I and its induced edges, i.e. $V_i \cap V_j = I$ for all $i, j \in \{1, \cdots, k\}$. In this version of the problem the set of optional edges induces a complete k-partite graph and hence this also generalizes probe graphs. This generalized version can be solved in polynomial time for comparability and permutation graphs [9]. We conjecture that it can be solved in polynomial time for interval graphs.

References

1. Berry, A., Golumbic, M.C., Lipshteyn, M.: Recognizing chordal probe graphs and cycle-bicolorable graphs. SIAM J. Discret. Math. 21(3), 573–591 (2007)
2. Booth, K.S., Lueker, G.S.: Testing for the consecutive ones property, interval graphs, and graph planarity using PQ-tree algorithms. J. Comput. System Sci. 13, 335–379 (1976)
3. Brass, P., Cenek, E., Duncan, C.A., Efrat, A., Erten, C., Ismailescu, D.P., Kobourov, S.G., Lubiw, A., Mitchell, J.S.B.: On simultaneous planar graph embeddings. Comput. Geom. Theory Appl. 36(2), 117–130 (2007)
4. Estrella-Balderrama, A., Gassner, E., Junger, M., Percan, M., Schaefer, M., Schulz, M.: Simultaneous geometric graph embeddings. In: Hong, S.-H., Nishizeki, T., Quan, W. (eds.) GD 2007. LNCS, vol. 4875, pp. 280–290. Springer, Heidelberg (2008)
5. Chandler, D.B., Chang, M., Kloks, T., Liu, J., Peng, S.: Partitioned probe comparability graphs. Theor. Comput. Sci. 396(1-3), 212–222 (2008)
6. Golumbic, M.C.: Algorithmic Graph Theory And Perfect Graphs. Academic Press, New York (1980)
7. Golumbic, M.C., Kaplan, H., Shamir, R.: Graph sandwich problems. J. Algorithms 19(3), 449–473 (1995)
8. Golumbic, M.C., Lipshteyn, M.: Chordal probe graphs. Discrete Appl. Math. 143(1-3), 221–237 (2004)
9. Jampani, K.R., Lubiw, A.: The simultaneous representation problem for chordal, comparability and permutation graphs. In: WADS 2009, pp. 387–398 (2009)
10. Jampani, K.R., Lubiw, A.: Simultaneous Interval Graphs. Arxiv 2010(2010)
11. Johnson, J.L., Spinrad, J.P.: A polynomial time recognition algorithm for probe interval graphs. In: SODA 2001, Philadelphia, PA, USA, pp. 477–486 (2001)
12. McConnell, R.M., Nussbaum, Y.: Linear-time recognition of probe interval graphs. In: Fiat, A., Sanders, P. (eds.) ESA 2009. LNCS, vol. 5757, pp. 349–360. Springer, Heidelberg (2009)
13. McConnell, R.M., Spinrad, J.P.: Construction of probe interval models. In: SODA 2002, Philadelphia, PA, USA, pp. 866–875 (2002)
14. McMorris, F.R., Wang, C., Zhang, P.: On probe interval graphs. Discrete Appl. Math. 88(1-3), 315–324 (1998)
15. Tanenbaum, P.J.: Simultaneous intersection representation of pairs of graphs. J. Graph Theory. 32(2), 171–190 (1999)

Unbalanced Graph Partitioning

Angsheng Li[1,*] and Peng Zhang[2,**]

[1] State Key Laboratory of Computer Science, Institute of Software,
Chinese Academy of Sciences, Beijing 100190, China
angsheng@ios.ac.cn
[2] School of Computer Science and Technology, Shandong University,
Jinan 250101, China
algzhang@sdu.edu.cn

Abstract. We investigate the unbalanced cut problems. A cut (A, B) is called *unbalanced* if the size of its smaller side is at most k (called k-size) or exactly k (called Ek-size), where k is an input parameter. An s-t cut (A, B) is called unbalanced if its s-side is of k-size or Ek-size. We consider three types of unbalanced cut problems, in which the quality of a cut is measured with respect to the capacity, the sparsity, and the conductance, respectively.

We show that even if the input graph is restricted to be a tree, the Ek-Sparsest Cut problem (to find an Ek-size cut with the minimum sparsity) is still **NP**-hard. We give a bicriteria approximation algorithm for the k-Sparsest Cut problem (to find a k-size cut with the minimum sparsity), which outputs a cut whose sparsity is at most $O(\log n)$ times the optimum and whose smaller side has size at most $O(\log n)k$. As a consequence, this leads to a $(O(\log n), O(\log n))$-approximation algorithm for the Min k-Conductance problem (to find a k-size cut with the minimum conductance). We also prove that the Min k-Size s-t Cut problem is **NP**-hard and give an $O(\log n)$-approximation algorithm for it.

1 Introduction

Cut problems have been fundamental and long-standing problems in combinatorial optimization and approximation algorithms. Given an undirected graph $G = (V, E)$ with nonnegative edge capacities $\{c(e)\}$, a cut (A, B) of graph G is the partition (A, B) of the vertex set V (that is, $A \neq \emptyset$, $B \neq \emptyset$, $A \cup B = V$ and $A \cap B = \emptyset$). Given a partition (A, B), it is convenient to view the cut (A, B) as the set of edges having one endpoint in A and the other endpoint in B. The capacity $c(A, B)$ of a cut (A, B), as usual, is the sum of capacities of edges in

* Supported by the hundred talent program of the Chinese Academy of Sciences, and the grand challenge program, *Network Algorithms and Digital Information*, Institute of Software, Chinese Academy of Sciences.
** Corresponding author. Supported by NSFC 60970003, the StarTrack Program of Microsoft Research Asia, and China Postdoctoral Science Foundation 20080441144, 200902562. Part of this work was done when the author visited Microsoft Research Asia.

O. Cheong, K.-Y. Chwa, and K. Park (Eds.): ISAAC 2010, Part I, LNCS 6506, pp. 218–229, 2010.

the cut. Note that if every edge in the graph has unit capacity, the capacity of a cut is simply the number of edges in the cut.

A cut is called *unbalanced* if its smaller side is size-bounded. We distinguish two types of unbalanced cuts. A cut (A, B) is a *k-size cut* if the size of its smaller side is at most k. A cut (A, B) is an *Ek-size cut* if the size of its smaller side is exactly k. For convenience, given an unbalanced cut (A, B), we always assume that A is the side with smaller size between A and B. In this article, we study three types of unbalanced cut problems, i.e., the min capacity unbalanced cut problem, the min sparsity unbalanced cut problem, and the min conductance unbalanced cut problem.

The k-Sparsest Cut Problem (k-SC). The input of the problem consists of an undirected graph $G = (V, E)$ with nonnegative capacities $\{c(e)\}$ defined on edges, a terminal pair set $D = \{(s_1, t_1), (s_2, t_2), \ldots, (s_q, t_q)\}$, and a positive integer k. For every terminal pair (s_i, t_i), there is a nonnegative demand $\text{dem}(i)$. The problem asks to find a k-size cut (A, B) such that its sparsity $s(A, B)$ is minimized. Given a cut (A, B), its *sparsity* $s(A, B)$ is defined as $\frac{c(A,B)}{\text{dem}(A,B)}$, where $\text{dem}(A, B) = \sum_{i: |A \cap \{s_i, t_i\}|=1} \text{dem}(i)$ is the total demands separated by the cut (A, B). If every vertex pair in $V(G)$ forms a terminal pair with unit demand, then the problem is called *uniform*. Note that in the uniform version $\text{dem}(A, B)$ is just $|A| \cdot |B|$.

Given an undirected graph G with nonnegative edge capacities and a cut (A, B) of G, the *conductance* of cut (A, B) is defined [11] by $\phi(A, B) = \frac{c(A,B)}{\min\{\text{vol}(A), \text{vol}(B)\}}$. Given any vertex subset $S \subseteq V(G)$, its *volume* $\text{vol}(S)$ is defined as $\sum_{u \in S} \sum_{v \in V} c(u, v)$, assuming that $c(u, v) = 0$ if $(u, v) \notin E(G)$. If for a vertex $u \in V(G)$, let's define $d(u) = \sum_{e \in \delta(u)} c(e)$, then the volume $\text{vol}(S)$ is just $\sum_{u \in S} d(u)$. Notice that if every edge in G has unit capacity, then $\text{vol}(S)$ is exactly the sum of degrees of vertices in S.

The Min k-Conductance Problem. In the problem we are given an undirected graph $G = (V, E)$ with nonnegative edge capacities and an integer $k > 0$. We are asked to find a k-size cut (A, B) whose conductance is minimized.

The Min k-Size Cut Problem. We are given an undirected graph $G = (V, E)$ with nonnegative capacities defined on edges, and a positive integer k. The problem asks to find a k-size cut (A, B) with minimized capacity.

Given two distinguished vertices s and t of graph G, where s is known as the source and t the sink, an *s-t cut* (A, B) is a cut which separates s and t. For the convenience, when we talk about an *s-t* cut (A, B), we always mean that $s \in A$ and $t \in B$. A k-size (Ek-size, resp.) *s-t* cut of G is an *s-t* cut whose s-side has size at most k (exactly k, resp.).

The Min k-Size s-t Cut Problem. Given an undirected graph $G = (V, E)$ with nonnegative edge capacities, a source-sink pair (s, t), and a positive integer k, the Min k-Size s-t Cut problem asks to find a k-size s-t cut such that its capacity is minimized among all such cuts of G.

For all unbalanced cut problems given above, their Ek-versions are defined similarly in the obvious way. This gives the Ek-Sparsest Cut problem (Ek-SC), the Min Ek-Size Cut problem, and the Min Ek-Size s-t Cut problem. For a problem P considered in this article, we use P(T) to denote the corresponding problem restricted in trees. For example, Ek-SC(T) denotes the Ek-Sparsest Cut in Trees problem. As usual, given any graph G, let n denote the number of vertices in G, and m denote the number of edges in G. All graphs considered in this article are undirected.

1.1 Motivations

The unbalanced cut problems are natural variants and generalization of the classical cut problems. In applications, the unbalanced cut problems are closely related to the clustering and classification problems in social and information networks. In a social network, the vertices represent underlying social entities and the edges represent some sort of interaction between pairs of vertices. A community (a.k.a. cluster) in a social network is thought of as a set of vertices that there have many connections between its members. As the rapid development of internet, social networks get more and more attention from researchers coming from many different areas. Identifying communities in a network and analyzing the community structure of a network are fundamental key problems in the research on social networks and information networks.

The Min Conductance problem (to calculate $\phi(G) = \min_{\emptyset \neq A \subset V}\{\phi(A, \overline{A})\}$) is closely related to the Sparsest Cut problem. Finding a min conductance cut in social networks is a commonly used method in discovery of community structure of social networks. Whereas there are many works aiming at analyzing the whole community structure of a network (see [8,12]), it is natural to identify some particular communities from a network, e.g., the community having bounded size. In this way, Leskovec et $al.$ [11, Section 3.1] studied large number of social and information networks by experimentally solving the Min Ek-Conductance problem and discovered a surprising phenomenon, that the best-expressed network communities in large-scale social networks are rather small (of size up to roughly 100 nodes), and that the sizes are practically independent of the network size. It is an interesting open question to theoretically analyze or prove the discovery by [11]. To the best of our knowledge, there is no known approximation algorithm with guaranteed performance ratio for the Min Ek-Conductance problem. For this, we study the unbalanced sparsest cut problems and the unbalanced min cut problems, which we would hope to be a step towards understanding the $small$ $community$ $phenomenon$.

1.2 Our Results

We give a bicriteria approximation algorithm for the k-Sparsest Cut problem, which outputs a cut whose sparsity is at most $O(\log n)$ times the optimum and whose smaller side has size at most $O(\log n)k$. As a consequence, this leads to a $(O(\log n), O(\log n))$-approximation algorithm for the Min k-Conductance problem. That is, we can find a cut in polynomial time such that its conductance is at most $O(\log n)$ times the minimum conductance over all k-size cuts of the input graph, and its smaller side has size at most $O(\log n)k$. To the best of our knowledge, there are no approximation algorithmic results for k-Sparsest Cut and Min k-Conductance previously. To approximate k-SC, we first convert the constraint k on number of vertices to a constraint on edge capacities. This idea is inspired by the work of Armon and Zwick [1] to solve the Min k-size Cut problem. The conversion results in a budget version of the Sparsest Cut problem in graphs with two capacity functions on edges, which is solved by reducing it to a budgeted version of the maximum multicut problem.

We show that the Ek-Sparsest Cut problem is **NP**-hard even if the input instances are restricted to trees. The idea behind this is the following. Even if Ek-SC is restricted to trees, to minimize the sparsity over Ek-size cuts we still need to find an Ek-size cut which separates as many as possible demands between terminal pairs. Finding such a cut is similar to the **NP**-hard MAX CUT problem.

For the min capacity unbalanced cut problems, we prove that the Min k-Size s-t Cut problem is **NP**-hard. This is an interesting result by noticing that the Min k-Size Cut problem can be optimally solved in polynomial time [1]. We give improved approximation ratio $O(\log n)$ for Min Ek-Size Cut. Based on this result, we show that Min k-Size s-t Cut can be approximated within a factor of $O(\log n)$ in polynomial time. Our idea to approximate Min k-Size s-t Cut is inspired by the work of finding minimum s-t bisection in [5].

Räcke in his recent paper [14] gave an elegant tree decomposition method to route multiflow in an undirected graph with low congestion. All of our algorithmic results (except for that in trees) in this paper eventually rely on Räcke's result.

1.3   Related Works

To the best of our knowledge, our paper gives theoretical results for k-Sparsest Cut and Min k-Conductance at the first time. For the uniform Sparsest Cut problem and the Min Conductance problem, Arora, Rao and Vazirani [3] gave $O(\sqrt{\log n})$-approximation algorithms. For the general Sparsest Cut problem, Arora, Lee and Naor [2] showed that it can be approximated within a factor of $O(\sqrt{\log q} \log \log q)$. The works in [3] and [2] are a remarkable breakthrough in approximating Sparsest Cut.

Along the way of the budget versions of cut problems, Engelberg et al. [4] proposed the Budgeted Multicut problem and the Budgeted Multiway Cut problem. They gave a $(1 - \frac{1}{e}, O(\log^2 n \log \log n))$-approximation algorithm for the Budgeted Multicut problem and a constant factor approximation algorithm for the Budgeted Multiway Cut problem.

Both the Min Ek-Size Cut problem and the Min Ek-Size s-t Cut problem are **NP**-hard (please refer to the Minimum Cut into Equal-sized Subsets problem, which is known to be **NP**-hard [7]). For the Min Ek-Size Cut problem, Feige and Krauthgamer [5] gave an $O(\log^{1.5} n)$-approximation algorithm for general k

based on their approximation algorithm for the Min Bisection problem. In [6], Feige, Krauthgamer and Nissim considered approximating the Min Ek-Size Cut problem and the Min Ek-Size s-t Cut problem when k is relatively small using Karger's edge contraction technique [10]. For both of these two problems, they gave a polynomial time randomized approximation scheme when $k = O(\log n)$ and a randomized $O(\frac{k}{\log n})$-approximation algorithm when $k = \Omega(\log n)$. As pointed out by the authors [6], their $O(\frac{k}{\log n})$-approximation algorithm should be used only when k is slightly larger than $O(\log n)$.

For the Min k-Size Cut problem, Armon and Zwick [1] showed that it can be optimally solved in $O(n^6 \log n)$ time. Armon and Zwick's algorithm for Min k-Size Cut eventually relies on enumerating all the 2-approximate min cuts in an undirected graph, where a 2-approximate min cut is a cut whose capacity is at most two times that of the min cut. By the work of Nagamochi $et\ al.$ [13], the enumeration can be done in $O(n^6)$ time.

There are also two works that can be put in the theme of unbalanced cut problems. Hayrapetyan $et\ al.$ [9] studied the Min-Size Bounded-Capacity Cut problem (MinSBCC, for short), while Svitkina and Tardos [15] considered the Max-Size Bounded-Capacity Cut problem (MaxSBCC, for short). Hayrapetyan $et\ al.$ [9] showed that MinSBCC has interesting applications in the containment of epidemics and finding small communities in social networks. For both MinS-BCC and MaxSBCC, only bicriteria approximation algorithms are currently known [9,15].

2   The General Unbalanced Sparsest Cut Problems

2.1   Hardness Result

The (general) unbalanced Sparest Cut problems are more complicated than their uniform versions. For the Sparsest Cut problem (whether or not it is required to be unbalanced), it is helpful to describe its demands by a $demand\ graph$. Fix any instance of the Sparsest Cut problem including input graph G and terminal pair set D. The demand graph H for the instance is constructed as follows. The vertex set of graph H is simply $V(G)$. For every terminal pair $(s_i, t_i) \in D$, there is an undirected edge $e = (s_i, t_i)$ in $E(H)$ with demand $d(e) = d_i$. For the Sparsest Cut problem, the goal of finding a cut with minimized sparsity is actually to find a cut such that the total capacity in graph G crossing the cut is as small as possible, meanwhile the total demand in graph H crossing the cut is as large as possible.

As mentioned in the introduction, we distinguish two versions of the un-balanced Sparsest Cut problem, i.e., the Ek-Sparsest Cut problem and the k-Sparsest Cut problem. In the following we show that even restricted to trees, the Ek-Sparsest Cut problem (Ek-SC(T)) is still **NP**-hard. The idea behind the proof is as follows. Although the input graph is a tree in Ek-SC(T), the corre-sponding demand graph is not a tree in general. The task to find a cut in the demand graph that maximizes as far as possible the total separated demands is similar to that of the MAX CUT problem, which is known to be **NP**-complete [7].

Theorem 1. *The Ek-Sparsest Cut in Trees problem is* **NP**-*hard.*

Proof. We proceed by proving that the decision version of Ek-SC(T) is **NP**-complete. In this proof we slightly abuse notations to refer to Ek-SC(T) as both its optimization version and its decision version.

Given an instance (T, k, D, s) of Ek-SC(T) and a cut (A, B) of T, we can easily check in polynomial time whether (A, B) is an Ek-size cut and its sparsity is at most s. This means that Ek-SC(T) is in **NP**.

Then we shall reduce MAX CUT to Ek-SC(T). The instance $\mathcal{I}_M$ of MAX CUT consists of an undirected graph G and a positive integer c; and the problem asks whether there is a cut of G whose capacity is at least c. Note that in $\mathcal{I}_M$ every edge has unit capacity, so the capacity of a cut in the MAX CUT problem is actually the number of edges crossing it.

We construct the instance $\mathcal{I}_m = (T, k, D, s)$ of Ek-SC(T) as follows. Let $n = |V(G)|$. T is a tree of height 1 having $2n$ leaves and an additional vertex r as its root. T uses all the vertices in $V(G) = \{v_1, v_2, \cdots, v_n\}$ and n additional vertices in $U = \{u_1, u_2, \cdots, u_n\}$ as its leaves. Every edge in T has unit capacity. For every edge (v_i, v_j) in $E(G)$, there is a terminal pair (v_i, v_j) in D. That is, graph G is actually used as the demand graph of instance $\mathcal{I}_m$. Finally, let $k = n$ $(= \frac{1}{2}(|V(T)| - 1))$ and $s = \frac{n}{c}$. Obviously the construction can be done in polynomial time.

Denote by c^* the capacity of a max cut of instance $\mathcal{I}_M$, and by s^* the sparsity of a sparsest Ek-size cut of instance $\mathcal{I}_m$. We show that $c^* \geq c$ iff $s^* \leq \frac{n}{c}$, and hence the theorem holds.

(Sufficiency.) Suppose that (A', B') is a sparsest Ek-size cut of $\mathcal{I}_m$ satisfying $|A'| = k$ (recall that $k = n$) and $|B'| = n + 1$. We claim that $r \notin A'$. Otherwise A' has to contain $n - 1$ leaves and thus $c_T(A', B') = n + 1$. Because B' contains at most $n - 1$ vertices from $\{v_1, v_2, \cdots, v_n\}$ (otherwise $\text{dem}_T(A', B')$ would be zero, which is impossible), there are at least 2 vertices in B' from U. Suppose one of such two vertices is u. Define $A'' = A' \cup \{u\} - \{r\}$ and $B'' = V(T) - A''$. Since there is no any demand incident to u and r, we have that $\text{dem}_T(A'', B'') = \text{dem}_T(A', B')$. Moreover, $c_T(A'', B'') = n < c_T(A', B')$. So we have constructed an Ek-size cut (A'', B'') whose sparsity is strictly less than that of (A', B'), contradicting the optimality of (A', B').

Since $r \notin A'$, we have that $c_T(A', B') = n$. Because $s(A', B') = \frac{c_T(A', B')}{\text{dem}_T(A', B')} \leq \frac{n}{c}$, we have $\text{dem}_T(A', B') \geq c$. Define $A = A' \cap V(G)$ and $B = V(G) - A$. Then (A, B) is a cut of G satisfying that $c_G(A, B) = \text{dem}_T(A', B')$. So we conclude that $c^* \geq c_G(A, B) \geq c$.

(Necessity.) Suppose that (A, B) is a max cut of G with $c_G(A, B) = c^* \geq c$. We define $A' = A \cup \{u_1, u_2, \cdots, u_j\}$ and $B' = V(T) - A'$, where $j = n - |A|$. So we have that $c_T(A', B') = n$ and $\text{dem}_T(A', B') = c_G(A, B) = c^*$. This gives $s_T(A', B') = \frac{n}{c^*} \leq \frac{n}{c}$. □

Note that in the proof of Theorem 1 we have $k < \frac{1}{2}|V(T)|$ for the instance $\mathcal{I}_m$ of Ek-SC(T). In fact we can make as many as arbitrary polynomial (in n) copies of all the vertices in U, leading to that k becomes more far away from half of $|V(T)|$, and the Ek-sparsest cut (A', B') becomes more unbalanced.

2.2 Approximating k-Sparsest Cut

In this section we give a $(O(\log n), O(\log n))$-approximation algorithm for k-Sparsest Cut in Graphs. Let $\text{OPT}_{k-\text{SC}}$ denote the sparsity of an optimal cut for the k-Sparsest Cut problem. In polynomial time, our algorithm gives a cut (A, B) such that $s(A, B) \leq O(\log n)\text{OPT}_{k-\text{SC}}$ and $|A| \leq O(\log n)k$.

First we give approximation results for some related cut problems. We then use these results to design the bicriteria approximation algorithm for k-Sparsest Cut.

Budgeted Multicut in Graphs (BMC). The instance of BMC consists of a graph $G = (V, E)$ with edge capacity function, a nonnegative bound B, and a terminal pair set $D = \{(s_1, t_1), (s_2, t_2), \cdots, (s_q, t_q)\}$. For each terminal pair $(s_i, t_i) \in D$, there is a positive demand dem(i). The goal is to find a multicut $E' \subseteq E$ such that $c(E') \leq B$ and dem(E') is maximized.

For the BMC problem Engelberg *et al.* [4] gave a (α, β)-approximation algorithm, where $\alpha = 1 - \frac{1}{e}, \beta = O(\log^2 n \log \log n)$. In polynomial time, their algorithm finds a multicut E' such that dem(E') is at least $1 - \frac{1}{e}$ times the optimum and $c(E') \leq \beta B$. Using the tree decomposition technique of Räcke [14], we remark that the factor β can be improved to $O(\log n)$. The proof of Theorem 2 is omitted due to space limitation and will be given in the full version.

Theorem 2. *The Budgeted Multicut in Graphs problem can be approximated within a bi-factor of $(1 - \frac{1}{e}, O(\log n))$ in polynomial time.*

Multi-criteria Multicut (MCMC). In the MCMC problem, we are given a graph $G = (V, E)$. There are two capacity functions $c_1 \colon E \to \mathbf{R}^+$ and $c_2 \colon E \to \mathbf{R}^+$ defined for edges of graph G. We are also given two nonnegative bounds B_1 and B_2, and a terminal pair set $D = \{(s_1, t_1), (s_2, t_2), \cdots, (s_q, t_q)\}$. For each terminal pair $(s_i, t_i) \in D$, there is a positive demand dem(i). We are asked to find a multicut $E' \subseteq E$ such that $c_1(E') \leq B_1$, $c_2(E') \leq B_2$ and dem(E') is maximized.

Lemma 1. *Suppose that BMC admits (α, β)-approximation. Then there exists an approximation algorithm for MCMC which outputs in polynomial time a multicut E' such that dem(E') is at least α times the optimum, $c_1(E') \leq 2\beta B_1$ and $c_2(E') \leq 2\beta B_2$.*

Proof. We reduce MCMC to BMC by normalizing its capacity functions. Let $\mathcal{I} = (G, c_1, c_2, B_1, B_2, D, \text{dem})$ be an instance of MCMC. We construct the BMC instance $\mathcal{J} = (G, c, B, D, \text{dem})$ as follows. Define a new capacity function $c'_2(e) = \frac{B_1}{B_2} c_2(e)$, where $e \in E(G)$. In instance $\mathcal{J}$, let $c(e) = c_1(e) + c'_2(e)$ for every edge e, $B = 2B_1$.

Let E' be a (α, β)-approximate solution to instance $\mathcal{J}$, $\text{OPT}(\mathcal{I})$ and $\text{OPT}(\mathcal{J})$ be the optimal values for instance $\mathcal{I}$ and $\mathcal{J}$, respectively. Then we have $c_1(E') \leq$

$c(E') \leq \beta B = 2\beta B_1$, $c_2(E') = \frac{B_2}{B_1}c_2'(E') \leq \frac{B_2}{B_1}c(E') \leq \frac{B_2}{B_1}\beta B = 2\beta B_2$, and $\text{dem}(E') \geq \alpha\text{OPT}(\mathcal{J}) \geq \alpha\text{OPT}(\mathcal{I})$, where $\text{OPT}(\mathcal{J}) \geq \text{OPT}(\mathcal{I})$ holds since an optimal solution E^* ($c_1(E^*) \leq B_1$ and $c_2(E^*) \leq B_2$) to instance $\mathcal{I}$ is a feasible solution to instance $\mathcal{J}$ with the same solution value. □

Budgeted Sparsest Cut (BSC). The instance of BSC is the same as that of MCMC except that there is no bound B_1. The goal is to find a cut (A, B) such that $c_2(A, B) \leq B_2$ and (A, B)'s sparsity $s(A, B) = \frac{c_1(A,B)}{\text{dem}(A,B)}$ is minimized.

Lemma 2. *Let $E' \subseteq E(G)$ be a multicut on BSC instance $(G, c_1, c_2, B_2, D, \text{dem})$. Then there exists a cut (A, B) of G such that $\frac{c_1(A,B)}{\text{dem}(A,B)} \leq \frac{c_1(E')}{\text{dem}(E')}$ and $c_2(A, B) \leq c_2(E')$.*

Proof. Removing E' from graph G breaks G into several connected components, say $V_1, V_2, \cdots, V_r$. Then we have

$$\min_{1 \leq i \leq r}\left\{\frac{c_1(V_i, \overline{V_i})}{\text{dem}(V_i, \overline{V_i})}\right\} \leq \frac{\sum_i c_1(V_i, \overline{V_i})}{\sum_i \text{dem}(V_i, \overline{V_i})} = \frac{2 \cdot c_1(E')}{2 \cdot \text{dem}(E')} = \frac{c_1(E')}{\text{dem}(E')}$$

and

$$\forall 1 \leq i \leq r, c_2(V_i, \overline{V_i}) \leq c_2(E').$$

Letting (A, B) be the cut with minimized sparsity in $\{(V_i, \overline{V_i})\}$ completes the proof. □

Lemma 3. *Suppose that MCMC admits $(\gamma, \beta_1, \beta_2)$-approximation. Then there exists an approximation algorithm for BSC which outputs in polynomial time a cut (A, B) such that the sparsity of (A, B) is at most $\frac{1}{\gamma}(1 + \epsilon)\beta_1$ times the optimum and $c_2(A, B) \leq \beta_2 B_2$, where $\epsilon > 0$ is an arbitrary small constant.*

Proof. By adapting the idea of [4, Theorem 26], we reduce BSC to MCMC by the following algorithm.

> Algorithm for BSC.
> 1. **for** each integer i from $\lceil\log_{1+\epsilon}\min_{e \in E(G)}\{c_1(e)\}\rceil$ to $\lceil\log_{1+\epsilon}c_1(E(G))\rceil$ **do**
> 2.   Call the $(\gamma, \beta_1, \beta_2)$-approximation algorithm for MCMC on instance $\mathcal{J}_i = (G, c_1, c_2, B_1(i), B_2, D, \text{dem})$, where $B_1(i) = (1 + \epsilon)^i$. Let E_i be the multicut found by the algorithm.
> 3.   Get a cut (A_i, B_i) from E_i by Lemma 2.
> 4. **endfor**
> 5. **return** the cut with the minimum sparsity in $\{(A_i, B_i)\}$.

Suppose that (A^*, B^*) is an optimal cut to BSC with sparsity $s^* = \frac{c_1(A^*,B^*)}{\text{dem}(A^*,B^*)}$ and $c_2(A^*, B^*) \leq B_2$. Let $j = \lceil\log_{1+\epsilon}c_1(A^*, B^*)\rceil$. Then we have $B_1(j) = (1 + \epsilon)^j \leq (1 + \epsilon)c_1(A^*, B^*)$ and $B_1(j) \geq c_1(A^*, B^*)$.

The multicut E_j found by the algorithm for MCMC on instance $\mathcal{J}_j$ satisfies that $c_1(E_j) \leq \beta_1 B_1$, $c_2(E_j) \leq \beta_2 B_2$ and $\text{dem}(E_j) \geq \gamma \cdot d^*(j)$, where $d^*(j)$ is the optimum of instance $\mathcal{J}_j$.

So for the cut (A_j, B_j) corresponding to E_j we have

$$c_2(A_j, B_j) \leq c_2(E_j) \leq \beta_2 B_2$$

and

$$\frac{c_1(A_j, B_j)}{\text{dem}(A_j, B_j)} \leq \frac{c_1(E_j)}{\text{dem}(E_j)} \leq \frac{1}{\gamma}\beta_1 \frac{B_1(j)}{d^*(j)} \leq \frac{1}{\gamma}(1 + \epsilon)\beta_1 \frac{c_1(A^*, B^*)}{d^*(j)}$$
$$\leq \frac{1}{\gamma}(1 + \epsilon)\beta_1 \frac{c_1(A^*, B^*)}{\text{dem}(A^*, B^*)} = \frac{1}{\gamma}(1 + \epsilon)\beta_1 s^*,$$

where the last inequality holds since (A^*, B^*) is a feasible solution to the MCMC instance $\mathcal{J}_j$ and hence $d^*(j) \geq \text{dem}(A^*, B^*)$.   □

By Lemma 3, we know that BSC can be approximated within a bi-factor of (α', β') where $\alpha' = \frac{1}{\gamma}(1 + \epsilon)\beta_1$ and $\beta' = \beta_2$ for any small constant $\epsilon > 0$. We then use the algorithm for BSC in Lemma 3 to design the approximation algorithm $\mathcal{A}$ for k-SC. Let us call the approximation algorithm for Sparsest Cut in [2] the ALN algorithm, which is also used in Algorithm $\mathcal{A}$.

Algorithm $\mathcal{A}$ for k-SC.
1. **if** $k > \frac{n}{4\beta'-1}$ **then**
2.   Call the ALN algorithm for Sparsest Cut on instance $\mathcal{J}_{\text{SC}} = (G, c, D, \text{dem})$.
3. **else**
4.   Let $G^c = (V, E^c)$ be the complete graph corresponding to graph G. For every edge $e \in E^c$, define $c_1(e) = c(e)$ if $e \in E(G)$, and $c_1(e) = 0$ otherwise, and define $c_2(e) = 1$. Let $B_2 = k(n - k)$. This gives a BSC instance $\mathcal{J}_{\text{BSC}} = (G^c, c_1, c_2, B_2, D, \text{dem})$.
5.   Call the approximation algorithm in Lemma 3 on instance $\mathcal{J}_{\text{BSC}}$.
6. **endif**
7. **return** the cut (A, B) found in Step 2 or Step 5.

Theorem 3. *Algorithm $\mathcal{A}$ is a $(O(\log n), O(\log n))$-approximation algorithm for the k-Sparsest Cut problem. That is, in polynomial time, Algorithm $\mathcal{A}$ outputs a cut (A, B) such that the sparsity of (A, B) is at most $O(\log n)$ times the optimum and $|A| \leq O(\log n)k$.*

Proof. Let $s_k^* = \text{OPT}_{k-\text{SC}}$ be the sparsity of an optimal cut to the instance of k-SC.

First consider the case $k > \frac{n}{4\beta'-1}$. Let s^* be the sparsity of an optimal cut to Sparsest Cut instance $\mathcal{J}_{\text{SC}}$. It is obviously that $s^* \leq s_k^*$. By [2], $s(A, B) \leq O(\sqrt{\log q} \log \log q)s^* \leq O(\log n)s_k^*$. Since $k > \frac{n}{4\beta'-1}$, $|A| \leq \frac{n}{2} < \frac{1}{2}(4\beta' - 1)k$. By Lemmas 3, 1 and Theorem 2, we know $\beta' = \beta_2 = O(\log n)$. So the theorem is proved under the case $k > \frac{n}{4\beta'-1}$.

Next consider the case $k \leq \frac{n}{4\beta'-1}$.

Let s^* be the sparsity of an optimal cut to BSC instance $\mathcal{J}_{\text{BSC}}$. Suppose that (A^*, B^*) is an optimal solution to k-SC with $|A^*| \leq k$. Then in graph G^c,

$c_2(A^*, B^*) \leq k(n - k)$. So (A^*, B^*) is a feasible solution to BSC instance $\mathcal{J}_{\text{BSC}}$ and hence $s^* \leq s_{G^c}(A^*, B^*) = s_G(A^*, B^*) = s_k^*$. For the cut (A, B) found in Step 5 of Algorithm $\mathcal{A}$, we have $s_G(A, B) = \frac{c(A,B)}{\text{dem}(A,B)} = \frac{c_1(A,B)}{\text{dem}(A,B)} = s_{G^c}(A, B) \leq \alpha' s^* \leq \alpha' s_k^*$. By Lemmas 3, 1 and Theorem 2, we know $\alpha' = \frac{1}{\gamma}(1 + \epsilon)\beta_1 = \frac{2e}{e-1}(1 + \epsilon)\beta = O(\log n)$.

Then we analyze the unbalance property of (A, B). Since (A, B) is a solution to the BSC instance $\mathcal{J}_{\text{BSC}}$, $|A| \cdot |B| = c_2(A, B) \leq \beta' k(n - k)$. As before, suppose that A is the one of smaller size between the two parts of cut (A, B). We claim that $|A| \leq 2\beta' k = O(\log n)k$ when $k \leq \frac{n}{4\beta'-1}$. Suppose not and hence $|A| > 2\beta' k$. Then we have

$$
\begin{aligned}
c_2(A, B) &> 2\beta' k \cdot (n - 2\beta' k) \\
&= \beta' k(2n - 4\beta' k) \\
&= \beta' k(n - k + n - (4\beta' - 1)k) \\
&\geq \beta' k(n - k),
\end{aligned}
$$

which is a contradiction. This concludes the theorem. □

Theorem 4. *The Min k-Conductance problem can be approximated within a bi-factor of $(O(\log n), O(\log n))$. That is, we can calculate in polynomial time a cut (A, B) such that its conductance is at most $O(\log n)$ times ϕ_k^* and $|A| \leq O(\log n)k$, where ϕ_k^* is the minimum conductance over all k-size cuts of the problem instance.*

Proof. We show that the standard technique of approximating the Min Conductance problem via solving the Sparsest Cut problem [16] can be generalized to approximate Min k-Conductance.

Recall that the conductance $\phi(A, B) = \frac{c(A,B)}{\min\{\text{vol}(A),\text{vol}(B)\}}$. The symmetrized variant ϕ' of ϕ is defined as $\phi'(A, B) = \frac{c(A,B)}{\text{vol}(A) \cdot \text{vol}(B)}$. Fix any k-size cut (A, B). Without loss of generality, suppose $\text{vol}(A) \leq \text{vol}(B)$. Since $\frac{1}{2}\text{vol}(V) \leq \text{vol}(B) \leq \text{vol}(V)$, $\frac{1}{\text{vol}(V)}\phi(A, B) \leq \phi'(A, B) \leq \frac{2}{\text{vol}(V)}\phi(A, B)$. So up to constant factor of 2, approximating $\min_{|A| \leq k}\{\phi'(A, \overline{A})\}$ suffices to approximate ϕ_k^*.

For each vertex pair (u, v), define $\text{dem}(u, v) = d(u) \cdot d(v)$. Then $\text{vol}(A) \cdot \text{vol}(B) = \left(\sum_{u \in A} d(u)\right)\left(\sum_{v \in B} d(v)\right) = \sum_{u \in A}\sum_{v \in B} \text{dem}(u, v) = \text{dem}(A, B)$. So $\phi'(A, B) = s(A, B)$, and thus approximating $\min_{|A| \leq k}\{\phi'(A, \overline{A})\}$ is in fact equivalent to approximating the k-Sparsest Cut problem.

Let (A^*, B^*) is an optimal k-size cut to the Min k-Conductance problem. Compute an approximate k-sparsest cut (A, B) on the k-SC instance $\mathcal{J}_{k-\text{SC}} = (G, c, k, D, \text{dem})$ by Algorithm $\mathcal{A}$. Suppose that (A', B') is an optimal k-size cut to instance $\mathcal{J}_{k-\text{SC}}$. Let $\beta = O(\log n)$. Then we have

$$
s(A, B) \leq \beta \cdot s(A', B') \leq \beta \cdot s(A^*, B^*) \leq \frac{2\beta}{\text{vol}(V)}\phi(A^*, B^*) = \frac{2\beta}{\text{vol}(V)}\phi_k^*
$$

and

$$
s(A, B) = \phi'(A, B) \geq \frac{1}{\text{vol}(V)}\phi(A, B).
$$

Consequently, we have $\phi(A, B) \leq \text{vol}(V) \cdot s(A, B) \leq 2\beta \cdot \phi_k^*$. Since $|A| \leq O(\log n)k$, the theorem is concluded. □

Theorem 5. *The k-Sparsest Cut in Trees problem can be approximated within a bi-factor of $(3.17, 4)$. That is, there exists an approximation algorithm for the problem which in polynomial time outputs a cut (A, B) such that the sparsity of (A, B) is at most 3.17 times the optimum and $|A| \leq 4k$.*

Proof. The proof is similar to that of Theorem 3. We distinguish the two cases that $k > \frac{n}{7}$ and $k \leq \frac{n}{7}$. When $k > \frac{n}{7}$, we use the optimal cut (A, B) to the Sparsest Cut in Trees (SC(T)) instance (G, c, D, dem) as the solution to k-SC(T). It is well known that SC(T) can be optimally solved in polynomial time. Since $|A| \leq \frac{n}{2} < \frac{7}{2}k$, we get a $(1, 3.5)$-approximation to k-SC(T) under this case.

Then consider the case $k \leq \frac{n}{7}$. By [4], BMC(T) can be approximated within $1 - \frac{1}{e}$. So Lemma 1 implies that MCMC(T) can be approximated within $(1 - \frac{1}{e}, 2, 2)$, and Lemma 3 implies that BSC(T) can be approximated within $(\frac{2e}{e-1}(1+\epsilon), 2)$. Let s_k^* be the optimum of k-SC(T). By the same method as in Algorithm $\mathcal{A}$ we can get a cut (A, B) such that $s(A, B) \leq \frac{2e}{e-1}(1+\epsilon)s_k^* \leq 3.17s_k^*$ for sufficiently small ϵ and $|A| \leq 4k$. This shows k-SC(T) can be approximated within $(3.17, 4)$. The theorem follows. □

3   The Unbalanced Min Cut Problems

In this section, we mainly consider the unbalanced min s-t cut problems. These problems have potential applications in identifying a network community around a given node from social networks. We first prove that the Min k-Size s-t Cut problem is **NP**-hard. Then we show that the Min Ek-Size Cut problem admits $O(\log n)$-approximation. Finally, we show that the Min k-Size s-t Cut problem can be approximated within $O(\log n)$ by reducing it to the Min Ek-Size Cut problem.

Theorem 6. *The Min k-Size s-t Cut problem is **NP**-hard.*

Theorem 7. *The Min k-Size s-t Cut problem can be approximated within $O(\log n)$ in polynomial time.*

The proofs of Theorem 6 and Theorem 7 are omitted due to space limitation and will be given in the full version of the paper.

4   Conclusions

We investigate the unbalanced cut problems. We give $(O(\log n), O(\log n))$ approximation algorithms for the k-Sparsest Cut and Min k-Conductance problems, and an $O(\log n)$-approximation algorithm for the Min k-Size s-t Cut problem. For the Ek-Sparsest Cut problem, we prove that it is **NP**-hard even if the problem is restricted to trees. For the k-Sparsest Cut in Trees problem, we give a $(3.17, 4)$-approximation algorithm.

Two specific interesting open questions are: to decide the complexity of the k-Sparsest Cut in Trees problem, and to build approximation algorithms to the Ek-Sparsest Cut problem. In general, unbalanced graph partitioning is a widely area and still needs more research. In applications for social networks, it is also very interesting to study the unbalanced partitioning of power law graphs.

Acknowledgements

We are very grateful for helpful discussions with Wei Chen, Pinyan Lu and Yajun Wang (Theory Group, Microsoft Research Asia) on this topic.

References

1. Armon, A., Zwick, U.: Multicriteria Global Minimum Cuts. Algorithmica 46(1), 15–26 (2006)
2. Arora, S., Lee, J., Naor, A.: Euclidean distortion and the sparsest cut. In: Proc. of STOC, pp. 553–562 (2005)
3. Arora, S., Rao, S., Vazirani, U.: Expander flows, geometric embeddings and graph partitioning. In: Proc. of STOC, pp. 222–231 (2004)
4. Engelberg, R., Könemann, J., Leonardi, S., Naor, J. (Seffi): Cut problems in graphs with a budget constraint. Journal of Discrete Algorithms 5, 262–279 (2007)
5. Feige, U., Krauthgamer, R.: A polylogarithmic approximation of the minimum bisection. SIAM Review 48(1), 99–130 (2006); Preliminary version appears in: SIAM Journal on Computing, 31(4), 1090–1118 (2002)
6. Feige, U., Krauthgamer, R., Nissim, K.: On cutting a few vertices from a graph. Discrete Applied Mathematics 127, 643–649 (2003)
7. Garey, M., Johnson, D., Stockmeyer, L.: Some simplified NP-complete graph problems. Theoretical Computer Science 1, 237–267 (1976)
8. Girvan, M., Newman, M.E.J.: Community structure in social and biological networks. Proceedings of the National Academy of Sciences of the United States of America 99(12), 7821–7826 (2002)
9. Hayrapetyan, A., Kempe, D., Pál, M., Svitkina, Z.: Unbalanced Graph Cuts. In: Brodal, G.S., Leonardi, S. (eds.) ESA 2005. LNCS, vol. 3669, pp. 191–202. Springer, Heidelberg (2005)
10. Karger, D., Stein, C.: A new approach to the minimum cut problem. Journal of the ACM 43(4), 601–640 (1996)
11. Leskovec, J., Lang, K., Dasgupta, A., Mahoney, M.: Statistical properties of community structure in large social and information networks. In: Proc. of WWW, pp. 695–704 (2008)
12. Mann, C., Matula, D., Olinick, E.: The use of sparsest cut to reveal the hierarchical community structure of social networks. Social Networks 30, 223–234 (2008)
13. Nagamochi, H., Nishimura, K., Ibaraki, T.: Computing all small cuts in an undirected network. SIAM Journal on Discrete Mathematics 10(3), 469–481 (1997)
14. Räcke, H.: Optimal hierarchical decompositions for congestion minimization in networks. In: Proc. of STOC, pp. 255–264 (2008)
15. Svitkina, Z., Tardos, É.: Min-max multiway cut. In: Jansen, K., Khanna, S., Rolim, J.D.P., Ron, D. (eds.) RANDOM 2004 and APPROX 2004. LNCS, vol. 3122, pp. 207–218. Springer, Heidelberg (2004)
16. Vazirani, V.: Approximation Algorithms. Springer, Berlin (2001)

On the Intersection of Tolerance and Cocomparability Graphs

George B. Mertzios and Shmuel Zaks

Department of Computer Science, Technion, Haifa, Israel
{mertzios,zaks}@cs.technion.ac.il

Abstract. It has been conjectured by Golumbic and Monma in 1984 that the intersection of tolerance and cocomparability graphs coincides with bounded tolerance graphs. Since cocomparability graphs can be efficiently recognized, a positive answer to this conjecture in the general case would enable us to efficiently distinguish between tolerance and bounded tolerance graphs, although it is NP-complete to recognize each of these classes of graphs separately. The conjecture has been proved under some – rather strong – *structural* assumptions on the input graph; in particular, it has been proved for complements of trees, and later extended to complements of bipartite graphs, and these are the only known results so far. Furthermore, it is known that the intersection of tolerance and cocomparability graphs is contained in the class of trapezoid graphs. In this article we prove that the above conjecture is true for every graph G, whose tolerance representation satisfies a slight assumption; note here that this assumption concerns only the given tolerance *representation* R of G, rather than *any* structural property of G. This assumption on the representation is guaranteed by a wide variety of graph classes; for example, our results immediately imply the correctness of the conjecture for complements of triangle-free graphs (which also implies the above-mentioned correctness for complements of bipartite graphs). Our proofs are algorithmic, in the sense that, given a tolerance representation R of a graph G, we describe an algorithm to transform R into a bounded tolerance representation R^* of G. Furthermore, we conjecture that any minimal tolerance graph G that is not a bounded tolerance graph, has a tolerance representation with exactly one unbounded vertex. Our results imply the non-trivial result that, in order to prove the conjecture of Golumbic and Monma, it suffices to prove our conjecture. In addition, there already exists evidence in the literature that our conjecture is true.

Keywords: Tolerance graphs, cocomparability graphs, 3-dimensional intersection model, trapezoid graphs, parallelogram graphs.

1   Introduction

A simple undirected graph $G = (V, E)$ on n vertices is called a *tolerance* graph if there exists a collection $I = \{I_u \mid u \in V\}$ of closed intervals on the real line

O. Cheong, K.-Y. Chwa, and K. Park (Eds.): ISAAC 2010, Part I, LNCS 6506, pp. 230–240, 2010.

and a set $t = \{t_u \mid u \in V\}$ of positive numbers, such that for any two vertices $u, v \in V$, $uv \in E$ if and only if $|I_u \cap I_v| \geq \min\{t_u, t_v\}$. The pair $\langle I, t \rangle$ is called a *tolerance representation* of G. A vertex u of G is called a *bounded vertex* (in a certain tolerance representation $\langle I, t \rangle$ of G) if $t_u \leq |I_u|$; otherwise, u is called an *unbounded vertex* of G. If G has a tolerance representation $\langle I, t \rangle$ where all vertices are bounded, then G is called a *bounded tolerance* graph and $\langle I, t \rangle$ a *bounded tolerance representation* of G.

Tolerance graphs find numerous applications (in bioinformatics, constrained-based temporal reasoning, resource allocation, and scheduling problems, among others). Since their introduction in 1982 [9], these graphs have attracted many research efforts [2, 4, 7, 10–12, 15, 16], as they generalize in a natural way both interval and permutation graphs [9]; see [12] for a detailed survey.

Given an undirected graph $G = (V, E)$ and a vertex subset $M \subseteq V$, M is called a *module* in G, if for every $u, v \in M$ and every $x \in V \setminus M$, x is either adjacent in G to both u and v or to none of them. Note that $\emptyset$, V, and all singletons $\{v\}$, where $v \in V$, are trivial modules in G. A *comparability* graph is a graph which can be transitively oriented. A *cocomparability* graph is a graph whose complement is a comparability graph. A *trapezoid* (resp. *parallelogram* and *permutation*) graph is the intersection graph of trapezoids (resp. parallelograms and line segments) between two parallel lines L_1 and L_2 [8]. Such a representation with trapezoids (resp. parallelograms and line segments) is called a *trapezoid* (resp. *parallelogram* and *permutation*) *representation* of this graph. A graph is bounded tolerance if and only if it is a parallelogram graph [2]. Permutation graphs are a strict subset of parallelogram graphs [3]. Furthermore, parallelogram graphs are a strict subset of trapezoid graphs [19], and both are subsets of cocomparability graphs [8, 12]. On the other hand, not every tolerance graph is a cocomparability graph [8, 12].

Cocomparability graphs have received considerable attention in the literature, mainly due to their interesting structure that leads to efficient algorithms for several NP-hard problems, see e.g. [5, 6, 12, 14]. Furthermore, the intersection of the class of cocomparability graphs with other graph classes has interesting properties and coincides with other widely known graph classes. For instance, their intersection with chordal graphs is the class of interval graphs, while their intersection with comparability graphs is the class of permutation graphs [8]. These structural characterizations find also direct algorithmic implications to the recognition problem of interval and permutation graphs, respectively, since the class of cocomparability graphs can be recognized efficiently [8, 20]. In this context, the following conjecture has been made in 1984 [10]:

Conjecture 1 ([10]). The intersection of cocomparability graphs with tolerance graphs is exactly the class of bounded tolerance graphs.

Note that the inclusion in one direction is immediate: every bounded tolerance graph is a cocomparability graph [8, 12], as well as a tolerance graph by definition. Conjecture 1 has been proved for complements of trees [1], and later extended to complements of bipartite graphs [18], and these are the only known results so far. Furthermore, it has been proved that the intersection of tolerance and cocomparability graphs is contained in the class of trapezoid graphs [7].

Since cocomparability graphs can be efficiently recognized [20], a positive answer to Conjecture 1 would enable us to efficiently distinguish between tolerance and bounded tolerance graphs, although it is NP-complete to recognize both tolerance and bounded tolerance graphs [16]. Recently, an intersection model for general tolerance graphs has been presented in [15], given by 3D-parallelepipeds. This *parallelepiped representation* of tolerance graphs generalizes the parallelogram representation of bounded tolerance graphs; the main idea is to exploit the third dimension to capture the information given by unbounded tolerances.

Our contribution. In this article we prove that Conjecture 1 is true for every graph G, whose parallelepiped representation R satisfies a slight assumption (to be defined later). This assumption is guaranteed by a wide variety of graph classes; for example, our results immediately imply correctness of the conjecture for complements of triangle-free graphs (which also implies the above mentioned correctness for complements of trees [1] and complements of bipartite graphs [18]). Furthermore, we state a new conjecture regarding only the separating examples between tolerance and bounded tolerance graphs (cf. Conjecture 2). There already exists evidence in the literature that this conjecture is true [12]. Our results reduce Conjecture 1 to our conjecture; that is, the correctness of our conjecture implies the correctness of Conjecture 1.

Specifically, we state three conditions on the unbounded vertices of G (in the parallelepiped representation R). Condition 1 is that R has exactly one unbounded vertex. Condition 2 is that, for every unbounded vertex u of G (in R), there exists no unbounded vertex v of G whose neighborhood is strictly included in the neighborhood of u. Note that these two conditions concern only the parallelepiped representation R; furthermore, the second condition is weaker than the first one. Then, Condition 3 concerns also the position of the unbounded vertices in the trapezoid representation R_T, and it is weaker than the other two.

Assuming that this (weaker) Condition 3 holds, we algorithmically construct a parallelogram representation of G, thus proving that G is a bounded tolerance graph. The proof of correctness relies on the fact that G can be represented simultaneously by R and by R_T. The main idea is to iteratively "eliminate" the unbounded vertices of R. That is, assuming that the input representation R has $k \geq 1$ unbounded vertices, we choose an unbounded vertex u in R and construct a parallelepiped representation R^* of G with $k-1$ unbounded vertices; specifically, R^* has the same unbounded vertices as R except for u (which becomes bounded in R^*). As a milestone in the above construction of the representation R^*, we construct an induced subgraph G_0 of G that includes u, with the property that the vertex set of $G_0 \setminus \{u\}$ is a module in $G \setminus \{u\}$. The presented techniques are new and provide geometrical insight for the graphs that are both tolerance and cocomparability.

In order to state our Conjecture 2, we define a graph G to be a *minimally unbounded tolerance* graph, if G is tolerance but not bounded tolerance, while G becomes bounded tolerance if we remove any vertex of G.

Conjecture 2. Any minimally unbounded tolerance graph has a tolerance representation with exactly one unbounded vertex.

In other words, Conjecture 2 states that any minimally unbounded tolerance graph G has a tolerance representation (or equivalently, a parallelepiped representation) R that satisfies Condition 1 (stated above). Our results imply the non-trivial result that, in order to prove Conjecture 1, it suffices to prove Conjecture 2. In addition, there already exists evidence that Conjecture 2 is true, as to the best of our knowledge it is true for all known examples of minimally unbounded tolerance graphs in the literature (see e.g. [12]).

Organization of the paper. We first review in Section 2 some properties of tolerance and trapezoid graphs. Then we define the notion of a *projection representation* of a tolerance graph G, which is an alternative way to think about a parallelepiped representation of G. Furthermore, we introduce the *right* and *left border properties* of a vertex in a projection representation, which are crucial for our analysis. In Section 3 we prove our main results and we discuss how these results reduce Conjecture 1 to Conjecture 2. Finally, we discuss the presented results and further research in Section 4. Due to space limitations, the proofs are omitted here; a full version of the paper can be found in [17].

2   Definitions and Basic Properties

Notation. We consider in this article simple undirected graphs with no loops or multiple edges. In a graph $G = (V, E)$, the edge between vertices u and v is denoted by uv, and in this case u and v are called *adjacent* in G. Given a vertex subset $S \subseteq V$, $G[S]$ denotes the induced subgraph of G on the vertices in S. Whenever it is clear from the context, we may not distinguish between a vertex set S and the induced subgraph $G[S]$ of G. Furthermore, we denote for simplicity the induced subgraph $G[V \setminus S]$ by $G \setminus S$. Denote by $N(u) = \{v \in V \mid uv \in E\}$ the set of neighbors of a vertex u in G, and $N[u] = N(u) \cup \{u\}$. For any two sets A, B, we write $A \subseteq B$ if A is included in B, and $A \subset B$ if A is strictly included in B.

Consider a trapezoid graph $G = (V, E)$ and a trapezoid representation R_T of G, where vertex $u \in V$ corresponds to the trapezoid T_u in R_T. Since trapezoid graphs are also cocomparability graphs [8], we can define the partial order $(V, \ll_{R_T})$, such that $u \ll_{R_T} v$, or equivalently $T_u \ll_{R_T} T_v$, if and only if T_u lies completely to the left of T_v in R_T (and thus also $uv \notin E$). Note that there are several trapezoid representations of a particular trapezoid graph G. Given one such representation R_T, we can obtain another one R'_T by *vertical axis flipping* of R_T, i.e. R'_T is the mirror image of R_T along an imaginary line perpendicular to L_1 and L_2.

Let us now briefly review the parallelepiped representation model of tolerance graphs [15]. Consider a tolerance graph $G = (V, E)$ and let V_B and V_U denote the set of bounded and unbounded vertices of G (for a certain tolerance representation), respectively. Consider now two parallel lines L_1 and L_2 in the plane. For every vertex $u \in V$, consider a parallelogram $\overline{P}_u$ with two of its lines on L_1 and L_2, respectively, and ϕ_u be the (common) slope of the other two lines of $\overline{P}_u$ with L_1 and L_2. For every unbounded vertex $u \in V_U$, the parallelogram $\overline{P}_u$

is trivial, i.e. a line. In the model of [15], every bounded vertex $u \in V_B$ corresponds to the parallelepiped $P_u = \{(x, y, z) \mid (x, y) \in \overline{P}_u, 0 \le z \le \phi_u\}$ in the 3-dimensional space, while every unbounded vertex $u \in V_U$ corresponds to the line $P_u = \{(x, y, z) \mid (x, y) \in \overline{P}_u, z = \phi_u\}$. The resulting set $\{P_u \mid u \in V\}$ of parallelepipeds in the 3-dimensional space constitutes the *parallelepiped representation* of G. In this model, two vertices u, v are adjacent if and only if $P_u \cap P_v \ne \emptyset$. That is, R is an intersection model for G. For more details we refer to [15]. An example of a parallelepiped representation R is illustrated in Figure 1(a). This representation corresponds to the induced path $P_4 = (z, u, v, w)$ with four vertices (P_4 is a tolerance graph); in particular, vertex w is unbounded in R, while the vertices z, u, v are bounded in R.

Definition 1 ([15]). *An unbounded vertex $v \in V_U$ of a tolerance graph G is called* inevitable *(in a certain parallelepiped representation R), if making v a bounded vertex in R, i.e. if replacing P_v with $\{(x, y, z) \mid (x, y) \in P_v, 0 \le z \le \phi_v\}$, creates a new edge in G.*

Definition 2 ([15]). *A parallelepiped representation R of a tolerance graph G is called* canonical *if every unbounded vertex in R is inevitable.*

For example, the parallelepiped representation of Figure 1(a) is canonical, since w is the only unbounded vertex and it is inevitable. A canonical representation of a tolerance graph G always exists, and can be computed in $O(n \log n)$ time, given a parallelepiped representation of G, where n is the number of vertices of G [15].

Given a parallelepiped representation R of the tolerance graph G, we define now an alternative representation, as follows. Let $\overline{P}_u$ be the projection of P_u to the plane $z = 0$ for every $u \in V$. Then, for two bounded vertices u and v, $uv \in E$ if and only if $\overline{P}_u \cap \overline{P}_v \ne \emptyset$. Furthermore, for a bounded vertex v and an unbounded vertex u, $uv \in E$ if and only if $\overline{P}_u \cap \overline{P}_v \ne \emptyset$ and $\phi_v > \phi_u$. Moreover,

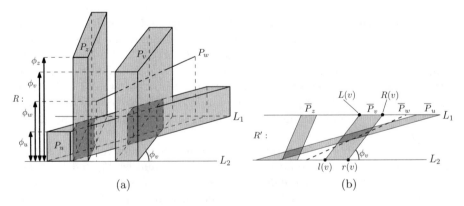

Fig. 1. (a) A parallelepiped representation R of the tolerance graph G and (b) the corresponding projection representation R' of G, where G is the induced path $P_4 = (z, u, v, w)$ with four vertices

two unbounded vertices u and v of G are never adjacent (even in the case where $\overline{P}_u$ intersects $\overline{P}_v$). In the following, we will call such a representation a *projection representation* of a tolerance graph. Note that $\overline{P}_u$ is a parallelogram (resp. a line segment) if u is bounded (resp. unbounded). The projection representation that corresponds to the parallelepiped representation of Figure 1(a) is presented in Figure 1(b). In the sequel, given a tolerance graph G, we will call a projection representation R of G a *canonical representation* of G, if R is the projection representation that is implied by a canonical parallelepiped representation of G. In the example of Figure 1, the projection representation R' is canonical, since the parallelepiped representation R is canonical as well.

Let R be a projection representation of a tolerance graph $G = (V, E)$. For every parallelogram $\overline{P}_u$ in R, where $u \in V$, we define by $l(u)$ and $r(u)$ (resp. $L(u)$ and $R(u)$) the lower (resp. upper) left and right endpoint of $\overline{P}_u$, respectively (cf. the parallelogram $\overline{P}_v$ in Figure 1(b)). Note that $l(u) = r(u)$ and $L(u) = R(u)$ for every unbounded vertex u. We assume throughout the paper w.l.o.g. that all endpoints and all slopes of the parallelograms in a projection representation are distinct [12, 13, 15]. For simplicity of the presentation, we will denote in the following $\overline{P}_u$ just by P_u in any projection representation.

Similarly to a trapezoid representation, we can define the relation $\ll_R$ also for a projection representation R. Namely, $P_u \ll_R P_v$ if and only if P_u lies completely to the left of P_v in R. Otherwise, if neither $P_u \ll_R P_v$ nor $P_v \ll_R P_u$, we will say that P_u *intersects* P_v in R, i.e. $P_u \cap P_v \neq \emptyset$ in R. Note that, for two vertices u and v of a tolerance graph $G = (V, E)$, P_u may intersect P_v in a projection representation R of G, although u is not adjacent to v in G, i.e. $uv \notin E$. Thus, a projection representation R of a tolerance graph G is *not* necessarily an intersection model for G.

In [11, 15] the *hovering set* of an unbounded vertex in a tolerance graph has been defined. According to these definitions, the hovering set depends on a particular representation of the tolerance graph. In the following, we extend this definition to the notion of *covering* vertices of an arbitrary graph G, which is independent of any representation of G.

Definition 3. *Let $G = (V, E)$ be an arbitrary graph and $u \in V$ be a vertex of G. Then,*
  - *the set $\mathcal{C}(u) = \{v \in V \setminus N[u] \mid N(u) \subseteq N(v)\}$ is the* covering set *of u, and every vertex $v \in \mathcal{C}(u)$ is a* covering vertex *of u,*
  - *$V_0(u)$ is the set of connected components of $G \setminus N[u]$ that have at least one covering vertex $v \in \mathcal{C}(u)$ of u.*

In the following, for simplicity of the presentation, we may not distinguish between the connected components of $V_0(u)$ and the vertex set of these components. In the next definition we introduce the notion of the right (resp. left) border property of a vertex u in a projection representation R of a tolerance graph G. This notion is of particular importance for the sequel of the paper.

Definition 4. *Let $G = (V, E)$ be a tolerance graph, u be an arbitrary vertex of G, and R be a projection representation of G. Then, u has the* right *(resp. left)* border property *in R, if there exists no pair of vertices $w \in N(u)$ and $x \in V_0(u)$, such that $P_w \ll_R P_x$ (resp. $P_x \ll_R P_w$).*

We denote in the sequel by TOLERANCE the class of tolerance graphs, and we use the corresponding notations for the classes of bounded tolerance, cocomparability, and trapezoid graphs. Let $G \in$ TOLERANCE $\cap$ COCOMPARABILITY. Then G is also a trapezoid graph [7]. Thus, since TRAPEZOID $\subseteq$ COCOMPARABILITY, it follows that TOLERANCE $\cap$ COCOMPARABILITY = TOLERANCE $\cap$ TRAPEZOID. Furthermore, clearly BOUNDED TOLERANCE $\subseteq$ (TOLERANCE $\cap$ TRAPEZOID), since BOUNDED TOLERANCE $\subseteq$ TOLERANCE and BOUNDED TOLERANCE $\subseteq$ TRAPEZOID. In the following we consider a graph $G \in$ (TOLERANCE $\cap$ TRAPEZOID) \ BOUNDED TOLERANCE, assuming that one exists, and our aim is to get to a contradiction; namely, to prove that (TOLERANCE $\cap$ TRAPEZOID) = BOUNDED TOLERANCE.

3 Main Results

In this section, we prove that for a graph $G \in$ (TOLERANCE $\cap$ TRAPEZOID) it follows that also $G \in$ BOUNDED TOLERANCE by making a slight assumption on the unbounded vertices of a projection representation R of G. In particular, we choose a certain unbounded vertex u in R and we "eliminate" u in R in the following sense: assuming that R has $k \geq 1$ unbounded vertices, we construct a projection representation R^* of G with $k - 1$ unbounded vertices, where all bounded vertices remain bounded and u is transformed to a bounded vertex. In Section 3.1 we deal with the case where the unbounded vertex u has the right or the left border property in R, while in Section 3.2 we deal with the case where u has neither the left nor the right border property in R. Finally we combine these two results in Section 3.3, in order to eliminate all k unbounded vertices in R, regardless of whether or not they have the right or left border property.

3.1 The Case Where u Has the Right or the Left Border Property

In this section we consider an arbitrary unbounded vertex u of G in the projection representation R, and we assume that u has the right or the left border property in R. Then, as we prove in the next theorem, there exists another projection representation R^* of G, in which u has been replaced by a bounded vertex.

Theorem 1. *Let $G = (V, E) \in$ (TOLERANCE $\cap$ TRAPEZOID) \ BOUNDED TOLERANCE with the smallest number of vertices. Let R be a projection representation of G with k unbounded vertices and u be an unbounded vertex in R. If u has the right or the left border property in R, then G has a projection representation R^* with $k - 1$ unbounded vertices.*

3.2 The Case Where u Has Neither the Left Nor the Right Border Property

In this section we consider a graph $G \in$ (TOLERANCE $\cap$ TRAPEZOID) \ BOUNDED TOLERANCE with the smallest number of vertices. Furthermore, let

R and R_T be a canonical projection and a trapezoid representation of G, respectively, and u be an unbounded vertex of G in R. We consider the case where G has no unbounded vertex in R with the right or the left border property (otherwise Theorem 1 can be applied). It can be proved that $V_0(u) \neq \emptyset$ and that $V_0(u)$ is connected (cf. [17]). Therefore, since u is not adjacent to any vertex of $V_0(u)$ by Definition 3, either all trapezoids of $V_0(u)$ lie to the left, or all to the right of T_u in R_T.

Consider first the case where all trapezoids of $V_0(u)$ lie to the *left* of T_u in R_T, i.e. $T_x \ll_{R_T} T_u$ for every $x \in V_0(u)$. It can be proved that $N(v) \neq N(u)$ for every unbounded vertex $v \neq u$ in R (cf. [17]). Denote by $Q_u = \{v \in V_U \mid N(v) \subset N(u)\}$ the set of unbounded vertices v of G in R, whose neighborhood set is included in the neighborhood set of u. Since no two unbounded vertices are adjacent, we can partition the set Q_u into the two subsets $Q_1(u) = \{v \in Q_u \mid T_v \ll_{R_T} T_u\}$ and $Q_2(u) = \{v \in Q_u \mid T_u \ll_{R_T} T_v\}$. Furthermore, it can be proved that $T_v \ll_{R_T} T_x \ll_{R_T} T_u$ for every $v \in Q_1(u)$ and every $x \in V_0(u)$ (cf. [17]). That is, $Q_1(u) = \{v \in Q_u \mid T_v \ll_{R_T} T_x$ for every $x \in V_0(u)\}$.

Consider now the case where all trapezoids of $V_0(u)$ lie to the *right* of T_u in R_T, i.e. $T_u \ll_{R_T} T_x$ for every $x \in V_0(u)$. Then, by performing vertical axis flipping of R_T, we partition similarly to the above the set Q_u into the sets $Q_1(u)$ and $Q_2(u)$. That is, in this (symmetric) case, the sets $Q_1(u)$ and $Q_2(u)$ will be $Q_1(u) = \{v \in Q_u \mid T_x \ll_{R_T} T_v$ for every $x \in V_0(u)\}$ and $Q_2(u) = \{v \in Q_u \mid T_v \ll_{R_T} T_u\}$.

We state now three conditions on G, regarding the unbounded vertices in R; the third one depends also on the representation R_T. Note that the second condition is weaker than the first one, while the third one is weaker than the other two. Then, we prove Theorem 2, assuming that the third condition holds. First, we introduce the notion of neighborhood maximality for unbounded vertices in a tolerance graph.

Definition 5. *Let G be a tolerance graph, R be a projection representation of G, and u be an unbounded vertex in R. Then, u is* unbounded-maximal *if there is no unbounded vertex v in R, such that $N(u) \subset N(v)$.*

Condition 1. *The projection representation R of G has exactly one unbounded vertex.*

Condition 2. *For every unbounded vertex u of G in R, $Q_u = \emptyset$; namely, all unbounded vertices are unbounded-maximal.*

Condition 3. *For every unbounded vertex u of G in R, $Q_2(u) = \emptyset$, i.e. $Q_u = Q_1(u)$.*

In the following of the section we assume that Condition 3 holds, which is weaker than Conditions 1 and 2. We present now the main theorem of this section. The proof of this theorem is based on the fact that G has simultaneously the two representations R and R_T.

Theorem 2. *Let* $G = (V, E) \in$ (TOLERANCE $\cap$ TRAPEZOID) \ BOUNDED TOLERANCE *with the smallest number of vertices. Let* R_T *be a trapezoid representation of* G *and* R *be a projection representation of* G *with* k *unbounded vertices. Then, assuming that Condition 3 holds, there exists a projection representation* R^* *of* G *with* $k - 1$ *unbounded vertices.*

3.3   The General Case

Recall now that TOLERANCE $\cap$ COCOMPARABILITY $=$ TOLERANCE $\cap$ TRAPEZOID [7]. The next main theorem follows by recursive application of Theorem 2.

Theorem 3. *Let* $G = (V, E) \in$ (TOLERANCE $\cap$ COCOMPARABILITY)*,* R_T *be a trapezoid representation of* G*, and* R *be a projection representation of* G*. Then, assuming that one of the Conditions 1, 2, or 3 holds,* G *is a bounded tolerance graph.*

As an immediate implication of Theorem 3, we prove in the next corollary that Conjecture 1 is true in particular for every graph G that has no three independent vertices a, b, c such that $N(a) \subset N(b) \subset N(c)$, since Condition 2 is guaranteed to be true for every such graph G. Therefore, in particular, the conjecture is also true for the complements of triangle-free graphs.

Corollary 1. *Let* $G = (V, E) \in$ (TOLERANCE $\cap$ COCOMPARABILITY)*. Suppose that there do not exist three independent vertices* $a, b, c \in V$ *such that* $N(a) \subset N(b) \subset N(c)$*. Then,* G *is also a bounded tolerance graph.*

Definition 6. *Let* $G \in$ TOLERANCE \ BOUNDED TOLERANCE*. If* $G \setminus \{u\}$ *is a bounded tolerance graph for every vertex of* G*, then* G *is a* minimally unbounded tolerance *graph.*

Assume now that Conjecture 1 is not true, and let G be a counterexample with the smallest number of vertices. Then, in particular, G is a minimally unbounded tolerance graph by Definition 6. Now, if our Conjecture 2 is true (see Section 1), then G has a projection representation R with exactly one unbounded vertex, i.e. R satisfies Condition 1. Thus, G is a bounded tolerance graph by Theorem 3, which is a contradiction, since G has been assumed to be a counterexample to Conjecture 1. Thus, we obtain the following theorem.

Theorem 4. *Conjecture 2 implies Conjecture 1.*

Therefore, in order to prove Conjecture 1, it suffices to prove Conjecture 2. In addition, there already exists evidence that this conjecture is true, as to the best of our knowledge all known examples of minimally unbounded tolerance graphs have a tolerance representation with exactly one unbounded vertex; for such examples, see e.g. [12].

4   Concluding Remarks and Open Problems

In this article we dealt with the over 25 years old conjecture of [10], which states that if a graph G is both tolerance and cocomparability, then it is also bounded tolerance. Specifically, we proved that the conjecture is true for every graph G, whose tolerance *representation* R of G satisfies a slight assumption, instead of making *any structural assumption* on G – as it was the case in all previously known results. Furthermore, we conjectured that any minimal graph G that is a tolerance but not a bounded tolerance graph, has a tolerance representation with exactly one unbounded vertex. Our results imply the non-trivial result that, in order to prove the conjecture of [10], it suffices to prove our conjecture. In addition, there already exists evidence in the literature that this conjecture is true [12].

References

1. Andreae, T., Hennig, U., Parra, A.: On a problem concerning tolerance graphs. Discrete Applied Mathematics 46(1), 73–78 (1993)
2. Bogart, K.P., Fishburn, P.C., Isaak, G., Langley, L.: Proper and unit tolerance graphs. Discrete Applied Mathematics 60(1-3), 99–117 (1995)
3. Brandstädt, A., Le, V.B., Spinrad, J.P.: Graph classes: a survey. SIAM, Philadelphia (1999)
4. Busch, A.H.: A characterization of triangle-free tolerance graphs. Discrete Applied Mathematics 154(3), 471–477 (2006)
5. Corneil, D.G., Olariu, S., Stewart, L.: LBFS orderings and cocomparability graphs. In: Proceedings of the Tenth Annual ACM-SIAM Symposium on Discrete Algorithms (SODA), pp. 883–884 (1999)
6. Deogun, J.S., Steiner, G.: Polynomial algorithms for hamiltonian cycle in cocomparability graphs. SIAM Journal on Computing 23(3), 520–552 (1994)
7. Felsner, S.: Tolerance graphs and orders. Journal of Graph Theory 28(3), 129–140 (1998)
8. Golumbic, M.C.: Algorithmic graph theory and perfect graphs. Annals of Discrete Mathematics, vol. 57. North-Holland Publishing Co., Amsterdam (2004)
9. Golumbic, M.C., Monma, C.L.: A generalization of interval graphs with tolerances. In: Proceedings of the 13th Southeastern Conference on Combinatorics, Graph Theory and Computing, Congressus Numerantium 35, pp. 321–331 (1982)
10. Golumbic, M.C., Monma, C.L., Trotter, W.T.: Tolerance graphs. Discrete Applied Mathematics 9(2), 157–170 (1984)
11. Golumbic, M.C., Siani, A.: Coloring algorithms for tolerance graphs: reasoning and scheduling with interval constraints. In: Proceedings of the Joint International Conferences on Artificial Intelligence, Automated Reasoning, and Symbolic Computation (AISC/Calculemus), pp. 196–207 (2002)
12. Golumbic, M.C., Trenk, A.N.: Tolerance Graphs. Cambridge studies in advanced mathematics (2004)
13. Isaak, G., Nyman, K.L., Trenk, A.N.: A hierarchy of classes of bounded bitolerance orders. Ars Combinatoria 69 (2003)
14. Kratsch, D., Stewart, L.: Domination on cocomparability graphs. SIAM Journal on Discrete Mathematics 6(3), 400–417 (1993)

15. Mertzios, G.B., Sau, I., Zaks, S.: A new intersection model and improved algorithms for tolerance graphs. SIAM Journal on Discrete Mathematics 23(4), 1800–1813 (2009)
16. Mertzios, G.B., Sau, I., Zaks, S.: The recognition of tolerance and bounded tolerance graphs. In: Proceedings of the 27th International Symposium on Theoretical Aspects of Computer Science (STACS), pp. 585–596 (2010)
17. Mertzios, G.B., Zaks, S.: The structure of the intersection of tolerance and cocomparability graphs. Technical Report AIB-2010-09, Department of Computer Science, RWTH Aachen University (May 2010),
http://sunsite.informatik.rwth-aachen.de/Publications/AIB/2010/
2010-09.pdf
18. Parra, A.: Eine Klasse von Graphen, in der jeder Toleranzgraph ein beschr"ankter Toleranzgraph ist. Abhandlungen aus dem Mathematischen Seminar der Universit"at Hamburg 64(1), 125–129 (1994)
19. Ryan, S.P.: Trapezoid order classification. Order 15, 341–354 (1998)
20. Spinrad, J.P.: Efficient graph representations. Fields Institute Monographs, vol. 19. AMS, Providence (2003)

Flows in One-Crossing-Minor-Free Graphs

Erin Chambers[1] and David Eppstein[2]

[1] Dept. of Mathematics and Computer Science, Saint Louis University
[2] Computer Science Department, University of California, Irvine

Abstract. We study the maximum flow problem in directed H-minor-free graphs where H can be drawn in the plane with one crossing. If a structural decomposition of the graph as a clique-sum of planar graphs and graphs of constant complexity is given, we show that a maximum flow can be computed in $O(n \log n)$ time. In particular, maximum flows in directed $K_{3,3}$-minor-free graphs and directed K_5-minor-free graphs can be computed in $O(n \log n)$ time without additional assumptions.

1 Introduction

Computing maximum flows is fundamental in algorithmic graph theory, and has many applications. Although flows can be computed in polynomial time for arbitrary graphs, it is of interest to find classes of graphs for which flows can be computed more quickly, and specialized algorithms are known for flows in planar graphs [3, 12, 16, 17, 20, 21, 23, 24, 26, 34], graphs of bounded genus [4, 5], graphs with small crossing number [18], and graphs of bounded treewidth [14].

Planar graphs, graphs of bounded genus, and graphs of bounded treewidth are *minor-closed graph families*, families of graphs closed under edge contractions and edge deletions. According to the Robertson–Seymour graph minor theorem [31], any minor-closed graph family can be described as the X-minor-free graphs, graphs that do not have as a minor any member of a finite set X of non-members of the family; for instance, the planar graphs are exactly the $\{K_5, K_{3,3}\}$-minor-free graphs [32]. In many cases the properties of a graph family are closely related to the properties of its excluded minors: for instance, the minor-closed graph families with bounded treewidth are exactly the families of X-minor-free graphs for which X includes at least one planar graph [28], and the families with bounded local treewidth (a functional relationship between the diameter of a graph and its treewidth) are exactly those for which X includes at least one *apex graph*, a graph that can be made planar by removing a single vertex [10]. If X includes a graph that can be drawn in the plane with a single pair of crossing edges, then the X-minor-free graphs have a structural decomposition as a *clique-sum* of smaller graphs that are either planar or have bounded treewidth [8,29]. In this last case we say that the family of X-minor-free graphs is *one-crossing-minor-free*; families of this type include the $K_{3,3}$-minor-free graphs and K_5-minor-free graphs, since $K_{3,3}$ and K_5 are one-crossing graphs (Figure 1).

O. Cheong, K.-Y. Chwa, and K. Park (Eds.): ISAAC 2010, Part I, LNCS 6506, pp. 241–252, 2010.

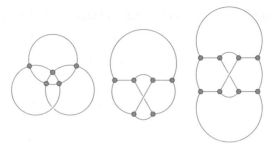

Fig. 1. One-crossing drawings of K_5 (left), $K_{3,3}$ (center), and the Wagner graph (right)

In this paper we consider flows in one-crossing-minor-free graph families. We provide $O(n \log n)$ algorithms to compute maximum flows in any directed H-minor-free graph, where H is a fixed one-crossing graph and where the structural decomposition of the graph is provided as part of the input. In the case of $K_{3,3}$-minor-free graphs and K_5-minor-free graphs, algorithms are known that can find such a decomposition efficiently [2, 27], and by combining our techniques with those known decomposition algorithms we provide an $O(n \log n)$ time algorithm for maximum flow in directed $K_{3,3}$-minor-free and K_5-minor-free graphs without requiring the decomposition to be part of the input.

Our main motivation for looking at flows in one-crossing-minor-free graphs is to try to make progress towards finding flows in arbitrary minor-closed graph families. Due to known separator theorems for minor-closed families [1], an algorithm of Johnson and Venkatesan [21] can be applied to find flows in any minor-closed family in time $O(n^{3/2} \log n)$, but this does not come close to the nearly-linear time bounds known for planar and bounded-genus graphs. Like one-crossing-minor-free graphs, graphs in more general minor-closed families have a structural decomposition in terms of bounded-genus surfaces, clique-sums, apexes (a constant number of vertices that can be adjacent to arbitrary subsets of each of the bounded-genus surfaces), and vortexes (bounded-treewidth graphs glued into faces of the bounded-genus surfaces) [7, 30]. However, these decompositions are greatly simplified in the one-crossing-minor-free case: the surfaces are planes and there are no apexes or vortexes. To handle the general case, we would need to combine clique-sums, bounded-genus, apexes, and vortexes. The problem of flows on bounded genus surfaces has been previously examined [4, 5] and the present work focuses on clique-sums, as these are the main feature in the structural decomposition for one-crossing-minor-free graphs. However, it remains unclear how to handle apexes and vortexes.

As an important tool in our results, we greatly simplify the *mimicking networks* of Hagerup et al [14] for multiterminal flow networks in the case of four terminals with a single source, leading to significantly reduced constant factors in the running time of algorithms that use this networks. Similar simplifications had been achieved in the undirected case [6] but to our knowledge our small directed mimicking network is novel.

2 Preliminaries

2.1 Flows and Cuts

A *flow network* is a graph $G = (V, E)$, where each edge $e \in E$ has an associated nonnegative real capacity c_e, along with two distinguished vertices s and t which are called the *source* and *sink*, respectively. A *flow* in this graph is a set of nonnegative real values f_e for each edge $e \in E$ such that $f_e \le c_e$ for every $e \in E$ and $\sum_{uv \in E} f_{uv} = \sum_{(vu) \in E} f_{vu}$ for every $v \in V - \{s, t\}$. The *value* of the flow is the amount of net flow going from s to t, or $\sum_{sv \in E} f_{sv}$.

A *cut* in a flow network is a set of edges separating s from t; the *capacity* of the cut is the sum of the capacities of all the edges that cross the cut in the direction from s to t. Clearly, the value of any flow is less than or equal to the value of any cut. The classic max-flow min-cut theorem states that the maximum possible flow from s to t is in fact equal to the minimum possible cut.

In our setting, we will need to know the maximum possible flow that can travel through a subgraph of the input graph; this will allow us to simplify the graph by removing the subgraph and replacing it with an equivalent (but much smaller) subgraph. To do this, we compute an *external flow*. In external flow networks, instead of a single source and sink, we have a set of terminals $Q = \{q_1, \ldots q_k\}$ where each terminal q_i has associated with it a number x_i; we require that $\sum_i x_i = 0$. If x_i is positive then it is interpreted as a supply, and if x_i is negative it is interpreted as a demand. (It may also be the case that x_i is zero, in which case x_i carries neither a supply nor a demand.) A *realizable external flow* is a set of k values $(x_1, \ldots, x_k)$ for $(q_1, \ldots, q_k)$, along with a flow f such that $\sum_{(q_i, v) \in E} f_{(q_i, v)} - \sum_{(v, q_i) \in E} f_{(v, q_i)} = x_i$ for all i. Basically, the flow remains balanced at every vertex in $V \setminus Q$, and the imbalance of flow at each vertex in $q_i \in Q$ is exactly x_i.

It will be helpful to define a special case of external flow networks, which we call *single-source external flow networks*. A single-source external flow network is, simply, an external flow network for which only q_1 may have a positive supply x_i; every other terminal has $x_i \le 0$ indicating that it is either a demand node or is inactive as a terminal. As before, a *realizable single-source external flow* is a realizable external flow subject to this constraint on the values of x_i.

We define $S \not\to T$, for sets of terminals S and T in an external flow network, to be the minimum value of a cut for which every terminal in S is on the source side of the cut and every terminal in T is on the sink side of the cut. We will further abbreviate the notation by writing strings of symbols instead of bracketed set notation for the sets of terminals on each side of the cut; e.g., $s \not\to abc$ should be interpreted as an abbreviation for $\{s\} \not\to \{a, b, c\}$.

2.2 Mimicking Networks

Let G be an external flow network or single-source external flow network with a fixed specification of the edge capacities and a fixed ordered set of terminals Q,

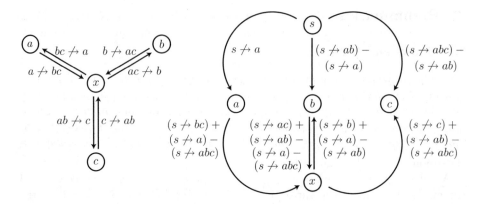

Fig. 2. Mimicking networks for an external flow network with three terminals a, b, and c (left), and for a single-source external flow network with four terminals s, a, b, and c (right). In the single-source mimicking network, a, b, and c are permuted if necessary so that $s \not\to a \geq \max\{s \not\to b, s \not\to c\}$ and $s \not\to ab \geq s \not\to ac$.

but where the supply and demand quantities x_i remain variable. A key ingredient for our technique, as formalized by Hagerup et al. [14], is the concept of a *mimicking network*, a network H that has the same terminals as G and has the same realizable external flows, but may have many fewer edges and vertices than G. In what amounts to a of separator based sparsification [11], Hagerup et al. solve flow problems on bounded-treewidth networks by repeatedly replacing subnetworks of the given network with smaller mimicking networks. Their construction of mimicking networks is based on an observation of Gale [13]:

Lemma 1. *[14] An external flow $(x_1, \ldots, x_k)$ is realizable in a network $G = (V, E)$ with terminals $Q = \{q_1, \ldots, q_k\}$ if and only if the following relations are satisfied:*

$$\sum_{i=1}^{k} x_i = 0 \tag{1}$$

$$\sum_{q_i \in S} x_i \leq (S \not\to (Q \setminus S)), \text{ for all } S \subseteq Q \text{ with } \emptyset \neq S \neq Q. \tag{2}$$

Essentially, this means that in order to understand the possible flow patterns in a subnetwork with k terminals, one needs only to know the $(2^k - 2)$ minimum cut values from a nonempty subset of terminals to its nonempty complement. If two networks have the same minimum cut values for each subset then they behave the same with respect to flows. Based on this observation, Hagerup et al. show that there exists a replacement network of at most 2^{2^k-2} vertices that behaves equivalently to any k-terminal network:

Lemma 2. *[14] Given any external flow network G with k terminals, there exists a flow network having at most 2^{2^k-2} vertices which has the same external flow value as G.*

Specifically, the mimicking network of Lemma 2 can be constructed from G by finding a set of $2^k - 2$ minimum cuts, one for each partition of the terminals into two nonempty subsets, and by collapsing subsets of vertices in G into a single supervertex whenever all vertices in the subset are on the same side of every cut. Note that the mimicking network is not necessarily a minor of G, as the collapsed subsets need not form connected subgraphs of G.

However, the size of the mimicking networks formed by Lemma 2, while constant, is large. Our algorithms will involve external flow networks with up to four terminals, and if we applied Lemma 2 directly we might get as many as 16384 vertices in our mimicking networks. It is possible to reduce the number of vertices in the construction of Hagerup et al. from a power of two to a Dedekind number by requiring nested partitions of the terminals to have nested cuts, but this would still lead to 168 vertices for the mimicking network of a four-terminal network. We describe in the full version arXiv:1007.1484 a simpler mimicking network of Chaudhuri et al. [6] for external flows with at most three terminals, and a new mimicking network for single source external flows with at most four terminals that are both much smaller, requiring at most one nonterminal vertex. These simplified mimicking networks are depicted in Figure 2. It is important for our techniques that the mimicking network for a three-terminal network is planar and has its three terminals in a single face of its planar embedding (more strongly, in fact, it is outerplanar); however, we do not rely on the planarity of the four-terminal mimicking network.

2.3   Structure of Minor Free Graphs

A *minor* of a graph G is a graph that can be formed from G by contracting and removing edges. A graph family F is *minor-closed* if every minor of a graph in F also belongs to F. If X is a finite set of graphs, the X-minor-free graphs are the graphs G such that no minor of G belongs to X; the X-minor-free graphs are obviously a minor-closed graph family, and (much less obviously) the Robertson–Seymour graph minor theorem [31] states that every minor-closed graph family has this form. If F is the family of X-minor-free graphs, then X is the set of *forbidden minors* for F; for instance, K_5 and $K_{3,3}$ are the forbidden minors for the planar graphs. We will abuse notation and abbreviate the $\{H\}$-minor-free graphs (where H is a single graph) as the H-minor-free graphs.

A *clique-sum* is an operation that combines two graphs by identifying the vertices in two equal-sized cliques in the two graphs, and then possibly removing some of the edges of the cliques. A *k-sum* is a clique-sum where all cliques have at most k vertices. More generally, we will say that a given graph is a k-sum of a collection of more than two graphs if it can be formed by repeatedly replacing pairs of graphs of the collection by their k-sum. Clique-sums are closely related to vertex-connectivity of graphs: for instance, the decomposition of a

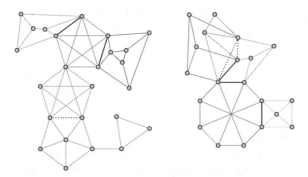

Fig. 3. A $K_{3,3}$-free graph expressed as a 2-sum of planar graphs and copies of K_5 (left), and a K_5-free graph expressed as a 3-sum of planar graphs and copies of the Wagner graph (right). The edges forming the cliques of the clique-sum operations are shown as solid black (if they remain in the final graph) or dashed (if they were removed after the clique-sum operation); the colors of the remaining edges identify the subgraphs entering into the clique-sum operations.

biconnected graph into triconnected components, formalized in the SPQR-tree, gives a representation of the graph as a 2-sum of triconnected graphs, cycles, and two-vertex multigraphs [9, 19, 25]. (See the full version for a description of SPQR-tree decompositions arXiv:1007.1484.)

Besides their relation to connectivity, clique-sums have also played an important role in describing the structure of minor-closed graph families since the proof by Wagner [32] that K_5-minor-free graphs are exactly the graphs that can be formed by 3-sums of planar graphs and the 8-vertex nonplanar Wagner graph, the graph formed by adding four edges connecting opposite pairs of vertices in an 8-cycle (Figure 3, right). Wagner [33] and D. W. Hall [15] also proved that $K_{3,3}$-minor-free graphs are exactly the 2-sums of planar graphs and the five-vertex complete graph K_5 (Figure 3, left). These two characterizations of H-minor-closed families can be generalized to any H-minor-free graph with the property that H can be drawn in the plane with only a single edge crossing, as $K_{3,3}$ and K_5 both can. In this case, the H-minor-free graphs can always be decomposed into 3-sums of planar graphs and graphs of bounded treewidth [29].

Algorithmically, a decomposition of a $K_{3,3}$-minor-free graph into a 2-sum of planar graphs and K_5 can be computed in linear time [2]. Essentially, this is simply a matter of constructing the SPQR tree and verifying that all triconnected components are either planar or K_5. As has been shown more recently, a decomposition of a K_5-minor-free graph into a 3-sum of planar graphs and the Wagner graph can also be constructed in linear time [22, 27]. Algorithmic versions of the generalized decomposition of graphs in any H-minor-free family (where H is a one-crossing graph) into 3-sums of planar graphs and graphs of bounded treewidth can also be solved in polynomial time [8], but the polynomial $O(n^4)$ is too high to be of use in flow algorithms, so in this case we will assume that a decomposition has been given to us as part of the input. In case a future development leads to linear time algorithms for decomposition of one-crossing-minor-free graphs, this assumption may be removed.

3   Flow Algorithm

Recall first that any H-minor free graph G (where H is a one-crossing graph) can be decomposed into 3-sums of planar and bounded treewidth graphs; we refer to these smaller graphs as *components* of the clique-sum decomposition. In order to handle the case that three or more components are glued together by a 3-sum at a single shared clique, we may represent the decomposition as a 2-colored tree, in which the vertices of one color represent the components of the decomposition and the vertices of the other color represent the cliques at which they are glued together. We may identify two distinguished components of the decomposition, one containing s and another containing t, where s and t are the two terminals of our given flow problem; if s or t is part of a clique on which multiple components are glued, then the distinguished component containing that terminal may be chosen arbitrarily among all components containing it. The two distinguished components are necessarily connected by a path in the clique-sum decomposition tree (possibly a path of length zero).

The vertices of G that belong to the cliques of the clique-sum decomposition, in effect, will form terminals in each component, so any flows into and out of components of the decomposition can be treated as an external flow computations. Our algorithm will iteratively replace each component in the decomposition with a mimicking network of constant size. However care is needed in ordering these computations, as some components may have more than a constant number of vertices that belong to cliques of the decomposition and we can only use mimicking networks for components that have a constant number of terminals. We perform this replacement in two phases: in the first phase, the components that are replaced do not belong to the path from s to t of the decomposition tree, and in the second phase the replaced components lie along this path.

Refining the decomposition. For technical reasons it is necessary to ensure that the planar components in our decomposition are only glued together by 3-sums along their faces, which we achieve by refining the given clique-sum decomposition tree. To do so, we make the following changes, each of which replaces one of the components in the decomposition by a clique-sum of smaller components:

- If any component of the decomposition is not biconnected, replace it with the 1-sum of its biconnected components, glued together by 1-sums at its articulation vertices.
- If any component of the decomposition is biconnected but not triconnected, find an SPQR tree representing it as a 2-sum of its triconnected components [9, 19, 25]; see the full version for more details arXiv:1007.1484.
- At this stage of refinement, each component is triconnected, and in particular each planar component has a unique planar embedding. Within each planar

component, find all the triangles at which 3-sums are glued, and compare them against the list of triangular faces of the graph. The gluing triangles that are not faces are separating triangles of the component, and may be used to partition each planar component into a 3-sum of smaller planar components.

All these refinements may be performed in linear time, and after this stage each triangle in each planar component of the decomposition forms a face of the component.

Simplification phase I: components off the terminal path. Next, we deal with all components that are *not* on the s to t path in the clique-sum decomposition tree. Our algorithm iteratively finds a component C_i that is a leaf in the clique-sum decomposition tree; let C_j be the component to which C_i is glued by a clique-sum, one step closer to the s–t path. Then C_i is connected to C_j and the rest of the graph via at most three vertices, so we can treat the flow calculation within that component as an external flow problem with at most three terminals. Since the subgraph is either planar or of bounded treewidth, we can in $O(n \log n)$ time compute the min-cut values of all 6 possible partitions of its terminals, and as shown in the full version, we can replace the component C_i with a mimicking network C_i' of at most four vertices. We then split into cases according to how C_i and C_j fit into the rest of the clique-sum decomposition tree:

- If C_j is a bounded-treewidth component, then C_i' can be glued directly into C_j by performing the clique-sum that connects these two components, forming a decomposition tree with one fewer component. The treewidth of the merged component is the maximum of 3 and the treewidth of C_j (using the fact that the mimicking network for C_i is at worst an oriented form of K_4), so after any number of repetitions of steps of this type the treewidth remains bounded.
- If C_j is a planar component, and C_i connects to C_j via one or two terminals or via a triangle of three terminals that is not used for any other 3-sum in the decomposition tree, then again C_i' can be immediately glued into C_j forming a decomposition tree with one fewer component. Due to the planar structure of our three-terminal mimicking networks and due to the fact that the gluing triangle is a face of C_j, this step preserves both the planarity of C_j and the property that all gluing triangles in the new larger planar component are still faces of their component.
- If C_j is a planar component and the gluing triangle of C_i is shared with another four-vertex replacement network C_h', then C_h' and C_i' may be merged into a single mimicking network. Then, if this merged component is the only component that uses that gluing triangle, it may be glued into C_j as in the previous case.
- In the remaining cases, we replace C_i by C_i' in the decomposition tree but do not merge it with C_j.

After all simplifications of this type have been performed, the remaining decomposition tree must have the structure of a path of components connecting s to t, where each of the gluing triangles in this path may also be connected to a single mimicking network off the path.

Simplification phase II: components along the terminal path. To complete the algorithm we will perform a similar sequence of replacements along the path from s to t. However, we must use a slightly different technique, since we may now need to deal with four terminals at a time: s and the three vertices in the 3-sum connecting the first two components in the s–t path. Additionally, after some of the replacements in this final stage of the algorithm, we will have to compute flows in networks that are neither planar nor of bounded treewidth, but that can be made planar by the removal of a constant number of edges; a result of Hochstein and Weihe [18] allows us to compute minimum cuts in these graphs in near-linear time. In this stage of the algorithm, as long as s and t belong to different components, we perform the following steps:

- Let C_i be the component containing s and T_i be the set of vertices in the clique-sum connecting C_i to the next component in the s–t path.
- Compute the minimum cut amounts between s and each nonempty subset of T_i, using the algorithm of Hochstein and Weihe [18] if C_i has been formed by adding a constant number of edges to a planar component, and the algorithm of Hagerup et al. [14] if C_i is a bounded-treewidth component.
- Replace C_i by a mimicking network C_i' for single-source external flows (as described in the full version) using the computed cut amounts.
- Glue C_i', and (if it exists) the other mimicking network sharing the same gluing triangle, into the next component in the s–t path, forming a path with one fewer component. If the next component was planar, it becomes a graph formed from a planar graph by adding a constant number of edges, while if it had bounded treewidth, its new treewidth is again bounded by the maximum of its old treewidth and 3.

Eventually this simplification will leave us with a planar or bounded treewidth graph containing both s and t, and we can compute the maximum flow between them directly in near linear time.

Reversing the simplifications and constructing a flow. We then reverse the sequence of simplifications we have performed, replacing each mimicking network with the larger network it replaced; for each such replacement we perform a single flow computation to find valid flow amounts forming the same external flow as in the mimicking network. At the end of this reversal of the simplification process, we will have a correct maximum flow in the original network that we were given as input.

We summarize our results as a theorem.

Theorem 1. *If we are given as input a decomposition of a directed flow network into a 3-clique-sum of planar and bounded-treewidth components, then in*

$O(n \log n)$ *time we may compute a maximum flow between any two terminals* s *and* t *of the network.*

Corollary 1. *For any fixed one-crossing graph* H, *maximum flows in directed* H-*minor-free flow networks may be computed in* $O(n \log n)$ *time once a clique-sum decomposition of the network has been found.*

In the case of K_5-free and $K_{3,3}$-free graphs, we can find a clique-sum decomposition in $O(n)$ time [2, 27]. The case of $K_{3,3}$-free graphs is particularly simple, since it uses only 2-sums, and therefore involves simpler mimicking networks.

Corollary 2. *We may compute maximum flows between any two terminals in a* K_5-*free or* $K_{3,3}$-*free directed flow network in* $O(n \log n)$ *time.*

4   Conclusions

We have shown how to find flow in near linear time when the input graph is a clique-sum of planar and bounded tree-width graphs by using mimicking networks to iteratively simplify the graph. This technique allows us to use known near linear algorithms in each bounded tree-width or planar component of the decomposition.

There is no added generality in considering 4-sums of planar graphs in place of 3-sums (any 4-sum involving a planar graph can be rearranged into a combination of 3-sums), but our methods immediately generalize to 2-sums of bounded-genus graphs and bounded-treewidth graphs. Flow computation in 3-sums of bounded genus graphs is more problematic due to the possible existence of non-facial non-separating triangles.

The larger goal, however, is computing flow quickly in more arbitrary minor-free families of graphs. Since flow can be computed efficiently in bounded genus and bounded tree-width graphs, the primary remaining open questions are those of computing flow in graphs with vortices or apexes, since these are the relevant building blocks for more general minor free families. Even the case of a planar graph with two apexes would be of great interest, since it could be used to solve a generalization of the maximum flow problem on planar graphs in which the flow is allowed to have multiple sources and sinks.

References

1. Alon, N., Seymour, P., Thomas, R.: A separator theorem for graphs with an excluded minor and its applications. In: Proc. 22nd ACM Symp. on Theory of Computing (STOC 1990), pp. 293–299 (1990),
   doi: http://doi.acm.org/10.1145/100216.100254
2. Asano, T.: An approach to the subgraph homeomorphism problem. Theoretical Computer Science 38(2-3), 249–267 (1985), doi:10.1016/0304-3975(85)90222-1
3. Borradaile, G., Klein, P.: An $O(n \log n)$ algorithm for maximum st-flow in a directed planar graph. Journal of the ACM 56(2):Art. No. 9 (2009),
   doi:10.1145/1502793.1502798,
   http://www.math.uwaterloo.ca/~glencora/downloads/maxflow-full.pdf

4. Chambers, E.W., Erickson, J., Nayyeri, A.: Homology flows, cohomology cuts. In: Proc. 41st ACM Symposium on Theory of Computing (STOC 2009), pp. 273–282 (2009), doi:10.1145/1536414.1536453

5. Chambers, E.W., Erickson, J., Nayyeri, A.: Minimum cuts and shortest homologous cycles. In: Proc. 25th ACM Symposium on Computational Geometry (SoCG 2009), pp. 377–385 (2009), doi:10.1145/1542362.1542426

6. Chaudhuri, S., Subrahmanyam, K.V., Wagner, F., Zaroliagis, C.D.: Computing mimicking networks. Algorithmica 26(1), 31–49 (2000), doi:10.1007/s004539910003

7. Demaine, E.D., Hajiaghayi, M., Ichi Kawarabayashi, K.: Algorithmic Graph Minor Theory: Decomposition, Approximation, and Coloring. In: Proc. 46th IEEE Symp. on Foundations of Computer Science (FOCS 2005), pp. 637–646 (2005), http://erikdemaine.org/papers/Decomposition_FOCS2005/

8. Demaine, E.D., Hajiaghayi, M., Thilikos, D.M.: 1.5-Approximation for treewidth of graphs excluding a graph with one crossing as a minor. In: Jansen, K., Leonardi, S., Vazirani, V.V. (eds.) APPROX 2002. LNCS, vol. 2462, pp. 67–80. Springer, Heidelberg (2002), doi:10.1007/3-540-45753-4_8

9. Di Battista, G., Tamassia, R.: Incremental planarity testing. In: Proc. 30th Annual Symposium on Foundations of Computer Science (FOCS 1989), pp. 436–441 (1989), doi:10.1109/SFCS.1989.63515

10. Eppstein, D.: Diameter and treewidth in minor-closed graph families. Algorithmica 27, 275–291 (2000), doi:10.1007/s004530010020, arXiv:math.CO/9907126

11. Eppstein, D., Galil, Z., Italiano, G.F., Spencer, T.H.: Separator based sparsification for dynamic planar graph algorithms. In: Proc. 25th ACM Symp. on Theory of Computing (STOC 1993), pp. 208–217 (1993), doi:10.1145/167088.167159

12. Ford, L.R., Fulkerson, D.R.: Maximal flow through a network. Canadian Journal of Mathematics 8, 399–404 (1956)

13. Gale, D.: A theorem on flows in networks. Pacific Journal of Mathematics 7, 1073–1082 (1957)

14. Hagerup, T., Katajainen, J., Nishimura, N., Ragde, P.: Characterizing multiterminal flow networks and computing flows in networks of small treewidth. Journal of Computer and System Sciences 57(3), 366–375 (1998), doi:10.1006/jcss.1998.1592

15. Hall, D.W.: A note on primitive skew curves. Bulletin of the American Mathematical Society 49(12), 935–936 (1943), doi:10.1090/S0002-9904-1943-08065-2

16. Hassin, R.: Maximum flow in (s,t) planar networks. Information Processing Letters 13(3), 107 (1981), http://www.math.tau.ac.il/~hassin/planarst.pdf

17. Hassin, R., Johnson, D.B.: An O($n log^2 n$) algorithm for maximum flow in undirected planar networks. SIAM J. Comput. 14(3), 612–624 (1985), doi:10.1137/0214045

18. Hochstein, J.M., Weihe, K.: Maximum s-t-flow with k crossings in O($k^3 n log n$) time. In: Proc. 18th ACM–SIAM Symposium on Discrete Algorithms (SODA 2007), pp. 843–847 (2007), http://portal.acm.org/citation.cfm?id=1283383.1283473

19. Hopcroft, J., Tarjan, R.: Dividing a graph into triconnected components. SIAM J. Comput. 2(3), 135–158 (1973), doi:10.1137/0202012

20. Itai, A., Shiloach, Y.: Maximum flow in planar networks. SIAM J. Comput. 8(2), 135–150 (1979), doi:10.1137/0208012

21. Johnson, D.B., Venkatesan, S.: Using divide and conquer to find flows in directed planar networks in O($n^{3/2} log n$) time. In: Proc 20th Allerton Conference on Communication, Control, and Computing, pp. 898–905 (1982)

22. Kézdy, A., McGuinness, P.: Sequential and parallel algorithms to find a K_5 minor. In: Proc. 3rd ACM-SIAM Symposium on Discrete Algorithms (SODA 1992), pp. 345–356 (1992), http://portal.acm.org/citation.cfm?id=139404.139475

23. Khuller, S., Naor, J.: Flow in planar graphs with vertex capacities. Algorithmica 11(3), 200–225 (1994), doi:10.1007/BF01240733

24. Khuller, S., Naor, J., Klein, P.: The lattice structure of flow in planar graphs. SIAM J. Discrete Math. 6(3), 477–490 (1993), doi:10.1137/0406038

25. Mac Lane, S.: A structural characterization of planar combinatorial graphs. Duke Mathematical Journal 3(3), 460–472 (1937), doi:10.1215/S0012-7094-37-00336-3

26. Miller, G.L., Naor, J.: Flow in planar graphs with multiple sources and sinks. SIAM J. Comput 24(5), 1002–1017 (1995), doi:10.1137/S0097539789162997

27. Reed, B., Li, Z.: Optimization and recognition for K_5-minor free graphs in linear time. In: Laber, E.S., Bornstein, C., Nogueira, L.T., Faria, L. (eds.) LATIN 2008. LNCS, vol. 4957, Springer, Heidelberg (2008), doi:10.1007/978-3-540-78773-0_18

28. Robertson, N., Seymour, P.: Graph minors. V. Excluding a planar graph. Journal of Combinatorial Theory, Series B 41(1), 92–114 (1986), doi:10.1016/0095-8956(86)90030-4

29. Robertson, N., Seymour, P.: Excluding a graph with one crossing. In: Graph Structure Theory: Proc. AMS–IMS–SIAM Joint Summer Research Conference on Graph Minors. Contemporary Mathematics, vol. 147, pp. 669–675. AMS, Providence (1993)

30. Robertson, N., Seymour, P.: Graph minors. XVI. Excluding a non-planar graph. Journal of Combinatorial Theory, Series B 89(1), 43–76 (2003), doi:10.1016/S0095-8956(03)00042-X

31. Robertson, N., Seymour, P.: Graph Minors. XX. Wagner's conjecture. Journal of Combinatorial Theory, Series B 92(2), 325–357 (2004), doi:10.1016/j.jctb.2004.08.001

32. Wagner, K.: Über eine Eigenschaft der ebenen Komplexe. Mathematische Annalen 114(1), 570–590 (1937), doi:10.1007/BF01594196

33. Wagner, K.: Über eine Erweiterung des Satzes von Kuratowski. Deutsche Mathematik 2, 280–285 (1937)

34. Weihe, K.: Maximum (s,t)-flows in planar networks in $O(|V| \log |V|)$ time. Journal of Computer and System Sciences 55(3), 454–476 (1997), doi:10.1006/jcss.1997.1538

From Holant to #CSP and Back: Dichotomy for Holant^c Problems

Jin-Yi Cai[1,*], Sangxia Huang[2,**], and Pinyan Lu[3]

[1] University of Wisconsin-Madison and Peking University
jyc@cs.wisc.edu
[2] Kungliga Tekniska högskolan (KTH)
sangxia@kth.se
[3] Microsoft Research Asia
pinyanl@microsoft.com

Abstract. We explore the intricate interdependent relationship among counting problems, considered from three frameworks for such problems: Holant Problems, counting CSP and weighted H-colorings. We consider these problems for general complex valued functions that take Boolean inputs. We show that results from one framework can be used to derive results in another, and this happens in both directions. Holographic reductions discover an underlying unity, which is only revealed when these counting problems are investigated in the complex domain $\mathbb{C}$. We prove three complexity dichotomy theorems, leading to a general theorem for Holant^c problems. This is the natural class of Holant problems where one can assign constants 0 or 1. More specifically, given any signature grid on $G = (V, E)$ over a set $\mathscr{F}$ of symmetric functions, we completely classify the complexity to be in P or #P-hard, according to $\mathscr{F}$, of

$$\sum_{\sigma:E\to\{0,1\}} \prod_{v\in V} f_v(\sigma\,|_{E(v)}),$$

where $f_v \in \mathscr{F} \cup \{\mathbf{0}, \mathbf{1}\}$ ($\mathbf{0}, \mathbf{1}$ are the unary constant 0, 1 functions). Not only is holographic reduction the main tool, but also the final dichotomy is naturally stated in the language of holographic transformations. The proof goes through another dichotomy theorem on Boolean complex weighted #CSP.

1   Introduction

In order to study the complexity of counting problems, several interesting frameworks have been proposed. One is called counting Constraint Satisfaction Problems (#CSP) [1–3, 13, 17]. Another well studied framework is called H-coloring or Graph Homomorphism, which can be viewed as a special case of #CSP problems [4, 5, 14–16, 19, 20]. Recently, we proposed a new refined framework called

---

* Supported by NSF CCF-0830488 and CCF-0511679.
** Work partly done when the author was an undergraduate student at Shanghai Jiao Tong University and an intern student at Microsoft Research Asia.

O. Cheong, K.-Y. Chwa, and K. Park (Eds.): ISAAC 2010, Part I, LNCS 6506, pp. 253–265, 2010.
© Springer-Verlag Berlin Heidelberg 2010

Holant Problems [8, 10] inspired by Valiant's Holographic Algorithms [25, 26]. One reason such frameworks are interesting is because the language is *expressive* enough so that they can express many natural counting problems, while *specific* enough so that we can prove *dichotomy theorems* (i.e., every problem in the class is either in P or #P-hard) [11]. By a theorem of Ladner, if P $\neq$ NP, or P $\neq$ P$^{\#P}$, then such a dichotomy for NP, or for #P, is *false*. Many natural counting problems can be expressed in all three frameworks. This includes counting the number of vertex covers, the number of k-colorings in a graph, and many others. However, some natural and important counting problems, such as counting the number of perfect matchings in a graph, *cannot* be expressed as a graph homomorphism function [18], but can be naturally expressed as a Holant Problem. Both #CSP and Graph Homomorphisms can be viewed as special cases of Holant Problems. The Holant framework of counting problems makes a finer complexity classification. A rich mathematical structure is uncovered in the Holant framework regarding the complexity of counting problems, which is sometimes difficult even to state in #CSP. This is particularly true when we apply holographic reductions [8, 25, 26].

We give a brief description of the Holant framework.[1] A *signature grid* $\Omega = (G, \mathscr{F}, \pi)$ is a tuple, where $G = (V, E)$ is a graph, and π labels each $v \in V$ with a function $f_v \in \mathscr{F}$. We consider all edge assignments (in this paper 0-1 assignments). An assignment σ for every $e \in E$ gives an evaluation $\prod_{v \in V} f_v(\sigma \mid_{E(v)})$, where $E(v)$ denotes the incident edges of v, and $\sigma \mid_{E(v)}$ denotes the restriction of σ to $E(v)$. The counting problem on the instance Ω is to compute

$$\text{Holant}_\Omega = \sum_\sigma \prod_{v \in V} f_v(\sigma \mid_{E(v)}).$$

For example, consider the PERFECT MATCHING problem on G. This problem corresponds to attaching the EXACT-ONE function at every vertex of G, and then consider all 0-1 edge assignments. In this case, Holant$_\Omega$ counts the number of perfect matchings. If we use the AT-MOST-ONE function at every vertex, then we count all (not necessarily perfect) matchings. We use the notation Holant($\mathscr{F}$) to denote the class of Holant problems where all functions are given by $\mathscr{F}$.

To see that Holant is a more expressive framework, we show that #CSP can be simulated by Holant. Represent an instance of a #CSP problem by a bipartite graph where left hand side (LHS) are labeled by variables and right hand side (RHS) are labeled by constraint functions. Now the signature grid Ω on this bipartite graph is as follows: Every variable node on LHS is labeled with an EQUALITY function, every constraint node on RHS is labeled with the given constraint. Then Holant$_\Omega$ is exactly the answer to the counting CSP problem. In effect, the EQUALITY function on a node in LHS forces the incident edges to take the same value; this effectively reduces to a vertex assignment on LHS as in #CSP. We can show that #CSP is equivalent to Holant problems where

---

[1] The term Holant was first used by Valiant in [25]. It denotes a sum which is a special case (corresponding to PERFECT MATCHING) of Holant$_\Omega$ in the definition here [8, 10]. The term Holant emphasizes its relationship with holographic transformations.

EQUALITY functions of k variables, for arbitrary k (denoted by $=_k$), are freely and implicitly available as constraints. However, this process cannot be reversed in general. While #CSP is the same as adding all $=_k$ to Holant, the effect of making them freely available is non-trivial. From the lens of holographic transformations, $=_3$ is a full-fledged non-degenerate symmetric function of arity 3.

Starting from the Holant framework, rather than assuming EQUALITY functions are free, one can consider new classes of counting problems which are difficult to express as #CSP problems. One such class, called Holant* Problems [10], is the class of Holant Problems where all unary functions are freely available. If we allow only two special unary functions **0** and **1** as freely available, then we obtain the family of counting problems called Holant^c Problems, which is even more appealing. This is the class of all Holant Problems (on Boolean variables) where one can set any particular edge (variable) to 0 or 1 in an input graph.

Previously a dichotomy theorem was proved for Holant*($\mathscr{F}$), where $\mathscr{F}$ is any set of complex-valued symmetric functions [10]. It was used to prove a dichotomy theorem for #CSP in [10]. For Holant^c($\mathscr{F}$) we were able to prove a dichotomy theorem valid only real-valued functions [10]. In this paper we manage to traverse in the other direction, going from #CSP to Holant Problems. First we establish a dichotomy theorem for a special Holant class. Second we prove a more general dichotomy for bipartite Holant Problems. Finally by going through #CSP, we prove a dichotomy theorem for complex-valued Holant^c Problems. Now we describe our results in more detail.

A symmetric function $f : \{0,1\}^k \rightarrow \mathbb{C}$ will be written as $[f_0, f_1, \ldots, f_k]$, where f_j is the value of f on inputs of Hamming weight j. Our first main result (in Section 3) is a dichotomy theorem for Holant($\mathscr{F}$), where $\mathscr{F}$ contains a single ternary function $[x_0, x_1, x_2, x_3]$. More generally, we can apply holographic reductions to prove a dichotomy theorem for Holant($[y_0, y_1, y_2]|[x_0, x_1, x_2, x_3]$) defined on 2-3 regular bipartite graphs. Here the notation indicates that every vertex of degree 2 on LHS has label $[y_0, y_1, y_2]$ and every vertex of degree 3 on RHS has label $[x_0, x_1, x_2, x_3]$. This is the foundation of the remaining two dichotomy results in this paper. Previously we proved a dichotomy theorem for Holant($[y_0, y_1, y_2]|[x_0, x_1, x_2, x_3]$), when all x_i, y_j take values in $\{0,1\}$ [8]. Kowalczyk extended this to $\{-1, 0, 1\}$ in [21]. In [9], we gave a dichotomy theorem for Holant($[y_0, y_1, y_2]|[1, 0, 0, 1]$), where y_0, y_1, y_3 take arbitrary real values. Finally this last result was extended to arbitrary complex numbers [22]. Our result here is built upon these results, especially [22].

Our second result (Section 4) is a dichotomy theorem, under a mild condition, for bipartite Holant problems Holant($\mathscr{F}_1|\mathscr{F}_2$) (see Sec. 2 for definitions). Under this mild condition, we first use holographic reductions to transform it to Holant($\mathscr{F}_1'|\mathscr{F}_2'$), where we transform some non-degenerate function $[x_0, x_1, x_2, x_3] \in \mathscr{F}_2$ to the EQUALITY function ($=_3$) $= [1, 0, 0, 1] \in \mathscr{F}_2'$. Then we prove that we can "realize" the binary EQUALITY function ($=_2$) $= [1, 0, 1]$ in the left side and reduce the problem to #CSP($\mathscr{F}_1' \cup \mathscr{F}_2'$). This is a new proof approach. Previously in [10], we reduced a #CSP problem to a Holant problem and obtained results for #CSP. Here, we go the opposite way, using results for #CSP to prove dichotomy theorems for Holant problems. This is made possible by our complete dichotomy theorem for Boolean complex weighted #CSP [10]. We note that proving this over $\mathbb{C}$ is crucial, as holographic reductions naturally go beyond $\mathbb{R}$. We also note that our dichotomy theorem here does not require the functions in $\mathscr{F}_1$ or $\mathscr{F}_2$ to be symmetric. This will be useful in the future.

Our third and main result, also the initial motivation of this work, is a dichotomy theorem for symmetric complex Holant^c problems. This improves our previous result in [10]. We made a conjecture in [10] about the dichotomy theorem of Holant^c for symmetric complex functions. It turns out that this conjecture is not correct as stated. For example, Holant$^c([1, 0, i, 0])$ is tractable (according to our new theorem), but not included in the tractable cases by the conjecture. After isolating these new tractable cases we prove *everything else* is #P-hard. Generally speaking, non-trivial and previously unknown tractable cases are what make dichotomy theorems particularly interesting, but at the same time make them more difficult to prove (especially for hardness proofs, which must "carve out" exactly what's left). The proof approach here is also different from that of [10]. In [10], the idea is to interpolate all unary functions and then use the results for Holant* Problems. Here we first prove that we can realize some non-degenerate ternary function, for which we can use the result of our first dichotomy theorem. Then we use our second dichotomy theorem to further reduce the problem to #CSP and obtain a dichotomy theorem for Holant^c.

The study of Holant Problems is strongly influenced by the development of holographic algorithms [7, 8, 25, 26]. Holographic reduction is a primary technique in the proof of these dichotomies, for both the tractability part and the hardness part. More than that—and this seems to be the first instance— holographic reduction even provides the correct language for the *statements* of these dichotomies. Without using holographic reductions, it is not easy to even fully describe what are the tractable cases in the dichotomy theorem. Another interesting observation is that by employing holographic reductions, complex numbers appear naturally and in an essential way. Even if one is only interested in integer or real valued counting problems, in the complex domain $\mathbb{C}$ the picture becomes whole. "It has been written that the shortest and best way between two truths of the real domain often passes through the imaginary one." —*Jacques Hadamard.*

2   Preliminaries

Our functions take values in $\mathbb{C}$ by default. Strictly speaking complexity results should be restricted to computable numbers in the Turing model; but it is more convenient to express this over $\mathbb{C}$. We say a problem is tractable if it is computable in P. The framework of Holant Problems is defined for functions mapping any $[q]^k \to \mathbb{C}$ for a finite q. Our results in this paper are for the Boolean case $q = 2$.

Let $\mathscr{F}$ be a set of such functions. A *signature grid* $\Omega = (G, \mathscr{F}, \pi)$ is a tuple, where $G = (V, E)$ is a graph, and $\pi : V \to \mathscr{F}$ labels each $v \in V$

with a function $f_v \in \mathscr{F}$. The Holant problem on instance Ω is to compute $\text{Holant}_\Omega = \sum_{\sigma:E\to\{0,1\}} \prod_{v\in V} f_v(\sigma \mid_{E(v)})$, a sum over all 0-1 edge assignments, of the products of the function evaluations at each vertex. Here $f_v(\sigma \mid_{E(v)})$ denotes the value of f_v evaluated using the restriction of σ to the incident edges $E(v)$ of v. A function f_v can be represented as a truth table. It will be more convenient to denote it as a vector in $\mathbb{C}^{2^{\deg(v)}}$, or a tensor in $(\mathbb{C}^2)^{\otimes \deg(v)}$, when we perform holographic tranformations. We also call it a *signature*. We denote by $=_k$ the EQUALITY signature of arity k. A symmetric function f on k Boolean variables can be expressed by $[f_0, f_1, \dots, f_k]$, where f_j is the value of f on inputs of Hamming weight j. Thus, for example, $\mathbf{0} = [1,0]$, $\mathbf{1} = [0,1]$ and $(=_k) = [1,0,\dots,0,1]$ (with $(k-1)$ 0's).

Definition 1. *Given a set of signatures $\mathscr{F}$, we define* $\text{Holant}(\mathscr{F})$:
  Input: A signature grid $\Omega = (G, \mathscr{F}, \pi)$;
  Output: Holant_Ω.

We would like to characterize the complexity of Holant problems in terms of its signature set $\mathscr{F}$. Some special families of Holant problems have already been widely studied. For example, if $\mathscr{F}$ contains all EQUALITY signatures $\{=_1, =_2, =_3, \dots\}$, then this is exactly the weighted #CSP problem. In [10], we also introduced the following two special families of Holant problems by assuming some signatures are freely available.

Definition 2. *Let $\mathscr{U}$ denote the set of all unary signatures. Then* $\text{Holant}^*(\mathscr{F}) = \text{Holant}(\mathscr{F} \cup \mathscr{U})$.

Definition 3. *For any set of signatures $\mathscr{F}$,* $\text{Holant}^c(\mathscr{F}) = \text{Holant}(\mathscr{F} \cup \{\mathbf{0}, \mathbf{1}\})$.

Replacing a signature $f \in \mathscr{F}$ by a constant multiple cf, where $c \neq 0$, does not change the complexity of $\text{Holant}(\mathscr{F})$. It introduces a global factor to Holant_Ω.
  An important property of a signature is whether it is degenerate.

Definition 4. *A signature is degenerate iff it is a tensor product of unary signatures. In particular, a symmetric signature in $\mathscr{F}$ is degenerate iff it can be expressed as $\lambda[x,y]^{\otimes k}$.*

We use $\mathscr{A}$ to denote the set of functions which has the form $\chi_{[AX=0]} \cdot i^{\sum_{j=1}^{n} \langle \alpha_j, X \rangle}$, where $i = \sqrt{-1}$, $X = (x_1, x_2, \dots, x_k, 1)$, A is matrix over $\mathbb{F}_2$, α_j is a vector over $\mathbb{F}_2$, and χ is a 0-1 indicator function such that $\chi_{[AX=0]}$ is 1 iff $AX = 0$.
  We use $\mathscr{P}$ to denote the set of functions which can be expressed as a product of unary functions, binary equality functions ($[1,0,1]$) and binary disequality functions ($[0,1,0]$).

Theorem 1. *[10] Suppose $\mathscr{F}$ is a set of functions mapping Boolean inputs to complex numbers. If $\mathscr{F} \subseteq \mathscr{A}$ or $\mathscr{F} \subseteq \mathscr{P}$, then #CSP($\mathscr{F}$) is computable in polynomial time. Otherwise, #CSP($\mathscr{F}$) is #P-hard.*

To introduce the idea of holographic reductions, it is convenient to consider bipartite graphs. This is without loss of generality. For any general graph, we can make it bipartite by adding an additional vertex on each edge, and giving each new vertex the EQUALITY function $=_2$ on 2 inputs.

We use $\text{Holant}(\mathscr{G}|\mathscr{R})$ to denote all counting problems, expressed as Holant problems on bipartite graphs $H = (U, V, E)$, where each signature for a vertex in U or V is from $\mathscr{G}$ or $\mathscr{R}$, respectively. An input instance for the bipartite Holant problem is a bipartite signature grid and is denoted as $\Omega = (H, \mathscr{G}|\mathscr{R}, \pi)$. Signatures in $\mathscr{G}$ are denoted by column vectors (or contravariant tensors); signatures in $\mathscr{R}$ are denoted by row vectors (or covariant tensors).

One can perform (contravariant and covariant) tensor transformations on the signatures. We will define a simple version of holographic reductions, which are invertible. Suppose $\text{Holant}(\mathscr{G}|\mathscr{R})$ and $\text{Holant}(\mathscr{G}'|\mathscr{R}')$ are two Holant problems defined for the same family of graphs, and $T \in \mathbf{GL}_2(\mathbb{C})$. We say that there is an (invertible) holographic reduction from $\text{Holant}(\mathscr{G}|\mathscr{R})$ to $\text{Holant}(\mathscr{G}'|\mathscr{R}')$, if the *contravariant* transformation $G' = T^{\otimes g}G$ and the *covariant* transformation $R = R'T^{\otimes r}$ map $G \in \mathscr{G}$ to $G' \in \mathscr{G}'$ and $R \in \mathscr{R}$ to $R' \in \mathscr{R}'$, and vice versa, where G and R have arity g and r respectively. (Notice the reversal of directions when the transformation $T^{\otimes n}$ is applied. This is the meaning of *contravariance* and *covariance*.) Suppose there is a holographic reduction from $\#\mathscr{G}|\mathscr{R}$ to $\#\mathscr{G}'|\mathscr{R}'$ mapping signature grid Ω to Ω', then $\text{Holant}_\Omega = \text{Holant}_{\Omega'}$. In particular, for invertible holographic reductions from $\text{Holant}(\mathscr{G}|\mathscr{R})$ to $\text{Holant}(\mathscr{G}'|\mathscr{R}')$, one problem is in P iff the other one is in P, and similarly one problem is #P-hard iff the other one is also #P-hard.

In the study of Holant problems, we will often transfer between bipartite and non-bipartite settings. When this does not cause confusion, we do not distinguish signatures between column vectors (or contravariant tensors) and row vectors (or covariant tensors). Whenever we write a transformation as $T^{\otimes n}F$ or $T\mathscr{F}$, we view the signatures as column vectors (or contravariant tensors); whenever we write a transformation as $FT^{\otimes n}$ or $\mathscr{F}T$, we view the signatures as row vectors (or covariant tensors).

3   Dichotomy Theorem for Ternary Signatures

In this section, we consider the complexity of $\text{Holant}([x_0, x_1, x_2, x_3])$. It is trivially tractable if $[x_0, x_1, x_2, x_3]$ is degenerate, so in the following we always assume that it is non-degenerate. Similar to that in [10], we classify the sequence $[x_0, x_1, x_2, x_3]$ into one of the following three categories (with the convention that $\alpha^0 = 1$, and $k\alpha^{k-1} = 0$ if $k = 0$, even when $\alpha = 0$): (1) $x_k = \alpha_1^{3-k}\alpha_2^k + \beta_1^{3-k}\beta_2^k$, where $\det \begin{bmatrix} \alpha_1 & \beta_1 \\ \alpha_2 & \beta_2 \end{bmatrix} \neq 0$; (2) $x_k = Ak\alpha^{k-1} + B\alpha^k$, where $A \neq 0$; (3) $x_k = A(3-k)\alpha^{2-k} + B\alpha^{3-k}$, where $A \neq 0$. We call the first category as the *generic* case, the second and third one as the *double-root* case.

For the *generic* case, we can apply a holographic reduction using $T = \begin{bmatrix} \alpha_1 & \beta_1 \\ \alpha_2 & \beta_2 \end{bmatrix}$, and have $\text{Holant}([x_0, x_1, x_2, x_3]) \equiv_T \text{Holant}([y_0, y_1, y_2]|[1, 0, 0, 1])$, where $[y_0, y_1, y_2] = [1, 0, 1]T^{\otimes 2}$. (We note that $[x_0, x_1, x_2, x_3] = T^{\otimes 3}[1, 0, 0, 1]$.) Therefore we only need to give a dichotomy for $\text{Holant}([y_0, y_1, y_2]|[1, 0, 0, 1])$, which has been proved in [22]; we quote the theorem here.

Theorem 2. ([22]) *The problem* Holant($[y_0, y_1, y_2]|[1, 0, 0, 1]$) *is #P-hard for all* $y_0, y_1, y_2 \in \mathbb{C}$ *except in the following cases, for which the problem is in P: (1)* $y_1^2 = y_0 y_2$; *(2)* $y_0^{12} = y_1^{12}$ *and* $y_0 y_2 = -y_1^2$ ($y_1 \neq 0$); *(3)* $y_1 = 0$; *and (4)* $y_0 = y_2 = 0$.

For the *double-root* case, we have the following lemma.

Lemma 1. *Let* $x_k = Ak\alpha^{k-1} + B\alpha^k$, *where* $A \neq 0$ *and* $k = 0, 1, 2, 3$. *Unless* $\alpha^2 = -1$, Holant($[x_0, x_1, x_2, x_3]$) *is #P-hard. On the other hand, if* $\alpha = \pm i$, *then the problem is in P.*

Proof. If $\alpha = \pm i$, the signature $[x_0, x_1, x_2, x_3]$ satisfies the recurrence relation $x_{k+2} = \alpha x_{k+1} + x_k$, where $k = 0, 1$. This is a generalized Fibonacci signature (see [8]). Thus it is in P by holographic algorithms [8] using Fibonacci gates.

Now we assume that $\alpha \neq \pm i$. We first apply an *orthogonal* holographic transformation. The crucial observation is that we can view Holant($[x_0, x_1, x_2, x_3]$) as the bipartite Holant($[1, 0, 1]|[x_0, x_1, x_2, x_3]$) and an orthogonal transformation $T \in \mathbf{O}_2(\mathbb{C})$ keeps $(=_2) = [1, 0, 1]$ invariant: $[1, 0, 1]T^{\otimes 2} = [1, 0, 1]$. By a suitable orthogonal transformation T, we can transform $[x_0, x_1, x_2, x_3]$ to $[v, 1, 0, 0]$ for some $v \in \mathbb{C}$, up to a scalar. (Details are in the full paper [6].) So the complexity of Holant($[x_0, x_1, x_2, x_3]$) is the same as Holant($[v, 1, 0, 0]$).

Next we prove that Holant($[v, 1, 0, 0]$) is #P-hard for all $v \in \mathbb{C}$. First, for $v = 0$, Holant($[0, 1, 0, 0]$) is #P-hard, because it is the problem of counting all perfect matchings on 3-regular graphs [12]. Second, let $v \neq 0$. We can realize $[v^3 + 3v, v^2 + 1, v, 1]$ by connecting three $[v, 1, 0, 0]$'s as a triangle, so it is enough to prove that Holant($[v^3 + 3v, v^2 + 1, v, 1]$) is #P-hard. In tensor product notation this signature is $\frac{1}{2}\left(\begin{bmatrix} v+1 \\ 1 \end{bmatrix}^{\otimes 3} + \begin{bmatrix} v-1 \\ 1 \end{bmatrix}^{\otimes 3} \right)$. Then

$$\text{Holant}([v^3 + 3v, v^2 + 1, v, 1]) \equiv_T \text{Holant}([1, 0, 1]|[v^3 + 3v, v^2 + 1, v, 1])$$
$$\equiv_T \text{Holant}([v^2 + 2v + 2, v^2, v^2 - 2v + 2]|[1, 0, 0, 1])$$

where the second step is a holographic reduction using $\begin{bmatrix} v+1 & v-1 \\ 1 & 1 \end{bmatrix}$. We can apply Theorem 2 to Holant($[v^2 + 2v + 2, v^2, v^2 - 2v + 2]|[1, 0, 0, 1]$). Checking against the four exceptions we find that they are all impossible. Therefore Holant($[v^3 + 3v, v^2 + 1, v, 1]$) is #P-hard, and so is Holant($[v, 1, 0, 0]$) for all $v \in \mathbb{C}$.

By Theorem 2 and Lemma 1, we have a complete dichotomy theorem for Holant ($[x_0, x_1, x_2, x_3]$) and for bipartite Holant($[y_0, y_1, y_2]|[x_0, x_1, x_2, x_3]$).

Theorem 3. Holant($[x_0, x_1, x_2, x_3]$) *is #P-hard unless* $[x_0, x_1, x_2, x_3]$ *satisfies one of the following conditions, in which case the problem is in P:*

1. $[x_0, x_1, x_2, x_3]$ *is degenerate;*
2. *There is a* 2×2 *matrix* T *such that* $[x_0, x_1, x_2, x_3] = T^{\otimes 3}[1, 0, 0, 1]$ *and* $[1, 0, 1]T^{\otimes 2}$ *is in* $\mathscr{A} \cup \mathscr{P}$;
3. *For* $\alpha \in \{2i, -2i\}$, $x_2 + \alpha x_1 - x_0 = 0$ *and* $x_3 + \alpha x_2 - x_1 = 0$.

Theorem 4. Holant($[y_0, y_1, y_2] | [x_0, x_1, x_2, x_3]$) *is #P-hard unless* $[x_0, x_1, x_2, x_3]$ *and* $[y_0, y_1, y_2]$ *satisfy one of the following conditions, in which case the problem is in P:*

1. $[x_0, x_1, \dot{x}_2, x_3]$ *is degenerate;*
2. *There is a* 2×2 *matrix* T *such that* $[x_0, x_1, x_2, x_3] = T^{\otimes 3}[1, 0, 0, 1]$ *and* $[y_0, y_1, y_2]T^{\otimes 2}$ *is in* $\mathscr{A} \cup \mathscr{P}$;
3. *There is a* 2×2 *matrix* T *such that* $[x_0, x_1, x_2, x_3] = T^{\otimes 3}[1, 1, 0, 0]$ *and* $[y_0, y_1, y_2]T^{\otimes 2}$ *is of form* $[0, *, *]$;
4. *There is a* 2×2 *matrix* T *such that* $[x_0, x_1, x_2, x_3] = T^{\otimes 3}[0, 0, 1, 1]$ *and* $[y_0, y_1, y_2]T^{\otimes 2}$ *is of form* $[*, *, 0]$.

4   Reductions between Holant and #CSP

In this section, we extend the dichotomies in Section 3 for a single ternary signature to a set of signatures. We will give a dichotomy for Holant($[x_0, x_1, x_2, x_3] \cup \mathscr{F}$), or more generally for Holant($[y_0, y_1, y_2] \cup \mathscr{G}_1 | [x_0, x_1, x_2, x_3] \cup \mathscr{G}_2$), where $[y_0, y_1, y_2]$ and $[x_0, x_1, x_2, x_3]$ are non-degenerate. In this section, we focus on the generic case of $[x_0, x_1, x_2, x_3]$, and the double root case will be handled in the next section in Lemma 3. For the generic case, we can apply a holographic reduction to transform $[x_0, x_1, x_2, x_3]$ to $[1, 0, 0, 1]$. Therefore we only need to give a dichotomy for Holant problems of the form Holant($[y_0, y_1, y_2] \cup \mathscr{G}_1 | [1, 0, 0, 1] \cup \mathscr{G}_2$), where $[y_0, y_1, y_2]$ is non-degenerate. We make one more observation: for any $T \in \mathscr{T}_3 \triangleq \left\{ \begin{bmatrix} 1 & 0 \\ 0 & 1 \end{bmatrix}, \begin{bmatrix} 1 & 0 \\ 0 & \omega \end{bmatrix}, \begin{bmatrix} 1 & 0 \\ 0 & \omega^2 \end{bmatrix} \right\}$, where $\omega = \omega_3 = e^{2\pi i/3}$, we have

$$\text{Holant}([y_0, y_1, y_2] | [1, 0, 0, 1] \cup \mathscr{F}) \equiv_T \text{Holant}([y_0, y_1, y_2]T^{\otimes 2} | [1, 0, 0, 1] \cup T^{-1}\mathscr{F}).$$

As a result, we can normalize $[y_0, y_1, y_2]$ by a holographic reduction with any $T \in \mathscr{T}_3$. In particular, we call a symmetric binary signature $[y_0, y_1, y_2]$ *normalized* if $y_0 = 0$ or it is not the case that y_2 is y_0 times a t-th primitive root of unity, and $t = 3t'$ where $\gcd(t', 3) = 1$. We can always normalize $[y_0, y_1, y_2]$ by applying a transformation $\begin{bmatrix} 1 & 0 \\ 0 & \omega^k \end{bmatrix} \in \mathscr{T}_3$. So in the following, we only deal with normalized $[y_0, y_1, y_2]$. In one case, we also need to normalize a unary signature $[x_0, x_1]$, namely $x_0 = 0$ or x_1 is not a multiple of x_0 by a t-th primitive root of unity, and $t = 3t'$ where $\gcd(t', 3) = 1$. Again we can normalize the unary signature by a suitable $T \in \mathscr{T}_3$.

Theorem 5. *Let* $[y_0, y_1, y_2]$ *be a normalized and non-degenerate signature. And in the case of* $y_0 = y_2 = 0$, *we further assume that* $\mathscr{G}_1$ *contains a unary signature* $[a, b]$, *which is normalized and* $ab \neq 0$. *Then*

$$\text{Holant}([y_0, y_1, y_2] \cup \mathscr{G}_1 | [1, 0, 0, 1] \cup \mathscr{G}_2) \equiv_T \#\text{CSP}([y_0, y_1, y_2] \cup \mathscr{G}_1 \cup \mathscr{G}_2).$$

Thus, Holant($[y_0, y_1, y_2] \cup \mathscr{G}_1 | [1, 0, 0, 1] \cup \mathscr{G}_2$) *is #P-hard unless* $[y_0, y_1, y_2] \cup \mathscr{G}_1 \cup \mathscr{G}_2 \subseteq \mathscr{P}$ *or* $[y_0, y_1, y_2] \cup \mathscr{G}_1 \cup \mathscr{G}_2 \subseteq \mathscr{A}$, *in which cases the problem is in P.*

This dichotomy is an important reduction step in the proof of our dichotomy theorem for Holant^c. It is also interesting in its own right as a connection between Holant and #CSP. The assumption on signature normalization in the statement of the theorem is without loss of generality. For a non-normalized signature, we can first normalize it and then apply the dichotomy criterion. (Note that when $y_0 = y_2 = 0$ the normalization on $[a, b]$ keeps $[y_0, y_1, y_2]$ normalized.) The additional assumption of the existence of a unary signature $[a, b]$ with $ab \neq 0$ circumvents a technical difficulty, and finds a circuitous route to the proof of our main dichotomy theorem for Holant^c. For Holant^c, the needed unary signature will be produced from $[1, 0]$ and $[0, 1]$. We also note that we do not require the signatures in $\mathscr{G}_1$ and $\mathscr{G}_2$ to be symmetric.

One direction in Theorem 5, from Holant to #CSP, is straightforward. Thus our main claim is a reduction from #CSP to these bipartite Holant problems. Start with $\#\mathrm{CSP}([y_0, y_1, y_2] \cup \mathscr{G}_1 \cup \mathscr{G}_2) \equiv_T \mathrm{Holant}([y_0, y_1, y_2] \cup \mathscr{G}_1 \cup \mathscr{G}_2 | \{=_k : k \geq 1\})$. The approach is to construct the binary equality $[1, 0, 1] = (=_2)$ in LHS in the Holant problem. As soon as we have $[1, 0, 1]$ in LHS, together with $[1, 0, 0, 1] = (=_3)$ in RHS, we get equality gates of all arities $(=_k)$ in RHS. Then with the help of $[1, 0, 1]$ in LHS we can transfer $\mathscr{G}_2$ to LHS.

If the problem $\mathrm{Holant}([y_0, y_1, y_2] | [1, 0, 0, 1])$ is already #P-hard, then for any $\mathscr{G}_1$ and $\mathscr{G}_2$, it is #P-hard. So we only need to consider the cases, where $\mathrm{Holant}([y_0, y_1, y_2] | [1, 0, 0, 1])$ is not #P-hard. For this, we again use Theorem 2 from [22]. The first tractable case $y_1^2 = y_0 y_2$ is degenerate, which does not apply here. The remaining three tractable cases are proved separately and the proofs can be found in the full paper [6].

5   Dichotomy Theorem for Complex Holant^c Problems

In this section, we prove our main result, a dichotomy theorem for Holant^c problems with complex valued symmetric signatures over Boolean variables, which is stated as Theorem 6. The proof crucially uses the dichotomies proved in the previous two sections. We first prove in Lemma 2 that we can always realize a non-degenerate ternary signature except in some trivial cases. With this non-degenerate ternary signature, we can immediately prove #P-hardness if it is not of one of the tractable cases in Theorem 3. For tractable ternary signatures, we use Theorem 5 to extend the dichotomy theorem to the whole signature set. In Theorem 5, we only considered the generic case of the ternary function. The double-root case is handled here in Lemma 3.

Lemma 2. *Given any set of symmetric signatures $\mathscr{F}$ which contains $[1, 0]$ and $[0, 1]$, we can construct a non-degenerate symmetric ternary signature $X = [x_0, x_1, x_2, x_3]$, except in the following two trivial cases:*

1. *Any non-degenerate signature in $\mathscr{F}$ is of arity at most 2;*
2. *In $\mathscr{F}$, all unary signatures are of form $[x, 0]$ or $[0, x]$; all binary signatures are of form $[x, 0, y]$ or $[0, x, 0]$; and all signatures of arity greater than 2 are of form $[x, 0, \ldots, 0, y]$.*

Proof (Sketch). Suppose Case 1 does not hold, and let $X \triangleq [x_0, x_1, \ldots, x_m] \in \mathscr{F}$ be a non-degenerate signature of arity at least 3. It must be that all ternary sub-signatures are degenerate, otherwise we are done. Then we can show that X must be of the form $[x_0, 0, \ldots, 0, x_m]$, where $x_0 x_m \neq 0$. If we have a unary signature, or a unary sub-signature of a binary signature, of the form $[a, b]$ ($ab \neq 0$), we can connect this signature to $m - 3$ dangling edges of X to get a non-degenerate ternary signature $[x, 0, 0, y]$, and we are done. Otherwise, we are in Case 2.

We next consider the double root case for a non-degenerate $X = [x_0, x_1, x_2, x_3]$. By Lemma 1, Holant(X) is already #P-hard unless the double eigenvalue is i or $-i$. Then, $x_{k+2} + \alpha x_{k+1} - x_k = 0$ for $k = 0, 1$, where $\alpha = \pm 2i$.

Lemma 3. *Let* $X = [x_0, x_1, x_2, x_3]$ *be a non-degenerate complex signature satisfying* $x_{k+2} + \alpha x_{k+1} - x_k = 0$ *for* $k = 0, 1$, *where* $\alpha = \pm 2i$. *Let* $Y = [y_0, y_1, y_2]$ *be a non-degenerate binary signature. Then Holant($Y|X$) is #P-hard unless* $y_2 + \alpha y_1 - y_0 = 0$ *(in which case Holant($\{X, Y\}$) is in P by Fibonacci gates).*

Proof (Sketch). We prove this result for $\alpha = -2i$. The other case is similar.

We have $X = T^{\otimes 3}[1, 1, 0, 0]^T$, where $T = \begin{bmatrix} 1 & \frac{B-1}{3} \\ i & A + \frac{B-1}{3}i \end{bmatrix}$, $A \neq 0$. By expressing $\begin{bmatrix} y_0 & y_1 \\ y_1 & y_2 \end{bmatrix} = T_0^T T_0$, which is always possible for some non-singular $T_0 = \begin{bmatrix} a & c \\ b & d \end{bmatrix}$, we have $Y = [1, 0, 1]T_0^{\otimes 2}$. Thus we apply a holographic reduction and have Holant($Y|X$) $\equiv_T$ Holant($[1, 0, 1]|(T_0 T)^{\otimes 3}[1, 1, 0, 0]^T$). Next, we try to use an orthogonal matrix to transform $T_0 T$ to be upper-triangular. We show that we can do this, except for the tractable cases. This leads to a reduction from Holant($[v, 1, 0, 0]$) to Holant($[1, 0, 1]|(T_0 T)^{\otimes 3}[1, 1, 0, 0]^T$) and therefore to Holant($Y|X$), for some v. By Lemma 1, Holant($[v, 1, 0, 0]$) is #P-hard.

Theorem 6. *Let* $\mathscr{F}$ *be a set of complex symmetric signatures. Holant$^c(\mathscr{F})$ is #P-hard unless* $\mathscr{F}$ *satisfies one of the following conditions, in which case it is tractable:*

1. *Holant$^*(\mathscr{F})$ is tractable (for which we have an effective dichotomy in [10]);*
2. *There exists a* $T \in \mathscr{T}$ *such that* $\mathscr{F} \subseteq T\mathscr{A}$, *where*

$$\mathscr{T} \triangleq \{T \mid [1, 0, 1]T^{\otimes 2}, [1, 0]T, [0, 1]T \in \mathscr{A}\}$$

Proof. First of all, if $\mathscr{F}$ is an exceptional case of Lemma 2, we know that Holant$^*(\mathscr{F})$ is tractable and we are done. Now we can assume that we can construct a non-degenerate symmetric ternary signature $X = [x_0, x_1, x_2, x_3]$ and the problem is equivalent to Holant$^c(\mathscr{F} \cup \{X\})$. As discussed in Section 3, there are three categories for X and we only need to consider the first two: (1) $x_k = \alpha_1^{3-k}\alpha_2^k + \beta_1^{3-k}\beta_2^k$; (2) $x_k = Ak\alpha^{k-1} + B\alpha^k$, where $A \neq 0$.
Case 1: $x_k = \alpha_1^{3-k}\alpha_2^k + \beta_1^{3-k}\beta_2^k$. In this case, $X = T^{\otimes 3}[1, 0, 0, 1]^T$, where $T = \begin{bmatrix} \alpha_1 & \beta_1 \\ \alpha_2 & \beta_2 \end{bmatrix}$. (Note that we can replace T by $T\begin{bmatrix} 1 & 0 \\ 0 & \omega^j \end{bmatrix}$, $0 \leq j \leq 2$, and $X = T^{\otimes 3}[1, 0, 0, 1]^T$ still holds.) So we have the following reduction chain,

$$\text{Holant}^c(\mathscr{F}) \equiv_T \text{Holant}^c(\mathscr{F} \cup \{X\}) \equiv_T \text{Holant}(\mathscr{F} \cup \{X, [1, 0], [0, 1]\})$$
$$\equiv_T \text{Holant}(\{[1, 0, 1], [1, 0], [0, 1]\}|\mathscr{F} \cup \{X\})$$
$$\equiv_T \text{Holant}(\{[1, 0, 1]T^{\otimes 2}, [1, 0]T, [0, 1]T\}|[1, 0, 0, 1] \cup T^{-1}\mathscr{F}).$$

Since $[1, 0, 1]T^{\otimes 2}$ is a non-degenerate binary signature, we can apply Theorem 5. (We replace T by $T\begin{bmatrix} 1 & 0 \\ 0 & \omega^j \end{bmatrix}$, $0 \le j \le 2$, to normalize $[1, 0, 1]T^{\otimes 2}$, if needed.) We need to verify is that when $[1, 0, 1]T^{\otimes 2} = [\alpha_1^2 + \alpha_2^2, \alpha_1\beta_1 + \alpha_2\beta_2, \beta_1^2 + \beta_2^2]$ is of the form $[0, *, 0]$, at least one of $[1, 0]T = [\alpha_1, \beta_1]$ or $[0, 1]T = [\alpha_2, \beta_2]$ has both entries non-zero. If not, we would have $\alpha_1\beta_1 = 0$ and $\alpha_2\beta_2 = 0$, which implies that $[1, 0, 1]T^{\otimes 2} = [0, 0, 0]$, a contradiction. (We may again replace T by $T\begin{bmatrix} 1 & 0 \\ 0 & \omega^j \end{bmatrix}$, $0 \le j \le 2$, to normalize this unary, if needed, which does not conflict with the normalization of $[1, 0, 1]T^{\otimes 2}$.) Therefore, by Theorem 5, we know that the problem is #P-hard unless $[1, 0, 1]T^{\otimes 2} \cup T^{-1}\mathscr{F} \subseteq \mathscr{P}$ or $\{[1, 0, 1]T^{\otimes 2}, [1, 0]T, [0, 1]T\} \cup T^{-1}\mathscr{F} \subseteq \mathscr{A}$. In the first case, Holant$^*(\mathscr{F})$ is tractable; and the second case is equivalent to having $T \in \mathscr{T}$ satisfying $\mathscr{F} \subseteq T\mathscr{A}$.

Case 2: $x_k = Ak\alpha^{k-1} + B\alpha^k$, where $A \ne 0$. In this case, if $\alpha \ne \pm i$, the problem is #P-hard by Lemma 1 and we are done. Now we consider the case $\alpha = i$ (the case $\alpha = -i$ is similar). Consider the following Equation

$$z_{k+2} - 2iz_{k+1} - z_k = 0. \tag{1}$$

We note that $X = [x_0, x_1, x_2, x_3]$ satisfies this equation for $k = 0, 1$. If all non-degenerate signatures $Z = [z_0, z_1, \ldots, z_m]$ in $\mathscr{F}$ with arity $m \ge 2$ fulfill

Condition: Z satisfies Equation (1) for $k = 0, 1, \ldots, m - 2$
then this is the second tractable case in the Holant* dichotomy theorem in [10] and we are done. So suppose this is not the case, and $Z = [z_0, z_1, \ldots, z_m] \in \mathscr{F}$, for some $m \ge 2$, is a non-degenerate signature that does not satisfy this Condition. By Lemma 3, if any non-degenerate sub-signature $[z_k, z_{k+1}, z_{k+2}]$ does not satisfy Equation (1), then, together with X which does satisfy (1), we know that the problem is #P-hard and we are done. So we assume every non-degenerate sub-signature $[z_k, z_{k+1}, z_{k+2}]$ of Z satisfies (1). In particular $m \ge 3$, and there exists some binary sub-signature of Z that is degenerate and does not satisfy (1).

If all binary sub-signatures of Z are degenerate (but Z itself is not), we claim that Z has the form $[z_0, 0, \ldots, 0, z_m]$, where $z_0 z_m \ne 0$. Then we can produce some $[a, 0, 0, b]$, $ab \ne 0$, and reduce to Case 1. Otherwise, we can find a ternary sub-signature $[z_k, z_{k+1}, z_{k+2}, z_{k+3}]$ (or its reversal) where $[z_k, z_{k+1}, z_{k+2}]$ is degenerate and $[z_{k+1}, z_{k+2}, z_{k+3}]$ is non-degenerate and thus satisfies $-z_{k+1} - 2iz_{k+2} + z_{k+3} = 0$. Then either we have got an instance of Case 1 or we could prove #P-hardness directly by Lemma 1. (Details are in the full paper [6].)

References

1. Bulatov, A.: A dichotomy theorem for constraint satisfaction problems on a 3-element set. J. ACM 53(1), 66–120 (2006)
2. Bulatov, A.: The complexity of the counting constraint satisfaction problem. In: Aceto, L., Damgård, I., Goldberg, L.A., Halldórsson, M.M., Ingólfsdóttir, A., Walukiewicz, I. (eds.) ICALP 2008, Part I. LNCS, vol. 5125, pp. 646–661. Springer, Heidelberg (2008)

3. Bulatov, A., Dalmau, V.: Towards a dichotomy theorem for the counting constraint satisfaction problem. Inf. Comput. 205(5), 651–678 (2007)
4. Bulatov, A., Grohe, M.: The complexity of partition functions. Theor. Comput. Sci. 348(2-3), 148–186 (2005)
5. Cai, J.-Y., Chen, X., Lu, P.: Graph homomorphisms with complex values: A dichotomy theorem. In: Gavoille, C. (ed.) ICALP 2010, Part I. LNCS, vol. 6198, pp. 275–286. Springer, Heidelberg (2010)
6. Cai, J.-Y., Huang, S., Lu, P.: From Holant To #CSP And Back: Dichotomy For Holant^c Problems. CoRR, abs/1004.0803 (2010),
   http://arxiv.org/abs/1004.0803
7. Cai, J.-Y., Lu, P.: Holographic algorithms: from art to science. In: STOC 2007: Proceedings of the Thirty-Ninth Annual ACM Symposium on Theory of Computing, pp. 401–410 (2007)
8. Cai, J.-Y., Lu, P., Xia, M.: Holographic algorithms by fibonacci gates and holographic reductions for hardness. In: FOCS 2008: Proceedings of the 49th Annual IEEE Symposium on Foundations of Computer Science, pp. 644–653 (2008)
9. Cai, J.-Y., Lu, P., Xia, M.: A Computational Proof of Complexity of Some Restricted Counting Problems. In: Chen, J., Cooper, S.B. (eds.) TAMC 2009. LNCS, vol. 5532, pp. 138–149. Springer, Heidelberg (2009)
10. Cai, J.-Y., Lu, P., Xia, M.: Holant Problems and Counting CSP. In: STOC 2009: Proceedings of the Thirty-Ninth Annual ACM Symposium on Theory of computing, pp. 715–724 (2009)
11. Creignou, N., Khanna, N., Sudan, M.: Complexity classifications of boolean constraint satisfaction problems. SIAM Monographs on Discrete Mathematics and Applications (2001)
12. Dagum, P., Luby, M.: Approximating the permanent of graphs with large factors. Theor. Comput. Sci. 102, 283–305 (1992)
13. Dyer, M.E., Goldberg, L.A., Jerrum, M.: The complexity of weighted boolean #CSP. CoRR, abs/0704.3683 (2007)
14. Dyer, M.E., Goldberg, L.A., Paterson, M.: On counting homomorphisms to directed acyclic graphs. J. ACM 54(6) (2007)
15. Dyer, M.E., Greenhill, C.S.: The complexity of counting graph homomorphisms (extended abstract). In: Proceedings of SODA, pp. 246–255 (2000)
16. Dyer, M.E., Greenhill, C.S.: Corrigendum: The complexity of counting graph homomorphisms. Random Struct. Algorithms 25(3), 346–352 (2004)
17. Feder, T., Vardi, M.Y.: The computational structure of monotone monadic snp and constraint satisfaction: A study through datalog and group theory. SIAM J. Comput. 28(1), 57–104 (1998)
18. Freedman, M., Lovász, L., Schrijver, A.: Reflection positivity, rank connectivity, and homomorphism of graphs. J. AMS 20, 37–51 (2007)
19. Goldberg, L.A., Grohe, M., Jerrum, M., Thurley, M.: A complexity dichotomy for partition functions with mixed signs. In: Proceedings of STACS, pp. 493–504 (2009), CoRR, abs/0804.1932 (2008)
20. Hell, P., Nešetřil, J.: On the complexity of h-coloring. Journal of Combinatorial Theory, Series B 48(1), 92–110 (1990)
21. Kowalczyk, M.: Classification of a Class of Counting Problems Using Holographic Reductions. In: Ngo, H.Q. (ed.) COCOON 2009. LNCS, vol. 5609, pp. 472–485. Springer, Heidelberg (2009)
22. Kowalczyk, M., Cai, J.-Y.: Holant Problems for Regular Graphs with Complex Edge Functions. In: Proceedings of STACS, pp. 525–536 (2010)

23. Vadhan, S.P.: The complexity of counting in sparse, regular, and planar graphs. SIAM J. Comput. 31(2), 398–427 (2001)
24. Valiant, L.G.: The complexity of enumeration and reliability problems. SIAM J. Comput. 8(3), 410–421 (1979)
25. Valiant, L.G.: Holographic algorithms. In: FOCS 2004: Proceedings of the 45th Annual IEEE Symposium on Foundations of Computer Science, pp. 306–315 (2004); SIAM J. Comput. 37(5), 1565–1594 (2008)
26. Valiant, L.G.: Accidental algorthims. In: FOCS 2006: Proceedings of the 47th Annual IEEE Symposium on Foundations of Computer Science, pp. 509–517 (2006)

Computing Sparse Multiples of Polynomials*

Mark Giesbrecht, Daniel S. Roche, and Hrushikesh Tilak

Cheriton School of Computer Science, University of Waterloo
{mwg,droche,htilak}@cs.uwaterloo.ca
http://www.cs.uwaterloo.ca/

Abstract. We consider the problem of finding a sparse multiple of a polynomial. Given $f \in \mathsf{F}[x]$ of degree d, and a desired sparsity t, our goal is to determine if there exists a multiple $h \in \mathsf{F}[x]$ of f such that h has at most t non-zero terms, and if so, to find such an h. When $\mathsf{F} = \mathbb{Q}$ and t is constant, we give a polynomial-time algorithm in d and the size of coefficients in h. When F is a finite field, we show that the problem is at least as hard as determining the multiplicative order of elements in an extension field of F (a problem thought to have complexity similar to that of factoring integers), and this lower bound is tight when $t = 2$.[1]

1 Introduction

Let F be a field, which will later be specified either to be the rational numbers ($\mathbb{Q}$) or a finite field with q elements ($\mathbb{F}_q$). We say a polynomial $h \in \mathsf{F}[x]$ is *t-sparse* (or *has sparsity t*) if it has at most t nonzero coefficients in the standard power basis; that is, h can be written in the form

$$h = h_1 x^{d_1} + h_2 x^{d_2} + \cdots + h_t x^{d_t} \quad \text{for } h_1, \ldots, h_t \in \mathsf{F} \text{ and } d_1, \ldots, d_t \in \mathbb{N}. \quad (1.1)$$

Sparse polynomials have a compact representation as a sequence of coefficient-degree pairs $(h_1, d_1), \ldots, (h_t, d_t)$, which allow representation and manipulation of very high degree polynomials. Let $f \in \mathsf{F}[x]$ have degree d. We examine the computation a t-sparse multiple of f. That is, we wish to determine if there exist $g, h \in \mathsf{F}[x]$ such that $fg = h$ and h has prescribed sparsity t, and if so, to find such an h. We do not attempt to find g, as it may have a superpolynomial number of terms even though h has a compact representation (see Theorem 3.6).

Sparse multiples over finite fields have cryptographic applications. Their computation is used in correlation attacks on LFSR-based stream ciphers (Aimani and von zur Gathen, 2007; Didier and Laigle-Chapuy, 2007). The security of the TCHo cryptosystem is also based on the conjectured computational hardness of sparsest multiple computation over $\mathbb{F}_2[x]$ (Aumasson et al., 2007); our results provide further evidence that this is in fact a computationally difficult problem.

---

* The authors would like to thank the Natural Sciences and Engineering Research Council of Canada (NSERC), and MITACS.

[1] Proofs of all statements in this paper may be found at
http://arxiv.org/abs/1009.3214

O. Cheong, K.-Y. Chwa, and K. Park (Eds.): ISAAC 2010, Part I, LNCS 6506, pp. 266–278, 2010.

Sparse multiples can be useful for extension field arithmetic (Brent and Zimmermann, 2003) and designing interleavers for error-correcting codes (Sadjadpour et al., 2001). The linear algebra formulation in Section 2 relates to finding the minimum distance of a binary linear code (Berlekamp et al., 1978; Vardy, 1997) as well as "sparsifications" of linear systems (Egner and Minkwitz, 1998).

One of our original motivations was to understand the complexity of sparse polynomial *implicitization* over $\mathbb{Q}$ or $\mathbb{R}$: Given a list of zeros of a (typically multivariate) function, we wish find a sparse polynomial with those zeros (see, e.g., Emiris and Kotsireas (2005)). Our work here can be thought of as a univariate version of implicitization, though a reduction from the multi- to univariate case via Kronecker-like substitutions seems quite feasible.

In general, we assume that the desired sparsity t is a constant. This seems reasonable given that over a finite field, even for $t = 2$, the problem is probably computationally hard (Theorem 5.1). In fact, we have reason to conjecture that the problem is intractable over $\mathbb{Q}$ or $\mathbb{F}_q$ when t is a parameter. Our algorithms are exponential in t but polynomial in the other input parameters when $t \in O(1)$.

Over $\mathbb{Q}[x]$, the analysis must consider coefficient size, and we will count machine word operations in our algorithms to account for coefficient growth. We follow the conventions of Lenstra (1999) and define the *height* of a polynomial: Let $f \in \mathbb{Q}[x]$ and $r \in \mathbb{Q}$ the least positive rational number such that $rf \in \mathbb{Z}[x]$. If $rf = \sum_i a_i x^{d_i}$ with each $a_i \in \mathbb{Z}$, then the *height* of f, written $\mathcal{H}(f)$, is $\max_i |a_i|$.

We examine variants of the sparse multiple problem over $\mathbb{F}_q$ and $\mathbb{Q}$. Since every polynomial in $\mathbb{F}_q$ has a 2-sparse multiple of high degree, given $f \in \mathbb{F}_q[x]$ and $n \in \mathbb{N}$ we consider the problem of finding a t-sparse multiple of f with degree at most n. For input $f \in \mathbb{Q}[x]$ of degree d, we consider algorithms which seek t-sparse multiples of height bounded above by an additional input value $c \in \mathbb{N}$. We present algorithms requiring time polynomial in d and $\log c$.

The remainder of the paper is structured as follows.

In Section 2, we consider the straightforward linear algebra formulation of the sparse multiple problem. This is useful over $\mathbb{Q}[x]$ once a bound on the output degree is derived, and also allows us to bound the output size. In addition, it connects our problems with related NP-complete coding theory problems.

In Section 3 we consider the problem of finding the least-degree binomial multiple of a rational polynomial. A polynomial-time algorithm in the size of the input is given which completely resolves the question in this case. This works despite the fact that we show polynomials with binomial multiples whose degrees and heights are both exponential in the input size!

In Section 4 we consider the more general problem of finding a t-sparse multiple of an input $f \in \mathbb{Q}[x]$ without repeated cyclotomic factors. We present a polynomial-time algorithm in the size of f and a given height bound.

Section 5 shows that, even for $t = 2$, finding a t-sparse multiple of a polynomial $f \in \mathbb{F}_q[x]$ is at least as hard as finding multiplicative orders in an extension of $\mathbb{F}_q$ (a problem thought to be computationally difficult). This lower bound is shown to be tight for binomial multiples.

Open questions and avenues for future research are discussed in Section 6.

2 Linear Algebra Formulation

The sparsest multiple problem can be formulated using linear algebra. This requires specifying bounds on degree, height and sparsity; later some of these parameters will be otherwise determined. This approach also highlights the connection to some problems from coding theory. We exhibit a randomized algorithm for finding a t-sparse multiple of a rational polynomial of degree d, given bounds c and n on the height and degree of the multiple respectively. When $t \in O(1)$, the algorithm runs in time polynomial in n and $\log \mathcal{H}(f)$ and returns the desired output with high probability. We also conjecture the intractability of some of these problems, based on similar problems in coding theory. Finally, we show that the problem of finding the sparsest vector in an integer lattice is NP-complete, which was conjectured by Egner and Minkwitz (1998).

Suppose R is a principal ideal domain, and basic ring operations have unit cost. Let $f \in \mathsf{R}[x]$ have degree d and $n \in \mathbb{N}$ be given. Suppose $g, h \in \mathsf{R}[x]$ have degrees $n - d$ and n respectively, with $f = \sum_0^d f_i x^i$, $g = \sum_0^{n-d} g_i x^i$ and $h = \sum_0^n h_i x^i$. The coefficients in equation $fg = h$ satisfy the linear system (2.1).

$$
\underbrace{
\begin{bmatrix}
f_0 & & & \\
f_1 & \ddots & & \\
\vdots & \ddots & f_0 & \\
f_d & \ddots & f_1 & \\
& \ddots & \vdots & \\
& & f_d &
\end{bmatrix}
}_{A_{f,n}}
\underbrace{
\begin{bmatrix}
g_0 \\
g_1 \\
\vdots \\
g_{n-d}
\end{bmatrix}
}_{v_g}
=
\underbrace{
\begin{bmatrix}
h_0 \\
h_1 \\
\vdots \\
h_n
\end{bmatrix}
}_{v_h}
\quad (2.1)
\qquad
\underbrace{
\begin{bmatrix}
1 & \alpha_1 & \cdots & \alpha_1^n \\
1 & \alpha_2 & \cdots & \alpha_2^n \\
\vdots & \vdots & \vdots & \vdots \\
1 & \alpha_d & \cdots & \alpha_d^n
\end{bmatrix}
}_{A_n(\alpha_1,\ldots,\alpha_d)}
\begin{bmatrix}
h_0 \\
h_1 \\
\vdots \\
h_n
\end{bmatrix}
= 0 \quad (2.2)
$$

Thus, a multiple of f of degree at most n and sparsity at most t corresponds to a vector with at most t nonzero coefficients (i.e., a t-sparse vector) in the linear span of $A_{f,n}$. If $f \in \mathsf{R}[x]$ is squarefree and has roots $\{\alpha_1, \ldots, \alpha_d\}$, possibly over a finite extension of R, then (2.2) also holds. Thus t-sparse multiples of a squarefree f correspond to t-sparse R-vectors in the nullspace of $A_n(\alpha_1, \ldots, \alpha_d)$.

2.1 Finding Bounded-Height Bounded-Degree Sparse Multiples

We now present an algorithm to find the sparsest bounded-degree, bounded-height multiple $h \in \mathbb{Q}[x]$ of an input $f \in \mathbb{Q}[x]$. Since $\mathcal{H}$ is invariant under scaling, we may assume that $f, g, h \in \mathbb{Z}[x]$.

The basic idea is the following: Having fixed the positions at which the multiple h has nonzero coefficients, finding low-height multiple is reduced to finding the nonzero vector with smallest l_∞ norm in the image of a small lattice.

Let $I = \{i_1, \ldots, i_t\}$ be a t-subset of $\{0, \ldots, n\}$, and $A_{f,n}^I \in \mathbb{Z}^{(n+1-t)\times(n-d+1)}$ the matrix $A_{f,n}$ with rows $i_1, \ldots, i_t$ removed. Denote by $B_{f,n}^I \in \mathbb{Z}^{t\times(n-d+1)}$ the matrix consisting of the removed rows $i_1, \ldots, i_t$ of the matrix $A_{f,n}$. Existence of a t-sparse multiple $h = h_{i_1} x^{i_1} + h_{i_2} x^{i_2} + \cdots + h_{i_t} x^{i_t}$ of input f is equivalent to the existence of a vector v_g such that $A_{f,n}^I \cdot v_g = \mathbf{0}$ and $B_{f,n}^I \cdot v_g = [h_{i_1}, \ldots, h_{i_t}]^T$.

Algorithm 2.1. Bounded-Degree Bounded-Height Sparsest Multiple

---

Input: $f \in \mathbb{Z}[x]$ and $t, n, c \in \mathbb{N}$
Output: A t-sparse multiple of f with $\deg(h) \leq n$ and $\mathcal{H}(h) \leq c$, or "NONE"

1 **for** $s = 2, 3, \ldots, t$ **do**
2   **foreach** s-subset $I = (1, i_2, \ldots, i_s)$ of $\{1, 2, \ldots, n\}$ **do**
3   Compute matrices $A_{f,n}^I$ and $B_{f,n}^I$ as defined above
4   **if** $A_{f,n}^I$ does not have full column rank **then**
5   Compute matrix $C_{f,n}^I$ whose columns span the nullspace of $A_{f,n}^I$
6           $\mathbf{h} \leftarrow$ shortest l_∞ vector in the column lattice of $B_{f,n}^I \cdot C_{f,n}^I$
7   **if** $l_\infty(\mathbf{h}) \leq c$ **then return** $h_1 + h_2 x^{i_2} + \cdots + h_t x^{i_t}$

8 **return** "NONE"

Now let $C_{f,n}^I$ be a matrix whose columns span the nullspace of the matrix $A_{f,n}^I$. Since $A_{f,n}$ has full column rank, the nullspace of $A_{f,n}^I$ has dimension $s \leq t$, and hence $C_{f,n}^I \in \mathbb{Z}^{(n-d+1) \times s}$. Thus, a t-sparse multiple $h = h_{i_1} x^{i_1} + \cdots + h_{i_t} x^{i_t}$ of f exists if and only if there exists a $v \in \mathbb{Z}^t$ such that

$$B_{f,n}^I \cdot C_{f,n}^I \cdot v = [h_{i_1}, \ldots, h_{i_t}]^T. \tag{2.3}$$

Note that $B_{f,n}^I \cdot C_{f,n}^I \in \mathbb{Z}^{t \times s}$.

Algorithm 2.1 outlines this approach. The following lemma uses the Smith normal form to show that Step 5 can be computed efficiently.

Lemma 2.1. Given $T \in \mathbb{Z}^{k \times \ell}$ with $k \geq \ell$ and nullspace of dimension s, we can compute a $V \in \mathbb{Z}^{s \times \ell}$ such that the image of V equals the nullspace of T. The algorithm requires $O(k\ell^2 s \log \|T\|)$ bit operations (ignoring logarithmic factors).

Lemma 2.2 shows that Step 6 can be performed efficiently. The proof employs the algorithm of (Ajtai et al., 2001) to find all shortest vectors in the l_2 norm within an approximation factor of $\sqrt{t}$. The shortest l_∞ vector must be among the computed set.

Lemma 2.2. The shortest l_∞ vector in the image of a matrix $U \in \mathbb{Z}^{t \times t}$ can be computed by a randomized algorithm using $2^{O(t \log t)} \cdot \|U\|^{O(1)}$ bit operations.

The correctness and efficiency of Algorithm 2.1 can be summarized as follows.

Theorem 2.3. Algorithm 2.1 correctly computes a t-sparse multiple h of f of degree n and height c, if it exists, with $(\log \mathcal{H}(f))^{O(1)} \cdot n^{O(t)} \cdot 2^{O(t \log t)}$ bit operations. The sparsity s of h is minimal over all multiples with degree less than n and height less than c, and $\deg h$ is minimal over all such s-sparse multiples.

2.2 Relationship to NP-Hard Problems

Note that the above algorithms require time exponential in t, and are only polynomial-time for constant t. It is natural to ask whether there are efficient algorithms which require time polynomial in t. We conjecture this problem is

NP-complete, and point out two relatively recent results of Vardy (1997) and Guruswami and Vardy (2005) on related problems that are known to be hard.

The formulation (2.2) seeks the sparsest vector in the nullspace of a (structured) matrix. For an unstructured matrix over finite fields, this is the problem of finding the minimum distance of a linear code, shown by Vardy (1997) to be NP-complete. The same problem over integers translates into finding the sparsest vector in an integer lattice. It was posed as an open problem in Egner and Minkwitz (1998). Techniques similar to Vardy (1997) prove that this problem is also NP-complete over the integers:

Theorem 2.4. *The problem* SparseLatticeVector *of computing the vector with the least Hamming weight in an integer lattice specified by its basis is* NP *complete.*

Of course, the problem may be easier for structured matrices as in (2.2) However, Guruswami and Vardy (2005) show that maximum likelihood decoding of cyclic codes, which seeks sparse solutions to systems of equations of similar structure to (2.2), is also NP-complete. They do require the freedom to choose a right-hand-side vector, whereas we insist on a sparse vector in the nullspace. While these two results certainly do not prove that the bounded-degree sparsest multiple problem is NP-complete, they support our conjecture that it is.

3   Binomial Multiples over $\mathbb{Q}$

In this section we completely solve the problem of determining if there exists a binomial multiple of a rational input polynomial (i.e., of sparsity $t = 2$). That is, given input $f \in \mathbb{Q}[x]$ of degree d, we determine if there exists a binomial multiple $h = x^m - a \in \mathbb{Q}[x]$ of f, and if so, find such an h with minimal degree. The constant coefficient a will be given as a pair $(r, e) \in \mathbb{Q} \times \mathbb{N}$ representing $r^e \in \mathbb{Q}$. The algorithm requires a number of bit operations which is polynomial in d and $\log \mathcal{H}(f)$. No a priori bounds on the degree or height of h are required. We show that m may be exponential in d, and $\log a$ may be exponential in $\log \mathcal{H}(f)$, and give a family of polynomials with these properties.

Algorithm 3.1 begins by factoring the given polynomial $f \in \mathbb{Q}[x]$ into irreducible factors (using, e.g., the algorithm of Lenstra et al. (1982)). We then show how to find a binomial multiple of each irreducible factor, and finally provide a combining strategy for the different multiples.

The following theorem of Risman (1976) characterizes binomial multiples of irreducible polynomials. Let ϕ be Euler's totient function, the number of positive integers less than or equal to n which are coprime to n.

Fact 3.1 (Risman (1976), Corollary 2.2). *Let* $f \in \mathbb{Q}[x]$ *be irreducible of degree d. Suppose the least-degree binomial multiple (if one exists) is of degree m. Then there exist $n, t \in \mathbb{N}$ with $n \mid d$ and $\phi(t) \mid d$ such that $m = n \cdot t$.*

Combining   Fact   3.1   with   number-theoretic   bounds   from Rosser and Schoenfeld (1962), we obtain the following explicit upper bound on the maximum degree of a binomial multiple of an irreducible polynomial.

Algorithm 3.1. Lowest degree Binomial Multiple of a Rational Polynomial

Input: $f \in \mathbb{Q}[x]$

Output: The lowest degree binomial multiple $h \in \mathbb{Q}[x]$ of f, or "NONE"

1 Factor f into irreducible factors: $f = x^b f_1 f_2 \cdots f_u$

2 **if** f *is not squarefree* **then return** "NONE"

3 **for** $i = 1, 2, 3, \ldots, u$ **do**

4 $\quad$ $m_i \leftarrow$ least $k \in \{d_i = \deg f_i, \ldots, (\lceil 3d_i \ln \ln d_i \rceil + 7)d_i\}$ s.t. x^k rem $f_i \in \mathbb{Q}$

5 $\quad$ **if** *no such* m_i *is found* **then return** "NONE"

6 $\quad$ **else** $r_i \leftarrow x^{m_i}$ rem f_i

7 $m \leftarrow \operatorname{lcm}(m_1, \ldots, m_u)$

8 **foreach** *2-subset* $\{i, j\} \subseteq \{1, \ldots, u\}$ **do**

9 $\quad$ **if** $|r_i|^{m_j} \neq |r_j|^{m_i}$ **then return** "NONE"

10 $\quad$ **else if** $sign(r_i^{m/m_i}) \neq sign(r_j^{m/m_j})$ **then** $m \leftarrow 2 \cdot \operatorname{lcm}(m_1, \ldots, m_u)$

11 **return** $x^b(x^m - r_1^{m/m_1})$, with r_1 and m/m_1 given separately

Theorem 3.2. *Let $f \in \mathbb{Q}[x]$ be irreducible of degree d. If a binomial multiple of f exists, and has minimal degree m, then $m \leq d \cdot (\lceil 3d \ln \ln d \rceil + 7)$.*

The above theorem ensures that for an irreducible f_i, Step 4 of Algorithm 3.1 computes the least-degree binomial multiple $x^{m_i} - r_i$ if it exists, and otherwise correctly reports failure. It clearly runs in polynomial time.

Assume the factorization of f is as computed in Step 1, and moreover f is squarefree (otherwise it cannot have a binomial multiple). If any factor does not have a binomial multiple, neither can the product. If every irreducible factor does have a binomial multiple, Step 4 computes the one with the least degree. The following relates the degrees of the minimal binomial multiple of the input polynomial to those of its irreducible factors.

Lemma 3.3. *Let $f \in \mathbb{Q}[x]$ be such that $f = f_1 \cdots f_u \in \mathbb{Q}[x]$ for distinct, irreducible $f_1, \ldots, f_u \in \mathbb{Q}[x]$. Let $f_i \mid (x^{m_i} - r_i)$ for minimal $m_i \in \mathbb{N}$ and $r_i \in \mathbb{Q}$, and let $f \mid (x^m - r)$ for $r \in \mathbb{Q}$. Then $\operatorname{lcm}(m_1, \ldots, m_u) \mid m$.*

The key step in the proof of Lemma 3.3 is showing that if any m_i does not divide m, then a binomial multiple of f_i with degree $(m \bmod m_i)$ can be constructed, contradicting the minimality of m_i.

Lemma 3.4. *For a polynomial $f \in \mathbb{Q}[x]$ factored into distinct irreducible factors $f = f_1 f_2 \ldots f_u$, with $f_i \mid (x^{m_i} - r_i)$ for $r_i \in \mathbb{Q}$ and minimal such m_i, a binomial multiple of f exists if and only if $|r_i|^{m_j} = |r_j|^{m_i}$ for every pair $1 \leq i, j \leq u$. If a binomial multiple exists, the least-degree binomial multiple of f is $x^m - r_i^{m/m_i}$ such that m either equals the least common multiple of the m_i or twice that number. It can be efficiently checked which of these cases holds.*

The following comes directly from the previous lemma and the fact that Algorithm 3.1 performs polynomially many arithmetic operations.

Theorem 3.5. *Given a polynomial $f \in \mathbb{Q}[x]$, Algorithm 3.1 outputs the least-degree binomial multiple $x^m - r_i^{m/m_i}$ (with r_i and m/m_i output separately) if one exists or correctly reports the lack of a binomial multiple otherwise. Furthermore, it runs in deterministic time $(d + \mathcal{H}(f))^{O(1)}$.*

The constant coefficient of the binomial multiple cannot be output in standard form, but must remain an unevaluated power. Polynomials exist whose minimal binomial multiples have exponentially sized degrees and heights.

Theorem 3.6. *For any $d \geq 841$ there exists a polynomial $f \in \mathbb{Z}[x]$ of degree at most $d \log d$ and height $\mathcal{H}(f) \leq \exp(2d \log d)$ whose minimal binomial multiple $x^m - a$ is such that $m > \exp(\sqrt{d})$ and $\mathcal{H}(a) > 2^{\exp(\sqrt{d})}$.*

4 t-Sparse Multiples over $\mathbb{Q}$

We examine the problem of computing t-sparse multiples of rational polynomials, for any fixed positive integer t. As with other types of polynomial computations, it seems that cyclotomic polynomials behave quite differently from cyclotomic-free ones. Accordingly, we first examine the case that our input polynomial f consists only of cyclotomic or cyclotomic-free factors. Then we see how to combine them, in the case that none of the cyclotomic factors are repeated.

Specifically, we will show that, given any rational polynomial f which does not have repeated cyclotomic factors, and a height bound $c \in \mathbb{N}$, we can compute a sparsest multiple of f with height at most c, or conclude that none exists, in time polynomial in the size of f and $\log c$ (but exponential in t).

First, notice that multiplying a polynomial by a power of x does not affect the sparsity, and so without loss of generality we may assume all polynomials are relatively prime to x; we call such polynomials *non-original* since they do not pass through the origin.

4.1 The Cyclotomic Case

Suppose the input polynomial f is a product of cyclotomic factors, and write the complete factorization of f as

$$f = \Phi_{i_1}^{e_i} \cdot \Phi_{i_2}^{e_2} \cdots \Phi_{i_k}^{e_k}. \tag{4.1}$$

Now let $m = \mathrm{lcm}(i_1, \ldots, i_k)$. Then m is the least integer such that $\Phi_{i_1} \cdots \Phi_{i_k}$ divides $x^m - 1$. Let $\ell = \max_i e_i$, the maximum multiplicity of any factor of f. This means that $(x^m - 1)^\ell$ is an $(\ell + 1)$-sparse multiple of f. To prove that this is in fact a sparsest multiple of f, we first require the following simple lemma. Here and for the remainder, for a univariate polynomial $f \in \mathsf{F}[x]$, we denote by f' the first derivative with respect to x, that is, $\frac{\mathrm{d}}{\mathrm{d}x} f$.

Lemma 4.1. *Let $h \in \mathbb{Q}[x]$ be a t-sparse and non-original polynomial, and write $h = a_1 + a_2 x^{d_2} + \cdots + a_t x^{d_t}$. Assume the complete factorization of h over $\mathbb{Q}[x]$ is $h = a_t h_1^{e_1} \cdots h_k^{e_k}$, with each h_i monic and irreducible. Then $\max_i e_i \leq t - 1$,*

An immediate consequence is the following:

Corollary 4.2. *Let $f \in \mathbb{Q}[x]$ be a product of cyclotomic factors, written as in (4.1). Then*

$$h = (x^{\mathrm{lcm}(i_1,\ldots,i_k)} - 1)^{\max_i e_i}$$

is a sparsest multiple of f.

4.2   The Cyclotomic-Free Case

We say a polynomial $f \in \mathbb{Q}[x]$ is *cyclotomic-free* if it contains no cyclotomic factors. Here we will show that a sparsest multiple of a cyclotomic-free polynomial must have degree bounded by a polynomial in the size of the input and output.

First we need the following elementary lemma.

Lemma 4.3. *Suppose $f, h \in \mathbb{Q}[x]$ with f irreducible, and k is a positive integer. Then $f^k | h$ if and only if $f|h$ and $f^{k-1}|h'$.*

The following technical lemma provides the basis for our degree bound on the sparsest multiple of a non-cyclotomic polynomial. The proof, omitted, is by induction on k, using the "gap theorem" from (Lenstra, 1999) in the base case.

Lemma 4.4. *Let $f, h_1, h_2, \ldots, h_\ell \in \mathbb{Q}[x]$ be non-original polynomials, where f is irreducible and non-cyclotomic with degree d, and each h_i satisfies $\deg h_i \leq u$ and $\mathcal{H}(h_i) \leq c$. Also let $k, m_1, m_2, \ldots, m_\ell$ be positive integers such that*

$$f^k | (h_1 x^{m_1} + h_2 x^{m_2} + \cdots + h_\ell x^{m_\ell}).$$

Then f^k divides each h_i whenever every "gap length", for $1 \leq i < \ell$, satisfies

$$m_{i+1} - m_i - \deg h_i \geq \frac{1}{2} d \cdot \ln^3(3d) \cdot \ln \left(u^{k-1} c \, (t-1) \right). \tag{4.2}$$

Our main tool in proving that Algorithm 2.1 is useful for computing the sparsest multiple of a rational polynomial, given only a bound c on the height, in polynomial time in the size of f and $\log c$, is the following degree bound on the sparsest height-bounded multiple of a rational polynomial. The proof follows from Lemma 4.4, observing that if the degree of h is sufficiently large compared to the sparsity t, then h must have at least one large gap.

Theorem 4.5. *Let $f \in \mathbb{Q}[x]$ with $\deg f = d$ be cyclotomic-free, and let $t, c \in \mathbb{N}$ such that f has a nonzero t-sparse multiple with height at most c. Denote by n the smallest degree of any such multiple of f. Then n satisfies*

$$n \leq 2(t-1)B \ln B, \tag{4.3}$$

where B is the formula polynomially bounded by d, $\log c$, and $\log t$ defined as

$$B = \frac{1}{2} d^2 \cdot \ln^3(3d) \cdot \ln \left(\hat{c} \, (t-1)^d \right), \tag{4.4}$$

and $\hat{c} = \max(c, 35)$.

In order to compute the sparsest multiple of a rational polynomial with no cyclotomic or repeated factors, we can therefore simply call Algorithm 2.1 with the given height bound c and degree bound as specified in (4.3).

Algorithm 4.1. Rational Sparsest Multiple

Input: Bounds $t, c \in \mathbb{N}$ and $f \in \mathbb{Q}[x]$ a non-original polynomial of degree d with no repeated cyclotomic factors

Output: t-sparse multiple h of f with $\mathcal{H}(h) \leq c$, or "NONE"

1 Factor f as $f = \Phi_{i_1} \cdot \Phi_{i_2} \cdots \Phi_{i_k} \cdot f_D$, where f_D is cyclotomic-free
2 $n \leftarrow$ degree bound from (4.3)
3 $\hat{h} \leftarrow$ sparsest multiple of f_D with $\mathcal{H}(\hat{h}) \leq c$ and $\deg \hat{h} \leq n$, using Algorithm 2.1
4 $\tilde{h} \leftarrow$ sparsest multiple of f with $\mathcal{H}(h) \leq c$ and $\deg h \leq n$, using Algorithm 2.1
5 **if** $\hat{h} =$ "NONE" *and* $\tilde{h} =$ "NONE" **then return** "NONE"
6 **else if** $\hat{h} =$ "NONE" *or* sparsity$(\tilde{h}) \leq 2 \cdot$ sparsity$(\hat{h})$ **then return** $\tilde{h}$
7 $m \leftarrow \mathrm{lcm}\{i_1, i_2, \ldots, i_k\}$
8 **return** $\hat{h} \cdot (x^m - 1)$

4.3   Handling Cyclotomic Factors

Suppose f is any non-original rational polynomial with no repeated cyclotomic factors. Factor f as $f = f_C \cdot f_D$, where f_C is a squarefree product of cyclotomics and f_D is cyclotomic-free. Write the factorization of f_C as $f_C = \Phi_{i_1} \cdots \Phi_{i_k}$, where Φ_n is the nth cyclotomic polynomial. Since every i^{th} root of unity is also a $(mi)^{\text{th}}$ root of unity for any $m \in \mathbb{N}$, f_C must divide the binomial $x^{\mathrm{lcm}\{i_1, \ldots, i_k\}} - 1$, which is in fact the sparsest multiple of f_C and clearly has minimal height.

Then we will show that the sparsest height-bounded multiple of f is either of small degree, or is equal to the sparsest height-bounded multiple of f_D times the binomial multiple of f_C specified above. Algorithm 4.1 uses this fact to compute the sparsest multiple of any such f.

Theorem 4.6. *Let $f \in \mathbb{Q}[x]$ be a degree-d non-original polynomial with no repeated cyclotomic factors. Given f and integers c and t, Algorithm 4.1 correctly computes a t-sparse multiple h of f satisfying $\mathcal{H}(h) \leq c$, if one exists. The sparsity of h will be minimal over all multiples with height at most c. The cost of the algorithm is $(d \log c)^{O(t)} \cdot 2^{O(t \log t)} \cdot (\log \mathcal{H}(f))^{O(1)}$.*

The main idea in the proof is that, if the sparsest multiple of f has very high degree, then it can be written as $h = h_1 + h_2 x^m$, and both h_1 and h_2 must be sparse multiples of f_D, the cyclotomic-free part of f.

4.4   An Example

Say we want to find a sparsest multiple of the following polynomial over $\mathbb{Q}[x]$.

$$f = x^{10} - 5x^9 + 10x^8 - 8x^7 + 7x^6 - 4x^5 + 4x^4 + x^3 + x^2 - 2x + 4.$$

First factor using (Lenstra et al., 1982) and identify cyclotomic factors:

$$f = \underbrace{(x^2 - x + 1)}_{\Phi_6} \cdot \underbrace{(x^4 - x^3 + x^2 - x + 1)}_{\Phi_{10}} \cdot \underbrace{(x^4 - 3x^3 + x^2 + 6x + 4)}_{f_D}.$$

Next, we calculate a degree bound from Theorem 4.5. Unfortunately, this bound is not very tight (despite being polynomial in the output size); using $t = 10$, $c = 1000$, and f given above, the bound is $n \leq 11\,195\,728$. So for this example, we will use the smaller (but artificial) bound of $n \leq 20$.

The next step is to calculate the sparsest multiples of both f_D and f with degree at most 20 and height at most 1000. Using Algorithm 2.1, these are

$$\hat{h} = x^{12} + 259x^6 + 64.$$
$$\tilde{h} = x^{11} - 3x^{10} + 12x^8 - 9x^7 + 10x^6 - 4x^5 + 9x^4 + 3x^3 + 8.$$

Since the sparsity of $\hat{h}$ is less than half that of $\tilde{h}$, a sparsest multiple is

$$h = (x^{12} + 259x^6 + 64) \cdot (x^{\mathrm{lcm}(6,10)} - 1) = x^{42} + 259x^{36} + 64x^{30} - x^{12} - 259x^6 - 64.$$

5   Sparse Multiples over $\mathbb{F}_q$

We prove that for any constant t, finding the minimal degree t-sparse multiple of an $f \in \mathbb{F}_q[x]$ is harder than finding orders of elements in $\mathbb{F}_{q^e}$. Order finding is reducible to integer factorization and to discrete logarithm, but reductions in the other direction are not known for finite fields (Adleman and McCurley, 1994). However, a fast algorithm for order finding in finite fields would give an efficient procedure for computing primitive elements, "one of the most important unsolved and notoriously hard problems in the computational theory of finite fields" (von zur Gathen and Shparlinski, 1999).

Formal problem definitions are as follows:

SpMul$_{\mathbb{F}_q}^{(t)}(f, n)$: Given a polynomial $f \in \mathbb{F}_q[x]$ and an integer $n \in \mathbb{N}$, determine if there exists a (nonzero) 2-sparse multiple $h \in \mathbb{F}_q[x]$ of f with $\deg h \leq n$.

Order$_{\mathbb{F}_{q^e}}(a, n)$: Given an element $a \in \mathbb{F}_{q^e}^*$ and an integer $n < q^e$, determine if there exists a positive integer $m \leq n$ such that $a^m = 1$.

The problem **Order**$_{\mathbb{F}_{q^e}}(a, n)$ is well-studied (see for instance Meijer (1996)), and has been used as a primitive in several cryptographic schemes. Note that an algorithm to solve **Order**$_{\mathbb{F}_{q^e}}(a, n)$ will allow us to determine the *multiplicative order* of any $a \in \mathbb{F}_{q^e}^*$ (the smallest nonzero m such that $a^m = 1$) with essentially the same cost (up to a factor of $O(e \log q)$) by using binary search.

The reduction from **Order**$_{\mathbb{F}_{q^e}}(a, n)$ to **SpMul**$_{\mathbb{F}_q}^{(t)}(f, n)$ works as follows: Given an instance of **Order**$_{\mathbb{F}_{q^e}}(a, n)$, we first check if the order o_a of a is less than t by brute-force. Otherwise, we construct the minimal polynomial g_{a^i} for each $a^0, a^1, a^2, \ldots, a^{t-1}$. We only keep distinct g_{a^i}, and call the product of these distinct polynomials $f_{a,t}$. We then run the **SpMul**$_{\mathbb{F}_q}^{(t)}(f, n)$ subroutine to search for the existence of a degree n, t-sparse multiple of the polynomial $f_{a,t}$.

Theorem 5.1. *Let $a \in \mathbb{F}_q$ be an element of order at least t. Then the least degree t-sparse multiple of $f_{a,t}$ is $x^{o_a} - 1$ where o_a is the order of a.*

Algorithm 5.1. Least degree binomial multiple of f over $\mathbb{F}_q$

Input: $f \in \mathbb{F}_q[x]$
Output: The least degree binomial multiple h of f

1 Factor $f = x^b f_1^{e_1} \cdot f_2^{e_2} \cdots f_\ell^{e_\ell}$ for irreducible $f_1, \ldots, f_\ell \in \mathbb{F}_q[x]$, and set $d_i \leftarrow \deg f_i$
2 **for** $i = 1, 2, \ldots, \ell$ **do**
3 $\quad$ $a_i \leftarrow x \in \mathbb{F}_q[x]/(f_i)$, a root of f_i in the extension $\mathbb{F}_{q^{d_i}}$
4 $\quad$ Calculate o_i, the order of a_i in $\mathbb{F}_q[x]/(f_i)$.
5 $n_1 \leftarrow \mathrm{lcm}(\{o_i/\gcd(o_i, q-1)\})$ for all i such that $d_i > 1$
6 $n_2 \leftarrow \mathrm{lcm}(\{order(a_i/a_j)\})$ over all $1 \le i, j \le u$
7 $n \leftarrow \mathrm{lcm}(n_1, n_2)$
8 $\tilde{h} \leftarrow (x^n - a_1^n)$
9 $e \leftarrow \lceil \log_p \max e_i \rceil$, the smallest e such that $p^e \ge e_i$ for all i
10 **return** $h = x^b(x^n - a_1^n)^{p^e}$

The proof uses the null vector formulation of (2.2) in Section 2 to reason that low-degree t-sparse multiples correspond to weight t null vectors of a certain matrix. It is similar to the construction of BCH codes of specified design distance. Of cryptographic interest is that fact that these order-finding polynomials are frequent enough in $\mathbb{F}_q[x]$ so that the reduction also holds in the average case.

Next we give a probabilistic algorithm for finding the least degree binomial multiple for polynomials $f \in \mathbb{F}_q$. This algorithm makes repeated calls to an **Order**$_{\mathbb{F}_{q^e}}(a, n)$ (defined in the previous section) subroutine. Combined with the hardness result of the previous section (with $t=2$), this precisely characterizes the complexity of finding least-degree binomial multiples in terms of the complexity of **Order**$_{\mathbb{F}_{q^e}}(a, n)$.

Algorithm 5.1 solves the binomial multiple problem in $\mathbb{F}_q$ by making calls to an **Order**$_{\mathbb{F}_{q^e}}(a, n)$ procedure that computes the order of elements in extension fields of $\mathbb{F}_q$. Thus **SpMul**$^{(2)}_{\mathbb{F}_q}(f)$ reduces to **Order**$_{\mathbb{F}_{q^e}}(a, n)$ in probabilistic polynomial time. Construction of an irreducible polynomial (required for finite field arithmetic) as well as the factoring step in the algorithm make it probabilistic.

Theorem 5.2. *Given $f \in \mathbb{F}_q[x]$ of degree d, Algorithm 5.1 correctly computes a binomial multiple h of f with least degree. It uses at most d^2 calls to a routine for order finding in $\mathbb{F}_{q^e}$, for various $e \le d$, and $d^{O(1)}$ other operations in $\mathbb{F}_q$. It is probabilistic of the Las Vegas type.*

The proof of correctness (omitted here) works in three steps. The algorithm is first shown to work correctly for irreducible polynomials, then for squarefree polynomials and finally for all polynomials.

6   Conclusions and Open Questions

We have presented an efficient algorithm to compute the least-degree binomial multiple of any rational polynomial. We can also compute t-sparse multiples of rational polynomials that do not have repeated cyclotomic factors, for any fixed t, and given a bound on the height of the multiple.

We have also shown that, even for fixed t, finding a t-sparse multiple of a degree-d polynomial over $\mathbb{F}_q[x]$ is at least as hard as finding the orders of elements in $\mathbb{F}_{q^d}$. In the $t = 2$ case, there is also a probabilistic reduction in the other direction, so that computing binomial multiples of degree-d polynomials over $\mathbb{F}_q[x]$ probabilisticly reduces to order finding in $\mathbb{F}_{q^d}$.

Several important questions remain unanswered. Although we have an unconditional algorithm to compute binomial multiples of rational polynomials, computing t-sparse multiples for fixed $t \geq 3$ requires an *a priori* height bound on the output as well as the requirement that the input contains no repeated cyclotomic factors. Removing these restrictions would be desirable (if possible).

Regarding lower bounds, we know that computing t-sparse multiples over finite fields is at least as hard as order finding, a result which is tight (up to randomization) for $t = 2$, but for larger t we believe the problem is even harder. Specifically, we suspect that computing t-sparse multiples is NP-complete over both $\mathbb{Q}$ and $\mathbb{F}_q$, when t is a parameter in the input.

References

Adleman, L.M., McCurley, K.S.: Open problems in number-theoretic complexity. II. In: Huang, M.-D.A., Adleman, L.M. (eds.) ANTS 1994. LNCS, vol. 877, pp. 291–322. Springer, Heidelberg (1994)

El Aimani, L., von zur Gathen, J.: Finding low weight polynomial multiples using lattices. Cryptology ePrint Archive, Report 2007/423 (2007), http://eprint.iacr.org/2007/423.pdf

Ajtai, M., Kumar, R., Sivakumar, D.: A sieve algorithm for the shortest lattice vector problem. In: Symp. Theory of Computing (STOC 2001), pp. 601–610 (2001)

Aumasson, J.-P., Finiasz, M., Meier, W., Vaudenay, S.: TCHo: a hardware-oriented trapdoor cipher. In: Pieprzyk, J., Ghodosi, H., Dawson, E. (eds.) ACISP 2007. LNCS, vol. 4586, pp. 184–199. Springer, Heidelberg (2007)

Berlekamp, E.R., McEliece, R.J., van Tilborg, H.C.: On the inherent intractability of certain coding problems. IEEE Transactions on Information Theory 24(3) (1978)

Brent, R.P., Zimmermann, P.: Algorithms for finding almost irreducible and almost primitive trinomials. In: Primes and Misdemeanours: Lectures in Honour of the Sixtieth Birthday of Hugh Cowie Williams, Fields Institute, p. 212 (2003)

Didier, F., Laigle-Chapuy, Y.: Finding low-weight polynomial multiples using discrete logarithms. In: Proc. IEEE International Symposium on Information Theory (ISIT 2007), pp. 1036–1040 (2007)

Egner, S., Minkwitz, T.: Sparsification of rectangular matrices. J. Symb. Comput. 26(2), 135–149 (1998)

Emiris, I.Z., Kotsireas, I.S.: Implicitization exploiting sparseness. In: Geometric and Algorithmic Aspects of Computer-Aided Design and Manufacturing. DIMACS Ser. Discrete Math. Theoret. Comput. Sci., vol. 67, pp. 281–297 (2005)

von zur Gathen, J., Shparlinski, I.: Constructing elements of large order in finite fields. In: Fossorier, M.P.C., Imai, H., Lin, S., Poli, A. (eds.) AAECC 1999. LNCS, vol. 1719, pp. 730–730. Springer, Heidelberg (1999)

Guruswami, V., Vardy, A.: Maximum-likelihood decoding of Reed-Solomon codes is NP-hard. In: SODA 2005: Proceedings of the Sixteenth Annual ACM-SIAM symposium on Discrete Algorithms, pp. 470–478 (2005)

Lenstra, A.K., Lenstra Jr., H.W., Lovász, L.: Factoring polynomials with rational coefficients. Math. Ann. 261(4), 515–534 (1982)

Lenstra Jr., H.W.: Finding small degree factors of lacunary polynomials. In: Number
Theory in Progress, vol. 1, pp. 267–276. De Gruyter, Berlin (1999)

Meijer, A.R.: Groups, factoring, and cryptography. Math. Mag. 69(2), 103–109 (1996)

Risman, L.J.: On the order and degree of solutions to pure equations. Proc. Amer.
Math. Soc. 55(2), 261–266 (1976)

Rosser, J.B., Schoenfeld, L.: Approximate formulas for some functions of prime numbers. Ill. J. Math. 6, 64–94 (1962)

Sadjadpour, H.R., Sloane, N.J.A., Salehi, M., Nebe, G.: Interleaver design for turbo
codes. IEEE J. Selected Areas in Communications 19(5), 831–837 (2001)

Vardy, A.: The intractability of computing the minimum distance of a code. IEEE
Transactions on Information Theory 43(6), 1757–1766 (1997)

Fractal Parallelism:
Solving SAT in Bounded Space and Time

Denys Duchier, Jérôme Durand-Lose*, and Maxime Senot

LIFO, Université d'Orléans,
B.P. 6759, F-45067 ORLÉANS Cedex 2
Jerome.Durand-Lose@univ-orleans.fr

Abstract. Abstract geometrical computation can solve NP-complete problems efficiently: any boolean constraint satisfaction problem, instance of SAT, can be solved in bounded space and time with simple geometrical constructions involving only drawing parallel lines on a Euclidean space-time plane. Complexity as the maximal length of a sequence of consecutive segments is quadratic. The geometrical algorithm achieves massive parallelism: an exponential number of cases are explored simultaneously. The construction relies on a fractal pattern and requires the same amount of space and time independently of the SAT formula.

Keywords: Abstract geometrical computation; Signal machine; Fractal; SAT; Massive parallelism; Model of computation.

1 Introduction

SAT, the problem of determining the satisfiability of propositional formulae, is the poster-child of combinatorial complexity and the natural representative of the classical time complexity class NP [Cook, 1971, Levin, 1973]. As such, it is a natural challenge to consider when investigating new computing machinery (quantum, NDA, membrane, hyperbolic spaces...) [Sosík, 2003, Margenstern and Morita, 2001]. In this paper, we show that *signal machines*, through fractal parallelization, are capable of solving SAT in bounded space and time, and thus by NP-completeness of SAT, signal machine can solve any NP-problem *i.e.* hard problems according to classical models like Turing-machine. We also offer a more pertinent notion of complexity, namely *depth*, which is quadratic for our proposed construction for SAT.

The geometrical context proposed here is the following: dimensionless *particles/signals* move uniformly on the real axis. When a set of particles collide, they are replaced by a new set of particles according to a chosen collection of *collision rules*. By adjoining a temporal dimension to the continuous space-line, we can visualize the chronology of motions and collisions of these particles as a *space-time diagram* (in which both space and time are continuous). Since particles have constant speed, their trajectories in the diagram consist of line segments.

---

* Corresponding author.

O. Cheong, K.-Y. Chwa, and K. Park (Eds.): ISAAC 2010, Part I, LNCS 6506, pp. 279–290, 2010.

Models of computation, conventional or not, are frequently based on mathematical idealizations of physical concepts and investigate the consequences, on computational power, of such abstractions (quantum, membrane, closed time-like curves, black holes...) [Păun, 2001, Brun, 2003, Etesi and Németi, 2002]. However, oftentimes, the idealization is such that it must be interpreted either as allowing information to have infinite density (e.g. an oracle), or to be transmitted at infinite speed (global clock, no spatial extension...). On this issue, the model of signal machines stands in contradistinction with other abstract models of computation: it respects the principle of causality, density and speed of information are finite, as are the sets of objects manipulated. Nonetheless, it remains a resolutely abstract model with no apriori ambition to be physically realizable, and it deals with theoretical issues such as computational power.

Signal machines are Turing-universal [Durand-Lose, 2005] and allows to do analog computation by a systematic use of the continuity of space and time [Durand-Lose, 2008, 2009a,b]. Other *geometrical models of computation* exist and allow to compute: colored universe [Jacopini and Sontacchi, 1990], geometric machines [Huckenbeck, 1989], piece-wise constant derivative systems [Asarin and Maler, 1995, Bournez, 1997], optical machines [Naughton and Woods, 2001]...

Most of the work to date in this domain, called *abstract geometrical computation* (AGC), has dealt with the simulation of sequential computations even though the model, seen as a continuous extension of cellular automata, is inherently parallel. (The connexion with CA is briefly illustrated on Fig. 1) In the present paper, we describe a massively parallel evaluation of all possible valuations for a given propositional formula. This is the first time that parallelism is really used in AGC and that an algorithm is proposed to solve directly hard problems (without simulating another model).

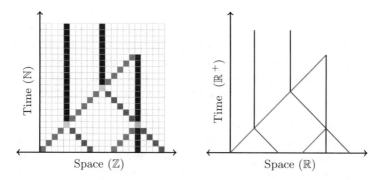

Fig. 1. From cellular automata to signal machines

To achieve this, we follow a fractal pattern to a depth of n (for n propositional variables) in order to partition the space in 2^n regions corresponding to the 2^n possible valuations of the formula. We call the resulting geometrical construction the *combinatorial comb* of propositional assignments. With a signal machine, such an exponential construction fits in bounded space and time regardless of the number of variables.

This constant time has to be understood in the context of continuous space and time. Feynman famously remarked that "there's plenty of room at the bottom." This is especially the case here where, by scaling things down, room can be provided anywhere. With proper handling, this also leads to unbounded acceleration [Durand-Lose, 2009a]. The fractal pattern provides a way to automatically scale down. The one implemented here is a recursive subdivision in halves.

Once the combinatorial comb is in place, it is used to implement a binary decision tree for evaluating the formula, where all branches are explored in parallel. Finally, all the results are collected and disjunctively aggregated to yield the final answer.

Signal machines are presented in Section 2. Sections 3 to 7 detail step by step our algorithm to solve SAT by geometrical computation: splitting the space, coding and generating the formula, broadcasting it, evaluating it and finalizing the answer by collecting the evaluations. Complexities are discussed in Section 8 and conclusion and remarks are gathered in Section 9.

2   Definitions

Signal machines are an extension of cellular automata from discrete time and space to continuous time and space. Dimensionless signals/particles move along the real line and rules describe what happens when they collide.

Signals. Each *signal* is an instance of a *meta-signal*. The associated meta-signal defines its *velocity* and what happen when signals meet. Figure 2 presents a very simple space-time diagram. Time is increasing upwards and the meta-signals are indicated as labels on the signals. Existing meta-signals are listed on the left of Fig. 2.

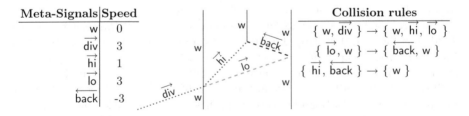

Meta-Signals	Speed	Collision rules
w	0	$\{\, w, \overrightarrow{div}\,\} \rightarrow \{\, w, \overrightarrow{hi}, \overrightarrow{lo}\,\}$
$\overrightarrow{div}$	3	$\{\, \overrightarrow{lo}, w\,\} \rightarrow \{\, \overleftarrow{back}, w\,\}$
$\overrightarrow{hi}$	1	$\{\, \overrightarrow{hi}, \overleftarrow{back}\,\} \rightarrow \{\, w\,\}$
$\overrightarrow{lo}$	3	
$\overleftarrow{back}$	-3	

Fig. 2. Geometrical algorithm for computing the middle

Generally, we use over-line arrows to indicate the direction of propagation of a meta-signal. For example, $\overleftarrow{a}$ and $\overrightarrow{a}$ denotes two different meta-signals; but as can be expected, they have similar uses and behaviors. Similarly b_r and b_l are different; both are stationary, but one is meant to be the version for right and the other for left.

Collision rules. When a set of signals collide, they are replaced by a new set of signals according to a matching collision rule. A rule has the form:

$$\{\sigma_1, \ldots, \sigma_n\} \rightarrow \{\sigma'_1, \ldots, \sigma'_p\}$$

where all σ_i are meta-signals of distinct speeds as well as σ'_j (two signals cannot collide if they have the same speed and outcoming signals must have different speeds). A rule matches a set of colliding signals if its left-hand side is equal to the set of their meta-signals. By default, if there is no exactly matching rule for a collision, the behavior is defined to regenerate exactly the same meta-signals. In such a case, the collision is called *blank*. Collision rules can be deduced from space-time diagram as on Fig. 2. They are also listed on the right of this figure.

Signal machine. A signal machine is defined by a set of meta-signals, a set of collision rules, and and initial configuration, i.e. a set of particles placed on the real line. The evolution of a signal machine can be represented geometrically as a *space-time diagram*: space is always represented horizontally, and time vertically, growing upwards. The geometrical algorithm displayed in Fig. 2 computes the middle: the new w is located exactly half way between the initial two w.

3   Combinatorial Comb

In order to determine by brute force whether a propositional formula with n variables is satisfiable, 2^n cases must be considered. These cases can be recursively enumerated using a binary decision tree. In this section, we explain how to construct in parallel the full decision tree in constant space and time. This is done for a fixed formula, so that n is a constant, and the construction of the signal machine depends on it. In later sections we will use this tree to evaluate the formula.

The intuition is that the decision for variable x_i will be represented by a stationary signal: the space on the left should be interpreted as $x_i = \texttt{false}$, and the space on the right as $x_i = \texttt{true}$. Then we will similarly subdivide the spaces to the left and to the right, with stationary signals for x_{i+1}, and so on recursively for all variables as illustrated in Fig. 3(a).

Starting with two bounding signals w and an initiator $\overrightarrow{\text{start}}$, space is recursively divided as shown in Fig. 3(b). The first step works exactly as in Fig. 2, but then continues on to a depth of n: the counting is realized by using successively $\overrightarrow{m_0}$, $\overrightarrow{m_1}$, $\overrightarrow{m_2}$ ... The necessary rules and meta-signals are summarized in Tab. 1.

Since each level of the tree is half the height of the previous one, the full tree can be constructed in bounded time regardless of its size. Also, note that the bottom level of the tree is not x_n but b_r and b_l. These are used both to evaluate the formula and to aggregate the results as explained later.

4   Formula Encoding

In this section, we will explain how to represent the formula as a set of signals. This is illustrated with the following example:

$$\phi = (x_1 \vee \neg x_2) \wedge x_3$$

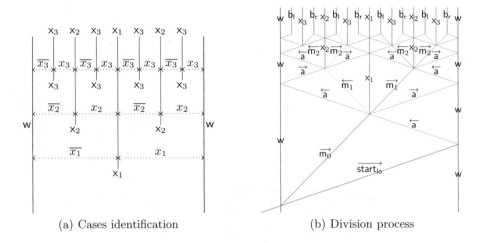

(a) Cases identification   (b) Division process

Fig. 3. Combinatorial comb

Table 1. Meta-Signals and collision rules to build the comb

Meta-Signal	Speed	Collision rules
start, start_{lo}, $\overrightarrow{a}$	3	$\{ \text{start}, w \} \rightarrow \{ w, \overrightarrow{\text{start}_{lo}}, \overrightarrow{m_0} \}$
$\overrightarrow{m_0}$, $\overrightarrow{m_1}$, $\overrightarrow{m_2}$ ...	1	$\{ \overrightarrow{\text{start}_{lo}}, w \} \rightarrow \{ \overleftarrow{a}, w \}$
x_1, x_2, x_3 ...	0	$\{ w, \overleftarrow{a} \} \rightarrow \{ w, \overrightarrow{a} \}$
$\overleftarrow{m_0}$, $\overleftarrow{m_1}$, $\overleftarrow{m_2}$ ...	-1	$\{ \overrightarrow{a}, w \} \rightarrow \{ \overleftarrow{a}, w \}$
$\overleftarrow{a}$	-3	$\{ \overrightarrow{m_i}, \overleftarrow{a} \} \rightarrow \{ \overleftarrow{a}, \overrightarrow{m_{i+1}}, x_i, \overleftarrow{m_{i+1}}, \overrightarrow{a} \}$
b_l, b_r	0	$\{ \overrightarrow{a}, \overleftarrow{m_i} \} \rightarrow \{ \overleftarrow{a}, \overleftarrow{m_{i+1}}, x_i, \overrightarrow{m_{i+1}}, \overrightarrow{a} \}$
		$\{ \overrightarrow{m_n}, \overleftarrow{a} \} \rightarrow \{ b_r \}$
		$\{ \overrightarrow{a}, \overleftarrow{m_n} \} \rightarrow \{ b_l \}$

A formula can be viewed as a tree whose nodes are labeled by symbols (connectives and variables). The evaluation of the formula for a given assignment is a bottom-up process that percolates from the leaves toward the root. In order to model that process, we shall represent each node of the tree by a signal. In Fig. 4(a), each node is additionally decorated with a *path* from the root uniquely identifying its position in the tree: thus we are able to conveniently distinguish multiple occurrences of the same symbol. These decorated symbols provide convenient names for the required meta-signals (see Fig. 4(b)). Thus a formula of size l requires the definition of $2l$ meta-signals.

The signals for all subformulae are sent along parallel trajectories and form a *beam*. They are stacked in the diagram in order of nesting, inner-most subformulae first. This order is important for the process of percolation that will take place at the end.

The process can be initiated by just 3 signals as shown in Fig. 4(c). The delay between the two signals from the left, $\overrightarrow{m_0}$ and ϕ_R, controls the width of the beam. Since space is continuous, this width can be made as small as desired.

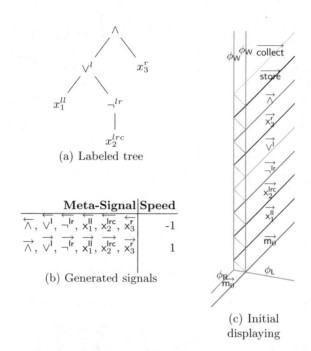

(a) Labeled tree

Meta-Signal	Speed
$\overleftarrow{\wedge}$, $\overleftarrow{\vee^l}$, $\overleftarrow{\neg^{lr}}$, $\overleftarrow{x_1^{ll}}$, $\overleftarrow{x_2^{lrc}}$, $\overleftarrow{x_3^r}$	-1
$\overrightarrow{\wedge}$, $\overrightarrow{\vee^l}$, $\overrightarrow{\neg^{lr}}$, $\overrightarrow{x_1^{ll}}$, $\overrightarrow{x_2^{lrc}}$, $\overrightarrow{x_3^r}$	1

(b) Generated signals

(c) Initial
displaying

Fig. 4. Compiling the formula

5   Propagating the Beam

The formula's beam is now propagated down the decision tree. For each decision point, the beam is duplicated: one part goes through, the other is reflected. Thus, by construction, every branch of the beam tree encounters a decision point for every variable at least once. If the beam is sufficiently narrow, the guarantee becomes "exactly once," as shown in Fig. 5(a). Although we lack space for a detailed explanation (see [Duchier et al., 2010, App. A] for proofs), it can easily be verified that emitting ϕ_L from the origin with a speed of $1 - 7/(3k \cdot 2^{n+2})$ is more than sufficient, where k is the number of signals in the beam and n is the number of variables in the formula. This rational number can be computed in time at most quadratic in the size of the formula.

When the beam encounters a decision point (a stationary signal for a variable x_i), then a split occurs producing two branches. Except for the sign of their velocity, most signals remain identical in both branches; most, except those corresponding to occurrences of x_i: those become `false` in the left branch and `true` in the right branch. Fig. 5(b) shows the beam intersecting the decision signal for variable x_1. Note how the incident signal $\overrightarrow{x_1^{ll}}$ becomes $\overleftarrow{f^{ll}}$ on the left and $\overrightarrow{t^{ll}}$ on the right; the path decoration is preserved since, as we shall see, it is essential later for the percolation process. This is achieved by the collision rule:

$$\{\overrightarrow{x_1^{ll}}, x_1\} \rightarrow \{\overleftarrow{f^{ll}}, x_1, \overrightarrow{t^{ll}}\}$$

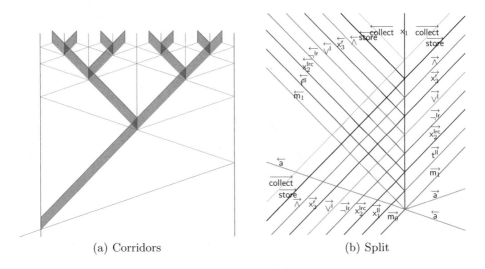

(a) Corridors   (b) Split

Fig. 5. Propagating the formula's beam

Since a decision point is encountered exactly once for each variable on each branch of the beam tree, at the bottom of the tree, all signals corresponding to occurrences of variables have been assigned a boolean value.

6   Evaluating the Formula

Remember how, at the very bottom of the decision tree, we added an extra division using signals b_l or b_r: their purpose is to initiate the percolation process. b_l is for starting the percolation process of a left branch, while b_r is for a right branch. Fig. 6 zooms on one case of our example: The invariant is that all signals that reach b_r have determined boolean values. When $\overrightarrow{t^{ll}}$ reaches b_r, it gets reflected as $\overleftarrow{T^{ll}}$. The change from lowercase to uppercase indicates that the subformula's signal is now able to interact with the signal of its parent connective. The stacking order ensures that reflected signals of subformulae will interact with the incoming signal of their parent connective before the latter reaches b_r. This enforces the invariant.

A connective is evaluated by colliding with the (uppercased) boolean signals of its arguments. For example, the disjunction collides with its first argument. Depending on its value, it becomes the one-argument function identity or the constant true. This is the way the rules of Tab. 2 should be understood.

Note how the path decorations are essential to ensure that the right subformulae interact with the right occurrences of connectives. Conjunctions and negations can be handled similarly. Finally, $\overrightarrow{store}$ projects the truth value of the formula's root on b_r where it is temporarilly stored until $\overrightarrow{collect}$ starts the aggregation of the results.

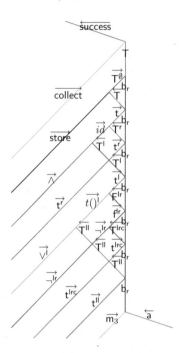

Fig. 6. Evaluation at the bottom of the comb

Table 2. Collision rules to evaluate the disjunction $\vee^l$

$$\{ \overrightarrow{\vee^l}, \overleftarrow{\mathsf{T}^{ll}} \} \to \{ \overrightarrow{t()^l} \} \qquad \{ \overrightarrow{t()^l}, \overleftarrow{\mathsf{T}^{lr}} \} \to \{ \overrightarrow{t^l} \} \qquad \{ \overrightarrow{id^l}, \overleftarrow{\mathsf{T}^{lr}} \} \to \{ \overrightarrow{t^l} \}$$
$$\{ \overrightarrow{\vee^l}, \overleftarrow{\mathsf{F}^{ll}} \} \to \{ \overrightarrow{id^l} \} \qquad \{ \overrightarrow{t()^l}, \overleftarrow{\mathsf{F}^{lr}} \} \to \{ \overrightarrow{t^l} \} \qquad \{ \overrightarrow{id^l}, \overleftarrow{\mathsf{F}^{lr}} \} \to \{ \overrightarrow{f^l} \}$$

7   Collecting the Results

At the end of the propagation phase, the results of evaluating the formula for all possible assignments have been stored as stationary signals replacing the b_l and b_r signals. We must now compute the disjunction of all these results. This is the collection phase and it is initiated and carried out by signals $\overleftarrow{collect}$ and $\overrightarrow{collect}$ as illustrated in Fig. 7. The required collision rules are summarized in Tab. 3.

Table 3. Collection rules

$$\{ B, \overrightarrow{collect} \} \to \{ \overrightarrow{B} \} \qquad \{ \overrightarrow{collect}, x_i \} \to \{ \overleftarrow{collect}, \mathsf{L}, \overrightarrow{collect} \}$$
$$\{ B, \overleftarrow{collect} \} \to \{ \overleftarrow{B} \} \qquad \{ x_i, \overleftarrow{collect} \} \to \{ \overleftarrow{collect}, \mathsf{R}, \overrightarrow{collect} \}$$
$$\{ \overrightarrow{B_1}, \mathsf{R}, \overleftarrow{B_2} \} \to \{ \overleftarrow{B_3} \} \qquad \text{for } B, B_1, B_2, B_3 \in \{\mathsf{T},\mathsf{F}\}$$
$$\{ \overrightarrow{B_1}, \mathsf{L}, \overleftarrow{B_2} \} \to \{ \overrightarrow{B_3} \} \qquad \text{and } B_3 = B_1 \vee B_2$$

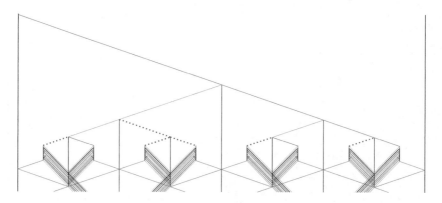

Fig. 7. Collecting the result

Putting it all together, we get the space-time diagram of Fig. 8. Although, it cannot be seen on the picture, four signals are emitted on the first collision (bottom left). Two have very close speeds so that when the signals for the formula are generated, the resulting beam is sufficiently narrow (see Fig. 4(c) for a zoom in on this part).

8   Complexities

We now turn to a crucial question: what is the complexity of our construction as a function of the size of the formula? What is a meaningful way to measure this complexity?

The width of the construction measures the space requirement: it is independent of the formula and can be fixed to any value we like. The height measures the time requirement: it is also independent of the formula because of the fractal construction and the continuity of space-time. If more variables are involved, the comb gains extra levels, but its height remains bounded by the fractal.

As a consequence, while width (space) and height (time) are the natural continuous extensions of traditional complexity measures used in the discrete universe of cellular automata, in the context of abstract geometrical computations, they loose all pertinence.

Instead we should regard our construction as a computational device transforming inputs into outputs. The inputs are given by the initial state of the signal machine at the bottom of the diagram. The output is the computed result that comes out at the top. The transformation is performed in parallel by many threads: a thread here is an ascending path through the diagram from an input to the output. The operations that are "performed" by the thread are all the collisions found along the path.

Thus, if we view the diagram as an acyclic graph of collisions (vertices) and signals (arcs), the time complexity can then be defined as the maximal length

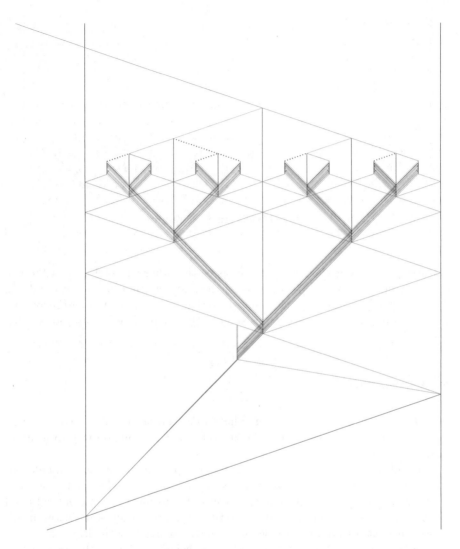

Fig. 8. The whole diagram

of a chain and the space complexity can be defined as the maximal length of an anti-chain.

Let t be the size of the formula and n the number of variables. At the bottom level of the comb, there is an anti-chain of length approximately $t2^n$. The space complexity is exponential.

Generation of the comb, initiation, propagation, evaluation and aggregation contribute along any path a number of collisions at most linear in the size of the formula. However, intersections of incident and reflected branches at every level add $O(nt)$ because there are $O(n)$ levels and the beam consists of $O(t)$ parallel signals. Thus the time complexity is $O(nt)$.

It should also be pointed out that the signal machine depends on the formula but the compilation of the formula into a rational signal machine is done in polynomial time, presicely in quadratic time (see [Duchier et al., 2010, App. B] for the schemes of compilation). The size of the generated signal machine is as follows. The number of meta-signals is linear in n for the comb and in t for the formula. The number of non blank collision rules is proportional to nt (each node of the formula is split on each variable). Counting the blank collision rules, it sums up to t^2. There are only seven distinct speeds: -6, -3, -1, 0, 1, 3 and one special rational value for the initiation.

9   Conclusion

In this article, we have shown how to achieve massive parallelism with signal machines, by means of a fractal pattern. We call this *fractal parallelism* and it is a novel contribution in the field of abstract geometrical computation.

Our approach is able to solve SAT, and thus any NP-problem, in bounded space and time by a methodic use of continuity. It does so while respecting the principle that everywhere the density of information is finite and its speed is bounded; a principle typically not considered by other abstract models of computation.

The complexity is not hidden inside the compilation of the machine nor in the initial configuration. Admittedly, the "magic" rational velocity used to control the narrowness of the beam constitutes an infelicity of presentation as it is the only one that depends on the formula. It can be eliminated using a slightly more involved beam-narrowing technique, but that extension is beyond the scope of the present article.

Since, clearly, time and space are no longer appropriate measures of complexity, we have also proposed to replace them respectively by the maximum length of a chain and an anti-chain in the space-time diagram regarded as a directed acyclic graph. According to these new definitions, our construction has exponential space complexity and quadratic time complexity. The compilation of formulae into signal machines can be done uniformly in quadratic time by a single classical machine.

We are currently furthering this research along two axes. First, we are considering how to tackle other complexity classes such as PSPACE, #P or EXP-TIME using abstract geometrical computation. Second, we would like to design a generic signal machine for SAT, *i.e.* a single machine solving any instance of SAT, where the formula is merely compiled into an initial configuration.

References

Asarin, E., Maler, O.: Achilles and the Tortoise climbing up the arithmetical hierarchy. In: Thiagarajan, P.S. (ed.) FSTTCS 1995. LNCS, vol. 1026, pp. 471–483. Springer, Heidelberg (1995)

Bournez, O.: Some bounds on the computational power of piecewise constant derivative systems. In: Degano, P., Gorrieri, R., Marchetti-Spaccamela, A. (eds.) ICALP 1997. LNCS, vol. 1256, pp. 143–153. Springer, Heidelberg (1997)

Brun, T.A.: Computers with closed timelike curves can solve hard problems efficiently. Foundations of Physics Letters 16, 245–253 (2003)

Cook, S.A.: The complexity of theorem proving procedures. In: 3rd Symposium on Theory of Computing (STOC 1971), pp. 151–158. ACM, New York (1971)

Duchier, D., Durand-Lose, J., Senot, M.: Fractal parallelism: solving SAT in bounded space and time (extended version). Research Report RR-2010-08, LIFO, Université d'Orléans (2010),
http://www.univ-orleans.fr/lifo/rapports.php

Durand-Lose, J.: Abstract geometrical computation: Turing computing ability and undecidability. In: Cooper, S.B., Löwe, B., Torenvliet, L. (eds.) CiE 2005. LNCS, vol. 3526, pp. 106–116. Springer, Heidelberg (2005)

Durand-Lose, J.: Abstract geometrical computation with accumulations: Beyond the Blum, Shub and Smale model. In: Beckmann, A., Dimitracopoulos, C., Löwe, B. (eds.) Logic and Theory of Algorithms, 4th Conf. Computability in Europe (CiE 2008) (abstracts and extended abstracts of unpublished papers), pp. 107–116. University of Athens (2008)

Durand-Lose, J.: Abstract geometrical computation 3: Black holes for classical and analog computing. Nat. Comput. 8(3), 455–572 (2009a)

Durand-Lose, J.: Abstract geometrical computation and computable analysis. In: Costa, J.F., Dershowitz, N. (eds.) UC 2009. LNCS, vol. 5715, pp. 158–167. Springer, Heidelberg (2009)

Etesi, G., Németi, I.: Non-turing computations via Malament-Hogarth space-time. International Journal of Theoret. Physics 41, 341–370 (2002)

Huckenbeck, U.: Euclidian geometry in terms of automata theory. Theoret. Comp. Sci. 68(1), 71–87 (1989)

Jacopini, G., Sontacchi, G.: Reversible parallel computation: an evolving space-model. Theoret. Comp. Sci. 73(1), 1–46 (1990)

Levin, L.: Universal search problems. In: Problems of Information Transmission, pp. 265–266 (1973)

Margenstern, M., Morita, K.: NP problems are tractable in the space of cellular automata in the hyperbolic plane. Theor. Comp. Sci. 259(1-2), 99–128 (2001)

Naughton, T.J., Woods, D.: On the computational power of a continuous-space optical model of computation. In: Margenstern, M., Rogozhin, Y. (eds.) MCU 2001. LNCS, vol. 2055, pp. 288–299. Springer, Heidelberg (2001)

Păun, G.: P systems with active membranes: Attacking NP-Complete problems. Journal of Automata, Languages and Combinatorics 6(1), 75–90 (2001)

Sosík, P.: Solving a PSPACE-Complete problem by P-systems with active membranes. In: Cavaliere, M., Martín-Vide, C., Păun, G. (eds.) Brainstorming Week on Membrane Computing, pp. 305–312. Universidad Rovira i Virgili, Tarragona (2003)

Interpretation of Stream Programs: Characterizing Type 2 Polynomial Time Complexity

Hugo Férée[1], Emmanuel Hainry[2,5],
Mathieu Hoyrup[3,5], and Romain Péchoux[4,5]

[1] ENS Lyon, 46 allée d'Italie, 69364 Lyon cedex 07, France
[2] Université Henri Poincaré, Nancy-Université, France
[3] INRIA Nancy - Grand Est, Villers-lès-Nancy, France
[4] Université Nancy 2, Nancy-Université, France
[5] LORIA, BP 239, 54506 Vandœuvre-lès-Nancy cedex, France
hugo.feree@ens-lyon.fr, {hainry,hoyrup,pechoux}@loria.fr

Abstract. We study polynomial time complexity of type 2 functionals. For that purpose, we introduce a first order functional stream language. We give criteria, named well-founded, on such programs relying on second order interpretation that characterize two variants of type 2 polynomial complexity including the Basic Feasible Functions (BFF). These characterizations provide a new insight on the complexity of stream programs. Finally, we adapt these results to functions over the reals, a particular case of type 2 functions, and we provide a characterization of polynomial time complexity in Recursive Analysis.

1 Introduction

Stream languages including lazy functional languages like Haskell allows the programmer to represent functionals, functions over functions. From this perspective, they can be understood as a way to simulate type 2 functions. There are many works in the literature that study computability and (polynomial time) complexity of such functions [5,14]. The implicit computational complexity (ICC) community has proposed characterizations of such complexity classes using function algebra and types [9,16,8] or recently as a logic [15] . These results are reminiscent of former characterizations of type 1 polynomial time functions [4,2,12] that led to other ICC works using polynomial interpretations.

Polynomial interpretations [13,11] are a well-known tool used in the termination analysis of first order functional programs for several decades. Variants, like sup-interpretations and quasi-interpretations [3], that allow the programmer to perform program complexity analysis have emerged in the last ten years. One of their drawbacks is that such tools are restricted to first order programs on inductive data types. The paper [7] was a first attempt to adapt such a tool to co-inductive data types and, more precisely, to stream programs. In this paper, we provide a second order variation of this interpretation methodology that fits to stream computation.

O. Cheong, K.-Y. Chwa, and K. Park (Eds.): ISAAC 2010, Part I, LNCS 6506, pp. 291–303, 2010.

It allows us to characterize exactly the set of functions computable in polynomial time by Unary Oracle Turing Machine (UOTM), that is functions computable by machines including oracles where the oracle has only unary input. It can also be used in order to characterize the set of functions computable in polynomial time by Oracle Turing Machine (OTM), that is shown to be equivalent to the BFF algebra in [9].

The first characterization has two advantages. First, it gives a new and intuitive notion of stream computational complexity in terms of Turing Machine. Second, it shows that this natural class can be easily captured using an adaptation of the interpretation tool. Using this tool we can analyze functions of this class in an easier way (based on the premise that it is practically easier to write a first order functional program on streams than the corresponding Turing Machine). The drawback is that the tool suffers from the same problem as polynomial interpretation: the difficulty to automatically synthesize the interpretation of a given program (see [1]).

The latter characterization gives a natural view of a well-know complexity class BFF, just by changing the interpretation codomain: indeed we use power towers instead of polynomials in the interpretation of a stream argument. It illustrates that the first characterization on UOTM is natural and flexible because it can be easily adapted to other complexity classes. Finally, it can be interpreted as a negative result showing that the BFF class, whose purpose is to study functions from $\mathbb{N} \to \mathbb{N}$, is surprisingly not well-suited to describe stream polynomial complexity (because of the power tower).

We also go one step further showing that these tools can be adapted to characterize the complexity of functions computing over reals defined in Recursive Analysis [10]. This approach is a first attempt to study the complexity of such functions through static analysis methods.

Outline of the paper. The paper is organized as follows. In section 2, we briefly recall the notion of (Unary) Oracle Turing Machine and its complexity. In section 3, we introduce a first order stream language. In section 4, we define the interpretation tools and a criterion on stream programs. We show our main characterization relying on the criterion in section 5. In a last section, this characterization is adapted to functions computing over reals.

2   Polynomial Time Oracle Turing Machines

In this section, we recall the notion of Oracle Turing Machine, used by Kapron and Cook in their characterization of Basic Poly-time functionals (BFF) [9], and we give a variant, Unary Oracle Turing Machine, more related to stream computations.

Definition 1 (Oracle Turing Machine). *An Oracle Turing Machine (denoted OTM) $\mathcal{M}$ with k oracles (where oracles are functions from $\mathbb{N}$ to $\mathbb{N}$) and l input tapes is a Turing machine with, for each oracle, a state, one query tape and one answer tape.*

If $\mathcal{M}$ is used with oracles $F_1, \ldots F_k : \mathbb{N} \to \mathbb{N}$, then on the oracle state $i \in \{1, \ldots, k\}$, $F_i(x)$ is written on the corresponding answer tape, whenever x is the content of the corresponding query tape.

We introduce now the notion of Unary OTM that are more related to stream computations as accessing the n-th element takes at least n steps (whereas it takes $\log(n)$ steps in OTM. See example 1 for details).

Definition 2 (Unary Oracle Turing Machine). *A Unary Oracle Turing Machine (denoted UOTM) is an OTM where numbers are written using unary notation on the query tape, i.e. on the oracle state i, $F_i(|x|)$ is written on the corresponding answer tape, whenever x is the content of the corresponding query tape.*

Definition 3 (Size of function). *The size $|F| : \mathbb{N} \to \mathbb{N}$ of a function $F : \mathbb{N} \to \mathbb{N}$ is defined by:*

$$|F|(n) = \max_{k \leq n} |F(k)|$$

where $|F(k)|$ represents the size of the binary representation of $F(k)$.

Definition 4 (Second order polynomial). *A second order polynomial is a polynomial generated by the following grammar:*

$$P := c \mid X \mid P + P \mid P \times P \mid Y\langle P \rangle$$

where X represents a first order variable, Y a second order one and c a constant in $\mathbb{N}$.

In the following, $P(Y_1, \ldots, Y_k, X_1, \ldots, X_l)$ will denote a second order polynomial where each Y_i represents a second order variable, and each X_i a first order variable.

Definition 5 (Polynomial running time). *The cost of a transition is:*

- *$|F|(|x|)$, if the machine is in a query state of the oracle F on input query x;[1]*
- *1 otherwise.*

An OTM $\mathcal{M}$ operates in time $T : (\mathbb{N} \to \mathbb{N})^k \to \mathbb{N}^l \to \mathbb{N}$ if for all inputs $x_1, \ldots x_l : \mathbb{N}$ and $F_1, \ldots F_k : \mathbb{N} \to \mathbb{N}$, the sum of the transition costs before $\mathcal{M}$ halts on these inputs is less than $T(|F_1|, \ldots, |F_k|, |x_1|, \ldots, |x_l|)$.

A function $G : (\mathbb{N} \to \mathbb{N})^k \to \mathbb{N}^l \to \mathbb{N}$ is OTM computable (resp. UOTM computable) in polynomial time if there exists a second order polynomial P such that $G(F_1, \ldots, F_k, x_1, \ldots x_l)$ is computed by an OTM (resp. UOTM) in time $P(|F_1|, \ldots, |F_k|, |x_1|, \ldots, |x_l|)$ on inputs $x_1, \ldots, x_l$ and oracles $F_1, \ldots, F_k$.

The set of polynomial time UOTM computable functions is strictly included in the set of polynomial time OTM computable functions (proved to be equal to the BFF algebra in [9]):

---

[1] This definition is equivalent to that of [9] which considers $|F|(x)$ but where $|F|$ is the maximum for $k \leq |n|$. It has the advantage to be uniform for UOTM and OTM.

Example 1. The function $G : (\mathbb{N} \to \mathbb{N}) \times \mathbb{N} \to \mathbb{N}$ defined by $G(F, x) = F(|x|)$ is UOTM computable in polynomial time (its running time is bounded by $2 \times (|x| + |F|(|x|))$, that is the cost to copy x on the query tape and to query the oracle). However $H(F, x) = F(x)$ is not but is in BFF.

3   First Order Stream Language

Syntax. We define a simple first order functional language with streams. This small language can be seen as a strict subset of a lazy functional language like Haskell. We denote by $\mathcal{F}$ the set of function symbols, $\mathcal{C}$ the set of constructor symbols and $\mathcal{X}$ the set of variable names. Programs in our language are lists of definitions $\mathcal{D}$ given by the following grammar:

$$p ::= x \mid c\ p_1\ \dots\ p_n \mid p : y \text{ (Patterns)}$$
$$e ::= x \mid t\ e_1\ \dots\ e_n \text{ (Expressions)}$$
$$d ::= f\ p_1\ \dots\ p_n = e \text{ (Definitions)}$$

where $x, y \in \mathcal{X}, t \in \mathcal{C} \cup \mathcal{F}, c \in \mathcal{C} \setminus \{:\}$ and $f \in \mathcal{F}$ and c, t and f are symbols of arity n.

Throughout the paper, we call closed expression any expression without variables.

The stream constructor $: \in \mathcal{C}$ is a special infix constructor of arity 2. In a stream expression $hd : tl$, hd is called the head and tl is called the tail (of the stream).

In a definition $f\ p_1\ \dots\ p_n = e$, all the variables of e appear in the patterns p_i. Moreover patterns are non overlapping and each variable appears at most once in the left-hand side. It entails that programs are confluent. In a program, we suppose that all pattern matchings are exhaustive. Finally, we only allow patterns of depth 1 for the stream constructor (*i.e.* only variables appear in the tail of a stream pattern). This is not restrictive since a program with higher pattern matching depth can be easily transformed into a program of this form using extra function symbols and definitions.

Type system. Programs contain inductive types that will be denoted by Tau throughout the paper. Unary integers are defined by data Nat = 0 | Nat +1, given that $0, +1 \in \mathcal{C}$. Consequently, each constructor symbol comes with a typed signature and we will use the notation c :: T to denote that the constructor symbol c has type T. For example, we have 0 :: Nat and +1 :: Nat $\to$ Nat.

Programs contain co-inductive types defined by data [Tau] = Tau : [Tau] for each inductive type Tau. This is a distinction with Haskell, where streams are defined to be both finite and infinite lists, but not a restriction since finite lists may be defined in this language and since we are only interested in showing properties of total functions (*i.e.* an infinite stream represents a total function).

Each function symbol f comes with a typed signature that we restrict to be either f :: $[\text{Tau}]^k \to \text{Tau}^l \to \text{Tau}$ or f :: $[\text{Tau}]^k \to \text{Tau}^l \to [\text{Tau}]$, with $k, l \geq 0$.

Throughout the paper, we will only consider well-typed programs where the left-hand side and the right-hand side of a definition can be given the same type using the following simple rules with $A, A_i \in \{\texttt{Tau}, \texttt{[Tau]}\}$:

$$\frac{}{\texttt{x :: A}}\ x \in \mathcal{X} \qquad \frac{\texttt{t :: A}_1 \to \ldots \to \texttt{A}_n \to \texttt{A} \qquad \forall i \in \{1, n\},\ \texttt{e}_i \texttt{ :: A}_i}{\texttt{t e}_1 \ldots \texttt{e}_n \texttt{ :: A}}\ t \in \mathcal{C} \cup \mathcal{F}$$

Semantics. Let lazy values and strict values be defined by:

$$\texttt{lv} ::= \texttt{e}_1 : \texttt{e}_2 \text{ (Lazy value)}$$
$$\texttt{v} ::= \texttt{c v}_1 \ \ldots \ \texttt{v}_n \text{ (Strict value)}$$

where $\texttt{e}_1, \texttt{e}_2$ are closed expressions and $\texttt{c}$ belongs to $\mathcal{C} \setminus \{:\}$. Lazy values are expressions with the constructor symbol : at the top level whereas strict values are expressions where only constructor symbols occur and are used to deal with fully evaluated elements.

Moreover, let $\mathfrak{S}$ represent the set of substitutions σ that map variables to expressions. As usual the result of applying the substitution σ to an expression $\texttt{e}$ is denoted $\sigma(\texttt{e})$.

The derivation rules are defined by:

$$\frac{(\texttt{f p}_1 \ \ldots \ \texttt{p}_n = \texttt{e}) \in \mathcal{D} \qquad \sigma \in \mathfrak{S} \qquad \forall i \in \{1, ..., n\},\ \sigma(\texttt{p}_i) = \texttt{e}_i}{\texttt{f e}_1 \ \ldots \ \texttt{e}_n \to \sigma(\texttt{e})}\ (d)$$

$$\frac{\texttt{e}_i \to \texttt{e}'_i \qquad t \in \mathcal{F} \cup \mathcal{C} \setminus \{:\}}{\texttt{t e}_1 \ \ldots \ \texttt{e}_i \ \ldots \ \texttt{e}_n \to \texttt{t e}_1 \ \ldots \ \texttt{e}'_i \ \ldots \ \texttt{e}_n}\ (t) \qquad \frac{\texttt{e} \to \texttt{e}'}{\texttt{e} : \texttt{e}_0 \to \texttt{e}' : \texttt{e}_0}\ (:)$$

We will write $\texttt{e} \to^n \texttt{e}'$ if there exist expressions $\texttt{e}_1, \ldots, \texttt{e}_{n-1}$ such that $\texttt{e} \to \texttt{e}_1 \cdots \to \texttt{e}_{n-1} \to \texttt{e}'$. Let $\to^*$ denote the transitive and reflexive closure of $\to$. We write $\texttt{e} \to_! \texttt{e}'$ if $\texttt{e}$ is normalizing to the expression $\texttt{e}'$, *i.e.* $\texttt{e} \to^* \texttt{e}'$ and there is no $\texttt{e}''$ such that $\texttt{e}' \to \texttt{e}''$. We can show easily wrt the derivation rules (and because definitions are exhaustive) that given a closed expression $\texttt{e}$, if $\texttt{e} \to_! \texttt{e}'$ and $\texttt{e} ::$ Tau then $\texttt{e}'$ is a strict value, whereas if $\texttt{e} \to_! \texttt{e}'$ and $\texttt{e} ::$ [Tau] then $\texttt{e}'$ is a lazy value. Indeed the (t) rule allows the reduction of an expression under a function or constructor symbol whereas the (:) rule only allows reduction of a stream head (this is why we do not allow stream patterns of depth greater than 1 in a definition).

These reduction rules are not deterministic but we could define a lazy call-by-need strategy to mimic Haskell's semantic.

4   Second Order Polynomial Interpretations

In the following, we call a positive functional any function in $((\mathbb{N} \to \mathbb{N})^k \times \mathbb{N}^l) \to T$ with $k, l \in \mathbb{N}$ and $T \in \{\mathbb{N}, \mathbb{N} \to \mathbb{N}\}$. Given a positive functional $F : ((\mathbb{N} \to \mathbb{N})^k \times \mathbb{N}^l) \to T$, the arity of F is $k + l$.

Let $>$ denote the usual ordering on $\mathbb{N}$ and $\mathbb{N} \to \mathbb{N}$, *i.e.* given $F, G : \mathbb{N} \to \mathbb{N}$, $F > G$ if $\forall n \in \mathbb{N} \setminus \{0\}, F(n) > G(n)$. We extend this ordering to positive functionals of arity l by: $F > G$ if $\forall x_1 \ldots x_l \in \{\mathbb{N} \setminus \{0\}, \mathbb{N} \to^\uparrow \mathbb{N}\}$, $F(x_1, \ldots, x_l) > G(x_1, \ldots, x_l)$, where $\mathbb{N} \to^\uparrow \mathbb{N}$ is the set of increasing functions on positive integers.

Definition 6 (Monotonic positive functionals). *A positive functional F of arity n is monotonic if $\forall i \in \{1, n\}, \forall x_i > x_i',\ F(\ldots, x_i, \ldots) > F(\ldots, x_i', \ldots)$, where $x_i, x_i' \in \{\mathbb{N} \setminus \{0\}, \mathbb{N} \to^\uparrow \mathbb{N}\}$.*

Definition 7. *The interpretation of a program is a total mapping of the function and constructor symbols to monotonic positive functionals. The type of the interpretation is inductively defined by the type of the corresponding symbol:*

- *a symbol* t *of type* Tau *has interpretation $(\!|t|\!)$ in $\mathbb{N}$*
- *a symbol* t *of type* [Tau] *has interpretation $(\!|t|\!)$ in $\mathbb{N} \to \mathbb{N}$*
- *a symbol* t *of type* A -> B *has interpretation $(\!|t|\!)$ in $T_A \to T_B$, where T_A and T_B are the types of the interpretations of the symbols of type* A *and, respectively, type* B.

We fix the interpretation of each constructor symbol by:

- $(\!|c|\!)(X_1, \ldots, X_n) = X_1 + \ldots X_n + 1$ *if* $c \in \mathcal{C} \setminus \{:\}$ *is of arity n*
- $(\!|:|\!)(X, Y)(Z + 1) = 1 + X + Y \langle Z \rangle^2$
- $(\!|:|\!)(X, Y)(0) = X$

Once the interpretation of each function and constructor symbol is fixed, we can define the interpretation of any expression by structural induction (notice that we preserve the previous correspondence between the type of the expression and the type of its interpretation):

- $(\!|x|\!) = X$ *if* x *is a variable of type* Tau, *i.e. we associate a unique first order variable X in $\mathbb{N}$ to each* $x \in \mathcal{X}$ *of type* Tau.
- $(\!|y|\!)(Z) = Y \langle Z \rangle$ *if* y *is a variable of type* [Tau], *i.e. we associate a unique second order variable $Y : \mathbb{N} \to \mathbb{N}$ to each* $y \in \mathcal{X}$ *of type* [Tau].
- $(\!|t\ e_1\ \ldots\ e_n|\!) = (\!|t|\!)((\!|e_1|\!), \ldots, (\!|e_n|\!))$ *if* $t \in \mathcal{C} \cup \mathcal{F}$

Consequently, the interpretation $(\!|\ |\!)$ maps any expression to a functional (of the interpretation of its free variables).

The interpretation of a program is polynomial if each function symbol is interpreted by a second order polynomial.

Example 2. The stream constructor : has type Tau $\to$ [Tau] $\to$ [Tau]. Consequently, its interpretation $(\!|:|\!)$ has type $(\mathbb{N} \times (\mathbb{N} \to \mathbb{N})) \to (\mathbb{N} \to \mathbb{N})^3$. Considering the expression p : (q : r), with p, q, r $\in \mathcal{X}$, we obtain that:

$$(\!|p : (q : r)|\!) = (\!|:|\!)((\!|p|\!), (\!|q : r|\!)) = (\!|:|\!)((\!|p|\!), (\!|:|\!)((\!|q|\!), (\!|r|\!))) = (\!|:|\!)(P, (\!|:|\!)(Q, R)) = F(R, P, Q)$$

where $F \in ((\mathbb{N} \to \mathbb{N}) \times \mathbb{N}^2) \to (\mathbb{N} \to \mathbb{N})$ is the positive functional such that:

---

[2] By abuse of notation, we will consider that the interpretation of : is also a positive functional.

[3] We will use the cartesian product instead of the arrow for the argument types of a function symbol in the following.

- $F(R,P,Q)(Z+2) = 1 + P + (\!|:\!|)(Q,R)(Z+1) = 2 + P + Q + R(Z)$
- $F(R,P,Q)(1) = 1 + P + (\!|:\!|)(Q,R)(0) = 1 + P + Q$
- $F(R,P,Q)(0) = P$

Lemma 1. *The interpretation of an expression* e *defines a positive functional in the interpretations of its free variables.*

Definition 8 (Well-founded polynomial interpretation). *The interpretation of a program is well-founded if for each definition* f $p_1 \ldots p_n$ = e $\in \mathcal{D}$,

$$(\!|f\ p_1 \ldots p_n|\!) > (\!|e|\!)$$

By extension, a program is well-founded (polynomial) if it admits a well-founded (polynomial) interpretation.

The following programs are examples of well-founded polynomial programs:

Example 3. The sum and product over unary integers:

```
plus :: Nat -> Nat -> Nat        mult :: Nat -> Nat -> Nat
plus 0 b = b                     mult 0 b = 0
plus (a+1) b = (plus a b)+1      mult (a+1) b = plus b (mult a b)
```

They admit the following well-founded interpretation[4] $(\!|plus|\!)(X_1,X_2) = 2 \times X_1 + X_2$, $(\!|mult|\!)(X_1,X_2) = 3 \times X_1 \times X_2$. Indeed, we check that the following inequalities are satisfied:

- $(\!|plus\ 0\ b|\!) = 2 + B > B = (\!|b|\!)$
- $(\!|plus\ (a+1)\ b|\!) = 2A + 2 + B > 2A + B + 1 = (\!|(plus\ a\ b)+1|\!)$
- $(\!|mult\ 0\ b|\!) = 3 \times (\!|0|\!) \times (\!|b|\!) = 3 \times B > 1 = (\!|0|\!)$
- $(\!|mult\ (a+1)\ b|\!) = 3 \times A \times B + 3 \times B > 2 \times B + 3 \times A \times B = (\!|plus\ b\ (mult\ a\ b)|\!)$

s !! ncomputes the $(n+1)^{th}$ element of the stream s:

```
!! :: [Tau] -> Nat -> Tau
(h:t) !! (n+1) = t !! n
(h:t) !! 0 = h
```

and admits a well-founded interpretation $(\!|!!|\!)$ in $((\mathbb{N} \to \mathbb{N}) \times \mathbb{N}) \to \mathbb{N}$ defined by $(\!|!!|\!)(Y,N) = Y\langle N \rangle$. Indeed, we check that:

- $(\!|(h:t)\ !!\ (n+1)|\!) = (\!|h:t|\!)((\!|n|\!) + 1) = 1 + (\!|h|\!) + (\!|t|\!)((\!|n|\!)) > (\!|t|\!)((\!|n|\!)) = (\!|t\ !!\ n|\!)$
- $(\!|(h:t)\ !!\ 0|\!) = (\!|h:t|\!)((\!|0|\!)) = (\!|h:t|\!)(1) = 1 + (\!|h|\!) + (\!|t|\!)(0) > (\!|h|\!)$

In the same way, we let the reader check that tln, which drops the first $n+1$ elements of a stream, admits the well-founded interpretation $(\!|tln|\!)$ of type $((\mathbb{N} \to \mathbb{N}) \times \mathbb{N}) \to (\mathbb{N} \to \mathbb{N})$ defined by $(\!|tln|\!)(Y,N)(Z) = Y\langle N + Z + 1 \rangle$.

---

[4] On programs without streams, well-founded polynomial interpretations correspond exactly to polynomial interpretations.

```
tln :: [Tau] -> Nat -> [Tau]
tln (h:t) (n+1) = tln t n
tln (h:t) 0 = t
```

Indeed, for the first rule, we just check that $(\!|\texttt{tln (h:t) (n+1)}|\!) > (\!|\texttt{tln t n}|\!)$, that is $\forall Z \in \mathbb{N} \setminus \{0\}$, $(\!|\texttt{tln (h:t) (n+1)}|\!)(Z) > (\!|\texttt{tln t n}|\!)(Z)$.

Lemma 2. *If* e *is an expression of a program with a well-founded interpretation* $(\!| \; |\!)$ *and* $e \rightarrow e'$, *then* $(\!|e|\!) > (\!|e'|\!)$.

Corollary 1. *Given a closed expression* e :: Tau *of a program having a well-founded interpretation* $(\!| \; |\!)$, *if* $e \rightarrow^n e'$ *then* $n \leq (\!|e|\!)$, *i.e. every reduction chain starting from an expression* e *of a well-founded program has its length bounded by* $(\!|e|\!)$.

Corollary 2. *Given a closed expression* e :: [Tau] *of a program having a well-founded interpretation* $(\!| \; |\!)$, *if* $e \texttt{ !! } k \rightarrow^n e'$ *then* $n \leq (\!|e \texttt{ !! } k|\!) = {}^5(\!|e|\!)((\!|k|\!)) = (\!|e|\!)(k+1)$, *i.e. at most* $(\!|e|\!)(k+1)$ *reduction steps are needed to compute the* k^{th} *element of a stream* e.

Productive streams are defined in the literature [6] as terms weakly normalizing to infinite lists, which is in our case equivalent to: a stream s is productive if for all n :: Nat, s !! n evaluates to a strict value.

Corollary 3. *Each stream expression of a program with a well-founded interpretation is productive.*

Corollary 4. *Given a function symbol* f :: $[\texttt{Tau}]^k \rightarrow \texttt{Tau}^l \rightarrow \texttt{Tau}$ *of a program with a well-founded polynomial interpretation* $(\!| \; |\!)$, *there is a second order polynomial* P *such that if* f $e_1 \ldots e_{k+1} \rightarrow^n_! v$ *then* $n \leq P((\!|e_1|\!), \ldots, (\!|e_{k+1}|\!))$, *for all closed expressions* $e_1, \ldots, e_{k+1}$.

The following lemma shows that in a well-founded program, the number of evaluated stream elements is bounded.

Lemma 3. *Given a function symbol* f :: $[\texttt{Tau}]^k \rightarrow \texttt{Tau}^l \rightarrow \texttt{Tau}$ *of a program having a well-founded interpretation* $(\!| \; |\!)$, *and closed expressions* $e_1, \ldots, e_l$ :: Tau, $d_1, \ldots, d_k$ :: [Tau], *if* f $d_1 \ldots d_k e_1 \ldots e_l \rightarrow_! v$ *and* $\forall n :: \texttt{Nat}$, $d_i \texttt{ !! } n \rightarrow_! v_i^n$ *then for all closed expressions* $d'_1 \ldots d'_k$ :: [Tau] *satisfying* $d'_i \texttt{ !! } n \rightarrow_! v_i^n$, $\forall n \leq (\!|\texttt{f } d_1 \ldots d_k e_1 \ldots e_l|\!)$, *we have* f $d'_1 \ldots d'_k e_1 \ldots e_l \rightarrow_! v$.

5   Characterizations of Polynomial Time

In this section, we provide a characterization of polynomial time UOTM computable functions using interpretations. We also provide a characterization of Basic Feasible Functionals using the same methodology.

Note that for the sake of simplicity, we implicitly consider that the inductive type Tau has an encoding in $\mathbb{N}$.

---

[5] Using the well-founded interpretation of example 3.

Theorem 1. *A function* $F : ((\mathbb{N} \to \mathbb{N})^k \to \mathbb{N}^l) \to \mathbb{N}$ *which is computable in polynomial time by a UOTM if and only if there exists a program* f *computing* F, *of type* $[\mathtt{Tau}]^k \to \mathtt{Tau}^l \to \mathtt{Tau}$ *admits a polynomial well-founded interpretation.*

To prove this theorem, we will show in lemma 4 that second order polynomials can be computed by programs having well-founded polynomial interpretations. We will then use this result to get soundness in lemma 5. Completeness (lemma 7) consists in computing a bound on the number of entries to read to compute an element of the output stream and then to the computation by a classical Turing machine.

Lemma 4. *Every second order polynomial can be computed by a well-founded polynomial program.*

Lemma 5 (Soundness). *Every polynomial time UOTM computable function can be computed by a well-founded polynomial program.*

PROOF
Let $f : ((\mathbb{N} \to \mathbb{N})^k \to \mathbb{N}^l) \to \mathbb{N}$ be a function computed by a UOTM $\mathcal{M}$ in time P, with P a second order polynomial. Without loss of generality, we will assume that $k = l = 1$. The idea of this proof is to write a program $\mathtt{f_0}$ giving the output of $\mathcal{M}$ after t steps, and to use lemma 4 to simulate the computation of P.

Let $\mathtt{f_0}$ be the function symbol describing the execution of $\mathcal{M}$:

$$\mathtt{f_0} \; :: \; \mathtt{[Bin]} \; \text{->} \; \mathtt{Nat} \; \text{->} \; \mathtt{Nat} \; \text{->} \; \mathtt{Bin}^8 \; \text{->} \; \mathtt{Bin}$$

where $\mathtt{Bin} = \mathtt{Nil} \mid \mathtt{0\ Bin} \mid \mathtt{1\ Bin}$.

The arguments of $\mathtt{f_0}$ represent respectively the input stream, the number of steps $\mathtt{t}$ the machine is allowed to compute, the state and the 4 tapes (each tape is represented by 2 binary numbers as illustrated in figure 1). The output will correspond to the content of the output tape after $\mathtt{t}$ steps.

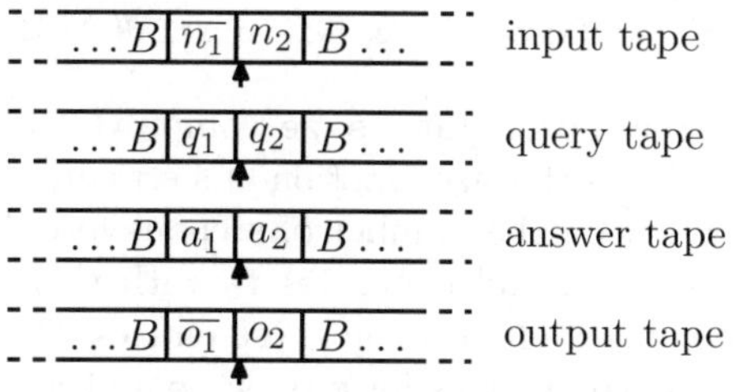

Fig. 1. Encoding of the content of the tapes of an OTM (or UOTM). $\bar{x}$ represents the mirror of the word x and the symbol $\uparrow$ represents the positions of the heads.

The function symbol $\mathtt{f_0}$ is defined recursively in its second argument:

– if the timer is 0, then we output the content of the output tape (after its head):

$\mathtt{f_0} \; \mathtt{s} \; \mathtt{0} \; \mathtt{q} \; \mathtt{n_1} \; \mathtt{n_2} \; \mathtt{q_1} \; \mathtt{q_2} \; \mathtt{a_1} \; \mathtt{a_2} \; \mathtt{o_1} \; \mathtt{o_2} \; = \; \mathtt{o_2}$

- for each transition of $\mathcal{M}$, we write a definition:

  `f₀ s (t+1) q n₁ n₂ q₁ q₂ a₁ a₂ o₁ o₂ = f₀ s t q' n'₁ n'₂ q'₁ q'₂ a'₁ a'₂ o'₁ o'₂`

  where n_1 and n_2 represent the input tape before the transition and n'_1 and n'_2 represent the input tape after the transition, the motion and writing of the head being taken into account, and so on for the other tapes.

Since the transition function is well described by a set of such definitions, the function f_0 produces the content of o_2 (*i.e.* the content of the output tape) after t steps on entry t and configuration $\mathcal{C}$ (*i.e.* the state and the representations of the tapes).

f_0 admits a well-founded polynomial interpretation $(\!|f_0|\!)$. Indeed, in each definition, the state can only increase by a constant, the length of the numbers representing the various tapes cannot increase by more than 1. The answer tape $(\!|a_2|\!)$ can undergo an important increase: when querying, it can increase by $(\!|s|\!)((\!|q_2|\!))$, that is the interpretation of the input stream taken in the interpretation of the query.

Then, $(\!|f_0|\!)(Y, T, Q, N_1, N_2, Q_1, Q_2, A_1, A_2, O_1, O_2)$ can be defined by $(T+1) \times (Y\langle Q_2\rangle + 1) + Q + N_1 + N_2 + Q_1 + A_1 + A_2 + O_1 + O_2$, which provides a well-founded polynomial interpretation. Lemma 4 shows how we can implement the polynomial P by a program p, and give it a polynomial well-founded interpretation. Finally, consider the programs `size`, `max`, `maxsize` and f_1 defined below:

```
size :: Bin -> Bin              max :: Nat -> Nat -> Nat
size Nil = 0                     max 0 n = n
size (0 x) = (size x)+1         max n 0 = n
size (1 x) = (size x)+1         max (n+1) (k+1) = (max n k)+1
```

```
maxsize :: [Bin] -> Nat -> Nat
maxsize (h:t) 0 = size h
maxsize (h:t) (n+1) = max (maxsize t n) (size h)
```

```
f₁ :: [Bin] -> Bin -> Bin
f₁ s n = f₀ s (p (maxsize s) (size n)) q₀ Nil n Nil Nil Nil Nil Nil Nil
```

where q_0 is the index of the initial state. `size` computes the size of a binary number, and `maxsize` computes the size function of a stream of binary numbers. f_1 computes an upper bound on the number of steps before $\mathcal{M}$ halts on entry n with oracle s (*i.e.* $P(|s|, |n|)$), and computes f_0 with this time bound. The output is then the value computed by M on these entries. Define the following well-founded polynomial interpretations for `max`, `size` and `maxsize`:

- $(\!|size|\!)(X) = 2X$
- $(\!|max|\!)(X_1, X_2) = X_1 + X_2$
- $(\!|maxsize|\!)(Y, X) = 2 \times Y\langle X\rangle$

Finally f_1 admits a well-founded polynomial interpretation since it is defined by composition of programs with well-founded polynomial interpretations.   □

The previous lemma also gives a hint for the completeness proof:

Corollary 5. *Any function* $f : \mathbb{N} \to \mathbb{N}$ *computable in polynomial time by a Turing Machine can be implemented by a stream program having a well-founded polynomial interpretation.*

Lemma 6. *If P is a second-order polynomial, then the function:*

$$F_1, \ldots, F_k, x_1, \ldots x_l \mapsto 2^{P(|F_1|, \ldots, |F_k|, |x_1|, \ldots, |x_l|)} - 1$$

is computable in polynomial time by a UOTM.

Lemma 7 (Completeness). *If a program* f *of type* [Tau]^k $\to$ Tau^l $\to$ Tau *admits a well-founded polynomial interpretation, then it computes a function* $f : (\mathbb{N} \to \mathbb{N})^k \to \mathbb{N}^l \to \mathbb{N}$ *which is computable in polynomial time by a UOTM.*

PROOF

Lemma 6 shows that given some inputs and oracles, a UOTM can compute $(\!|f|\!)$ applied on their sizes and get a unary integer N in polynomial time. According to lemma 3, the Haskell-like program needs at most the first N values of each oracle. Then, we can build a UOTM which queries all these values (in time $\sum_{i \leq N} |f|(N)$, which is polynomial in the size of the inputs and the size of the oracles) and computes f on these finite inputs: we can convert the program f into a program working on finite lists (which will also have polynomial time complexity), and according to corollary 5, this program can be computed in polynomial time by a (classical) Turing Machine.   □

Similarly, we can obtain a characterization of BFF, that is functions computable in polynomial time by OTM. Instead of using second order polynomials, we will use a larger set of second order functions named exp-poly.

Definition 9 (exp-poly). *We call exp-poly the set of functions generated by the following grammar:*

$$EP := P \mid EP + EP \mid EP \times EP \mid Y\langle 2^{EP}\rangle$$

where P denotes a first order polynomial and Y a second order variable.

The interpretation of a program is exp-poly if each symbol is interpreted by an exp-poly function.

Theorem 2. *BFF is exactly the set of functions that can be computed by programs that admit a well-founded exp-poly interpretation.*

6   Link with Polynomial Time Computable Real Functions

We show in this section that our complexity results can be adapted to real functions.

Up to now, we have considered stream programs as type 2 functionals in their own rights. However, type 2 functionals can be used to represent real functions.

Indeed Recursive Analysis models computation on reals as computation on converging sequences of rational numbers [17,10].

We will require a given convergence speed to be able to compute effectively. A real x is represented by a sequence $(q_n) \in \mathbb{Q}^{\mathbb{N}}$ if $\forall i \in \mathbb{N}, \|x - q_i\| < 2^{-i}$. This will be denoted by $(q_n) \rightsquigarrow x$. A function $f : \mathbb{R} \rightarrow \mathbb{R}$ will be said to be computed by a machine $\mathcal{M}$ if

$$[!t](q_n) \rightsquigarrow x \Rightarrow (\mathcal{M}(q_n)) \rightsquigarrow f(x). \tag{1}$$

Hence a computable real function will be computed by programs of type `[Q] -> [Q]` in our stream language, where `Q` is an inductive type describing the set of rationals $\mathbb{Q}$. Only programs encoding machines verifying the implication (1) will make sense in this framework. Following [10], we can define polynomial complexity of real functions using polynomial time UOTM computable functions.

Proposition 1. *If a program* `[Q] -> [Q]` *with a well-founded polynomial interpretation computes a real function on compact* $\mathbb{K}$, *then this function is computable in polynomial time.*

Proposition 2. *Any polynomial-time computable real function (defined over* $\mathbb{K}$*) can be implemented by a well-founded polynomial program.*

References

1. Amadio, R.M.: Synthesis of max-plus quasi-interpretations. Fundamenta Informaticae 65(1), 29–60 (2005)
2. Bellantoni, S., Cook, S.A.: A new recursion-theoretic characterization of the polytime functions. Computational complexity 2(2), 97–110 (1992)
3. Bonfante, G., Marion, J.Y., Moyen, J.Y.: Quasi-interpretations. Theor. Comput. Sci. (to appear)
4. Cobham, A.: The Intrinsic Computational Difficulty of Functions. In: Logic, methodology and philosophy of science III, p. 24. North-Holland Pub. Co., Amsterdam (1965)
5. Constable, R.L.: Type two computational complexity. In: Proc. 5th annual ACM STOC, pp. 108–121 (1973)
6. Endrullis, J., Grabmayer, C., Hendriks, D., Isihara, A., Klop, J.W.: Productivity of stream definitions. Theor. Comput. Sci. 411(4-5), 765–782 (2010)
7. Gaboardi, M., Péchoux, R.: Upper Bounds on Stream I/O Using Semantic Interpretations. In: Grädel, E., Kahle, R. (eds.) CSL 2009. LNCS, vol. 5771, pp. 271–286. Springer, Heidelberg (2009)
8. Irwin, R.J., Royer, J.S., Kapron, B.M.: On characterizations of the basic feasible functionals (Part I). J. Funct. Program. 11(1), 117–153 (2001)
9. Kapron, B.M., Cook, S.A.: A new characterization of type-2 feasibility. SIAM Journal on Computing 25(1), 117–132 (1996)
10. Ko, K.I.: Complexity theory of real functions. Birkhauser Boston Inc., Cambridge (1991)
11. Lankford, D.: On proving term rewriting systems are noetherien. Tech. Rep. (1979)
12. Leivant, D., Marion, J.Y.: Lambda calculus characterizations of poly-time. In: Typed Lambda Calculi and Applications, pp. 274–288 (1993)

13. Manna, Z., Ness, S.: On the termination of Markov algorithms. In: Third Hawaii International Conference on System Science, pp. 789–792 (1970)
14. Mehlhorn, K.: Polynomial and abstract subrecursive classes. In: Proceedings of the Sixth Annual ACM Symposium on Theory of Computing, pp. 96–109. ACM, New York (1974)
15. Ramyaa, R., Leivant, D.: Feasible functions over co-inductive data. In: WoLLIC, pp. 191–203 (2010)
16. Seth, A.: Turing machine characterizations of feasible functionals of all finite types. In: Feasible Mathematics II, pp. 407–428 (1995)
17. Weihrauch, K.: Computable analysis: an introduction. Springer, Heidelberg (2000)

New Upper Bounds on the Average PTF Density of Boolean Functions

Kazuyuki Amano

Dept of Comp Sci, Gunma Univ, Tenjin 1-5-1, Kiryu, Gunma 376-8515, Japan
amano@cs.gunma-u.ac.jp

Abstract. A Boolean function $f : \{1, -1\}^n \rightarrow \{1, -1\}$ is said to be sign-represented by a real polynomial $p : \mathbb{R}^n \rightarrow \mathbb{R}$ if $\mathrm{sgn}(p(x)) = f(x)$ for all $x \in \{1, -1\}^n$. The PTF density of f is the minimum number of monomials in a polynomial that sign-represents f. It is well known that every n-variable Boolean function has PTF density at most 2^n. However, in general, less monomials are enough. In this paper, we present a method that reduces the problem of upper bounding the average PTF density of n-variable Boolean functions to the computation of (some modified version of) average PTF density of k-variable Boolean functions for small k. By using this method, we show that almost all n-variable Boolean functions have PTF density at most $(0.617)2^n$, which is the best upper bound so far.

1 Introduction and Overview

This paper deals with the expressive power of real polynomials for representing Boolean functions. Let $f : \{1, -1\}^n \rightarrow \{1, -1\}$ be a Boolean function on n variables and let $p : \mathbb{R}^n \rightarrow \mathbb{R}$ be a real polynomial. We say that p sign-represents f if $\mathrm{sgn}(p(x)) = f(x)$ for all $x \in \{1, -1\}^n$. The complexity (i.e., the minimum degree or the number of monomials needed to represent a given Boolean function) of such a representation has been extensively investigated especially in complexity theory and in learning theory (see e.g., [6,12,8,9,13,14] and the references therein).

In this paper, we focus on the PTF *density* of Boolean functions, which is defined as the minimum number of monomials with non-zero coefficient in a polynomial that sign-represents a given Boolean function. It is classically known that every Boolean function on n variables can be sign-represented by a polynomial with 2^n monomials. However, in general, less monomials are enough. For example, it is easy exercise to show that every two-variable Boolean function can be sign-represented by a polynomial with at most three monomials, not four. Note that the PTF density is depending on the choice of the domain of Boolean functions. In this paper, we exclusively consider the case $\{1, -1\}^n$.

In spite of a long history of investigations, there still is a large gap between the upper and lower bounds on the PTF density of Boolean functions. For the lower bounds, Saks [12, Theorem 2.27] noted that the result of Cover [3] implies that almost all Boolean functions on n variables have PTF density at least $(0.11)2^n$.

O. Cheong, K.-Y. Chwa, and K. Park (Eds.): ISAAC 2010, Part I, LNCS 6506, pp. 304–315, 2010.

To this date, this is the best known lower bound on the PTF density even for the worst case.

For the upper bounds, Gotsman [4] proved that every Boolean function has PTF density at most $2^n - 2^{n/2} + 1$ by a simple harmonic analysis. O'Donnell and Servedio improved this by showing that every Boolean function has PTF density at most $(1 - \frac{1}{O(n)})2^n$ [8]. They also proved a stronger statement that every fixed set of $(1 - \frac{1}{O(n)})2^n$ monomials serve as the support of a PTF for almost all Boolean functions on n variables.

Recently, Oztop [10] (see also [11]) made a breakthrough by improving the upper bound to $(0.75)2^n$, which is the first result saying that a constant ratio, namely, $3/4$, of all monomials are always enough to represent a Boolean function. This is remarkable not only the bound itself but also the elegance of the proof. In fact, the proof uses only relatively simple linear algebra (that we will sketch in Section 3.1).

Table 1. The known bounds on the PTF density of Boolean functions. The right bottom is shown in this paper.

	Lower Bound	Upper Bound
All Functions	$(0.11)2^n$	$(0.75)2^n$
Almost All Functions	$(0.11)2^n$	$(0.617)2^n$

In this paper, we reveal that a natural extension of Oztop's method can yield a better upper bound on the *average* PTF density of Boolean functions. Intuitively, we show that the problem of upper bounding the average density of n-variable Boolean functions can be reduced to a problem of computing (some modified version of) average density of k-variable Boolean functions for small k (Theorem 8). After showing this, we do a computation of this quantity to show that almost all n-variable Boolean functions have PTF density at most $(0.617)2^n$. These are the main contributions of this paper. The known bounds on the PTF density of Boolean functions are summarized in Table 1.

Our method has an interesting feature. As is expected, an upper bound is improved by increasing the computational effort. In fact, the examination of two-variable functions gives the upper bound of $(0.688)2^n$, that for three-variable functions gives the upper bound of $(0.649)2^n$, and that for four-variable functions gives the upper bound of $(0.617)2^n$ which is the best we have obtained. If we have more computational resource (or more sophisticated algorithm to compute this quantity), it is quite conceivable that our bound will further be improved. This would certainly be helpful to *guess* the right constant c such that $c2^n$ is the PTF density of (almost) all n-variable Boolean functions. We will discuss this further in the final section of this paper.

The organization of this paper is as follows: In Section 2, we introduce the notations and definitions. Main results are described in Section 3. In Section 3.1, we review the proof technique developed by Oztop [10,11], and in the following sections (Sections 3.2 and 3.3), we extend their method to obtain a better upper bound on the average PTF density. Finally, we close the paper with some concluding remarks in Section 4.

2   Preliminaries

In most of this paper, we use $\{1, -1\}$ to represent Boolean values. The false or 0 is represented by 1, and the true or 1 is represented by -1.

Definition 1. *Let $f : \{1, -1\}^n \to \{1, -1\}$ be a Boolean function and $p : \mathbb{R}^n \to \mathbb{R}$ be a real polynomial. We say that p sign-represents f, if $p(x) \neq 0$ for all $x \in \{1, -1\}^n$ and $sgn(p(x)) = f(x)$ for all $x \in \{1, -1\}^n$. We also say that p is a polynomial threshold function (PTF, in short) for f. The support of p is a set of monomials with non-zero coefficient in p and the density of p is the size of the support of p. For a Boolean function f, the PTF density of f is the smallest density of a polynomial that sign-represents f.*

Since $x^2 = 1$ for $x \in \{1, -1\}$, we can assume without loss of generality that p is a multilinear polynomial and so p is a linear combination of all 2^n monomials over $x_1, \ldots, x_n$ which we will denote by $\mathcal{M}^n$.

It is very useful to writing this in vector notations. We follow the notation by Oztop [10,11].

For $n \geq 1$, let $\mathbf{D}^n$ be a Hadamard matrix of order 2^n defined as

$$\mathbf{D}^1 = \begin{pmatrix} 1 & 1 \\ 1 & -1 \end{pmatrix}, \quad \mathbf{D}^n = \begin{pmatrix} \mathbf{D}^{n-1} & \mathbf{D}^{n-1} \\ \mathbf{D}^{n-1} & -\mathbf{D}^{n-1} \end{pmatrix} \text{(for } n \geq 2).$$

The well known identities $\mathbf{D}^n\mathbf{D}^n = 2^n\mathbf{I}$ and $(\mathbf{D}^n)^{-1} = 2^{-n}\mathbf{D}^n$ are very useful. Each column of $\mathbf{D}^n$ is indexed by a monomial in $\mathcal{M}^n$ in the ordering of $1, x_1, x_2, x_2x_1, x_3, x_3x_1, x_3x_2, x_3x_2x_1, \ldots, x_nx_{n-1} \ldots x_1$ from the leftmost column. For a polynomial $p = \sum_{i=1}^{2^n} a_i m_i$ where $m_i = \prod_{j \in S_i} x_j$ with $S_i \subseteq \{1, \ldots, n\}$, the coefficient vector $\mathbf{a} = (a_1, \ldots, a_{2^n})$ is called the *spectrum* of p where we use the same ordering for monomials as above.

Then the column vector $\mathbf{D}^n\mathbf{a}$ represents the values of $p(x)$ where the assignments to $(x_n, x_{n-1}, \ldots, x_1)$ are ordered as $00 \ldots 00, 00 \cdots 01, 00 \cdots 10, 00 \cdots 11, \ldots, 11 \cdots 11$ (where 0's represent 1 and 1's represent -1). For a Boolean function f on n variables, let $\mathbf{f}$ denote the column vector of length 2^n whose elements are the values of $f(x)$ for all x. We call $\mathbf{f}$ as the *vector representation* of f.

In this notation, p sign-represents f iff $\mathbf{Y}\mathbf{D}^n\mathbf{a} > \mathbf{0}$, where $\mathbf{Y} = diag(\mathbf{f})$. If this is the case, we are allowed to say that $\mathbf{a}$ sign-represents f. The density of p is the number of non-zero elements of $\mathbf{a}$. In the following, we sometimes refer to a monomial $\prod_{i \in S} x_i$ as its characteristic vector $x_n x_{n-1} \cdots x_1 \in \{0, 1\}^n$ with $x_i = 1$ iff $i \in S$.

3   Upper Bounds on PTF Density

3.1   Basics

Recently, Oztop [10] established the $(0.75)2^n$ upper bound on the PTF density of *every* n-variable Boolean function. Since we will extend its proof method for obtaining the average-case upper bound on the PTF density, we include the sketch of the proof here.

The following simple fact is extremely useful.

Fact 2. *([11, Theorem 1]) Let $\mathbf{f}$ be the vector representation of a Boolean function on n variables. The set of solutions of the inequality $\mathrm{diag}(\mathbf{f})\mathbf{D}^n\mathbf{a} > \mathbf{0}$ is all positive linear combinations of the columns of $\mathbf{D}^n\mathrm{diag}(\mathbf{f})$.*

Proof. $\mathrm{diag}(\mathbf{f})\mathbf{D}^n\mathbf{a} > \mathbf{0}$ iff $\exists \mathbf{k} > \mathbf{0}[\mathrm{diag}(\mathbf{f})\mathbf{D}^n\mathbf{a} = \mathbf{k}]$ iff $\exists \mathbf{k} > \mathbf{0}[\mathbf{a} = \frac{1}{2^n}\mathbf{D}^n\mathrm{diag}(\mathbf{f})\mathbf{k}]$. $\square$

Theorem 3. *[10] For any Boolean function on n variables, there exists a sign-representing polynomial with at most $2^n - 2^n/4$ monomials.*

Proof. (sketch) For an n-variable Boolean function f, let $\mathbf{a}$ denote the (column) vector of the coefficients of a polynomial that sign-represents f. Let $\mathbf{f}$ denote the vector representation of f. We partition $\mathbf{a}$ and $\mathbf{f}$ into $\mathbf{a}_0, \mathbf{a}_1$ and $\mathbf{f}_0, \mathbf{f}_1$ of equal length, respectively. We can write as

$$\mathrm{diag}\begin{pmatrix}\mathbf{f}_0\\\mathbf{f}_1\end{pmatrix}\begin{pmatrix}\mathbf{D}^{n-1} & \mathbf{D}^{n-1}\\\mathbf{D}^{n-1} & -\mathbf{D}^{n-1}\end{pmatrix}\begin{pmatrix}\mathbf{a}_0\\\mathbf{a}_1\end{pmatrix} > \mathbf{0}.$$

This is equivalent to

$$\mathrm{diag}(\mathbf{f}_0)\mathbf{D}^{n-1}(\mathbf{a}_0 + \mathbf{a}_1) > \mathbf{0},$$
$$\mathrm{diag}(\mathbf{f}_1)\mathbf{D}^{n-1}(\mathbf{a}_0 - \mathbf{a}_1) > \mathbf{0}. \tag{1}$$

By Fact 2, this is equivalent to

$$(\mathbf{a}_0 + \mathbf{a}_1)^{\mathrm{T}} = 2\mathbf{k}_0\mathbf{Y}_0,$$
$$(\mathbf{a}_0 - \mathbf{a}_1)^{\mathrm{T}} = 2\mathbf{k}_1\mathbf{Y}_1. \tag{2}$$

for some row vectors $\mathbf{k}_0 > \mathbf{0}$ and $\mathbf{k}_1 > \mathbf{0}$, where $\mathbf{Y}_i$ $(i = 0, 1)$ denotes $2^{n-1} \times 2^{n-1}$ matrix $\mathrm{diag}(\mathbf{f}_i)\mathbf{D}^{n-1}$.

Let $\mathbf{Z}_0$ (resp, $\mathbf{Z}_1$) be a matrix consisting of all rows of $\mathbf{Y}_0$ indexed by $x \in \{0,1\}^{n-1}$ such that $f_0(x) = f_1(x)$ $(f_0(x) \neq f_1(x)$, resp). Then Eq. (2) can be written as

$$(\mathbf{a}_0 + \mathbf{a}_1)^{\mathrm{T}} = 2\mathbf{k}_{0,0}\mathbf{Z}_0 + 2\mathbf{k}_{0,1}\mathbf{Z}_1,$$
$$(\mathbf{a}_0 - \mathbf{a}_1)^{\mathrm{T}} = 2\mathbf{k}_{1,0}\mathbf{Z}_0 - 2\mathbf{k}_{1,1}\mathbf{Z}_1.$$

where $\mathbf{k}_{i,j} > 0$ $(i, j = 0, 1)$ is a suitable partition of $\mathbf{k}_i$ $(i = 0, 1)$. By solving this for $\mathbf{a}_0$ and $\mathbf{a}_1$, we have

$$\mathbf{a}_0^{\mathrm{T}} = (\mathbf{k}_{0,0} + \mathbf{k}_{1,0})\mathbf{Z}_0 + (\mathbf{k}_{0,1} - \mathbf{k}_{1,1})\mathbf{Z}_1,$$
$$\mathbf{a}_1^{\mathrm{T}} = (\mathbf{k}_{0,0} - \mathbf{k}_{1,0})\mathbf{Z}_0 + (\mathbf{k}_{0,1} + \mathbf{k}_{1,1})\mathbf{Z}_1.$$

Let z_0 and z_1 denote the number of rows in $\mathbf{Z}_0$ and $\mathbf{Z}_1$, respectively. Note that $z_0 + z_1 = 2^{n-1}$. If $z_1 \geq 2^{n-2}$, then for any $\mathbf{k}_{0,0}$ and $\mathbf{k}_{1,0}$, we can zero z_1 components of $\mathbf{a}_0$ by an appropriate setting of $\mathbf{k}_{0,1}$ and $\mathbf{k}_{1,1}$ since $\mathbf{Z}_1$ has full rank (this is because every rows in $\mathbf{D}^{n-1}$ are linearly independent). If $z_1 < 2^{n-2}$, which implies $z_0 > 2^{n-2}$, then for any $\mathbf{k}_{0,1}$ and $\mathbf{k}_{1,1}$, we can zero z_0 components of $\mathbf{a}_1$ by an appropriate setting of $\mathbf{k}_{0,0}$ and $\mathbf{k}_{1,0}$. This means that a polynomial given by coefficient $\mathbf{a} = (\mathbf{a}_0, \mathbf{a}_1)$ sign-represents f and has at most $2^n - 2^n/4$ non-zero elements. $\qquad\square$

As we see, this proof is based on a decomposition of f into two subfunctions $f|_{x_n=1}$ and $f|_{x_n=-1}$. It is quite natural to ask what happens if we decompose f by fixing two or more variables. Actually, this question is the starting point of this work.

Obviously, if $\mathsf{Prob}_x[f|_{x_n=1}(x) = f|_{x_n=-1}(x)]$ is far apart from $1/2$, then the above proof gives a sign-representing polynomial for f with significantly fewer monomials. The ratio to the all monomials is approaching to 50% (from 75%) when the probability is approaching to 0 or 1. However, for a random function, this probability is close to $1/2$ almost surely.

3.2   Decompose by Two Variables

In this section, we reveal that if we consider a two-variable decomposition then the upper bound of Theorem 3 can be improved for the *average* case.

Theorem 4. *Let $\epsilon > 0$ be an arbitrary constant. Then there is a constant $c > 0$ (depending on ϵ) such that all but a 2^{-c2^n} fraction of n-variable Boolean functions can be sign-represented by a polynomial with at most $(\frac{11}{16} + \epsilon)2^n = (0.6875 + \epsilon)2^n$ monomials.*

Proof. Let f be a Boolean function on n variables and $\mathbf{f}$ be its vector representation. Let p be a polynomial that sign-represents f and $\mathbf{a}$ be its coefficient vector. We partition $\mathbf{f}$ into four parts of equal length $\mathbf{f}_{00}$, $\mathbf{f}_{01}$, $\mathbf{f}_{10}$ and $\mathbf{f}_{11}$. Note that $\mathbf{f}_{i,j}$ is the vector representation of the function $f|_{x_n=i, x_{n-1}=j}$ (here we consider the input is $\{0,1\}$ instead of $\{+1, -1\}$), which we will denote $f_{i,j}$.

For a while, we follow the outline of the proof of Theorem 3 taking into account that we have partitioned $\mathbf{f}$ and $\mathbf{a}$ into four parts instead of two. In place of Eq. (2), we have

$$(\mathbf{a}_{00} + \mathbf{a}_{01} + \mathbf{a}_{10} + \mathbf{a}_{11})^{\mathrm{T}} = 4\mathbf{k}_{00}\mathbf{Y}_{00},$$
$$(\mathbf{a}_{00} - \mathbf{a}_{01} + \mathbf{a}_{10} - \mathbf{a}_{11})^{\mathrm{T}} = 4\mathbf{k}_{01}\mathbf{Y}_{01},$$
$$(\mathbf{a}_{00} + \mathbf{a}_{01} - \mathbf{a}_{10} - \mathbf{a}_{11})^{\mathrm{T}} = 4\mathbf{k}_{10}\mathbf{Y}_{10},$$
$$(\mathbf{a}_{00} - \mathbf{a}_{01} - \mathbf{a}_{10} + \mathbf{a}_{11})^{\mathrm{T}} = 4\mathbf{k}_{11}\mathbf{Y}_{11},$$

for some row vectors $\mathbf{k}_s > \mathbf{0}$ ($s \in \{0,1\}^2$), where $\mathbf{Y}_s$ ($s \in \{0,1\}^2$) denotes $2^{n-2} \times 2^{n-2}$ matrix $\mathsf{diag}(\mathbf{f}_s)\mathbf{D}^{n-2}$.

This time, we decompose $\mathbf{Y}_{00}$ into eight matrices $\mathbf{Z}_p$ ($p \in \{0,1\}^3$) as follows: For each $x \in \{0,1\}^{n-2}$, we assign a binary string p_x of length three to x in the following manner. If $f_{00}(x) = f_{01}(x)$, then the first bit of p_x is 0, and is 1 otherwise. The second bit is 0 (1, resp.), if $f_{00}(x) = f_{10}(x)$ ($f_{00}(x) \neq f_{10}(x)$, resp.), and the third bit is 0 (1, resp.), if $f_{00}(x) = f_{11}(x)$ ($f_{00}(x) \neq f_{11}(x)$, resp.). For each $p \in \{0,1\}^3$, let $\mathbf{Z}_p$ be a matrix consisting of all rows of $\mathbf{Y}_{00}$ indexed by x with $p_x = p$. For example, $\mathbf{Z}_{011}$ is consisting of all rows of $\mathbf{Y}_{00}$ indexed by $x \in \{0,1\}^{n-2}$ such that $(f_{00}(x), f_{01}(x), f_{10}(x), f_{11}(x)) = (1,1,-1,-1)$ or $(-1,-1,1,1)$.

Then we have

$$(\mathbf{a}_{00} + \mathbf{a}_{01} + \mathbf{a}_{10} + \mathbf{a}_{11})^{\mathrm{T}} = 4 \sum_{p \in \{0,1\}^3} \mathbf{k}_{00,p} \mathbf{Z}_p,$$

$$(\mathbf{a}_{00} - \mathbf{a}_{01} + \mathbf{a}_{10} - \mathbf{a}_{11})^{\mathrm{T}} = 4 \sum_{p \in \{0,1\}^3} (-1)^{p_1} \mathbf{k}_{01,p} \mathbf{Z}_p,$$

$$(\mathbf{a}_{00} + \mathbf{a}_{01} - \mathbf{a}_{10} - \mathbf{a}_{11})^{\mathrm{T}} = 4 \sum_{p \in \{0,1\}^3} (-1)^{p_2} \mathbf{k}_{10,p} \mathbf{Z}_p,$$

$$(\mathbf{a}_{00} - \mathbf{a}_{01} - \mathbf{a}_{10} + \mathbf{a}_{11})^{\mathrm{T}} = 4 \sum_{p \in \{0,1\}^3} (-1)^{p_3} \mathbf{k}_{11,p} \mathbf{Z}_p,$$

where the positive vectors $\mathbf{k}_{s,p} > \mathbf{0}$ ($s \in \{0,1\}^2, p \in \{0,1\}^3$) are a suitable partition of $\mathbf{k}_s$, and p_i ($i \in \{1,2,3\}$) denotes the i-th bit of p. By solving this, we have

$$\mathbf{a}_{00}^{\mathrm{T}} = \sum_{p \in \{0,1\}^3} \sum_{s \in \{0,1\}^2} \mathsf{sgn}(00, s, p) \mathbf{k}_{s,p} \mathbf{Z}_p,$$

$$\mathbf{a}_{01}^{\mathrm{T}} = \sum_{p \in \{0,1\}^3} \sum_{s \in \{0,1\}^2} \mathsf{sgn}(01, s, p) \mathbf{k}_{s,p} \mathbf{Z}_p,$$

$$\mathbf{a}_{10}^{\mathrm{T}} = \sum_{p \in \{0,1\}^3} \sum_{s \in \{0,1\}^2} \mathsf{sgn}(10, s, p) \mathbf{k}_{s,p} \mathbf{Z}_p,$$

$$\mathbf{a}_{11}^{\mathrm{T}} = \sum_{p \in \{0,1\}^3} \sum_{s \in \{0,1\}^2} \mathsf{sgn}(11, s, p) \mathbf{k}_{s,p} \mathbf{Z}_p, \tag{3}$$

where $\mathsf{sgn}(t, s, p) \in \{+1, -1\}$ represents a sign of each term. In fact, $\mathsf{sgn}(t, s, p)$ turns out to be $(-1)^{|t \cdot s|} m(s, p)$ where $|t \cdot s|$ denotes the number of one's in the bitwise AND of t and s, and $m(s, p)$ is -1 if ($s = 01$ and $p_1 = 1$) or ($s = 10$ and $p_2 = 1$) or ($s = 11$ and $p_3 = 1$), and is 1 otherwise.

The key observation is the following: For $p \in \{0,1\}^3$, let $\mathbf{S}_p$ be a 4×4 matrix such that (i) each row is indexed by $t \in \{0,1\}^2$, (ii) each column is indexed by $s \in \{0,1\}^2$ and (iii) the value of the (t, s)-entry is $\mathsf{sgn}(t, s, p)$. Then we can observe that $\mathbf{S}_{000} = \mathbf{D}^2$, $\mathbf{S}_{001} = \mathbf{D}^2\mathsf{diag}(1, 1, 1, -1)$, $\mathbf{S}_{010} = \mathbf{D}^2\mathsf{diag}(1, 1, -1, 1)$, $\ldots$, $\mathbf{S}_{111} = \mathbf{D}^2\mathsf{diag}(1, -1, -1, -1)$. The matrix $\mathbf{S}_p$ is of the form $\mathbf{D}^2\mathsf{diag}(\mathbf{f}_2)$ for some 2-variable function f_2 with $f_2(0, 0) = 1$ (see Fig. 1).

$t\backslash p$	000	001	010	011	100	101	110	111
00	++++	+++−	++−+	++−−	+−++	+−+−	+−−+	+−−−
01	+−+−	+−++	+−−−	+−−+	+++−	++++	++−−	++−+
10	++−−	++−+	+++−	++++	+−−−	+−−+	+−+−	+−++
11	+−−+	+−−−	+−++	+−+−	++−+	++−−	++++	+++−

Fig. 1. The sign matrix $\mathbf{S}_p$. In the table, "+" represents $+1$ and "−" represents -1.

Our goal is to make zeros in $(\mathbf{a}_{00}, \mathbf{a}_{01}, \mathbf{a}_{10}, \mathbf{a}_{11})$ as many as possible by appropriately determining $\mathbf{k}_*$'s.

First we consider $\mathbf{a}_{00}$. Fix $\mathbf{k}_{*,000} > \mathbf{0}$ so that $\mathbf{k}_{00,000} + \mathbf{k}_{01,000} + \mathbf{k}_{10,000} + \mathbf{k}_{11,000}$ is all one vector. We can satisfy this by setting all elements in $\mathbf{k}_{*,000}$ to $1/4$. Then we have

$$\mathbf{a}_{00}^{\mathrm{T}} = (w_1, w_2, \ldots, w_{2^{n-2}}) + \sum_{p \in \{0,1\}^3 \backslash \{000\}} \sum_{s \in \{0,1\}^2} \mathsf{sgn}(00, s, p) \mathbf{k}_{s,p} \mathbf{Z}_p, \quad (4)$$

for some w_*'s. The important thing here is that there is at least one "−" entries in $\mathsf{sgn}(00, *, p)$ for every $p \in \{0,1\}^3 \backslash \{000\}$ (see the top line of Fig. 1). Note that this is not the case for $p = 000$.

For a matrix $\mathbf{M}$, let $r(\mathbf{M})$ denote the number of rows of $\mathbf{M}$. Let $\mathbf{Z}'$ be the $(2^{n-2} - r(\mathbf{Z}_{000})) \times 2^{n-2}$ matrix consisting of all rows in all $\mathbf{Z}_p$'s with $p \in \{0,1\}^3 \backslash \{000\}$. Now we consider the *reduced row echelon form*[1] of $\mathbf{Z}'$, which is denoted by $\tilde{\mathbf{Z}}'$. Since the rows of $\mathbf{Z}'$ is orthogonal, $\tilde{\mathbf{Z}}'$ has no all-zero rows.

Let $c(i)$ denote the index of a column (starting from 1) of the leading nonzero element of the i-th row of $\tilde{\mathbf{Z}}'$. Define the row vector $\mathbf{v}$ so that

$$\mathbf{v} = \sum_{i=1}^{r(\mathbf{Z}')} -w_{c(i)} \tilde{\mathbf{Z}}'_i,$$

where $\tilde{\mathbf{Z}}'_i$ denotes the i-th row of $\tilde{\mathbf{Z}}'$. Since $\mathbf{v}$ is a linear combination of the rows of $\mathbf{Z}'$, there is a real row vector $\boldsymbol{\alpha}$ such that $\boldsymbol{\alpha}\mathbf{Z}' = \mathbf{v}$. At this moment, $\mathbf{k}_{*,p} > \mathbf{0}$ $(p \in \{0,1\}^3 \backslash \{000\})$ are free parameters and $\mathsf{sgn}(00, *, p)$ has at least one "−" entries, we can choose $\mathbf{k}_{*,p}$'s so that the signed sum of $\mathbf{k}_{*,p}$'s is equal to $\boldsymbol{\alpha}$, which means that the second term in Eq. (4) is equal to $\mathbf{v}$. As a result, we can zero at least $r(\mathbf{Z}')$ elements in $\mathbf{a}_{00}$.

If the numbers of rows of $\mathbf{Z}_p$'s are all equal, then we could zero $\frac{1}{4} \cdot \frac{7}{8} = \frac{7}{32}$ fraction of elements in $\mathbf{a}$ so far. However, this is less than a quarter, which is guaranteed by Theorem 3. So we should work on an another portion of $\mathbf{a}$. In fact, we below show that we can zero $\frac{2}{8}$ fraction of elements in $\mathbf{a}_{01}$ and also $\frac{1}{8}$ fraction of elements in $\mathbf{a}_{10}$ (again, when the sizes of $\mathbf{Z}_p$'s are equal).

---

[1] A matrix is in *reduced row echelon form* if (i) All nonzero rows are above any rows of all zeros, (ii) The leading nonzero coefficient of a nonzero row is always strictly to the right of the leading nonzero coefficient of the row above it, and (iii) Every leading coefficient is 1, and is the only nonzero element in its column.

For example, consider an $x \in \{0,1\}^{n-2}$ such that the row of $\mathbf{Y}_{00}$ indexed by x is belonging to, say, $\mathbf{Z}_{011}$. Let α_x and $k_{s,x}$ ($s \in \{0,1\}^2$) denote the value of the corresponding element in $\boldsymbol{\alpha}$ and in $\mathbf{k}_{s,011}$, respectively. In making zero-elements in $\mathbf{a}_{00}$, we needed to satisfy that

$$\alpha_x = k_{00,x} + k_{01,x} - k_{10,x} - k_{11,x}, \tag{5}$$

for some $k_{00,x}, k_{01,x}, k_{10,x}, k_{11,x} > 0$ (see the entry for $(t,p) = (00, 011)$ in Fig. 1 to get the signs of k_*'s in Eq.(5)). This is always possible for every $\alpha_x \in \mathbb{R}$ since the signs of k_*'s in Eq. (5) are mixed. If we additionally apply a similar process to this dimension for $\mathbf{a}_{01}$, we should satisfy that

$$\beta_x = k_{00,x} - k_{01,x} - k_{10,x} + k_{11,x}, \tag{6}$$

for every fixed $\beta_x \in \mathbb{R}$ (see the entry for $(t,p) = (01, 011)$ in Fig. 1). The questions are: Is this always possible? If so, what for the third?

In order to discuss these questions, we introduce the following definition.

Definition 5. *Let $\mathbf{M}$ be an $N \times N$ matrix whose element is 1 or -1. Let $\mathbf{w} = (w_1, \ldots, w_N)^T$ be a column vector of length N. For a subset $S \subseteq \{1, \ldots, N\}$, we say that $\mathbf{M}$ is free on S if for any values of $w_i \in \mathbb{R}$ for all $i \in S$, there exists a column vector $\mathbf{k} > \mathbf{0}$ and $w_i \in \mathbb{R}$ for each $i \notin S$ such that $\mathbf{Mk} = \mathbf{w}$.*

Since our sign matrix is always of the form $\mathbf{D}^n \mathrm{diag}(\mathbf{f})$ for a Boolean function f, the following simple lemma is extremely useful to analyzing the performance of our method. Below we represents monomials by their characteristic vectors, and so $\mathcal{M}^n$ is mapped to $\{0,1\}^n$.

Lemma 6. *Let $\mathbf{f}$ be the vector representation of an n-variable Boolean function f. Then $\mathbf{D}^n \mathrm{diag}(\mathbf{f})$ is free on $S \subseteq \{0,1\}^n$ if f can be sign-represented by a polynomial with support $\{0,1\}^n - S$.*

Proof. Suppose that f can be sign-represented by a polynomial with support $\{0,1\}^n - S$. This means that there is a column vector $\mathbf{w}' = (w'_{00\cdots0}, \ldots, w'_{11\cdots1})^T$ with $w'_i = 0$ for $i \in S$ such that $\mathrm{diag}(\mathbf{f})\mathbf{D}^n \mathbf{w}' = \mathbf{k}'$ for some $\mathbf{k}' > \mathbf{0}$. Let k' be the smallest absolute value of all elements in $\mathbf{k}'$.

Let $\mathbf{w}'' = (w''_{00\cdots0}, \ldots, w''_{11\cdots1})^T$ be an arbitrary vector satisfying $w''_i = 0$ for $i \notin S$. Put $\mathbf{k}'' = \mathrm{diag}(\mathbf{f})\mathbf{D}^n \mathbf{w}''$, and let k'' be the largest absolute value of all elements in $\mathbf{k}''$.

Let t be an arbitrary constant satisfying $t > k''/k'$. We define column vectors $\mathbf{w}$ and $\mathbf{k}$ as

$$\mathbf{w} = t\mathbf{w}' + \mathbf{w}'', \quad \mathbf{k} = t\mathbf{k}' + \mathbf{k}''.$$

Then we have

$$\mathrm{diag}(\mathbf{f})\mathbf{D}^n \mathbf{w} = t\mathbf{k}' + \mathbf{k}'' = \mathbf{k} > \mathbf{0}.$$

Since $[\mathrm{diag}(\mathbf{f})\mathbf{D}^n \mathbf{w} = \mathbf{k}] \iff [\mathbf{w} = 2^{-n}\mathbf{D}^n \mathrm{diag}(\mathbf{f})\mathbf{k} = (\mathbf{D}^n \mathrm{diag}(\mathbf{f}))(2^{-n}\mathbf{k})]$, the fact follows. □

We now go back to the aforementioned example (discussed just before Definition 5). In that example, the sign matrix is $\mathbf{D}^n\mathrm{diag}(1,1,-1,-1)$. This corresponds to the function x_2 which can be sign-represented by $p(x_1,x_2) = x_2$. Hence Lemma 6 guarantees that it is free on the first, second and fourth monomials. So we can always satisfy Eqs. (5) and (6) by an appropriate choice of $k_{*,x} > 0$.

By examining all eight functions on two variables with $f(1,1) = 1$, we observe that seven functions except the constant-one function are free on the first monomial, and two functions x_2 and $x_1 \oplus x_2$ whose vector representations are $(1,1,-1,-1)$ and $(1,-1,-1,1)$ are free on the first and second monomials. Note that $1 \oplus 1 = (-1) \oplus (-1) = 1$ and $(-1) \oplus 1 = 1 \oplus (-1) = -1$ since we map $\{0,1\}$ to $\{1,-1\}$. In addition, the function $x_1 \oplus x_2$ is free on the first, second and third monomials since it can be sign-represented by $p(x_1,x_2) = x_1 x_2$. This means that we can use a freedom of $\mathbf{k}_*$'s on dimensions corresponding to $\mathbf{Z}_{011}$ and $\mathbf{Z}_{110}$ for making zeros in $\mathbf{a}_{01}$ and that corresponding to $\mathbf{Z}_{110}$ for making zeros in $\mathbf{a}_{10}$.

Now we are ready to finish the proof of Theorem 4.

Consider the second equality in Eq.(3) concerning $\mathbf{a}_{01}$. Fix every $\mathbf{k}_{*,p} > \mathbf{0}$ arbitrary for $p \in \{001, 010, 100, 101, 111\}$ so that to satisfy the condition to make zeros in $\mathbf{a}_{00}$. This is always possible by the above argument. Recall that we have fixed $\mathbf{k}_{*,000}$'s and so $\mathbf{k}_{*,011}$'s and $\mathbf{k}_{*,110}$'s are remaining to be fixed. By using the freedom of dimensions in $\mathbf{k}_{*,011}$ and $\mathbf{k}_{*,110}$, we can zero $r(\mathbf{Z}_{011}) + r(\mathbf{Z}_{110})$ elements in $\mathbf{a}_{01}$ in an analogous way to make zeros in $\mathbf{a}_{00}$.

Finally, we consider the third equality in Eq.(3) concerning $\mathbf{a}_{10}$. Fix every $\mathbf{k}_{*,011} > \mathbf{0}$ so that to satisfy the conditions to be needed to make zeros in $\mathbf{a}_{00}$ and $\mathbf{a}_{01}$. The analogous procedure to the above can make zeros at least $r(\mathbf{Z}_{110})$ elements in $\mathbf{a}_{10}$.

In total, we can zero at least

$$r(\mathbf{Z}_{001}) + r(\mathbf{Z}_{010}) + 2r(\mathbf{Z}_{011}) + r(\mathbf{Z}_{100}) + r(\mathbf{Z}_{101}) + 3r(\mathbf{Z}_{110}) + r(\mathbf{Z}_{111})$$

elements in $\mathbf{a}$. The standard argument using the Chernoff bound (see below and e.g., [2, p.515] for the proof) shows that $r(\mathbf{Z}_p)$ is within $(1/8 \pm \epsilon)2^{n-2}$ for every p with probability $1 - 2^{-\Omega(\epsilon^2)2^n}$ when f is chosen uniformly. This completes the proof of Theorem 4.   □

Theorem 7. *(Chernoff Bound)   Let $X_1, X_2, \ldots, X_n$ be mutually independent random variables over $\{0,1\}$ and let $\mu = \sum_{i=1}^{n} E[X_i]$. Then for every $c > 0$,*

$$\mathsf{Prob}\left[\left|\sum_{i=1}^{n} X_i - \mu\right| \geq c\mu\right] \leq 2 \cdot e^{-\min\{c^2/4, c/2\}\mu}.$$

3.3   Finer Decompositions Yield Better Bounds

The proof in the last section can naturally be extended for finer decompositions.

Suppose that we partition a given function f into eight parts by fixing three variables and follow the proof in the last section. We should introduce $2^{2^3-1} = 2^7$ submatrices $\mathbf{Z}_p$. Then, in place of Fig. 1, we obtain 2^7 sign matrices $\mathbf{S}_p$ of size $2^3 \times 2^3$ which corresponds to all 3-variable Boolean functions with $f(1,1,1) = 1$.

Notice that we can also use another freedom to enlarge the number of zero elements in $\mathbf{a}$. In the above proof, we work on $\mathbf{a}_{00}, \mathbf{a}_{01}$ and $\mathbf{a}_{10}$ in this order. Indeed, we can consider *any* ordering of the partitions of $\mathbf{a}$, e.g., $\mathbf{a}_{11} \rightarrow \mathbf{a}_{10} \rightarrow \mathbf{a}_{01} \rightarrow \mathbf{a}_{00}$. In fact, the final bound will not be affected if we decompose by $k = 2$ variables. However, we see that this is not the case for $k \geq 3$. Once an ordering is given, the computation of the bound can be executed systematically using Lemma 6.

Recall that $\mathcal{M}^k$ denotes the all 2^k monomials on k variables. Let π be a mapping from $\{1, \ldots, 2^k\}$ to $\mathcal{M}^k$. The mapping π naturally represents the ordering of the monomials $\{\pi(1), \pi(2), \ldots, \pi(2^k)\}$. For a Boolean function f and an ordering of monomials π, the *freedom* of f with respect to π, denoted by $\mathsf{free}(f, \pi)$ is defined as the maximum t such that f can be sing-represented by a polynomial with monomials $\mathcal{M}^k - \{\pi(1), \pi(2), \ldots, \pi(t)\}$. For a positive integer k and an ordering π of $\mathcal{M}^k$, let $d(k, \pi)$ denote the average of $\mathsf{free}(f, \pi)/2^k$ over all k-variable Boolean functions f with $f(1, 1, \ldots, 1) = 1$ which we call the *average freedom* with respect to π. Note that the last condition ($f(1, 1, \ldots, 1) = 1$) can be removed without changing the value of $d(k, \pi)$ by symmetry.

The generalization of Theorem 4, which is the main theorem in this paper, can be stated as follows:

Theorem 8. *Let $\epsilon > 0$ an arbitrary constant. Let $k \geq 1$ be an integer and π be an ordering of $\mathcal{M}^k$. Then, there is a constant $c > 0$ (depending on ϵ and k) such that all but a 2^{-c2^n} fraction of n-variable Boolean functions have PTF density at most $(1 - d(k, \pi) + \epsilon)2^n$.* □

The proof is entirely analogous to the proof of Theorem 4, and is omitted.

It is a bit tedious but easy to verify that $d(2, \pi) = 5/16$ for every π of $\mathcal{M}^2$, which matches Theorem 4. The computation of $d(k, \pi)$ for $k \geq 3$ is done by a computer. We can see whether a given k-variable function f has a sign-representing polynomial with support S by checking the feasibility of the linear system $\mathsf{diag}(\mathbf{f})\mathbf{D}^k\mathbf{a} > \mathbf{0}$ where $a_i = 0$ for every $i \notin S$, which is an easy task by any linear programming package when k is small. The computation of $d(3, \pi)$ for every possible π is quite feasible by a standard PC. We use the GLPK package [5] in our experiments. We found that

$$d(3, \{1, x_1, x_2, x_1x_2, x_3, x_1x_3, x_2x_3, x_1x_2x_3\}) = 316/(2^7 \times 8) = 0.3085 \cdots,$$
$$d(3, \{1, x_1, x_2, x_3, x_1x_2, x_1x_3, x_2x_3, x_1x_2x_3\}) = 360/(2^7 \times 8) = 0.3515 \cdots,$$

and that all orderings of $\mathcal{M}^3$ are categorized into one of the above two. This gives the upper bound on the average PTF density of $(0.649)2^n$ which is better than Theorem 4.

The computation of $d(4, \pi)$ for all π seems out of reach. However, the result for $k = 3$ inspires the ordering

$$\pi_4 = \{1, x_1, \ldots, x_4, x_1x_2, \ldots, x_3x_4, x_1x_2x_3, \ldots, x_2x_3x_4, x_1x_2x_3x_4\},$$

would be a good candidate. We compute $d(4, \pi_4)$ (again by a computer) to find

$$d(4, \pi_4) = 195804/(2^{15} \times 16) = 0.3734 \cdots, \tag{7}$$

which gives the upper bound of $(0.627)2^n$. However, this is not the best.

We then computed the average freedom of 2000 randomly chosen orderings of $\mathcal{M}^4$, which took about two days on a standard PC. The best ordering we have found is

$$\pi_4' = \{1, x_1x_2x_3, x_1x_4, x_1x_3x_4, x_1x_2x_3x_4, x_3x_4, x_2x_4, x_3, x_1x_3,$$
$$x_1, x_1x_2, x_2x_3x_4, x_2x_3, x_2, x_4\},$$

which has the average freedom of

$$d(4, \pi_4') = 200964/(2^{15} \times 16) = 0.3833 \cdots . \tag{8}$$

Note that 8 (out of 2000) orderings have the same value. This immediately gives the following upper bound, which is the best we have obtained so far.

Corollary 9. *Almost all Boolean functions on n variables have PTF density at most $(0.617)2^n$.* □

We provide verifiable data for Eqs. (7) and (8) on the web page [1]. These are the lists of polynomial representations of all 4-variable functions with maximum freedom with respect to the designated ordering. The correctness of Eqs. (7) and (8) can be verified by hand in a several weeks, or by a computer in a few seconds. At the time of writing this article, we don't know whether π_4' is the best among all orderings of $\mathcal{M}^4$ or not.

4   Concluding Remarks

In this paper, we develop a method for proving an upper bound on the average PTF density of Boolean functions. Apparently, our method would yield a better upper bound if we have more computational resource (or more sophisticated algorithm for computing the average freedom). A random sampling experiment suggests that $1 - d(5, \pi_5)$ is around 0.598 where π_5 is an obvious extension of π_4. To see the limit of our method, or to characterize a good monomial ordering would be an interesting future work.

To improve the current best constant 0.75 in the maximum PTF density is also interesting. For small values of n, the maximum PTF density among all n-variable Boolean functions $\Pi(n)$ can be computed as $1, 3, 4$, and 9 for $n = 1, 2, 3$, and 4 (see e.g., [10]). This leads to a speculation that $\Pi(n) = (0.5)2^n$ if n is odd, and $\Pi(n) = (0.5)2^n + 1$ if n is even [10]. However, during the preparation of this paper, we experimentally found that $\Pi(5) = 11$, which is smaller than was expected.

The upper bound is verified by random search for every possible function (i.e., 616,126 representatives of NPN-equivalence classes [15, Sequence A00370]), and the lower bound is verified by exhaustive search for e.g., a function whose truth table is 000f3563 in hexadecimal notation. Based on this, we conjecture that the true constant is well below 0.5. In addition, we believe that the conjecture by O'Donnell and Servedio [8, Conjecture 23] stating that $\Pi(n) \leq (0.5)2^n$ for sufficiently large n would hold even if we replace "sufficiently large n" by "for every $n \geq 5$".

A final remark is on the implementation. As to the Oztop's method [10,11], our proof is constructive in a sense that we can easily make a computer program *executing* the proof of the upper bound. We made a program based on a two variable decomposition described in Section 3.2. For example, for 10^5 random functions on 10 variables, the average ratio of monomials with non-zero coefficients produced by the program is around 67.8% ($\sim$ 693.9 out of 2^{10}) which is close to the bound that theoretically proven (i.e., 68.8%).

References

1. Amano, K.: Supplemental data of the paper are available at
   http://www.cs.gunma-u.ac.jp/~amano/poly/index.html
2. Arora, S., Barak, B.: Computational Complexity: A Modern Approach. Cambridge University Press, Cambridge (2009)
3. Cover, T.: Geometrical and Statistical Properties of Systems of Linear Inequalities with Applications in Pattern Recognition. IEEE Trans. Electronic Computers EC-14(3), 326–334 (1965)
4. Gotsman, C.: On Boolean Functions, Polynomials and Algebraic Threshold Functions. Tech. Rep. TR-89-18, Department of Computer Science, Hebrew University (1989)
5. Makhorin, A.: The GLPK (GNU Linear Programming Kit) Package,
   http://www.gnu.org/software/glpk/
6. Minsky, M., Papert, S.: Perceptrons: An Introduction to Computational Geometry. MIT Press, Cambridge (1968)
7. Wegener, I.: The Complexity of Boolean Functions. Wiley-Teubner Series in Computer Science (1987)
8. O'Donnell, R., Servedio, R.: Extremal Properties of Polynomial Threshold Functions. J. Comput. Syst. Sci. 74(3), 298–312 (2008); Conference Version in Proc. of CCC 2003, pp. 3–12 (2003)
9. O'Donnell, R., Servedio, R.: New Degree Bounds for Polynomial Threshold Functions. In: Proc. of STOC 2003, pp. 325–334 (2003)
10. Oztop, E.: An Upper Bound on the Minimum Number of Monomials Required to Separate Dichotomies of $\{-1, 1\}^n$. Neural Computation 18(12), 3119–3138 (2006)
11. Oztop, E.: Sign-representation of Boolean Functions using a Small Number of Monomials. Neural Networks 22(7), 938–948 (2009)
12. Saks, M.E.: Slicing the Hypercubes. Surveys in Combinatorics, pp. 211–255. Cambridge University Press, Cambridge (1993)
13. Sherstov, A.A.: The Intersection of Two Halfspaces has High Threshold Degree. In: Proc. of FOCS 2009, pp. 343–362 (2009)
14. Sherstov, A.A.: Optimal Bounds for Sign-representing the Intersection of Two Halfspaces by Polynomials. In: Proc. of STOC 2010, pp. 523–532 (2010)
15. The On-Line Encyclopedia of Integer Sequences (2010), Published electronically at http://oeis.org

An Optimal Algorithm for Computing Angle-Constrained Spanners*

Paz Carmi[1] and Michiel Smid[2]

[1] Department of Computer Science, Ben-Gurion University, Beer-Sheva, Israel
[2] School of Computer Science, Carleton University, Ottawa, Canada

Abstract. Let S be a set of n points in $\mathbb{R}^d$. A graph $G = (S, E)$ is called a t-spanner for S, if for any two points p and q in S, the shortest-path distance in G between p and q is at most $t|pq|$, where $|pq|$ denotes the Euclidean distance between p and q. The graph G is called θ-angle-constrained, if any two distinct edges sharing an endpoint make an angle of at least θ. It is shown that, for any θ with $0 < \theta < \pi/3$, a θ-angle-constrained t-spanner can be computed in $O(n \log n)$ time, where t depends only on θ.

1 Introduction

Let S be a set of n points in $\mathbb{R}^d$ and let $G = (S, E)$ be a graph with vertex set S, in which the length (or weight) of every edge $\{p, q\}$ is equal to the Euclidean distance $|pq|$ between p and q. The length of a path in G is defined to be the sum of the lengths of the edges on the path. For any two points p and q in S, we denote by $\delta_G(p, q)$ the minimum length of any path in G between p and q. For a real number $t \geq 1$, G is a *t-spanner* for S, if $\delta_G(p, q) \leq t|pq|$ for any two points p and q of S.

The problem of efficiently constructing spanners for a given point set has been well-studied. For any set S of n points in $\mathbb{R}^d$ and any constant $t > 1$, a t-spanner for S whose maximum degree only depends on t can be computed in $O(n \log n)$ time (see Arya and Smid [3]); observe that such a spanner is sparse in the sense that it consists of $O(n)$ edges. For overviews of the main results for geometric spanners, we refer to Eppstein [9] and Narasimhan and Smid [14].

Geometric spanners have received much attention in the wireless network community; see, e.g., the book [11] by Li. For a positive real number θ, we say that the graph $G = (S, E)$ is *θ-angle-constrained*, if for any two distinct edges $\{p, q\}$ and $\{p, r\}$ in E, The angle $\angle(pq, pr)$ between them is at least θ. On page 238 in his book, Li mentions that a wireless network being an angle-constrained spanner is a desirable property, because it reduces signal interference and receiving power cost when directional antennae are used, and it guarantees short paths between any pair of nodes. In this paper, we consider the problem of computing angle-constrained spanners for point sets in $\mathbb{R}^d$.

* This work was supported by the Natural Sciences and Engineering Research Council of Canada.

O. Cheong, K.-Y. Chwa, and K. Park (Eds.): ISAAC 2010, Part I, LNCS 6506, pp. 316–327, 2010.

The "path-greedy" algorithm of Althöfer *et al.* [1] is a well-known algorithm for constructing a t-spanner. Soares [16] has shown that this spanner is θ-angle-constrained, where θ depends on t. However, the fastest known algorithm for computing the greedy spanner has a running time of $O(n^2 \log n)$; see Bose *et al.* [4]. In [15], Salowe presents an "angle-greedy" algorithm that constructs an angle-constrained spanner; this algorithm, however, also has a running time of $O(n^2 \log n)$. The "gap-greedy" algorithm of Arya and Smid [3] can be modified so that it constructs an angle-constrained spanner in $O(n \log^d n)$ time. Finally, none of the known $O(n \log n)$–time algorithms that construct spanners whose maximum degree is bounded by a constant produces a graph that is angle-constrained. In this paper, we prove the following result:

Theorem 1. *Let S be a set of n points in $\mathbb{R}^d$ and let θ and ϵ be two real constants such that $0 < \theta < \pi/3$ and $0 < \epsilon < (\pi - 3\theta)/(21 + \pi)$. In $O(n \log n)$ time, a θ-angle-constrained t-spanner for S can be computed, where*

$$t = \max\left(1 + \epsilon, \frac{1 + \sqrt{2(1 + \epsilon)(1 - \cos(\theta + 7\epsilon)) + \epsilon^2}}{2\cos(\theta + 7\epsilon) - 1 - \epsilon}\right).$$

Our construction will be based on a combination of the spanner based on the well-separated pair decomposition of Callahan and Kosaraju [5,6] (see also Section 2) and ideas that have been used in analyzing the Θ-graph spanner of Clarkson [8] and Keil and Gutwin [10]. As we will show in Section 3, this combination leads to a simple and sufficient condition for a graph being a spanner. In order to satisfy this condition, our algorithms will use simplicial cones, which are described in Section 4. In Section 5, we will present a simple algorithm that constructs an angle-constrained graph that satisfies the condition in Section 3; thus, it produces an angle-constrained spanner. The running time of this algorithm is, however, $O(n \log^{d-1} n)$. Moreover, it does not work in the algebraic computation-tree model. In Section 6, we will show that the algorithm can be modified such that it works in the algebraic computation-tree model and its running time is $O(n \log n)$. The main ingredient of this final algorithm is the use the "dumbbell trees" of Arya *et al.* [2].

2 Well-Separated Pairs

For any point set A in $\mathbb{R}^d$, we denote its *bounding box* by $R(A)$. Let $s > 0$ be a real number, which we call the *separation constant*. Two point sets A and B in $\mathbb{R}^d$ are *well-separated* with respect to s, if there exist two balls of the same radius, say, ρ, one ball containing $R(A)$ and the other ball containing $R(B)$, such that the distance between the balls is at least $s\rho$.

Lemma 1. *Let A and B be two sets of points that are well-separated with respect to s, let a, a', and a'' be points in $R(A)$, and let b, b', and b'' be points in $R(B)$. Then the following inequalities hold:*

$$|aa'| \leq (2/s)|a''b''|, |bb'| \leq (2/s)|a''b''|, |ab| \leq (1 + 4/s)|a'b'|.$$

Let S be a set of n points in $\mathbb{R}^d$. A *well-separated pair decomposition (WSPD)* of S is a sequence $\{A_1, B_1\}, \ldots, \{A_m, B_m\}$ of well-separated pairs of subsets of S, such that, for any two distinct points p and q in S, there is a unique index i such that $p \in A_i$ and $q \in B_i$ or $p \in B_i$ and $q \in A_i$. We will refer to the number m of pairs as the *size* of the WSPD.

Callahan and Kosaraju [6] have shown that a WSPD of size $O(n)$ can be computed in $O(n \log n)$ time. Their algorithm uses the so-called fair-split tree, which is a binary tree storing the points of S at its leaves. For each pair $\{A_i, B_i\}$ in the WSPD, there are two nodes u and v in this tree such that A_i is the set of all points stored in u's subtree and B_i is the set of all points stored in v's subtree. It follows from their construction that $A_i = S \cap R(A_i)$ and $B_i = S \cap R(B_i)$.

3   A Sufficient Condition for Being a Spanner

In this section, we introduce a general property which implies that a geometric graph is a spanner. The property is based on a combination of the WSDP-spanner of [5] and techniques that have been used in the analysis of the Θ-graph spanner of [8,10]. We fix real numbers α, λ, ϵ, and s, such that

$$0 < \alpha < \pi/3, \ \lambda \geq 1, \ 0 < \epsilon < 2\cos\alpha - 1, \ \text{and} \ s > 8\lambda. \tag{1}$$

Let S be a set of n points in $\mathbb{R}^d$ and consider a WSPD $\{A_1, B_1\}, \ldots, \{A_m, B_m\}$ of S with separation constant s. Let $G = (S, E)$ be a graph with vertex set S. For any i with $1 \leq i \leq m$, consider the following three properties P.1, P.2, and P.3:

P.1: For every point p in A_i, the edge set E contains an edge $\{p, r\}$ such that for every point q in B_i, $\angle(pq, pr) \leq \alpha$ and $|pr| \leq (1 + \epsilon)|pq|$.

P.2: For every point q in B_i, the edge set E contains an edge $\{q, r\}$ such that for every point p in A_i, $\angle(qp, qr) \leq \alpha$ and $|qr| \leq (1 + \epsilon)|pq|$.

P.3: Let ℓ_i be the distance between the centers of $R(A_i)$ and $R(B_i)$. The edge set E contains an edge $\{x, y\}$, such that for every point p in A_i and every point q in B_i, both $|px|$ and $|qy|$ are at most $(2\lambda/s)\ell_i$.

In words, property P.1 states that every point p in A_i has an edge $\{p, r\}$, such that the line segment pr takes us in the direction of B_i and does not take us too far beyond B_i. Property P.2 is symmetric to P.1. Finally, property P.3 states that there exists an edge $\{x, y\}$, where x is "close" to A_i and y is "close" to B_i.

Lemma 3 below states that the graph G is a spanner, provided for each pair $\{A_i, B_i\}$, at least one of P.1, P.2, and P.3 holds. The proof of this lemma will use the following technical geometric result:

Lemma 2. *Let t be a real number such that*

$$t \geq \max\left(1 + \epsilon, \frac{1 + \sqrt{2(1 + \epsilon)(1 - \cos\alpha) + \epsilon^2}}{2\cos\alpha - 1 - \epsilon}\right).$$

Let p, q, and r be three distinct points in $\mathbb{R}^d$, such that $\angle(pq, pr) \leq \alpha$ and $|pr| \leq (1 + \epsilon)|pq|$. Then $|rq| < |pq|$ and $|pr| + t|rq| \leq t|pq|$.

Lemma 3. *Assume that for each i with $1 \leq i \leq m$, at least one of the properties P.1, P.2, and P.3 is satisfied. Then for any real number t with*

$$t \geq \max \left(1 + \epsilon, \frac{1 + \sqrt{2(1+\epsilon)(1 - \cos \alpha) + \epsilon^2}}{2 \cos \alpha - 1 - \epsilon}, \frac{s + 8\lambda}{s - 8\lambda} \right),$$

the graph G is a t-spanner for S.

Proof. We have to show that $\delta_G(p, q) \leq t|pq|$ for all p and q in S. The proof is by induction on the rank of the distance $|pq|$ in the sorted sequence of distances in S. If $p = q$, then the claim obviously holds. Let $p \neq q$ and assume that $\delta_G(a, b) \leq t|ab|$ for all a and b in S with $|ab| < |pq|$. Let i be the index such that (i) $p \in A_i$ and $q \in B_i$ or (ii) $p \in B_i$ and $q \in A_i$. We may assume without loss of generality that (i) holds. If property P.1 or P.2 holds, then we use the induction hypothesis and Lemma 2. If property P.3 holds, then we use the induction hypothesis and Lemma 1. □

In the rest of this section, we give an informal description of our algorithm. Let θ be a real number with $0 < \theta < \pi/3$. We choose a small positive real number $\epsilon \ll \theta$ such that $\alpha := \theta + O(\epsilon) < \pi/3$. We also choose a real number $\lambda \geq 1$. Let $\mathcal{C}$ be a collection of cones that cover $\mathbb{R}^d$ such that each cone has its apex at the origin and angular diameter at most ϵ.

Consider a WSPD $\{A_1, B_1\}, \ldots, \{A_m, B_m\}$ of the point set S with separation constant s. For each i, let ℓ_i be the distance between the centers of $R(A_i)$ and $R(B_i)$. Let $R^\lambda(A_i)$ and $R^\lambda(B_i)$ be boxes of diameter $O(\lambda \ell_i/s)$ that contain $R(A_i)$ and $R(B_i)$, respectively.

We choose the separation constant s to be large enough such that for any point x in $R^\lambda(A_i)$ and any two points y and y' in $R^\lambda(B_i)$, $\angle(xy, xy') \leq \epsilon$. Thus, there are only a "few" cones C in $\mathcal{C}$ such that the box $R^\lambda(B_i)$ overlaps the translated cone $x + C$.

For each cone C in $\mathcal{C}$, we define $\overline{C}$ to be the union of all cones C' in $\mathcal{C}$ that make an angle of at most $\theta + O(\epsilon)$ with C.

Consider a pair $\{A_i, B_i\}$ in the WSPD. Let c be the center of $R(A_i)$ and let C be a cone in $\mathcal{C}$ such that $R(B_i)$ overlaps $c + C$. Consider the corresponding cone $\overline{C}$. Since $\epsilon \ll \theta$, the box $R^\lambda(B_i)$ is located "near" the center of $c + \overline{C}$: For any point $r \notin c + \overline{C}$ and any point $b \in R^\lambda(B_i)$, the angle between cr and cb is at least θ; for any point $r \in c + \overline{C}$ and any point $b \in R^\lambda(B_i)$, the angle between cr and cb is at most α.

Assume that the pairs in the WSPD have been sorted such that $\ell_1 \leq \ell_2 \leq \ldots \leq \ell_m$. The algorithm will start with an empty edge set E. Then it processes each pair in the WSPD. Consider the current pair $\{A_i, B_i\}$.

Let c be the center of $R(A_i)$ and let C be a cone in $\mathcal{C}$ such that $R(B_i)$ overlaps $c + C$. Let c' be the center of $R(B_i)$ and let C' be a cone in $\mathcal{C}$ such that $R(A_i)$ overlaps $c' + C'$. There are three possible cases:

First, if every point p in A_i is incident on some edge $\{p, r\}$ with $r \in p + \overline{C}$, then property P.1 holds for the pair $\{A_i, B_i\}$ and, thus, there is no need to add an additional edge to E.

Second, if every point q in B_i is incident on some edge $\{q,r\}$ with $r \in q + \overline{C'}$, then property P.2 holds for $\{A_i, B_i\}$ and, again, there is no need to add an additional edge to E.

Otherwise, we pick an arbitrary point x in $R^\lambda(A_i)$ that is not incident on any edge $\{x,r\}$ with $r \in x + \overline{C}$ and an arbitrary point y in $R^\lambda(B_i)$ that is not incident on any edge $\{y,r\}$ with $r \in y + \overline{C'}$ and add the edge $\{x,y\}$ to E. The addition of this edge guarantees that property P.3 holds for the pair $\{A_i, B_i\}$. Furthermore, the new edge $\{x,y\}$ makes an angle of at least θ with all edge in the old set E that are incident on x or y.

4 Simplicial Cones

Let V be a set of d linearly independent points in $\mathbb{R}^d$. The set

$$C = \left\{ \sum_{v \in V} \mu_v v : \mu_v \geq 0 \text{ for all } v \in V \right\}$$

is called a *simplicial cone* with *apex* at the origin 0. If we define r_v to be the infinite ray emanating from the origin and going through v, then this cone is equal to the convex hull of the rays r_v, where v ranges over all elements of V. Thus, C is bounded by d hyperplanes, each one containing the origin. The *angular diameter* of C is defined to be $\max\{\angle(0x, 0y) : x, y \in C \setminus \{0\}\}$.

We fix real numbers ϵ and λ such that $0 < \epsilon \leq \pi/2$ and $\lambda \geq 1$. Let S be a set of n points in $\mathbb{R}^d$, and consider a WSPD $\{A_1, B_1\}, \ldots, \{A_m, B_m\}$ of S with separation constant s, where

$$s \geq \max\left(8\lambda\sqrt{d}, \frac{4\lambda\sqrt{d}}{\sin \epsilon} \right). \tag{2}$$

For each i with $1 \leq i \leq m$, define ℓ_i to be the distance between the centers of the bounding boxes $R(A_i)$ and $R(B_i)$ of A_i and B_i, respectively.

For each i with $1 \leq i \leq m$, we assume that we are given boxes $R^\lambda(A_i)$ and $R^\lambda(B_i)$, both having diameter at most $2\lambda\sqrt{d}\ell_i/s$, that contain $R(A_i)$ and $R(B_i)$, respectively. Observe that, by Lemma 1, $R^\lambda(A_i)$ and $R^\lambda(B_i)$ exist.

Let $\mathcal{C}$ be a collection of simplicial cones that cover $\mathbb{R}^d$, such that each cone has its apex at the origin and angular diameter at most ϵ. Lukovszki [12] has shown how to obtain such a collection consisting of $O(1/\epsilon^{d-1})$ cones. (See also Chapter 5 in [14].) For each cone C in $\mathcal{C}$, we fix an arbitrary point y_C in $C \setminus \{0\}$.

Let $\theta > 0$ be a real number and define $\theta' = \theta + 3\epsilon$. For any cone C in $\mathcal{C}$, consider all cones C' in $\mathcal{C}$ for which there exists a point z in $C' \setminus \{0\}$ with $\angle(0y_C, 0z) \leq \theta'$. We define $\overline{C}$ to be the union of all these cones C'. Observe that, for any point r in $\overline{C} \setminus \{0\}$, $\angle(0y_C, 0r) \leq \theta' + \epsilon$.

We now state the properties about the cones C, the corresponding sets $\overline{C}$, and the WSPD that were mentioned at the end of Section 3.

Lemma 4. *Consider a pair $\{A_i, B_i\}$ in the WSPD, let x be a point in the box $R^\lambda(A_i)$, and let y and y' be points in the box $R^\lambda(B_i)$. Then $|xy| \geq \ell_i/2$ and $\angle(xy, xy') \leq \epsilon$.*

Lemma 5. *Consider a pair $\{A_i, B_i\}$ in the WSPD, let c be the center of the bounding box $R(A_i)$ of A_i, and let C be a cone in $\mathcal{C}$ such that $R(B_i)$ overlaps the translated cone $c + C$. Let b be a point in the intersection of $R(B_i)$ and $c + C$, and let x be a point in $R^\lambda(A_i)$. Then there exists a point b' in the translated cone $x + C$ such that $b' \neq x$ and $\angle(xb, xb') \leq \epsilon$.*

Lemma 6. *Consider a pair $\{A_i, B_i\}$ in the WSPD, let c be the center of the bounding box $R(A_i)$ of A_i, and let C be a cone in $\mathcal{C}$ such that $R(B_i)$ overlaps the translated cone $c + C$. Let x be a point in $R^\lambda(A_i)$, let y be a point in $R^\lambda(B_i)$, and let r be a point that is not contained in $x + \overline{C}$. Then $\angle(xy, xr) \geq \theta$.*

Lemma 7. *Consider a pair $\{A_i, B_i\}$ in the WSPD, let c be the center of the bounding box $R(A_i)$ of A_i, and let C be a cone in $\mathcal{C}$ such that $R(B_i)$ overlaps the translated cone $c + C$. Let x be a point in $R^\lambda(A_i)$, let y be a point in $R^\lambda(B_i)$, and let r be a point that is contained in $x + \overline{C}$. Then $\angle(xy, xr) \leq \theta + 7\epsilon$.*

5   A Preliminary Algorithm

The input to the algorithm is a set S of n points in $\mathbb{R}^d$ and two real constants θ and ϵ such that $0 < \theta < \pi/3$ and $0 < \epsilon < (\pi - 3\theta)/(21 + \pi)$.

Let $\mathcal{C}$ be the collection of $O(1/\epsilon^{d-1}) = O(1)$ simplicial cones of angular diameter at most ϵ; see Section 4. Recall how we defined $\overline{C}$ for every cone C in $\mathcal{C}$. Let

$$s = \max\left(8\sqrt{d}, \frac{4\sqrt{d}}{\sin \epsilon}, \frac{4}{\sqrt{1+\epsilon} - 1}, 8 + \frac{16}{\epsilon}\right).$$

Step 1: Compute a WSPD $\{A_1, B_1\}, \ldots, \{A_m, B_m\}$ for S with separation constant s, where $m = O(n)$. For each i with $1 \leq i \leq m$, let ℓ_i be the distance between the centers of the bounding boxes $R(A_i)$ and $R(B_i)$. Sort the pairs in the WSPD according to the values of ℓ_i. Renumber the pairs so that $\ell_1 \leq \ell_2 \leq \ldots \leq \ell_m$.

Step 2: Initialize an empty edge set E.

Step 3: Process the pairs in the WSPD in increasing order of their indices. Let $\{A_i, B_i\}$ be the current pair to be processed.

1. Let c be the center of $R(A_i)$, let C be a cone in $\mathcal{C}$ such that $R(B_i)$ overlaps the cone $c + C$, and let $L(A_i)$ be the set of all points p in A_i such that the current edge set E does not contain any edge $\{p, r\}$ with $r \in p + \overline{C}$.
2. Let c' be the center of $R(B_i)$, let C' be a cone in $\mathcal{C}$ such that $R(A_i)$ overlaps the cone $c' + C'$, and let $L(B_i)$ be the set of all points q in B_i such that the current edge set E does not contain any edge $\{q, r\}$ with $r \in q + \overline{C'}$.

3. If both $L(A_i)$ and $L(B_i)$ are non-empty, choose an arbitrary point x in $L(A_i)$ and an arbitrary point y in $L(B_i)$, and add the edge $\{x, y\}$ to E.

Step 4: Return the graph $G = (S, E)$.

The next two lemmas state that this algorithm returns an angle-constrained spanner. Their proofs use the results in Sections 3 and 4.

We define $\alpha = \theta + 7\epsilon$ and $\lambda = 1$. Then, the conditions in (1) are satisfied and, thus, the results in Section 3 can indeed be applied. We define, for each i with $1 \leq i \leq m$, $R^\lambda(A_i) = R(A_i)$ and $R^\lambda(B_i) = R(B_i)$. By Lemma 1, the diameters of $R(A_i)$ and $R(B_i)$ are at most $2\sqrt{d}\ell_i/s$. Also, the restriction on the separation constant s in (2) is satisfied. Thus, the results in Section 4 can be applied.

Lemma 8. *The graph $G = (S, E)$ that is returned by the above algorithm is θ-angle-constrained.*

Proof. Consider an edge $\{x, y\}$ that is added to the edge set E during the processing of the pair $\{A_i, B_i\}$. During the processing of this pair, the algorithm chooses a cone C in $\mathcal{C}$ such that $R(B_i)$ overlaps $c + C$, where c is the center of $R(A_i)$. It follows from the algorithm that $x \in L(A_i)$ and $y \in L(B_i)$ and, therefore, $x \in A_i$ and $y \in B_i$. Furthermore, just before $\{x, y\}$ was added to E, there was no edge $\{x, r\}$ in E with $r \in x + \overline{C}$. It follows from Lemma 6 that $\{x, y\}$ makes an angle of at least θ with all edges incident on x that were previously added to the edge set E. $\square$

Lemma 9. *The graph $G = (S, E)$ that is returned by the above algorithm is a t-spanner, where*

$$t = \max\left(1 + \epsilon, \frac{1 + \sqrt{2(1 + \epsilon)(1 - \cos(\theta + 7\epsilon)) + \epsilon^2}}{2\cos(\theta + 7\epsilon) - 1 - \epsilon}\right).$$

Proof. We will prove that the assumption in Lemma 3 holds. Since, by our choice of s, $(s + 8)/(s - 8) \leq 1 + \epsilon$, this will imply the lemma. Let i be an integer with $1 \leq i \leq m$ and consider the iteration in which the pair $\{A_i, B_i\}$ is processed. Consider the sets $L(A_i)$ and $L(B_i)$ in Step 3. There are three possible cases.

Case 1: $L(A_i) = \emptyset$. We will show that property P.1 holds for the pair $\{A_i, B_i\}$. Recall from the algorithm that c is the center of $R(A_i)$ and C is a cone in $\mathcal{C}$ such that $R(B_i)$ overlaps the cone $c + C$.

Let p be an arbitrary point in A_i. Since $L(A_i) = \emptyset$, the edge set E contains an edge $\{p, r\}$ with $r \in p + \overline{C}$. Let q be an arbitrary point in B_i. By Lemma 7, we have $\angle(pq, pr) \leq \theta + 7\epsilon = \alpha$.

It remains to show that $|pr| \leq (1 + \epsilon)|pq|$. Let j be the index such that the edge $\{p, r\}$ was added to E during the processing of the pair $\{A_j, B_j\}$. Since this pair was processed before $\{A_i, B_i\}$, we have $\ell_j \leq \ell_i$. It follows from the algorithm that (i) $p \in A_j$ and $r \in B_j$ or (ii) $p \in B_j$ and $r \in A_j$. Combining this with Lemma 1, we obtain $|pr| \leq (1 + 4/s)\ell_j \leq (1 + 4/s)\ell_i \leq (1 + 4/s)^2|pq|$. By our choice of s, we have $(1 + 4/s)^2 \leq 1 + \epsilon$.

Case 2: $L(B_i) = \emptyset$. Using a symmetric argument, property P.2 holds for the pair $\{A_i, B_i\}$.

Case 3: Both $L(A_i)$ and $L(B_i)$ are non-empty. We will show that property P.3 holds for the pair $\{A_i, B_i\}$. Consider the edge $\{x, y\}$ that is added to E during the processing of the pair $\{A_i, B_i\}$. Observe that $x \in A_i$ and $y \in B_i$. Since, by Lemma 1, both A_i and B_i have diameter at most $(2/s)\ell_i$, property P.3 is satisfied.   □

A naive implementation of the algorithm has a running time which is proportional to $\sum_{i=1}^{m}(|A_i| + |B_i|)$. This summation can be as large as $\Theta(n^2)$.

We can improve the running time by maintaining, for each cone C in $\mathcal{C}$, a range tree RT_C storing all points p of S for which the current edge set E does not contain any edge $\{p, r\}$ with $r \in p + \overline{C}$. At the start of Step 3, each tree RT_C stores all points of S. Consider the iteration in Step 3 in which the pair $\{A_i, B_i\}$ is processed. Deciding whether $L(A_i)$ and $L(B_i)$ are both non-empty can be done by performing two range emptiness queries: one in RT_C with the bounding box $R(A_i)$ of A_i and the other in $RT_{C'}$ with the bounding box $R(B_i)$ of B_i. If the algorithm adds an edge $\{x, y\}$ to the edge set E, then x is deleted from all range trees $RT_{C''}$ for which $y \in x + \overline{C''}$, and y is deleted from all range trees $RT_{C''}$ for which $x \in y + \overline{C''}$.

Using a result of Mehlhorn and Näher [13], we obtain an overall running time of $O(n \log^{d-1} n)$. This result has, however, two drawbacks. First, the space requirement is $O(n \log^{d-1} n)$. Second, the algorithms in [13] do not work in the algebraic computation-tree model. Thus, even though the running time is $O(n \log n)$ in the case when $d = 2$, the $\Omega(n \log n)$ lower bound of [7] on the time to compute any spanner does not apply.

6   An Optimal Algorithm

Arya *et al.* [2] define, for each WSPD-pair $\{A_i, B_i\}$, the *dumbbell* D_i to be the geometric object consisting of the bounding boxes $R(A_i)$ and $R(B_i)$, together with the line segment joining the centers of these boxes. The two boxes $R(A_i)$ and $R(B_i)$ are called the *heads* of the dumbbell. The *length* of the dumbbell D_i is defined to be the distance between the centers of its heads. Thus, using our previous notation, the length of D_i is equal to ℓ_i.

In the algorithm of Section 5, it is crucial that the pairs $\{A_i, B_i\}$ (or, equivalently, the dumbbells D_i) are processed in non-decreasing order of their lengths; see Case 1 in the proof of Lemma 9. Assume that for any two dumbbells, either their four heads are pairwise disjoint, or one dumbbell is completely contained in the head of the other dumbbell. Then we can store the dumbbells in a "nesting tree" such that, for every dumbbell D_i, all dumbbells that are completely contained in either of its heads are stored in the subtree of D_i. In particular, each dumbbell in the subtree of D_i has a length that is less than ℓ_i.

During the processing of the dumbbells, each node v for which (i) its dumbbell has not been processed but (ii) all dumbbells in its subtree have been processed, stores $O(1)$ lists: Let D_i be the dumbbell stored at v. Then, for every cone C in

$\mathcal{C}$, the node v stores two lists $L_A(v, C)$ and $L_B(v, C)$. The list $L_A(v, C)$ stores all points $p \in A_i$ such that (i) p is in some dumbbell stored in the subtree of v and (ii) the current graph does not have any edge $\{p, r\}$ with $r \in p + \overline{C}$. The list $L_B(v, C)$ stores a subset of B_i and is similarly defined.

These lists allow us to implement Step 3 of the algorithm in $O(n)$ time, because processing the dumbbells in sorted order implies a bottom-up traversal of the nesting tree.

Unfortunately, in general, the dumbbells do not have this nesting property. Arya *et al.* [2], however, have shown that the dumbbells can be partitioned into $O(1)$ groups, such that the nesting property "almost" holds for each group: Each group can be stored in a "dumbbell tree" such that the following holds: For any dumbbell D_i, all dumbbells in its subtree are much shorter than D_i and very close to D_i. Thus, even though there may be a point p in the subtree of D_i that is not contained in either of its heads, p is still close to one of the heads. This implies that, by using a value of λ that is larger than 1, we can still apply the results in Section 3.

6.1   Dumbbell Trees

We will follow the exposition in Chapter 11 of [14]; in particular, refer to Section 11.9. Let $s > 1$ be the separation constant and consider the WSPD $\{A_1, B_1\}, \ldots,$ $\{A_m, B_m\}$ of the point S and the corresponding set $\mathcal{D}$ of dumbbells $D_1, \ldots, D_m$. Let β and γ be real numbers such that

$$0 < \beta < \min\left(\frac{1}{2}, \frac{s}{\sqrt{d}(s + 4)}\right), \tag{3}$$

$$6\beta + \frac{8\sqrt{d}}{s} < \min\left(\gamma, \frac{s}{s + 4}\right), \tag{4}$$

and

$$\beta\left(1 + 2\gamma + \frac{2\sqrt{d}}{s}\right) \leq \gamma < 1. \tag{5}$$

Let R_0 be a large box that contains all dumbbells of $\mathcal{D}$. We define a *dummy dumbbell* D_0 whose heads are R_0 and a translated copy R'_0 of R_0 such that the distance between the centers of R_0 and R'_0 (and, thus, the length ℓ_0 of D_0) is equal to $1/\beta$ times the maximum length of any dumbbell in $\mathcal{D}$.

In $O(n \log n)$ time, the set $\mathcal{D}$ of dumbbells can be partitioned into $k = O(1)$ groups $\mathcal{D}_1, \ldots, \mathcal{D}_k$, each of which can be stored in a *dumbbell tree*. For any index ℓ with $1 \leq \ell \leq k$, each node in the dumbbell tree T_ℓ is either

1. a *dumbbell node*, in which case it stores a dumbbell of the set $\mathcal{D}_\ell \cup \{D_0\}$,
2. a *head node*, in which case it stores a head of some dumbbell in the set $\mathcal{D}_\ell \cup \{D_0\}$, or
3. a *leaf*, in which case it stores a point of S.

The tree T_ℓ has the following properties:

1. Each dumbbell in $\mathcal{D}_\ell \cup \{D_0\}$ is stored at a unique dumbbell node, each head of each dumbbell in this set is stored at a unique head node, and each point of S is stored at a unique leaf.
2. The root is a dumbbell node and stores the dummy dumbbell D_0.
3. Each dumbbell node storing a dumbbell D_i, has two children, which are head nodes storing the heads of D_i.
4. Each head node v storing a head R of a dumbbell D_i, has dumbbell nodes and leaves as children:
   (a) If a child is a dumbbell node storing a dumbbell D_j, then $\ell_j \leq \beta \ell_i$ and D_j is within distance $\gamma \ell_j$ of the head R.
   (b) If a child is a leaf storing a point p, then $p \in R$ and p is not contained in any of the heads that are stored in the proper subtree v.

We mention the following property, which is Lemma 11.8.2 in [14]:

Lemma 10. *Consider a dumbbell tree T_ℓ. Let p be a point of S, let u be the leaf in T_ℓ that stores p, and let v be a head node in T_ℓ whose head contains p. Then u is in the subtree of v.*

In the rest of this section, we will take

$$s \geq 32\sqrt{d}, \ \beta = \min\left(\frac{1}{6\sqrt{d}}, \frac{1}{4s}\right), \text{ and } \gamma = 1/2. \tag{6}$$

It is not difficult to verify that the conditions in (3), (4), and (5) are satisfied.

The following lemma states that for every head node v, all points in the subtree of v are "close" to the head stored at v. This will allow us to use the results of Section 3 with the value $\lambda = 2$.

Lemma 11. *Let v be a head node of a dumbbell tree, let R be the head stored at v, and let D_i be the dumbbell stored at the parent of v (thus, R is a head of D_i). Let S_v be the set of all points in S that are stored at the leaves of the subtree of v. Then*

1. *the diameter of S_v is at most $4\ell_i/s$,*
2. *the bounding box of S_v has diameter at most $4\sqrt{d}\ell_i/s$.*

6.2   The Algorithm

We are now ready to present the final algorithm. The input is a set S of n points in $\mathbb{R}^d$ and real constants θ and ϵ such that $0 < \theta < \pi/3$ and $0 < \epsilon < (\pi - 3\theta)/(21 + \pi)$.

Consider again the collection $\mathcal{C}$ of $O(1/\epsilon^{d-1}) = O(1)$ simplicial cones of angular diameter at most ϵ; see Section 4. Let

$$s = \max\left(32\sqrt{d}, \frac{8\sqrt{d}}{\sin \epsilon}, \frac{8}{\sqrt{1+\epsilon}-1}, 16 + \frac{32}{\epsilon}\right).$$

The algorithm starts by computing a WSPD $\{A_i, B_i\}$, $1 \le i \le m = O(n)$, for S with separation constant s, the corresponding dumbbells $D_1, \ldots, D_m$, and the dumbbell trees $T_1, \ldots, T_k$, where $k = O(1)$.

We assume that the pairs in the WSPD have been sorted so that $\ell_1 \le \ell_2 \le \ldots \le \ell_m$. As in Section 5, the algorithm processes the dumbbells in this order. Observe that if a dumbbell has been processed, all dumbbells in its subtree have also been processed. The algorithm will maintain the following invariant:

Invariant: Every head node v of every dumbbell tree stores lists $L(v, C)$, where C ranges over all cones in $\mathcal{C}$. Let S_v be the set of all points in S that are stored at the leaves of v's subtree. If (i) the dumbbell stored at the parent of v has not been processed but (ii) all dumbbells in the subtree of v have been processed, then the list $L(v, C)$ stores all points p in S_v for which the current edge set E does not contain any edge $\{p, r\}$ with $r \in p + \overline{C}$.

Let p be a point in S and let C be a cone in $\mathcal{C}$. Each dumbbell tree has at most one head node v satisfying (i) and (ii). Therefore, p is stored in at most $k = O(1)$ lists $L(v, C)$. We assume that these lists are connected by cross-pointers so that, if we know the position of p in one list $L(\cdot, C)$, we can access p in all lists $L(\cdot, C)$ in which it occurs, in $O(1)$ total time.

Initialization:

1. For every head node v of every dumbbell tree, and for every cone C in $\mathcal{C}$, initialize an empty list $L(v, C)$.
2. For every leaf w of every dumbbell tree, and for every cone C in $\mathcal{C}$, add the point p to the lists $L(v, C)$, where p is the point stored at w and v is the parent of w.
3. Initialize an empty edge set E.

Processing the pair $\{A_i, B_i\}$: Let u be the dumbbell node storing D_i, and let v and w be the children of u storing the heads $R(A_i)$ and $R(B_i)$ of D_i, respectively.

1. Let c be the center of $R(A_i)$ and let C be a cone in $\mathcal{C}$ such that $R(B_i)$ overlaps the cone $c + C$.
2. Let c' be the center of $R(B_i)$ and let C' be a cone in $\mathcal{C}$ such that $R(A_i)$ overlaps the cone $c' + C'$.
3. If both $L(v, C)$ and $L(w, C')$ are non-empty, do the following:
   (a) Choose an arbitrary point x in $L(v, C)$ and an arbitrary point y in $L(w, C')$, and add the edge $\{x, y\}$ to E.
   (b) For each cone C'' in $\mathcal{C}$ for which $y \in x + \overline{C''}$, delete x from all lists $L(\cdot, C'')$ in which it occurs.
   (c) For each cone C'' in $\mathcal{C}$ for which $x \in y + \overline{C''}$, delete y from all lists $L(\cdot, C'')$ in which it occurs.
4. Let u' be the parent of u. For each cone C'' in $\mathcal{C}$, concatenate the lists $L(v, C'')$, $L(w, C'')$, and $L(u', C'')$, and rename the resulting list as $L(u', C'')$.

After all pairs have been processed, the algorithm returns the graph $G = (S, E)$. The next lemma states that this graph is an angle-constrained spanner. The proof is similar to those of Lemmas 8 and 9, and they use the results of Sections 3 and 4.

Lemma 12. *The graph $G = (S, E)$ that is returned by the above algorithm is a θ-angle-constrained t-spanner, where*

$$t = \max \left(1 + \epsilon, \frac{1 + \sqrt{2(1 + \epsilon)(1 - \cos(\theta + 7\epsilon)) + \epsilon^2}}{2\cos(\theta + 7\epsilon) - 1 - \epsilon} \right).$$

References

1. Althöfer, I., Das, G., Dobkin, D.P., Joseph, D., Soares, J.: On sparse spanners of weighted graphs. Discrete & Computational Geometry 9, 81–100 (1993)
2. Arya, S., Das, G., Mount, D.M., Salowe, J.S., Smid, M.: Euclidean spanners: short, thin, and lanky. In: STOC, pp. 489–498 (1995)
3. Arya, S., Smid, M.: Efficient construction of a bounded-degree spanner with low weight. Algorithmica 17, 33–54 (1997)
4. Bose, P., Carmi, P., Farshi, M., Maheshwari, A., Smid, M.: Computing the greedy spanner in near-quadratic time. Algorithmica (to appear)
5. Callahan, P.B., Kosaraju, S.R.: Faster algorithms for some geometric graph problems in higher dimensions. In: SODA, pp. 291–300 (1993)
6. Callahan, P.B., Kosaraju, S.R.: A decomposition of multidimensional point sets with applications to k-nearest-neighbors and n-body potential fields. Journal of the ACM 42, 67–90 (1995)
7. Chen, D.Z., Das, G., Smid, M.: Lower bounds for computing geometric spanners and approximate shortest paths. Discrete Applied Mathematics 110, 151–167 (2001)
8. Clarkson, K.L.: Approximation algorithms for shortest path motion planning. In: STOC, pp. 56–65 (1987)
9. Eppstein, D.: Spanning trees and spanners. In: Handbook of Computational Geometry, pp. 425–461. Elsevier Science, Amsterdam (2000)
10. Keil, J.M., Gutwin, C.A.: Classes of graphs which approximate the complete Euclidean graph. Discrete & Computational Geometry 7, 13–28 (1992)
11. Li, X.-Y.: Wireless Ad Hoc and Sensor Networks. Cambridge University Press, Cambridge (2008)
12. Lukovszki, T.: New Results on Geometric Spanners and Their Applications. Ph.D. thesis, University of Paderborn, Germany (1999)
13. Mehlhorn, K., Näher, S.: Dynamic fractional cascading. Algorithmica 5, 215–241 (1990)
14. Narasimhan, G., Smid, M.: Geometric Spanner Networks. Cambridge University Press, Cambridge (2007)
15. Salowe, J.S.: Euclidean spanner graphs with degree four. Discrete Applied Mathematics 54, 55–66 (1994)
16. Soares, J.: Approximating Euclidean distances by small degree graphs. Discrete & Computational Geometry 11, 213–233 (1994)

Approximating Minimum Bending Energy Path in a Simple Corridor*

Jinhui Xu, Lei Xu, and Yulai Xie

Department of Computer Science and Engineering
State University of New York at Buffalo
Buffalo, NY 14260, USA
{jinhui,lxu,xie}@buffalo.edu

Abstract. In this paper, we consider the problem of computing a minimum bending energy path (or MinBEP) in a simple corridor. Given a simple 2D corridor C bounded by straight line segments and arcs of radius $2r$, the MinBEP problem is to compute a path P inside C and crossing two pre-specified points s and t located at each end of C so that the bending energy of P is minimized. For this problem, we first show how to lower bound the bending energy of an optimal curve with bounded curvature, and then use this lower bound to design a $(1+\epsilon)$-approximation algorithm for this restricted version of the MinBEP problem. Our algorithm is based on a number of interesting geometric observations and approximation techniques on smooth curves, and can be easily implemented for practical purpose. It is the first algorithm with a guaranteed performance ratio for the MinBEP problem.

1 Introduction

In this paper, we consider the following minimum bending energy path (MinBEP) problem: Given a simple corridor C bounded by the straight line segments and arcs of radius $2r$ (a precise definition of simple corridor is given later) and two points s and t at each end of C, compute a minimum energy path P crossing s and t and traversing the corridor C. The energy of a smooth path P is measured by its bending energy E_b.

The bending energy was suggested by Bernoulli in 1738 and studied intensively for the case of planar curves by Euler [2]. Given a smooth parameterized planar curve s, its bending energy is defined as the integral of the squared curvature over the length of the curve, i.e., $E_b = \int_s \kappa^2 ds$, where κ is the curvature of the curve at point s. Let $s(t) = (x(t), y(t))$ be the parametrization of an arbitrary point on a smooth curve. Then the curvature [14] of the curve at $s(t)$ is given by $\kappa(t) = \frac{\dot{x}(t)\ddot{y}(t) - \dot{y}(t)\ddot{x}(t)}{(\dot{x}(t)^2 + \dot{y}(t)^2)^{3/2}}$. Intuitively, the curvature of a smooth curve at point $s(t)$ is defined as the inverse of its minimum curvature radius (i.e., the radius of the osculating circle at $s(t)$). The MinBEP problem considered in this paper is

---

* The research of this work was supported in part by National Science Foundation (NSF) through a CAREER award CCF-0546509 and a grant IIS-0713489.

O. Cheong, K.-Y. Chwa, and K. Park (Eds.): ISAAC 2010, Part I, LNCS 6506, pp. 328–339, 2010.
© Springer-Verlag Berlin Heidelberg 2010

motivated by applications in intervational procedures for cardiovascular surgery [12,13]. The guidewire problem can be formulated as a MinBEP problem in 3D. In this paper we consider the 2D case. The framework of our approach is readily applicable to the 3D case.

Due to the wide use of elastic curves in many fields, the minimum energy path problem has been extensively studied in the past. Most of the studies have focused on computing closed or knotted elastic curves with prescribed boundary conditions and fixed length [8,6,3,9]. Almost all of them are based on numerical techniques. For unknotted curves, several results are known for solving the guidewire problem [12,13]. Bending energy has also been used in computational biology for understanding the cellular processes [5].

In this paper, we consider the problem of approximating the 2D MinBEP problem. To fix the problems associated with existing solutions (e.g., [12,13]), we expect that the computed path (1) has a bending energy close to optimal, (2) is smooth at every point, and (3) can be computed efficiently. To achieve these goals, our strategy is to first solve a constrained version of the problem. In this constrained version, we require that the to-be-computed path P have a bounded curvature κ_{max}. Once we obtain solutions to the curvature bounded MinBEP problem, we can then perform a binary search on the maximum curvature to find the best solution to the MinBEP problem. Thus our focus is on the curvature bounded minimum bending energy path problem.

Thus our goal for the MinBEP problem is to obtain a good approximation solution. We first use a set of recursive shortest paths and local features of the corridor to establish a lower bound on the energy of an optimal solution as well as the possible energy distribution. With this lower bound, we then place various types of grid points to discretize the continuous space and guess the positions of the optimal path. We show that it is possible to obtain a $(1 + \epsilon)$-approximation from the discretized space. Our algorithm runs in near quadratic time and can be easily implemented for practical purpose. To our best knowledge, this is the first quality guaranteed solution to the MinBEP problem.

2   Preliminaries

Let D be a disk of radius r. A pipe is a rectangle with a fixed width of $2r$ and a non-fixed length e. An elbow is a sector with a fixed radius of $2r$ and a non-fixed center angle of $\pi - \theta$ for some $0 \le \theta \le \pi$. A corridor is a set of alternatively appearing pipes and elbows concatenated in a way that the radius of each elbow completely overlaps with the side of a neighboring pipe with length $2r$. A corridor C is simple if every pipe or elbow does not intersect the interior of any other pipe or elbow, and non-simple otherwise (see Figure 1). Clearly, there is no hole in a such defined simple corridor. Each elbow together with its two adjacent pipes forms an elbow region ER (see Figure 2).

Let $\eta = \{v_1, v_2, \ldots, v_n\}$ be the centerline of C. It is easy to see that η is a simple 2D curve, where the edge between every pair of consecutive vertices v_i and v_{i+1} is either a straight line segment or an arc of radius r which alternatively

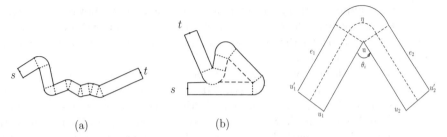

Fig. 1. (a) A simple corridor C; (b) A non-simple corri- **Fig. 2.** An elbow region
dor C' ER

appears along η. Thus C can also be viewed as the Minkowski sum $\eta \oplus D$ of η and D. The corridor C is bounded by straight line segments and arcs. Let C_u and C_l be the upper and lower boundary of C, separated from the two ends of C (Assume that the segment $\overline{v_1 v_n}$ is horizontally oriented.) Let $\{u_1, u_2, \dots, u_n\}$ and $\{w_1, w_2, \dots, w_n\}$ be the vertices of C_u and C_l respectively. Thus the edges between consecutive vertices in C_u and C_l are either segments or arcs. It is also easy to see that for each vertex in the boundary of C, if it is adjacent to two segments, then it is a reflex vertex (i.e., the inner angle is larger than π).

3   Lower Bound of the MinBEP

To design a quality guaranteed algorithm for the MinBEP problem, we need to know the minimum energy of an optimal solution, i.e., the lower bound of the Min-BEP path. We lower bound (1) the arc length of the portions of an optimal curve where the curvature is non-zero, (2) the curvatures at those portions. We first consider the problem of lower bounding (1). To estimate the curve length of (1), we consider the shortest path inside the corridor C. From [4], we know that the shortest path inside a simple polygon can be computed in linear time. (Note that although the corridor C is not a simple polygon, we can easily convert it into a simple polygon by replacing each arc with a convex polygonal curve around the arc.)

Let SP be the shortest path inside C with starting point s and ending point t. We assume that s and t are not visible to each other in C (since otherwise the segment $\overline{st}$ will be the trivial shortest path and also the minimum bending energy path in C). Clearly, SP consists of a set of straight line segments and has unbounded curvature. Let $V_{sp} = \{s_1, s_2, \dots, s_k\}$ be the set of interior vertices of SP (i.e., excluding s which is s_0 and t which is s_{k+1}) with the indices following their orders along SP.

Let $\acute{V}$ be the set of reflex vertices in the boundary ∂C of the corridor C. Clearly the boundary curve ∂C is non-differentiable at any point in $\acute{V}$. The following lemma shows that the shortest path SP makes turns only at reflex vertices.

Lemma 1. $V_{sp} \subseteq \acute{V}$.

Given any pair of consecutive vertices s_i and s_{i+1} on the shortest path SP, let s'_i and s'_{i+1} be the two closest (to $\overline{s_i s_{i+1}}$) intersection points of ∂C and the supporting line of segment $\overline{s_i s_{i+1}}$, where s'_i (or s'_{i+1}) is the intersection point closer to s_i (or s_{i+1}) than s_{i+1} (or s_i).

Let M be an arbitrary curvature-bounded smooth curve traversing the corridor C with the same starting and ending points s and t as SP, and κ_{max} be its maximum curvature. Let $M(t)$ be the parametrization of M with time t as the parameter, where $t_1 \leq t \leq t_2$ and $s = M(t_1)$ and $t = M(t_2)$ (i.e., we view the curve M as the loci of a point moving with unit speed).

Lemma 2. $\|M \cap \overline{s'_i s'_{i+1}}\| \geq 2$ for any $1 < i < k$ (i.e., M intersects $\overline{s'_i s'_{i+1}}$ at least twice and each intersection could be a consecutive portion of M overlapping with $\overline{s'_i s'_{i+1}}$).

Lemma 3. $\|M \cap \overline{ss'_1}\| \geq 2$ and $\|M \cap \overline{s'_k t}\| \geq 2$.

Let $s_i \in V_{sp}$ be any interior vertex of SP, and $\alpha_i = \angle \langle \overrightarrow{s_{i-1} s_i}, \overrightarrow{s_i s_{i+1}} \rangle$ be the minor angle between vectors $\overrightarrow{s_{i-1} s_i}$ and $\overrightarrow{s_i s_{i+1}}$. Clearly, $\angle \langle \overrightarrow{s_{i-1} s_i}, \overrightarrow{s_i s_{i+1}} \rangle = \pi - \angle s_{i-1} s_i s_{i+1}$. Let $r(s_i, s_{i-1})$ be the ray emitting from s_i and crossing s_{i-1}, and s'_{i-1} be the first intersection point of $r(s_i, s_{i-1}) \cap \partial C$ after passing s_{i-1}. Similarly, we have $r(s_i, s_{i+1})$ and s'_{i+1}. The two rays $r(s_i, s_{i-1})$ and $r(s_i, s_{i+1})$, together with the portion of ∂C on the opposite side of s_i form a closed region in C. We denote it as C_i, and call it the turning region of s_i. Let $e_{i,i-1}$ be the segment of $\overline{s_i s'_{i-1}}$ if s_i and s_{i-1} are on the same side of ∂C, and $\overline{s_i s_{i-1}}$ otherwise. Similarly, we have $e_{i,i+1}$. Clearly, $e_{i,i-1}$ and $e_{i,i+1}$ separate the turning region C_i from the remaining part of C.

Let x be any point in M with $x = M(t_x)$, $T_x = \dot{x} = M(t_x)$ be the unit tangent vector of M at point x (i.e., $\|T_x\| = \|\dot{x}\| = 1$), and κ_x be the curvature of M at x. To simplify our discussion, from now on we assume every vector is a unit vector (unless we specify otherwise).

Lemma 4. Let M be any planar smooth curve with non-zero curve length and $\triangle x_1 x_2 x_3$ be any triangle with $x_1 \in M$. Then there exists a point y in M but not in $\triangle x_1 x_2 x_3$.

Lemma 5. Let $s_i \in V_{sp}$ be any interior vertex of SP and M be any smooth curve in C (between s and t). There exist two points m_1 and m_2 in $M \cap C_i$ so that $\angle \langle T_{m_1}, T_{m_2} \rangle > \alpha_i$, where T_{m_1} and T_{m_2} are the unit tangent vectors of m_1 and m_2 respectively.

Next we consider the case that SP has only one or two interior points, i.e., $V_{sp} = \{s_1\}$ or $V_{sp} = \{s_1, s_2\}$.

Lemma 6. If $V_{sp} = \{s_1\}$ or $V_{sp} = \{s_1, s_2\}$, then $\exists m_1 \in M$, and $\exists m_2 \in M$, such that $\alpha_i < \angle \langle T_{m_1}, T_{m_2} \rangle$ for $i \in \{1, 2\}$.

Corollary 1. Let $s_i \in V_{sp}$ be any interior vertex of SP, and M be any smooth curve in C_i intersecting both $e_{i,i-1}$ and $e_{i,i+1}$ at a position other than s_i. Then there exist two points m_1 and m_2 in $M \cap C_i$ such that $\angle \langle T_{m_1}, T_{m_2} \rangle > \alpha_i$.

Lemma 7. *Let M be a smooth curve in C between s and t with maximum curvature of κ_{max}. Then the arc length of the portion of M with non-zero curvature in C_i is at least $\frac{\alpha_i}{\kappa_{max}}$.*

Let E_i be the bending energy of P in C_i. Then to lower bound E_i, we need to determine the curvature of this portion of P.

Lemma 8. *Among all smooth curves with an accumulated angle change of α_i, the curve with the same curvature at every point has the minimum bending energy.*

Definition 1. *An elbow in C is a short elbow if SP does not touch its reflex vertex u. An elbow region containing a short elbow is called a short elbow region.*

We have two cases about the turning region C_i: (A) C_i contains no short elbow, and (B) C_i contains at least one short elbow. We first consider case (A). As discussed above, C_i is bounded by segment $e_{i,i-1}$ and $e_{i,i+1}$. Let s''_{i-1} (s''_{i+1}) be the farthest endpoint of $e_{i,i-1}$ ($e_{i,i+1}$) away from s_i. Clearly, s''_{i-1} (s''_{i+1}) is same as s'_{i-1} (s'_{i+1}) if s_i and s_{i-1} (s_{i+1}) are on the same side of ∂C and s_{i-1} (s_{i+1}) otherwise. Thus we have four cases of C_i depending on whether $s''_{i-1} = s_{i-1}$ and $s''_{i+1} = s_{i+1}$. Let ER_i be the elbow region in C_i with reflex vertex $u = s_i$ and an outer angle of θ_i (see Figure 2). Let $\overline{uu_1}$ and $\overline{uu_2}$ be the two incident edges of u, e_1 and e_2 be the corresponding edges of $\overline{uu_1}$ and $\overline{uu_2}$ on the opposite side of ∂C, and u'_1 and u'_2 be the corresponding vertices of u_1 and u_2. Let $|e_i|$, $i \in \{1,2\}$, denote the length of e_i, and w denote the width of C. Clearly $|\overline{uu_1}| = |e_1|$, $|\overline{uu_2}| = |e_2|$, and $w = 2r$. To determine r_i, we need to first find B_{max} in C_i. Based on the relationships of B_{max} and ER_i, we have four subcases to consider.

A.1 B_{max} is tangent to both e_1 and e_2 at their interior points. B_{max} starts and ends at the two tangent points. In this case, s_i is incident to B_{max} due to the fact that B_{max} is the largest circular arc, the following conditions are satisfied. $\frac{w}{1-\sin\frac{\theta_i}{2}}\cos\frac{\theta_i}{2} < |e_1|$, and $\frac{w}{1-\sin\frac{\theta_i}{2}}\cos\frac{\theta_i}{2} < |e_2|$. Since $r_i = r_i \sin\frac{\theta_i}{2} + w$, we have

$$r_i = \frac{w}{1 - \sin\frac{\theta_i}{2}}. \tag{1}$$

A.2 B_{max} is tangent to e_1 only. In this case, s_i and s''_{i+1} are incident to B_{max}. B_{max} starts at the tangent point of e_1 and ends at s''_{i+1}.

A.3 B_{max} is tangent to e_2 only. This is the symmetric case of Case 2. In this case, s_i and s''_{i-1} are incident to B_{max}. B_{max} starts at s''_{i-1} and ends at the tangent point of e_2.

A.4 B_{max} is tangent to none of e_1 and e_2. This implies that s_i, s''_{i-1} and s''_{i+1} are all incident to B_{max}, and B_{max} starts at s''_{i-1} and ends at s''_{i+1}. Thus we have

$$r_i = \frac{1}{2\sin\theta_i}\sqrt{|e_{i,i-1}|^2 + |e_{i,i+1}|^2 - 2|e_{i,i-1}||e_{i,i+1}|\cos\theta_i}. \tag{2}$$

Since our goal is to lower bound E_i and $\cos \theta_i > -1$ (i.e., $0 < \theta_i < \pi$) in equation (2), we can choose

$$r_i = \frac{1}{2 \sin \theta_i} (|e_{i,i-1}| + |e_{i,i+1}|). \tag{3}$$

In each case of C_i, B_{max} is always inside C_i and the accumulative angle change of B_{max} is α_i (by Corollary 1). The following lemma follows directly from the above analysis.

Lemma 9. *In case (A), B_{max} has only the above 4 cases and r_i is a function of the local features (i.e., $w, \theta_i, |e_1|, e_2|, |e_{i,i-1}|$ and $|e_{i,i+1}|$) of C_i.*

Lemma 10. *r_i in case A.3 (or A.2) can be computed by using either equation (1) as in case A.1 or equation (3) as in case A.4.*

Lemma 11. *The optimal curve P in the turning region of C_i in case (A) has a bending energy*

$$E_i > E_i^l = \frac{\alpha_i}{\kappa_{max}} \kappa_i^2 = \frac{\alpha_i}{\kappa_{max}} \frac{1}{r_i^2}, \tag{4}$$

where E_i^l is the lower bound of E_i.

Now we consider case (B) (i.e., C_i contains at least one short elbow). We consider two subcases: (B.1) C_i contains exactly one short elbow, and (B.2) C_i contains more than one short elbows.

B.1: Let ER_k be the only short elbow region in C_i with reflex vertex k and k' be the middle point of the arc of ER_k. We first partition C_i, as well as C_{i+1}, into two parts by introducing the segment $\overline{kk'}$. Let C_i' and C_{i+1}' be the adjusted turning regions of s_i and s_{i+1} respectively. Obviously, $\overline{kk'}$ intersects the optimal curve P. Let p_k be the intersection point, and assume that the exact position of p_k is known in advance (in next section, we will show how to guess the positions of p_k). With p_k, we can then compute the shortest paths SP_{s_{i-1}, p_k} and $SP_{p_k, s_{i+1}}$ between s_{i-1} and p_k and between p_k and s_{i+1} respectively. This will give us more accurate α angles α_i', α_{i+1}' at s_i and s_{i+1}. Further, we can use a similar approach as in case (A) to derive better lower bounds of the bending energy for the portion of P in turning regions C_i' and C_{i+1}' respectively by using SP_{s_{i-1}, p_k} and $SP_{p_k, s_{i+1}}$. The only difference is that B_{max} ends (or starts) at p_k in C_i' (or C_{i+1}').

B.2: Let $\{ER_1, ER_2, \cdots, ER_m\}$ be the sequence of short elbow regions in C_i between s_{i-1} and s_{i+1}, with reflex vertices $\{k_1, k_2, \cdots, k_m\}$. Assume that we know p_m (same as p_k in case B.1) in advance. We can then compute shortest paths SP_{s_{i-1}, p_m}, $SP_{p_m, s_{i+1}}$, and $SP_{s_i, s_{i+1}}$ (if necessary), and adjust the turning regions for s_{i-1}, s_i and s_{i+1}. If there is some short elbow in any of the adjusted turning regions (based on the newly computed shortest paths), we can recursively apply the above strategy to adjust the turning regions and shortest paths until we eliminate all short elbows.

Lemma 12. *The bending energy of an optimal curve P is at least*

$$E_l = \frac{1}{2} \sum_{i=1}^{k} E_i^l. \tag{5}$$

where E_i^l is the lower bound on the bending energy of P in C_i.

4   Approximation Algorithm for Computing MinBEP

In this section, we present a $(1+\epsilon)$-approximation algorithm for the MinBEP problem with bounded curvature. Let E_b be the bending energy of an optimal curve P and E_a be the bending energy of the path P_A computed by our algorithm.

To achieve this, we first compute the shortest path SP in C and then isolate the task of approximating the optimal curve within each turning region (based on SP). For every turning region, we first sample a set of points as the possible positions for the optimal curve to either pass through or appear in their neighborhoods. Then we connect those sample points by smooth curves to form a network or graph G so that there exists at least one smooth curve in G which is geographically close (i.e. has a small Hausdorff or Fréchet distance) to the optimal curve P and differs in bending energy from P by only a small constant factor.

Lemma 13. *For any smooth curve P_c with bending energy E_c in the corridor C, after isotropic scaling with scale factor γ, the bending energy E_c' of the scaled curve P_c' is $\frac{1}{\gamma} E_c$.*

The above lemma suggests that isotropic scaling does not affect the performance ratio of an approximate path. Thus, to design a $(1 + \epsilon)$-approximation for the MinBEP problem, we can first choose an appropriate factor to scale the corridor C so that the width w becomes a large enough constant (depending on κ_{max}), and then compute a $(1 + \epsilon)$-approximating path P_A in the scaled corridor.

4.1   Sampling Points (Problem (1))

To sample the possible positions of the optimal curve P, we have two sub-problems to consider: (i) How to sample points inside a turning region C_i in case (A); (ii) How to sample points inside a turning region C_i in case (B).

To solve sub-problem (i), we first determine B_{max} which could span over at most three elbow regions. Let a and b be the two endpoints of B_{max}, and l_a and l_b be the two lines which cross a and b respectively and are orthogonal to the center line η. We denote the region of C bounded by l_a and l_b as φ_i (which could be either a subset or superset of C_i). From Lemma 11, we know that it is sufficient to focus only on φ_i when sampling points for C_i.

To sample points in φ_i, consider the elbow region ER_i in C_i with reflex point u (i.e., $s_i = u$). Let ER_i' be the extended elbow region of ER_i by enlarging or shortening the length of its two pipes so that φ_i is barely contained inside ER_i'. To simplify our discussion on sampling points for φ_i, we can first sample points in ER_i' and then throw away all points outside of φ_i to obtain the points for φ_i. Thus we will focus on ER_i' only.

Our idea for sampling points in ER_i' is to place a bended grid in ER_i' so that there always exists a set of grid points close to P. In a normal grid, the grid points are the intersections of a set of evenly spaced horizontal and vertical lines. In an elbow region, however, since the corridor is bended, our grid bends around the elbow. The "horizontal" lines in our grid are evenly spaced curves "parallel" to the center line η of the corridor C. The "vertical" lines in the two pipes are evenly spaced lines orthogonal to the boundary edges of the pipes, and the "vertical" lines in the elbow are evenly spaced rays emitted from the reflex point u. The intersections of the "horizontal" and "vertical" lines forms the set of grid points.

Let $\Delta s'$ be the distance between two neighboring "horizontal" lines and Δs be the maximum distance between two neighboring "vertical" lines. We choose $\Delta s' = \frac{2}{\kappa_{max}}$.

To determine the value for Δs, we only consider two types of C_i in cases A.1 and A.4, since case A.2 or A.3 can be reduced to one of them. More specifically, we let $\Delta s = \Delta sp$ for a pipe of ER_i' and $\Delta s = \Delta se$ for the elbow. Depending on the types of C_i, we have the following choices for Δs.

A.1 : $\Delta sp = \dfrac{w \cos \frac{\theta_i}{2}}{\kappa_{max} \alpha_i (1 - \sin \frac{\theta_i}{2})^3}$ and $\Delta se = \dfrac{w(\pi - \theta_i)}{\kappa_{max} \alpha_i (1 - \sin \frac{\theta_i}{2})^2}$.

A.4 : We have two subcases

A.4.1: If $|e_{i,i-1}| + |e_{i,i+1}| \leq \dfrac{2w}{1 - \sin \frac{\theta_i}{2}}$, then $\Delta sp = \dfrac{w}{\kappa_{max} \alpha_i \sin^2 \theta_i (1 - \sin \frac{\theta_i}{2})^3}$,

   and $\Delta se = \dfrac{w(\pi - \theta_i)}{\kappa_{max} \alpha_i \sin^2 \theta_i (1 - \sin \frac{\theta_i}{2})^2}$.

A.4.2: If $|e_{i,i-1}| + |e_{i,i+1}| > \dfrac{2w}{1 - \sin \frac{\theta_i}{2}}$, then $\Delta sp = \dfrac{(|e_{i,i-1}| + |e_{i,i+1}|)^3}{8 \kappa_{max} \alpha_i \sin^2 \theta_i w^2}$ and $\Delta se = \dfrac{(\pi - \theta_i)(|e_{i,i-1}| + |e_{i,i+1}|)^2}{4w \kappa_{max} \alpha_i \sin^2 \theta_i}$.

Our idea is to guess the position of P by placing grid points in the last short elbow ER_k in C_i (see case B.2). More specifically, let $\overline{kk'}$ be defined as in case B.1 in last section (i.e., the segment connecting the reflex vertex and the middle point of the arc). We place a set of grid points evenly spaced in $\overline{kk'}$ with a separating distance of $\Delta_s' = \frac{2}{\kappa_{max}}$. With the set of grid points, we can then recursively compute higher level shortest paths starting from (or ending at) each grid point to obtain the adjusted turning region as well as α_i angle for each reflex point in the set of shortest paths. For an adjusted turning region with multiple α_i values (from different grid points), we can use the same approach as in case (A) to obtain a set of densities (i.e., Δs) values of the grid and choose the densest one for its grid.

It is sufficient to find the grid point on $\overline{kk'}$ that maximizes α_i. (The maximum α_i means the densest grid points for the corresponding turning region.) Thus we choose the extreme point on $\overline{kk'}$ to compute α_i. Once we have this information, it can be reduced to case (A). In case (B.2), for each short elbow in C_i, we place

$\overline{kk'}$ with grid points as discussed above. For each $\overline{kk'}$, we compute the maximum α_i as in case (B.1) and thus choose a set of extreme grid points on each $\overline{kk'}$. Then sampling grid points in each adjusted turning region is one subcase of case (A). In this way, we could sample points inside a turning region C_i in case (B).

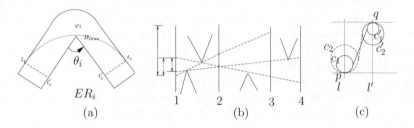

Fig. 3. (a)φ_i; (b) Weakly visible segments. (c)Connect circles by tangent line segments.

Our algorithm has to be able to produce a 0-energy path if P has 0 bending energy. We adopt two strategies to ensure this, (1) adding inter-elbow grid points and (2) adding inter-elbow edges in the graph G. We discuss our ideas on (1) next and (2) in next section.

Now we discuss how to generate inter-elbow grid points. This type of grid points is necessary. Let s_i and s_{i+1} be two consecutive interior vertices of the shortest path SP. Let $\{ER_1, ER_2, \cdots, ER_j\}$ be the sequence of short elbow regions between s_i and s_{i+1}, φ_i and φ_{i+1} be the corresponding grid region of s_i and s_{i+1}. If P has zero-curvature in those short elbow regions while the grid points are not dense enough in φ_i and φ_{i+1}, it is possible that the grid points in φ_i and φ_{i+1} are not visible to each other. Consequently, P_A will be forced to use a small amount of energy to pass each short elbow region and the total energy could be relatively large. Thus, it is difficult to guarantee the performance of P_A. Actually, this scenario may occur as long as any vertical line segment l_{v_i} in φ_i and any vertical line segment $l_{v_{i+1}}$ in φ_{i+1} are weakly visible to each other, but there is no grid point visible to each other. (Two segments l_{v_i} and $l_{v_{i+1}}$ are weakly visible to each other if there exists a pair of visible points on l_{v_i} and $l_{v_{i+1}}$ respectively.) Moreover, increasing the density of φ_i and φ_{i+1} does not solve the problem since the visible tunnel between l_{v_i} and $l_{v_{i+1}}$ could be very narrow (i.e., close to a line segment). To overcome this difficulty, our idea is to place inter-elbow grid points.

We first compute each shortest path map (using the algorithm in [11]) of the endpoints of l_{v_i} and endpoints of each vertical line l_k not in φ_i. For this purpose, we first convert C into a simple polygon C_P by replacing each arc with a convex polygonal curve around the arc. It is easy to see that expanding the corridor in this way does not affect the shortest maps of the endpoints of l_{v_i} and endpoints of l_k, since the shortest path between any pair of vertices will never go out of C. Then we determine the cusp vertices on the shortest path in constant time. Clipping l_{v_i} and l_k by the common tangent lines defined by these cusp vertices gives the portion e and e' (in constant time). Thus if the total number of vertical lines is $O(n)$, the worst running time of the above algorithm is $O(n^2)$. Moreover, since $\Delta s'$ is a constant, each vertical line l_{v_i} could be charged at most two inter-elbow grid points by simple amortized analysis even if the weakly visible portion is distinct for distinct pair of vertical lines. In Figure 3(b), there are four vertical line segments. Segment 1 is weakly visible by segment $2, 3, 4$ in different portions specified by dotted, dashed and solid arrow headed lines. We have the following lemma.

Lemma 14. *The total number of grid points is $O(n)$*

4.2 Connecting the Grid Points (Problem (2))

In this section, we discuss our idea on how to connect the set of grid points. To ensure the resulting path is smooth, for each grid point our idea is to first draw a set of circles tangent to the corresponding "horizontal" line (from both sides of the horizontal line) at the grid point with curvatures $0, \Delta \kappa, 2\Delta \kappa, \ldots, \kappa_{max}$ (the set of circles are said to be associated with the grid point), and then connect every pair of circles generated at different grid points by a tangent segment (i..e, a segment tangent to both circles) which is fully inside C.

To determine $\Delta \kappa$, consider Figure 3(c). Let l and l' be two consecutive "vertical" lines of the grid, and p be a grid point on the vertical line l associated with two circles c_1 and c_2 with curvature κ_1 and κ_2 respectively and satisfying the condition $|\kappa_1 - \kappa_2| \le \Delta \kappa$. Let q be a grid on l' associated with two circles c'_1 and c'_2 in a similar way. Let ΔE_{κ_1} (ΔE_{κ_2}) be the bending energy for a smooth curve using part of c_1 (c_2) and c'_1 (c'_2) to travel through the portion of C between l and l'. Let e_n be the total number of vertical lines of the grid passed by P in φ_i. Our idea is to choose $\Delta \kappa$ such that the following inequality holds.

$$e_n |\Delta E_{\kappa_1} - \Delta E_{\kappa_2}| \le \frac{1}{2} \epsilon E_i^l. \tag{6}$$

The idea behind the above inequality is to ensure that the difference of bending energy between any pair of smooth curves that travel through the region φ_i with a curvature difference no more than $\Delta \kappa$ is within a ϵ-factor. To find the exact value of $\Delta \kappa$, we consider case $A.1$ and $A.4$ since other cases are similar. In case $A.1$, it is easy to see that $e_n = (\frac{2w \cos \frac{\theta_i}{2}}{(1 - \sin \frac{\theta_i}{2}) \Delta sp} + \frac{w(\pi - \theta_i)}{\Delta se})$. Thus, $|\Delta E_{\kappa_1} - \Delta E_{\kappa_2}|$ satisfies

$$|\Delta E_{\kappa_1} - \Delta E_{\kappa_2}| \le 2(\frac{\kappa_1^2 \pi}{\kappa_1} - \frac{\kappa_2^2 \pi}{\kappa_2} + \frac{\kappa_1'^2 \pi}{\kappa_1'} - \frac{\kappa_2'^2 \pi}{\kappa_2'}) \le 4\Delta \kappa \pi. \tag{7}$$

By plugging (7) into (6), we have $\Delta \kappa \le \frac{\epsilon}{24\pi \kappa_{max}^2 w^2}$. Thus we can choose $\Delta \kappa = \frac{\epsilon}{24\pi \kappa_{max}^2 w^2}$. We obtain the same result for case $A.4$. The lemma below follows from Lemma 14 and the fact that $\Delta \kappa$ is a constant.

Lemma 15. *The total number of circles in C is $O(\frac{n}{\epsilon})$.*

Now we discuss our idea on how to connect the circles associated with different grid points. Let c_1 and c_2 be two circles associated with different grid points. c_1 and c_2 may be (1) in the same elbow region or (2) in different elbow regions. In (2), we have to add inter-elbow edges into G to ensure that the path we computed has 0 energy in the region if P has 0 energy in the same region. To connect c_1 and c_2, we first compute the four possible tangent segments between c_1 and c_2. Since only those tangent segments inside C are generated, we have to determine the visibility of the endpoints of each tangent segment. This implies that we need to check the visibility for the endpoints of every tangent segment.

To solve this problem more efficiently, we treat each pair of the endpoints (i.e., tangent points) as a visibility query, and build a data structure to answer those queries. Our idea is to use the optimal data structure developed in [15]. In this paper, a data structure was presented for solving the ray shooting problem inside a simple polygon PO with n vertices. In particular, it was shown that the data structure of size n can be preprocessed in $O(n \log n)$ time, and for each pair (q, u) of point q and direction u, it finds in $O(\log n)$ time the first edge of PO intersected by the ray emitting from q moving in the direction of u. To make use of this data structure, we first convert the corridor into a simple polygon C_P and then build the data structure for C_P. Thus, for each tangent segment, we can determine whether it is an interior tangent segment of C in $O(\log n)$ time.

The tangent points on each circle c_i partitions c_i into a linear number (in terms of the number of tangent points) of arcs. Some arcs may go outside of the corridor. Since each circle is drawn using the local feature of the elbow region, we can determine whether it is entirely inside C in $O(1)$ time.

Lemma 16. *The total number of interior arcs and interior tangent segments is $O(\frac{n^2}{\epsilon^2})$ in the worst case.*

With the set of interior tangent segments and arcs, we can now construct the graph G. In this graph, the vertices are the set of interior arcs. Each arc has an edge to its two neighboring arcs in the same circle. For two arcs in different circles, there is an edge between them if and only if the following two conditions are met: (i) There is an interior tangent segment connecting them (i.e., the tangent segment is between a pair of endpoints of the two arcs); (ii) The two arcs and the tangent segment together form a smooth curve. Thus there are two types of edges in G, tangent segments and edges connecting neighboring arcs in the same circle. Since the number of the second types of edges is no more than that of the arcs, the total number of edges is still bounded by $O(n^2/\epsilon^2)$. Also note that in the above visibility query computation, we treat s and t as circles of radius 0. Thus s and t are vertices of G and are connected to other vertices by tangent segments. The weight of each edge is 0 and the weight of each vertex is the bending energy of the corresponding arc. The lemma below follows directly from Lemma 16 and the above construction of the graph.

Lemma 17. *In graph G, $|V| = O(\frac{n^2}{\epsilon^2})$ and $|E| = O(\frac{n^2}{\epsilon^2})$ in the worst case. Furthermore, for any path in G, its corresponding path is a smooth curve in C.*

4.3 The Existence of a Smooth Curve with Similar Bending Energy (Problem 3)

Lemma 18. *There exists a curve P_s in G close to P with bending energy $E_s \leq (1 + \epsilon)E_b$, where E_b is the bending energy of P.*

The theorem below follows from all previous lemmas and the shortest path algorithm (e.g., Dijkstra's algorithm) running in $O(V \log V + E)$ time.

Theorem 1. *There exists a $(1 + \epsilon)$-approximation algorithm for the curvature bounded MinBEP problem with worst case running time $O((\frac{n}{\epsilon})^2 \log(\frac{n}{\epsilon}))$.*

Acknowledgments. The authors would like to thank Professor Joseph S.B. Mitchell for helpful discussion on this problem.

References

1. Dubins, L.: On Curves of Minimal Length with a Constraint on Average Curvature, and with Prescribed Initial and Terminal Positions and Tangents. American Journal of Mathematics 79, 497–516 (1957)
2. Euler, L.: Methodus inveniendi lineas curvas maximi minimive propriatate gaudentes, sive solutio problematis isoperimetrici lattisimo sensu accepti. Bousquet, Lausannae et Genevae, E65A. O.O. SER. I, vol. 24 (1744)
3. Freedman, M., He, Z., Wang, Z.: On the Mobius Energy of Knots and Unknots. Annals of Mathematics 139(1), 1–50 (1994)
4. Guibas, L., Hershberger, J., Leven, D., Sharir, M., Tarjan, R.: Linear Time Algorithm for Visibility and Shortest Path Problems Inside Simple Polygons. In: Proceedings of the Second Annual ACM-SIAM Symposium on Computational Geometry (1986)
5. Heekeren, R., Faas, F., Vliet, L.: Finding the Minimum-Cost Path Without Cutting Corners. In: Ersbøll, B.K., Pedersen, K.S. (eds.) SCIA 2007. LNCS, vol. 4522, pp. 263–272. Springer, Heidelberg (2007)
6. Ivey, T., Singer, D.: Knot Types, Homotopies and Stability of Closed Elastic Rods. Proceedings of the London Mathematical Society 79, 429–450 (1999)
7. Langer, J., Singer, D.: Knotted Elastic Curves in R^3. Journal of London Math. Society 30(2), 512–520 (1984)
8. von der Mosel, H.: Minimizing the Elastic Energy of Knots. Journal of Asymptotic Analysis 18(1-2), 49–65 (1998)
9. O'Hara, J.: Energy of a Knot. Topology 30, 241–247 (1991)
10. Pressley, A.: Elementary Differential Geometry. Springer, London (2001)
11. Sack, J.R., Urrutia, J.: Handbook of computational geometry, pp. 854–855. North-Holland Publishing Co., Amsterdam (2000)
12. Schafer, S., Singh, V., Hoffmann, K., Noël, P., Xu, J.: Planning image-guided endovascular interventions: guidewire simulation using shortest path algorithms. In: Proceedings of the SPIE, vol. 6509 (2007)
13. Schafer, S., Singh, V., Noël, P.B., Xu, J., Walczak, A., Hoffmann, K.R.: Real Time Endovascular Guidewire Position Simulation using Shortest Path Algorithms. International Journal of Computer Aided Radiaology and Surgery (IJCARS) 4, 597–608 (2009)
14. Thorpe, J.: Elementary Topic in Differential Geometry. Springer, Heidelberg (1979)
15. Chazelle, B., Guibas, L.J.: Visibility and intersection problems in plane geometry. Discrete and Computational Geometry 4(6), 551–581 (1989)

Analysis of an Iterated Local Search Algorithm for Vertex Coloring

Dirk Sudholt[1,2] and Christine Zarges[3]

[1] International Computer Science Institute, Berkeley, CA 94704, USA
[2] Cercia, University of Birmingham, Birmingham B15 2TT, UK
[3] Technische Universität Dortmund, 44221 Dortmund, Germany

Abstract. Hybridizations of evolutionary algorithms and local search are among the best-performing algorithms for vertex coloring. However, the theoretical knowledge about these algorithms is very limited and it is agreed that a solid theoretical foundation is needed. We consider an iterated local search algorithm that iteratively tries to improve a coloring by applying mutation followed by local search. We investigate the capabilities and the limitations of this approach using bounds on the expected number of iterations until an optimal or near-optimal coloring is found. This is done for two different mutation operators and for different graph classes: bipartite graphs, sparse random graphs, and planar graphs.

1 Introduction

The vertex coloring problem is one of the most fundamental and most difficult combinatorial problems. Given an undirected graph G one is looking for an assignment of colors to the vertices of G such that no two adjacent vertices share the same color and the number of different colors used is minimized. The problem has many applications in scheduling and for various allocation tasks.

Vertex coloring is known to be NP-hard even for planar graphs of maximum degree 4 (see, e. g., [1], page 91). It also remains NP-hard when certain properties of the graph are known. For instance, despite knowing that a graph is 3-colorable, it is still NP-hard to find a 4-coloring. More generally, coloring a k-colorable graph with $k^{(\log k)/25}$ colors is NP-hard for a large enough constant k [2]. It is therefore not surprising that the running time of the best known exact algorithms is exponential in the number of vertices.

This hardness has led many researchers and practitioners to resort to heuristic algorithms. This includes greedy algorithms, local search, simulated annealing, tabu search, evolutionary algorithms and ant colony optimization. A comprehensive survey is given by Galinier and Hertz [3]. These algorithms are also the method of choice in settings where it is impossible to design a custom-tailored algorithm; for instance, when the problem is only given as a black box [4,5].

Evolutionary algorithms maintain a population (multi-set) of candidate solutions and iteratively apply operators like mutation, crossover of multiple solutions, and selection. For vertex coloring hybridizations of evolutionary algorithms with local search strategies turned out to be highly successful [6,7,8,9]. These

O. Cheong, K.-Y. Chwa, and K. Park (Eds.): ISAAC 2010, Part I, LNCS 6506, pp. 340–352, 2010.

algorithms are often called *memetic algorithms*; in cases where the population only contains a single solution and no crossover is used one often speaks of *iterated local search* [10].

Despite many empirical studies, the theoretical foundation of hybrid evolutionary algorithms is still in its infancy. It is widely agreed that a solid theoretical foundation of heuristics is needed [4,5]. For plain evolutionary algorithms several theoretical analyses have appeared in the last 10–15 years, using methods from the analysis of randomized algorithms (see, e. g., the survey by Oliveto, He, and Yao [11]). Hybrid evolutionary algorithms are, in general, harder to analyze. Sudholt [12,13] analyzed the impact of the parametrization of local search in memetic algorithms. In [14] the author presented simply structured instances of combinatorial problems—Mincut, Knapsack, and Maxsat—where a hybrid algorithm finds global optima in expected polynomial time while many other heuristics need exponential time, with very high probability.

In this work we propose an iterated local search algorithm for vertex coloring and accompany it with theoretical results on its performance. The algorithm works on the set of feasible colorings. First, one out of two mutation operators is applied. These operators are based on so-called *Kempe chains* [15] and they are guaranteed to maintain feasibility. Then a local search is applied to the new solution, resulting in a *Grundy coloring*, that is, a coloring where every vertex has the smallest possible color. The new coloring replaces the previous coloring if it uses a larger number of smaller colors.

The precise algorithm and its components are formally defined in Section 2. In Sections 3, 4, and 5 we derive bounds on the asymptotic expected time until optimal or near-optimal colorings are found for several graph classes: bipartite (i. e. 2-colorable) graphs, sparse random graphs, and planar graphs. Our results show the capabilities and the limitations of iterated local search and contribute to a theoretical foundation of randomized search heuristics.

1.1   Related Work

Brockhoff [16] analyzed a simple evolutionary algorithm with a population of size 1 for vertex coloring. The search space contains feasible and infeasible colorings. The mutation operator selects a new color for each vertex independently with probability $1/n$, n the number of vertices. The new solution replaces the old one if the number of conflicting edges has not increased. This algorithm is able to color star graphs with two colors in an expected number of $O(n \log n)$ generations. However, it needs exponential time for 2-coloring binary trees.

Fischer and Wegener [17] considered a related problem as a test bed for evolutionary algorithms where the goal is to maximize the number of monochromatic edges. A simple evolutionary algorithm with populations size 1 finds a monochromatic coloring for an even ring in an expected number of $O(n^3)$ generations. In contrast, an evolutionary algorithm with population size 2 and crossover succeeds in only $O(n^2)$ generations, in expectation. Sudholt [18] presented a similar result: on binary trees the use of crossover can decrease an exponential expected time for finding a 2-coloring towards an expected number of $O(n^3)$ generations.

With a straightforward transformation, this result transfers to the problem of minimizing the number of conflicting edges when trying to 2-color binary trees.

Vertex coloring is also a very important topic in distributed computing as it is closely related to the problem of symmetry breaking. Results include 6-coloring planar graphs [19], 2-coloring bipartite graphs [20], and $(\Delta+1)$-coloring arbitrary graphs [21,22] where Δ is the maximum degree of the graph. This line of research relates to our investigations as in both cases iterative improvement strategies are used to find good or optimal colorings.

2   Preliminaries

Let $G = (V, E)$ denote an undirected graph with vertices V and edges E. Abbreviate $n := |V|$ and $m := |E|$. A coloring of G is an assignment of a color value from $\{1, \ldots, n\}$ for each vertex. Let $\deg(v)$ be the degree of a vertex v and $c(v)$ be its color in the current coloring.

There are several ways to formulate the optimization goal for a search heuristic based on iterative improvements. One approach is to allow feasible and infeasible colorings and to minimize the number of conflicts. The other approach is to restrict the search space to feasible colorings and to use operators that maintain feasibility throughout the search. In this study we follow the latter approach and only deal with feasible solutions, except for the initialization.

We first explain what we consider to be an improvement. The following relation among two solutions clarifies which solution should be preferred in the selection step. The basic idea is to guide a heuristic towards decreasing the number of conflicts for infeasible solutions. For feasible colorings we give hints towards gathering colors with small values.

Definition 1. *For x, y we say that x is better than y and write $x \succeq y$ iff*

- *The number of conflicting edges with x is smaller than with y or*
- *x and y have an equal number of conflicting edges and the following holds. Let $n_i(x)$ be the number of i-colored vertices in x, $1 \leq i \leq n$, and $n_i(y)$ be defined analogously. Then we require either $n_i(x) = n_i(y)$ for all i or $n_i(x) < n_i(y)$ for some i and $n_j(x) = n_j(y)$ for all $i < j \leq n$.*

Note that in this relation decreasing the number of vertices with the currently highest color (and not introducing yet a higher color) yields an improvement. If this number decreases to 0, the number of colors has decreased.

Next we describe the local search operator used in the sequel. A vertex v is called a *Grundy vertex* if v has the smallest color value not taken by any of its neighbors, formally $c(v) = \min\{i \in \{1, \ldots, n\} \mid \forall w \in \mathcal{N}(v) : c(w) \neq i\}$, where $\mathcal{N}(v)$ denotes the neighborhood of v. A coloring is called a *Grundy coloring* if all vertices are Grundy vertices [23]. Note that a Grundy coloring is always feasible. The following local search due to Hedetniemi, Jacobs, and Srimani [21] that we call *Grundy local search* always produces a Grundy coloring.

Algorithm 1. Grundy local search [21]

1: **while** the current coloring is not a Grundy coloring **do**
2:     Choose a non-Grundy vertex v.
3:     Set $c(v) := \min\{i \in \{1, \ldots, n\} \mid \forall w \in \mathcal{N}(v) \colon c(w) \neq i\}$.

In [21] it is proved that the algorithm finishes with a Grundy coloring within $n + 2m$ steps. The color of a vertex can only increase when it is recolored the first time and only in the presence of conflicts; afterwards or without conflicts, the color of a vertex can only decrease.

It is easy to see that the application of Grundy local search can never worsen a coloring. If y is the outcome of Grundy local search applied to x then $y \succeq x$. If x contains a non-Grundy node then y is strictly better, i.e., $y \succeq x$ and $x \not\succeq y$.

Our mutation operators are based on so-called *Kempe chain* [15] moves. The idea is to perform a chained sequence of operations that exchange two colors on a connected set of vertices. Such a set is described as follows.

Definition 2. *Consider a feasible coloring. A set $S \subseteq V$ is called an ij-set if S only contains vertices colored i or j. An ij-path is an ij-set whose vertices form a simple path, i.e., the colors i and j alternate on the path.*

By H_{ij} we denote the set of all vertices colored i or j in G. Then $H_j(v)$ is the connected component of $H_{c(v)j}$ that contains v.

The Kempe chain operator is applied to a vertex v and it exchanges the color of v (say i) with a specified color j. We restrict the choice of j to the set $\{1, \ldots, \deg(v) + 1\}$ since larger colors will be replaced in the following Grundy local search. In the connected component $H_j(v)$ the colors i and j of all vertices are exchanged. As no conflict within $H_j(v)$ is created and $H_j(v)$ is not neighbored to any vertex colored i or j, Kempe chains preserve feasibility.

Algorithm 2. Kempe chain

1: Choose $v \in V$ and $j \in \{1, \ldots, \deg(v) + 1\}$ uniformly at random.
2: **for** all $u \in H_j(v)$ **do**
3:     **if** $c(u) = c(v)$ **then** $c(u) := j$ **else** $c(u) := c(v)$.

We also define an alternative mutation operator called a *color elimination*. The idea is to eliminate a smaller color i in the neighborhood of a vertex v in one shot by trying to recolor all these vertices with another color j.

Algorithm 3. Color elimination

1: Choose $v \in V$ uniformly at random.
2: **if** $c(v) \geq 3$ **then**
3:     Choose $i, j \in \{1, \ldots, c(v) - 1\}$, $i \neq j$, uniformly at random.
4:     Let $v_1, \ldots, v_\ell$ enumerate all i-colored neighbors of v.
5:     **for** all $u \in H_j(v_1) \cup \cdots \cup H_j(v_\ell)$ **do**
6:         **if** $c(u) = i$ **then** $c(u) := j$ **else** $c(u) := i$.

A color elimination is, in fact, a union of several Kempe chain moves. If all j-colored neighbors $w_1, \ldots, w_r$ of v are not contained in $H_j(v_1) \cup \cdots \cup H_j(v_\ell)$, then

$v_1, \ldots, v_\ell$ as well as $w_1, \ldots, w_r$ are colored j and color i is eliminated from the neighborhood of v. Observe that the operator is not symmetric, i. e., swapping the values of the variables i and j in general leads to a different result.

Now we are ready to define our iterated local search algorithm, shortly called ILS. It repeatedly uses mutation followed by Grundy local search. At this point the mutation operator is not specified but regarded as a black box. In the initialization every vertex v receives a random color within $\{1, \ldots, \deg(v) + 1\}$; note that this range of colors is sufficient for every Grundy coloring.

Algorithm 4. Iterated local search (ILS)

1: Create an initial coloring by choosing for each vertex v a color in $\{1, \ldots, \deg(v)+1\}$ independently and uniformly at random.
2: Replace x by the result of Grundy local search applied to x.
3: **repeat forever**
4:     Let y be the result of a mutation operator applied to x.
5:     Let z be the outcome of Grundy Local Search applied to y.
6:     If $z \succeq x$ then $x := z$.

The *Grundy number* (also called Grundy chromatic number or First-Fit chromatic number) of a graph G, denoted by $\chi_G(G)$ is the maximum number of colors used by any Grundy coloring of G [24]. After the first iteration ILS finds a χ_G-coloring, regardless of the mutation operator used. This shows that the memetic algorithm inherits the good performance of the local search used.

As time measure we consider the number of generations, defined as the number of iterations in the outer loop of ILS. However, we keep in mind that Grundy local search runs up to $n + 2m$ iterations in its inner loop. This time must be accounted for when looking at more detailed notions of time.

3   Bipartite Graphs

We first investigate the performance of ILS with both mutation operators on bipartite graphs, i. e., a graph G whose vertices can be divided into two disjoint sets V_1 and V_2, such that for all edges $e = (u, v) \in E$ we have $u \in V_1$ and $v \in V_2$ or vice versa. W. l. o. g. assume $|V_1| \geq |V_2|$. Bipartite graphs are 2-colorable. We start with ILS with Kempe chains and see that it 2-colors a simple subclass of bipartite graphs, namely the even ring, efficiently.

Theorem 1. *On an even ring with n vertices ILS with Kempe chains finds a feasible 2-coloring in $3n/4$ expected generations.*

Proof. The initial Grundy local search yields a Grundy coloring with at most 3 colors. In such a coloring, every 3-colored vertex is surrounded by a 1-colored vertex and a 2-colored vertex. As the 2-colored vertex must be adjacent to a 1-colored vertex, the coloring is a concatenation of patterns 1-2-3, 1-3-2, and 1-2. As color 3 only appears once in patterns of length 3, the number of 3-colored vertices in any Grundy coloring is even and bounded by $n/3$. Recall that this number cannot increase due to the selection criterion.

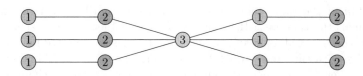

Fig. 1. Worst-case graph T_n for $n = 13$; the tree is rooted at the 3-colored vertex

We claim that every Kempe chain move recoloring 1- and 2-colored vertices reduces the number of 3-colored vertices by 2. Consider the component $H_{1,2}$ affected by the Kempe chain move and note that $H_{1,2}$ connects two 3-colored vertices u and v (if any). After the Kempe chain move, u and v each have monochromatic neighborhoods, hence they receive a smaller color in the following Grundy local search. This decreases the number of 3-colored vertices by 2.

Together, at most $n/6$ specific Kempe chain moves suffice to find a 2-coloring. The probability for one such move is at least 2/9 as the probability of selecting a 1- or 2-colored vertex is at least 2/3, and the probability to select j accordingly is 1/3. The expected number of generations is therefore bounded by $3n/4$.   □

While the even ring can be 2-colored efficiently by ILS with Kempe chains, this is not possible for bipartite graphs in general as already trees can lead to an exponential optimization time.

Theorem 2. *For every odd value of n there is a tree T_n with n vertices such that ILS with Kempe chains needs at least 2^{cn} generations, $c > 0$ an appropriate constant, for finding a 2-coloring, with probability $1 - 2^{-\Omega(n)}$.*

Proof. Choose T_n as a tree where the root has $N := (n-1)/2$ children and every child has exactly one leaf (cf. the example in Figure 1). We first prove that with probability $\Omega(1)$ after the first Grundy local search we have a feasible coloring where the root is 3-colored, all leaves and their parents have colors 1 or 2, and the number of 1-colored leaves is in between $N/8$ and $N - N/8$.

At initialization the probability that a specific leaf receives color 1 and its parent receives color 2 is 1/6. The expected number of leaves for which this holds is hence $N/6$. A symmetric argument applies to 2-colored leaves with 1-colored parents. By Chernoff bounds and the union bound, with probability $\Omega(1)$ the root receives a color at least 3, at least $N/8$ leaves are 1-colored with 2-colored parents, and at least $N/8$ leaves are 2-colored with 1-colored parents. All these leaves and their parents are Grundy vertices and will remain so until the end of local search. As all leaves and parents will be assigned colors 1 and 2, the root will eventually become 3-colored. This proves the claim.

After initialization each Kempe chain either creates search points that are worse than their parents or it exchanges the colors of a leaf and its parent. The latter either increases or decreases the number of 1-colored leaves by 1. The following Grundy local search will not alter this coloring unless all leaves have obtained the same color. The expected time until a 2-coloring is found hence equals the time until the number of 1-colored leaves has reached a value of either 0 or N. As all pairs of leaf and parent nodes are selected for the Kempe chain move with the same probability, this corresponds to an Ehrenfest model.

Given a generation where the number of 1-colored leaves is changed, whenever the number of 1-colored leaves is in $[1, N/8]$ the (conditional) probability of decreasing this number is at most $1/8$ and the (conditional) probability of increasing it is at least $7/8$. By the gambler's ruin problem, when starting with a value of $N/8$ the probability of reaching a value of 0 before returning to $N/8$ is $2^{-\Omega(n)}$. The situation at the other end of the scale is symmetric. By the union bound, the probability that either value 0 or value N is reached within 2^{cn} generations, $c > 0$ a sufficiently small constant, is still $2^{-\Omega(n)}$. □

We have seen that ILS with Kempe chains is not able to 2-color bipartite graphs efficiently as it is often not able to remove a certain color from the neighborhood of a node v which is necessary for assigning a smaller color to v. This is different for ILS with color eliminations.

Theorem 3. *On every bipartite graph $G = (V, E)$ ILS with color eliminations finds a feasible 2-coloring in $O(n^2 \log n)$ expected generations.*

Proof. Let V_1, V_2 with $|V_1| \geq |V_2|$ be the bipartition of G. The initial Grundy local search yields a Grundy coloring with at most $|V_1| + 1$ colors. Consider an arbitrary vertex v and two color values $i, j < c(v)$. Let $v_1, \ldots, v_\ell$ be all i-colored and $w_1, \ldots, w_r$ be all j-colored neighbors of v. Obviously, the color i can be eliminated from the neighborhood of v by a color elimination if $w_1, \ldots, w_r \notin H_j := H_j(v_1) \cup \ldots \cup H_j(v_\ell)$. This property holds for a feasible coloring on bipartite graphs: W. l. o. g. let $v \in V_1$ and thus $v_1, \ldots, v_\ell \in V_2$ (i-colored) and $w_1, \ldots, w_r \in V_2$ (j-colored). As the coloring is feasible only vertices $u_i \in V_1$ with $c(u_i) = j$ can be contained in H_j. Otherwise there would be an edge with both end points colored j. Hence, all possible color elimination moves will eliminate color i from the neighborhood of v.

Assume ℓ to be the highest color used in the current coloring. Eliminating the color i from the neighborhood of one of these vertices enables the following Grundy local search to reduce the number of ℓ-colored vertices by one by recoloring the vertex with color i. Since $i, j < \ell$ this is an improvement.

If there are $n_\ell \leq n$ ℓ-colored vertices the probability to select one of them for the color elimination is n_ℓ/n and the expected waiting time is n/n_ℓ. Thus, the expected time to eliminate color ℓ from the graph is at most $\sum_{k=1}^{n} n/k = n \cdot H(n) = O(n \log n)$. To obtain a 2-coloring we need to remove at most $|V_1| - 1 \leq n$ colors. Thus, the expected time to obtain a feasible 2-coloring is $O(n^2 \log n)$. □

Clearly, this upper bound may be rather pessimistic. We make this clear by considering the special case of complete bipartite graphs where already the initial Grundy local search is successful.

Theorem 4. *On complete bipartite graphs the initial Grundy local search finds a feasible 2-coloring.*

Proof. As G is a complete bipartite graph, a coloring is not feasible if V_1 and V_2 have at least one color in common. A Grundy vertex in V_1 can only be colored with color i if all colors $j < i$ exist in V_2 and no vertex in V_2 is i-colored.

We claim that every Grundy coloring uses at most 2 colors. Assume we have a feasible Grundy coloring with at least one i-colored vertex u and $i \geq 3$, w. l. o. g. $u \in V_1$. As u is a Grundy vertex there is at least one 1- and one 2-colored vertex w in V_2. As the coloring is feasible there cannot be a 1- or 2-colored vertex in V_1, contradicting the Grundy property of w and our assumption. □

4   Sparse Random Graphs

Now we investigate ILS with color eliminations on sparse random graphs in the $\mathcal{G}(n,p)$ model [25] with $p = c/n$ for some constant $c > 0$. The following arguments closely follow considerations by Witt [26] in an analysis of a local search-based algorithm for Vertex Cover. In the following let $\chi(G)$ denote the chromatic number, i. e., the smallest possible number of colors of a graph G.

Theorem 5. *For every constant $0 < c < 1$ the following holds. Construct a random graph G according to $\mathcal{G}(n, c/n)$ and then run ILS with color eliminations on G. With probability $1 - o(1)$ a $\chi(G)$-coloring is found within $O(n \log^3 n)$ generations.*

Proof. According to Bollobás [25, Corollary 5.8] and Witt [26,27] when constructing a graph in $\mathcal{G}(n, c/n)$ with $c < 1$ then with probability $1 - o(1)$ each connected component is a tree or a unicyclic graph, i. e., a graph that contains exactly one cycle, and contains at most $O(\log n)$ vertices and edges.

Assume G to have this property. Note that $\chi(G) = 2$ if each unicyclic component contains an even cycle and $\chi(G) = 3$ otherwise. Consider a vertex v and its i- and j-colored neighbors $v_1, \ldots, v_\ell$ and $w_1, \ldots, w_r$, respectively. Recall that the color i can be eliminated from the neighborhood of v by a color elimination if $w_1, \ldots, w_r \notin H_j(v_1) \cup \ldots \cup H_j(v_\ell)$. This property holds for components without odd cycles since these components are bipartite. For components with odd cycles we distinguish vertices on and outside the cycle. If v is not on the cycle, the property is fulfilled since all $H_j(v_i)$ are disjoint. Otherwise the component would contain more than one cycle. Thus, in these cases all color elimination moves will eliminate color i from the neighborhood of v.

If v is on the cycle, the property is not necessary fulfilled as the two neighbors of v on the cycle, say v_ℓ and v_r, may be connected via a $c(v_\ell)$-$c(v_r)$-path of even length. In this case it is not possible to remove the colors $c(v_\ell)$ and $c(v_r)$ with a color elimination. However, this only rules out the color elimination moves with $i \in \{c(v_\ell), c(v_r)\}$. Let ℓ be the highest color used in the current coloring. Since $\chi(G) = 3$, we can assume $\ell \geq 4$. Given v with $c(v) = \ell$ the probability to choose i and j appropriately is $(\ell - 3)/(\ell - 1) \geq 1/3$. Hence, in this case a color elimination move will eliminate color i from the neighborhood of v with constant probability.

The rest of the proof follows the argumentation of Theorem 3. Since all components have size $O(\log n)$, an initial coloring contains at most $O(\log n)$ colors. Thus, the expected time to obtain a feasible 3-coloring is $O(n \log^2 n)$ and by Markov's inequality this time is at most $O(n \log^3 n)$ with probability $1 - o(1)$. The theorem follows by taking the union bound for all considered events that have probability $o(1)$. $\qquad\square$

5   Planar Graphs

Balogh, Hartke, Liu, and Yu [24] showed that the Grundy number for any planar graph is bounded by $\log_{4/3}(n) + 8 - \log_{4/3}(7) \leq 2.41 \log(n) + 2$. We already know that this number of colors is achieved after the first call of local search.

Under certain conditions ILS can do with a constant number of colors. The following result is inspired by the proof of the Five Color Theorem (see Proposition 5.1.2 in [28]).

Theorem 6. *Let G be a planar graph with maximum degree at most 6. Then ILS with Kempe chains constructs a 5-coloring in $O(n \log n)$ expected generations.*

Proof. Fix a planar embedding for G and consider a vertex v of maximum color value. Note that this value is at most 7. If $c(v) \leq 5$ we already have found a 5-coloring. Otherwise v must have 5 or 6 neighbors, among which all colors from 1 to 5 must be present. Let v_1, v_2, v_3, v_4 be a disjoint set of neighbors of v in clockwise order of their appearance in the planar embedding such that these four vertices have pairwise different colors and these colors do not appear anywhere else in the neighborhood of v. Abbreviate $c_i := c(v_i)$ and note that as v has a maximal color, the c_i are all smaller than $c(v)$.

Assume w.l.o.g. that $c_1 < c_3$ and $c_2 < c_4$. Now, if there is no c_1-c_3-path between v_1 and v_3 a Kempe chain operation choosing v_3 and assigning color c_1 to it turns v into a non-Grundy node as c_3 is now a free color for v. The vertex v_3 remains a Grundy node as its color value has been decreased. The following Grundy local search cannot increase color values of vertices, so v will eventually be assigned a smaller color. If there is a c_1-c_3-path between v_1 and v_2 then there cannot be a c_2-c_4-path between v_2 and v_4 as these paths contain disjoint sets of colors and edges cannot cross in the planar embedding. Repeating the arguments with v_4 shows that also in this case v will receive a smaller color at the end of Grundy local search. In both cases the result will be accepted in the selection step as the number of vertices colored with the highest color has decreased.

It remains to estimate the time until a 5-coloring is obtained. The number of 7-colored vertices is non-increasing. The probability for a specific Kempe chain move is at least $1/(7n)$. If there are i 7-colored vertices the probability of reducing this number by at least 1 in one generation is at least $i/(7n)$. The expected waiting time until this happens is at most $7n/i$. Thus, the expected time until all 7-colored vertices have vanished is at most $\sum_{i=1}^{n} 7n/i = 7n \cdot H(n) = O(n \log n)$.

The same arguments then apply to the number of 6-colored vertices, leading to a bound of $O(n \log n)$ generations, in expectation.   □

The arguments from the proof of Theorem 6 break down in the case of higher-degree vertices. The question remains whether, using different arguments, ILS with Kempe chains can be proven to 5-color all planar graphs efficiently or whether ILS with color eliminations can do so. Surprisingly, the answer to both questions is no. Even if ILS is allowed to choose between both mutation operators in each generation, there are graphs that cannot always be 5-colored.

Theorem 7. *For each* $c \in \mathbb{N}$ *there is a 3-colorable graph* G_c *such that for ILS that in each generation chooses to perform either a Kempe chain or a color elimination the expected time until a* $(c - 1)$*-coloring is found is infinite.*

Proof. We construct G_c as $G_{c,c+1}$ according to the following procedure. The graph G_c is almost a tree, except for some cross-edges between siblings. Therefore, we use common notation for trees when describing G_c. Note that the size of G_c is exponential in c. Using a simple induction it is easy to check that G_c is 3-colorable. The procedure below describes a coloring x from which it is impossible to escape for ILS. Figure 2 shows an example for G_3.

Algorithm 5. Construction of $G_{c,k}$

1: Create a c-colored root r.
2: **for** $i = 1$ to $c - 1$ **do**
3:     **for** $j = 1$ to $c - 1$ **do**
4:         Create a new subtree $G_{i,c}$ with an i-colored root v_i and connect v_i to r.
5:         Create a new subtree $G_{j,c}$ with a j-colored root v_j and connect v_j to r.
6:         **if** $i \neq j$ **then** add the cross-edge $\{v_i, v_j\}$.
7: **for** $i = c + 1$ to $k - 1$ **do**
8:     Create a new subtree $G_{i,c}$ and connect its root to r.

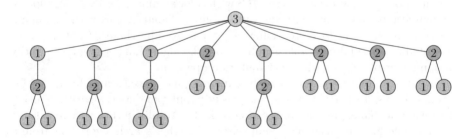

Fig. 2. The graph G_3 and its worst-case coloring

There is a positive probability that the coloring x will be chosen during initialization of ILS. Hence the claim follows if we can prove that neither a Kempe chain move nor a color elimination creates a different coloring that is accepted in the selection step.

For every c-colored vertex v all colors $1, \ldots, c - 1$ appear in v's neighborhood, which certifies that v is a Grundy vertex. Even stronger, if $k \geq c - 1$ is the largest color in the neighborhood of v then the neighborhood contains all color values up to $k - 1$. This implies that each Kempe chain move will again create a Grundy coloring as a Kempe chain preserves this property. In addition, each smaller color appears at least twice in the neighborhood of v. Larger color values appear only once. As this holds for the complete graph, each set H_{ij} contains strictly more vertices colored with the smaller color. (It is easy to check that this holds despite the cross-edges.) So every Kempe chain yields a Grundy coloring which is worse than x. We conclude that ILS cannot escape from x via Kempe chains.

Consider a color elimination w.r.t. a vertex v and colors $i, j < c(v)$. By construction v has two children v_i and v_j colored i and j, respectively. Due to the edge $\{v_i, v_j\}$ the color elimination will swap the colors of v_i and v_j and color i is not eliminated from the neighborhood of v. Hence v remains a Grundy vertex.

The resulting coloring is even worse than x. When v and its edges are removed from the graph, the graph is decomposed into connected components. A color elimination at v leads to at most one Kempe chain operation being performed in each component; hence all Kempe chain moves concern disjoint parts of G_c. Repeating the previous arguments on Kempe chains, in all components all vertices will be Grundy vertices (with respect to the original graph containing v) and all components will have strictly worsened with respect to the selection criterion. This also shows that color eliminations cannot help to escape from x. $\square$

6   Conclusions

We have proposed an iterated local search algorithm for vertex coloring and provided a theoretical analysis of its performance. Using color eliminations as mutation operator, the algorithm efficiently computes 2-colorings for bipartite graphs, while Kempe chain mutations can lead to exponential times, with overwhelming probability, even on trees. ILS with color eliminations also colors sparse random graphs optimally, with high probability. Planar graphs with maximum degree at most 6 are 5-colored efficiently using Kempe chains. Contrarily, we presented a class of 3-colorable graphs where the algorithm needs an arbitrarily large number of colors with either mutation operator.

For future work we suggest to analyze further randomized search heuristics for vertex coloring. In particular, the analysis of population-based algorithms using crossover is a challenging but interesting task. The results in this work may serve as a baseline for a performance comparison of further search algorithms from a theoretical perspective.

Acknowledgments. The authors would like to thank Benjamin Doerr and Christian Sohler for insightful discussions, Manouchehr Zaker for reference [24], and Carsten Witt for providing a full version of [26]. Dirk Sudholt was supported by a postdoctoral fellowship from the German Academic Exchange Service.

References

1. Wegener, I.: Complexity Theory—Exploring the Limits of Efficient Algorithms. Springer, Heidelberg (2005)
2. Khot, S.: Improved inapproximability results for MaxClique, chromatic number and approximate graph coloring. In: Proc. of FOCS 2001, pp. 600–609. IEEE, Los Alamitos (2001)
3. Galinier, P., Hertz, A.: A survey of local search methods for graph coloring. Computers and Operations Research 33(9), 2547–2562 (2006)
4. Wegener, I.: Towards a theory of randomized search heuristics. In: Rovan, B., Vojtáš, P. (eds.) MFCS 2003. LNCS, vol. 2747, pp. 125–141. Springer, Heidelberg (2003)
5. De Jong, K.: Evolutionary Computation—a unified approach. MIT Press, Cambridge (2006)
6. Culberson, J.C., Luo, F.: Exploring the k-colorable landscape with iterated greedy. DIMACS Series, vol. 26, pp. 245–284. AMS, Providence (1995)
7. Paquete, L., Stützle, T.: An experimental investigation of iterated local search for coloring graphs. In: Cagnoni, S., Gottlieb, J., Hart, E., Middendorf, M., Raidl, G.R. (eds.) EvoIASP 2002, EvoWorkshops 2002, EvoSTIM 2002, EvoCOP 2002, and EvoPlan 2002. LNCS, vol. 2279, pp. 122–131. Springer, Heidelberg (2002)
8. Chiarandini, M., Stützle, T.: An application of iterated local search to graph coloring. In: Proc. of the Computational Symposium on Graph Coloring and its Generalizations, pp. 112–125 (2002)
9. Galinier, P., Hao, J.K.: Hybrid evolutionary algorithms for graph coloring. Journal of Combinatorial Optimization 3(4), 379–397 (1999)
10. Lourenço, H.R., Martin, O., Stützle, T.: Iterated local search. In: Handbook of Metaheuristics, pp. 321–353. Kluwer Academic Publishers, Dordrecht (2002)
11. Oliveto, P.S., He, J., Yao, X.: Time complexity of evolutionary algorithms for combinatorial optimization: A decade of results. Int'l Journal of Automation and Computing 4(3), 281–293 (2007)
12. Sudholt, D.: Local search in evolutionary algorithms: the impact of the local search frequency. In: Asano, T. (ed.) ISAAC 2006. LNCS, vol. 4288, pp. 359–368. Springer, Heidelberg (2006)
13. Sudholt, D.: The impact of parametrization in memetic evolutionary algorithms. Theoretical Computer Science 410(26), 2511–2528 (2009)
14. Sudholt, D.: Hybridizing evolutionary algorithms with variable-depth search to overcome local optima. Algorithmica (to appear, 2010)
15. Jensen, T.R., Toft, B.: Graph coloring problems. Wiley Interscience, Hoboken (1995)
16. Brockhoff, D.: Randomisierte Suchheuristiken für das Graphfärbungsproblem. Master's thesis, Universität Dortmund (2005)
17. Fischer, S., Wegener, I.: The one-dimensional Ising model: Mutation versus recombination. Theoretical Computer Science 344(2-3), 208–225 (2005)
18. Sudholt, D.: Crossover is provably essential for the Ising model on trees. In: Proc. of GECCO 2005, pp. 1161–1167. ACM Press, New York (2005)
19. Ghosh, S., Karaata, M.H.: A self-stabilizing algorithm for coloring planar graphs. Distributed Computing 7(1), 55–59 (1993)
20. Kosowski, A., Kuszner, Ł.: Self-stabilizing algorithms for graph coloring with improved performance guarantees. In: Rutkowski, L., Tadeusiewicz, R., Zadeh, L.A., Żurada, J.M. (eds.) ICAISC 2006. LNCS (LNAI), vol. 4029, pp. 1150–1159. Springer, Heidelberg (2006)

21. Hedetniemi, S.T., Jacobs, D.P., Srimani, P.K.: Linear time self-stabilizing colorings. Information Processing Letters 87(5), 251–255 (2003)
22. Barenboim, L., Elkin, M.: Distributed $(\delta + 1)$-coloring in linear (in δ) time. In: Proc. of STOC 2009, pp. 111–120. ACM Press, New York (2009)
23. Zaker, M.: Inequalities for the Grundy chromatic number of graphs. Discrete Applied Mathematics 155(18), 2567–2572 (2007)
24. Balogh, J., Hartke, S., Liu, Q., Yu, G.: On the first-fit chromatic number of graphs. SIAM Journal of Discrete Mathematics 22, 887–900 (2008)
25. Bollobás, B.: Random Graphs. Cambridge University Press, Cambridge (2001)
26. Witt, C.: Greedy local search and vertex cover in sparse random graphs. In: Chen, J., Cooper, S.B. (eds.) TAMC 2009. LNCS, vol. 5532, pp. 410–419. Springer, Heidelberg (2009)
27. Witt, C.: Personal communication
28. Diestel, R.: Graph Theory. Springer, Heidelberg (2005)

Bounded Max-colorings of Graphs

Evripidis Bampis[1], Alexander Kononov[2], Giorgio Lucarelli[3], and Ioannis Milis[4]

[1] LIP6, Université Pierre et Marie Curie, France
Evripidis.Bampis@lip6.fr
[2] Sobolev Institute of Mathematics, pr Koptyuga 4, Novosibirsk, Russia
alvenko@math.nsc.ru
[3] LAMSADE, Université Paris-Dauphine and CNRS FRE 3234, France
lucarelli@lamsade.dauphine.fr
[4] Dept. of Informatics, Athens University of Economics and Business, Greece
milis@aueb.gr

Abstract. In a bounded max-coloring of a vertex/edge weighted graph, each color class is of cardinality at most b and of weight equal to the weight of the heaviest vertex/edge in this class. The bounded max-vertex/edge-coloring problems ask for such a coloring minimizing the sum of all color classes' weights. These problems generalize the well known max-coloring problems by taking into account the number of available resources (colors) in practical applications. In this paper we present complexity results and approximation algorithms for the bounded max-coloring problems on general graphs, bipartite graphs and trees.

1 Introduction

The *bounded max-vertex-coloring* (resp. *bounded max-edge-coloring*) problem takes as input a graph $G = (V, E)$, a weight function $w : V \to N$ (resp. $w : E \to N$) and an integer b; the question of this problem is to find a proper vertex- (resp. edge-) coloring of G, $\mathcal{C} = \{C_1, C_2, \ldots, C_k\}$, where each color C_i, $1 \le i \le k$, has weight $w_i = \max\{w(u) \mid u \in C_i\}$ (resp. $w_i = \max\{w(e) \mid e \in C_i\}$), cardinality $|C_i| \le b$, and the sum of colors' weights, $W = \sum_{i=1}^{k} w_i$, is minimized.

We shall denote the vertex and edge bounded max-coloring problems by $VC(w, b)$ and $EC(w, b)$, respectively. These problems, without the presence of the cardinality bound b, have been already addressed in the literature as *max-(vertex-)coloring* [21] and *max-edge-coloring* [6]; we denote them by $VC(w)$ and $EC(w)$, respectively. For unit weights our problems reduce to the *bounded vertex-coloring* [2] and *bounded edge-coloring* [1] problems, denoted here by $VC(b)$ and $EC(b)$, respectively. For both unbounded colors cardinalities and unit weights, we get the classical *vertex-coloring* (VC) and *edge-coloring* (EC) problems.

Any of the above generalizations of the edge-coloring problem, on a graph $G = (V, E)$, is equivalent to the corresponding vertex-coloring problem on the line graph, $L(G)$, of G (recall that each vertex of $L(G)$ represents an edge of G and any two vertices of $L(G)$ are adjacent if and only if their corresponding edges share a common endpoint in G). Thus, the results for any vertex-coloring

O. Cheong, K.-Y. Chwa, and K. Park (Eds.): ISAAC 2010, Part I, LNCS 6506, pp. 353–365, 2010.

problem on a graph G apply also to the corresponding edge-coloring problem on the graph $L(G)$ and *vice versa*, if both G and $L(G)$ are in the same graph class. This is true for general graphs and chains, but not for most special graph classes, including bipartite graphs and trees.

Motivation. Max-coloring problems have been well motivated in the literature. Max-vertex-coloring problems arise in the management of dedicated memories, organized as buffer pools, which is the case for wireless protocol stacks like GPRS or 3G [21,20]. Max-edge-coloring problems arise in switch based communication systems, like SS/TDMA [4,19], where messages are to be transmitted through direct connections established by an underlying network. Moreover, max-coloring problems correspond to scheduling jobs with conflicts into a batch scheduling environment [10,5].

In all applications mentioned above, context-related entities require their service by physical resources for a time interval. However, there exists in practice a natural constraint on the number of entities assigned the same resource or different resources at the same time. Indeed, the number of memory requests assigned the same buffer is determined by strict deadlines on their completion times, while the number of messages and jobs assigned, at the same time, to different channels and machines, respectively, is bounded by the number of the available resources. The existence of such a constraint motivates the bounded max-coloring problems.

Related Work. It is well known that for general graphs it is NP-hard to approximate the VC problem within a factor of $|V|^{1-\epsilon}$, for all $\epsilon > 0$, [22] and the EC problem within a factor less than $4/3$ [16]; for bipartite graphs both problems become polynomial.

The complexity of the VC(b) problem (known as Mutual Exclusion Scheduling [2]) on special graph classes has been extensively studied (see [14] and the references therein). It is polynomial for trees [17], but NP-complete for bipartite graphs even for three colors [3]. This last result implies also a $4/3$ inapproximability bound for the VC(b) problem on bipartite graphs.

The VC(w) problem is not approximable within a factor less than $8/7$ even for planar bipartite graphs, unless P=NP [10,20]. This bound has been attained for general bipartite graphs [8,20], while an $O(|V|/\log|V|)$-approximation algorithm for general graphs is known [10]. Although the complexity of the problem in trees is an open question, a PTAS for this case has been presented in [20,13]. Other results for the VC(w) problem on several graph classes have been also presented in [10,8,21,20,13,12,18].

The EC(b) problem is polynomial for bipartite graphs [4] as well as for general graphs if b is fixed [1]. Moreover, it is implied by the results in [14] that there is a $4/3$ approximation algorithm for the EC(b) problem on general graphs.

The EC(w) problem is not approximable within a factor less than $7/6$ even for cubic planar bipartite graphs with edge weights $w(e) \in \{1, 2, 3\}$, unless P=NP [8]. A simple greedy 2-approximation algorithm for general graphs has been proposed in [19]. Better than 2 ratios for bipartite graphs of moderate maximum degree have been also presented in [8,13,6]. The complexity of the EC(w) problem on trees remains also open, while a 3/2-approximation algorithm has been recently presented [6].

The VC(w, b) and EC(w, b) problems have been studied as parallel batch scheduling problems with compatibilities between jobs. In this context both problems have been shown to be polynomial for general graphs and $b = 2$ [5], while they have been studied for complements of several special graph classes.

In Table 1 we summarize the best known results for general graphs, bipartite graphs and trees together with our contribution.

Table 1. Known and ours (in bold) approximability results for bounded and/or max coloring problems. [1]The ratio H_b holds only if b is fixed. [2]Even the complexity of the problem is unknown.

Problem	General graphs		Bipartite graphs		Trees	
	Lower Bound	Upper Bound	Lower Bound	Upper Bound	Lower Bound	Upper Bound
VC(b)	$\|V\|^{1-\epsilon}$ [22]	$\mathbf{H_b}$ [1]	4/3 [3]	**4/3**	OPT [17]	
VC(w)		$O(\|V\|/\log \|V\|)$ [10]	8/7 [10,8,20]		open[2]	$PTAS$ [20,13]
VC(w,b)		$\mathbf{H_b}$ [1]	4/3 [3]	**17/11**	open[2]	**PTAS**
EC(b)	4/3 [16]	4/3 [1]	OPT [4]		OPT [4]	
EC(w)		2 [19]	7/6 [8]	2 [19]	open[2]	3/2 [6]
EC(w,b)		$\min\left\{\begin{array}{l}\mathbf{3 - 2/\sqrt{2b}}\\ \mathbf{H_b}^{(1)}\end{array}\right\}$	7/6 [8]	$\min\left\{\begin{array}{l}e\\ \mathbf{3 - 2/\sqrt{b}}\\ \mathbf{H_b}^{(1)}\end{array}\right\}$	**NP-complete**	**2**

Our results and organization of the paper. In this paper we deal with bounded max-coloring problems on general graphs, bipartite graphs and trees. Our interest in bipartite graphs and trees is two-fold. Despite their simplicity, these classes of graphs are important both from theoretical point of view but also from applications' perspective [19,20].

In the next section, we relate our problems with two well known problems, namely the *list coloring* and the *set cover* problems. We also introduce some useful notation. In Section 3, we deal with the VC(w, b) problem and we give a simple 2-approximation algorithm for bipartite graphs. As a byproduct, we show that this algorithm becomes a 4/3-approximation algorithm for the VC(b) problem, which matches the 4/3 inapproximability bound. Then, we present a generic scheme that we show to be a 17/11-approximation algorithm for bipartite graphs, while it becomes a PTAS for trees as well as for bipartite graphs when b is a fixed constant. In Section 4, we deal with the EC(w, b) problem and we present approximation algorithms of ratios $\min\{3 - 2/\sqrt{2b}, H_b\}$ for general graphs and $\min\{e, 3 - 2/\sqrt{b}, H_b\}$ for bipartite graphs, where $H_b = \sum_{i=1}^{b} \frac{1}{i}$ is the b-th harmonic number. More interestingly, we prove that the EC(w, b) problem on trees is NP-complete. Given that the complexity question of VC(w), EC(w) and VC(w, b) problems for trees remains open, this is the first max coloring problem on trees proven to be NP-hard. Finally, we propose a 2-approximation algorithm for the EC(w, b) problem on trees.

2   Preliminaries and Notation

We first establish a relation between ours and bounded list coloring problems.

BOUNDED LIST VERTEX (resp. EDGE) COLORING PROBLEM

INSTANCE: A graph $G = (V, E)$, a set of colors $C = \{C_1, C_2, \ldots, C_k\}$, a list of colors $\phi(u) \subseteq C$ for each $u \in V$ (resp. $\phi(e) \subseteq C$ for each $e \in E$), and integers b_i, $1 \leq i \leq k$.

QUESTION: Is there a k-coloring of G such that each vertex u (resp. edge e) is assigned a color in its list $\phi(u)$ (resp. $\phi(e)$) and every color C_i is used at most b_i times?

We denote the vertex and edge bounded list coloring problems by $\mathrm{VC}(\phi, b_i)$ and $\mathrm{EC}(\phi, b_i)$, respectively. In the next theorem we summarize some of the known results for these problems that we shall use in this paper.

Theorem 1

(i) The $\mathrm{VC}(\phi, b_i)$ problem is NP-complete even for chains, $|\phi(u)| \leq 2$, for all $u \in V$, and $b_i \leq 5$, $1 \leq i \leq k$ [11].

(ii) Both $\mathrm{VC}(\phi, b_i)$ and $\mathrm{EC}(\phi, b_i)$ problems are polynomial for trees if the number of colors k is fixed [15,9].

(iii) The $\mathrm{VC}(\phi, b_i)$ problem is polynomial for general graphs if $k = 2$ [15].

The next proposition follows easily by exhaustive transformations of an instance of the $\mathrm{VC}(w, b)$ and $\mathrm{EC}(w, b)$ problems to: (i) an instance of the $\mathrm{VC}(\phi, b)$ and $\mathrm{EC}(\phi, b)$ problems, respectively (and the use of Theorem 1(ii) for these instances), and (ii) instances of the set cover problem (and the use of Chvátal's algorithm [7] for these instances).

Proposition 1

(i) For a fixed number of colors k, both the $\mathrm{VC}(w, b)$ and $\mathrm{EC}(w, b)$ problems on trees are polynomial.

(ii) For a fixed bound b, there is an H_b-approximation algorithm for both the $\mathrm{VC}(w, b)$ and $\mathrm{EC}(w, b)$ problems on general graphs.

Our notation. Given a set S and a positive integer weight $w(s)$ for every element $s \in S$, we denote by $\langle S \rangle = \langle s_1, s_2, \ldots, s_{|S|} \rangle$ an ordering of S such that $w(s_1) \geq w(s_2) \geq \cdots \geq w(s_{|S|})$. For such an ordering of S and a positive integer b, let $k_S = \left\lceil \frac{|S|}{b} \right\rceil$. We define the *ordered b-partition* of S, denoted by $\mathcal{P}_S = \{S_1, S_2, \ldots, S_{k_S}\}$, to be the partition of S into k_S subsets, such that $S_i = \{s_j, s_{j+1}, \ldots, s_{\min\{j+b-1, |S|\}}\}$, $i = 1, 2, \ldots, k_S$, $j = (i-1)b + 1$. In other words, S_1 contains the b heaviest elements of S, S_2 contains the next b heaviest elements of S and so on; S_{k_S} contains the $|S| \bmod b$ lightest elements of S.

By $OPT = w_1^* + w_2^* + \cdots + w_{k^*}^*$, where $w_1^* \geq w_2^* \geq \cdots \geq w_{k^*}^*$, we denote the weight of an optimal solution to the $\mathrm{VC}(w, b)$ or $\mathrm{EC}(w, b)$ problem, where w_i^*, $1 \leq i \leq k^*$, is the weight of the i-th color class. By Δ we denote the maximum degree of a graph.

3   Bounded Max-vertex-Coloring

In this section we present approximation algorithms for the $VC(w, b)$ problem on bipartite graphs and trees.

3.1   A Simple Split Algorithm

Let $G = (U \cup V, E)$, $|U \cup V| = n$, be a vertex weighted bipartite graph. Our first algorithm colors the vertices of each class of G separately, by finding the ordered b-partitions of classes U and V. For the minimum number of colors k^* it holds that $k^* \geq \left\lceil \frac{|U|+|V|}{b} \right\rceil$ and, therefore, $k = \left\lceil \frac{|U|}{b} \right\rceil + \left\lceil \frac{|V|}{b} \right\rceil \leq \left\lceil \frac{|U|+|V|}{b} \right\rceil + 1 \leq k^* + 1$.

Algorithm Split
1. Let $\mathcal{P}_U = \{U_1, U_2, \ldots, U_{k_U}\}$ be the ordered b-partition of U;
2. Let $\mathcal{P}_V = \{V_1, V_2, \ldots, V_{k_V}\}$ be the ordered b-partition of V;
3. Return the coloring $\mathcal{C} = \mathcal{P}_U \cup \mathcal{P}_V$;

Theorem 2. *Algorithm* SPLIT *returns a solution of weight* $W \leq 2 \cdot w_1^* + w_2^* + \cdots + w_{k^*}^* \leq 2 \cdot OPT$ *for the* $VC(w, b)$ *problem in bipartite graphs.*

Proof. Let $\langle \mathcal{C} \rangle = \langle C_1, C_2, \ldots, C_k \rangle$ be the colors constructed by Algorithm SPLIT, that is $w_1 \geq w_2 \geq \cdots \geq w_k$. Assume, w.l.o.g., that U_x, $1 \leq x \leq k_U$, is the $i - th$ color in $\langle \mathcal{C} \rangle$. Let also u be the heaviest vertex of U_x, that is $w(u) = w_i$.

The ordered b-partition of U and V implies that the colors that appear before U_x in $\langle \mathcal{C} \rangle$ are the colors $U_1, U_2, \ldots, U_{x-1}$ and $V_1, V_2, \ldots, V_y$, $y = i - x$. The colors $U_1, U_2, \ldots, U_{x-1}$ are all of cardinality b and their $(x-1) \cdot b$ vertices are all of weight at least $w(u)$. The colors $V_1, V_2, \ldots, V_{y-1}$, are also all of cardinality b and their $(y-1) \cdot b$ vertices are all of weight at least the weight of the heaviest vertex of color V_y which is at least $w(u)$. Taking into account the vertex u itself it follows that there are in G at least $(x-1) \cdot b + [(y-1) \cdot b + 1] + 1 = (x+y-2) \cdot b + 2 = (i-2) \cdot b + 2$ vertices of weight at least $w(u) = w_i$. In an optimal solution, these vertices belong in at least $\left\lceil \frac{(i-2) \cdot b+2}{b} \right\rceil = (i - 1)$ colors, each one of weight at least w_i. Hence, $w_{i-1}^* \geq w_i$, $2 \leq i \leq k$. Clearly, $w_1 = w_1^*$, since both are equal to the weight of the heaviest vertex of the graph, and as $k \leq k^* + 1$, we obtain

$$W = \sum_{i=1}^k w_i = w_1^* + \sum_{i=2}^k w_i \leq w_1^* + \sum_{i=1}^{k-1} w_i^* \leq w_1^* + \sum_{i=1}^{k^*} w_i^* = 2 \cdot w_1^* + w_2^* + \cdots + w_{k^*}^* \leq 2 \cdot OPT. \qquad \square$$

The complexity of Algorithm SPLIT is dominated by the sorting needed to obtain the ordered b-partitions of U and V in Lines 1 and 2, that is $O(n \cdot \log n)$.

Algorithm SPLIT applies also to the $VC(b)$ problem on bipartite graphs. Moreover, the absence of weights in the $VC(b)$ problem allows a tight analysis with respect to the $\frac{4}{3}$ inapproximability bound. The proof is omitted.

Theorem 3. *There is a* $\frac{4}{3}$-*approximation algorithm for the* $VC(b)$ *problem on bipartite graphs.*

3.2   A Generic Scheme

To obtain our scheme we split a bipartite graph $G = (U \cup V, E)$, $|U \cup V| = n$, into two subgraphs $G_{1,j}$ and $G_{j+1,n}$ induced by the j heaviest and the $n - j$ lightest vertices of G, respectively (by convention, we consider $G_{1,0}$ as an empty subgraph). Our scheme depends on a parameter p such that all the vertices of G of weights $w_1^*, w_2^*, \ldots, w_{p-1}^*$ are in a subgraph $G_{1,j}$. This is always possible for some $j \le b(p-1)$, since each color of an optimal solution for G contains at most b vertices. In fact, for every j, $1 \le j \le b(p-1)$, we obtain a solution for the whole graph by concatenating an optimal solution of at most $p-1$ colors for $G_{1,j}$, if there is one, and the solution obtained by Algorithm SPLIT for $G_{j+1,n}$.

Algorithm Scheme(p)
1. Let $\langle U \cup V \rangle = \langle u_1, u_2, \ldots u_n \rangle$;
2. For $j = 0, 1, \ldots, b \cdot (p-1)$ do
3.   Split the graph into two vertex induced subgraphs:
      - $G_{1,j}$ induced by vertices $u_1, u_2, \ldots, u_j$
      - $G_{j+1,n}$ induced by vertices $u_{j+1}, u_{j+2}, \ldots, u_n$
4.   If there is a solution for $G_{1,j}$ with at most $p-1$ colors then
5.   Find an optimal solution for $G_{1,j}$ with at most $p-1$ colors;
6.   Run Algorithm SPLIT for $G_{j+1,n}$;
7.   Concatenate the two solutions found in Lines 5 and 6;
8. Return the best solution found;

Lemma 1. *Algorithm* SCHEME*(p) achieves a* $(1 + \frac{1}{H_p})$ *approximation ratio for the* VC(w, b) *problem.*

Proof. Consider the iteration j, $j \le b \cdot (p-1)$, of the algorithm where the weight of the heaviest vertex in $G_{j+1,n}$ is equal to the weight of the i-th color of an optimal solution, i.e. $w(u_{j+1}) = w_i^*$, $1 \le i \le p$.

The vertices of $G_{1,j}$ are a subset of those that appeared in the $i - 1$ heaviest colors of the optimal solution. Thus, an optimal solution for $G_{1,j}$ is of weight $OPT_{1,j} \le w_1^* + w_2^* + \cdots + w_{i-1}^*$.

The vertices of $G_{j+1,n}$ are a superset of those that appeared in the $k^* - (i-1)$ lightest colors of the optimal solution. The extra vertices of $G_{j+1,n}$ are of weight at most w_i^* and appear in an optimal solution in at most $i - 1$ colors. Thus, an optimal solution for $G_{j+1,n}$ is of weight $OPT_{j+1,n} \le w_i^* + w_{i+1}^* + \cdots + w_{k^*}^* + (i-1) \cdot w_i^* = i \cdot w_i^* + w_{i+1}^* + \cdots + w_{k^*}^*$. By Theorem 2, Algorithm SPLIT returns a solution for $G_{j+1,n}$ of weight $W_{j+1,n} \le (i+1) \cdot w_i^* + w_{i+1}^* + \cdots + w_{k^*}^*$.

Therefore, the solution found in this iteration j for the whole graph G is of weight $W_i = OPT_{1,j} + W_{j+1,n} \le w_1^* + w_2^* + \cdots + w_{i-1}^* + (i+1) \cdot w_i^* + w_{i+1}^* + \cdots + w_{k^*}^*$.

In all the iterations of the algorithm we obtain p such inequalities for W. By multiplying the i-th, $1 \le i \le p$, inequality by $\frac{1}{i \cdot (H_p + 1)}$ and adding up all of them, we have $\left(\sum_{i=1}^{p} \frac{1}{i \cdot (H_p + 1)} \right) \cdot W \le OPT$, that is $\frac{W}{OPT} \le \frac{H_p + 1}{H_p} = 1 + \frac{1}{H_p}$.   □

The complexity of the Algorithm SCHEME(p) is $O(bp(f(p) + n \log n))$, where $O(f(p))$ is the complexity of checking for the existence of solutions with at most

$p-1$ colors for $G_{1,j}$ and finding an optimal one among them, while $O(n \log n)$ is the complexity of Algorithm SPLIT. Algorithm SCHEME(1) coincides with Algorithm SPLIT. Algorithm SCHEME(2) has simply to check if the $j \leq b$ vertices of $G_{1,j}$ are independent from each other and, therefore, it derives a $\frac{5}{3}$ approximate solution in polynomial time. Algorithm SCHEME(3) has to check and find, a two color solution for $G_{1,j}$, if any. This can be done in polynomial time by Theorem 1(iii). Thus, Algorithm SCHEME(3) is a polynomial time $\frac{17}{11}$-approximation algorithm for the $\mathrm{VC}(w,b)$ problem on bipartite graphs.

However, when $p \geq 4$ and b is a part of the instance, finding an optimal solution in $G_{1,j}$ is an NP-hard problem (even for the $\mathrm{VC}(b)$ problem [3]). Hence, we consider that b is a fixed constant. In this case, we run an exhaustive algorithm for finding, if any, an optimal solution in $G_{1,j}$ of at most $p-1$ colors. The complexity of such an exhaustive algorithm is $O((p-1)^{b \cdot (p-1)})$ and thus, the complexity of Algorithm SCHEME(p), $p \geq 4$, becomes $O(bp^{bp} + n^2 \log n)$, since bp is $O(n)$. Choosing $\epsilon = \frac{1}{H_p}$, we get $p = O(2^{\frac{1}{\epsilon}})$. Consequently, for fixed b, we have a PTAS for the $\mathrm{VC}(w,b)$ problem on bipartite graphs, that is an approximation ratio of $1 + \frac{1}{H_p} = 1 + \epsilon$ within $O(b(2^{\frac{1}{\epsilon}})^{b2^{\frac{1}{\epsilon}}} + n^2 \log n)$ time.

Furthermore, in the particular case of trees, checking the existence of solutions with at most $p-1$ colors for $G_{1,j}$, and finding an optimal one among them, can be done, by Proposition 1, in polynomial time for fixed p. The complexity of our scheme in this case becomes $O(b2^{\frac{1}{\epsilon}}(n^{2^{\frac{1}{\epsilon}}} + n^2 \log n))$. Therefore, the following theorem holds.

Theorem 4. *For the* $\mathrm{VC}(w,b)$ *problem, Algorithm* SCHEME(p) *is a*
(i) polynomial time $\frac{17}{11}$*-approximation algorithm for bipartite graphs (for $p = 3$),*
(ii) PTAS for bipartite graphs if b is fixed,
(iii) PTAS for trees.

4 Bounded Max-edge-Coloring

In this section we deal with the approximability of the $\mathrm{EC}(w,b)$ problem on general graphs, bipartite graphs and trees. Moreover, we prove that the problem is NP-complete for trees.

4.1 General and Bipartite Graphs

We first give tight bounds on the number of colors in a solution to the $\mathrm{EC}(w,b)$ problem. In fact, our bounds apply to any *nice* solution $\langle \mathcal{C} \rangle = \langle C_1, C_2, \ldots, C_k \rangle$ to the $\mathrm{EC}(w,b)$ problem, that is a solution where each color C_i, $1 \leq i \leq k$, is of cardinality $|C_i| = b$ or C_i is maximal in the subgraph induced by the edges $\bigcup_{j=i}^{k} C_j$. It is easy to see that any solution to the $\mathrm{EC}(w,b)$ problem can be transformed into a nice one of the same total weight. The proof is quite technical and it is omitted.

Proposition 2. *For the number of colors k in any nice solution to the $EC(w,b)$ problem it holds that:*

$$\max\{\Delta, \left\lceil \frac{|E|}{b} \right\rceil\} \leq k \leq \begin{cases} \left\lceil \frac{|E|}{b} \right\rceil - \left\lceil \frac{\Delta^2}{2b} \right\rceil + (2\Delta - 1), & \text{for general graphs} \\ \left\lceil \frac{|E|}{b} \right\rceil - \lceil \frac{\Delta^2}{b} \rceil + (2\Delta - 1), & \text{for bipartite graphs} \end{cases}$$

We next adapt the greedy 2-approximation algorithm presented in [19] for the $EC(w)$ problem to the $EC(w,b)$ problem.

Algorithm Greedy

1. Let $\langle E \rangle = \langle e_1, e_2, \ldots, e_{|E|} \rangle$;
2. For $j = 1, 2, \ldots, |E|$ do
3.   Insert edge e_j in the first color of cardinality less than b which does not contain other edges adjacent to e_j;

The solution derived by Algorithm GREEDY is a nice one, since it is constructed in a first-fit manner. The analysis of this algorithm given in the next lemma is based on Proposition 2.

Lemma 2. *Algorithm GREEDY achieves approximation ratios of $(3 - \frac{2}{\sqrt{2b}})$, on general graphs, and $(3 - \frac{2}{\sqrt{b}})$, on bipartite graphs, for the $EC(w,b)$ problem.*

Proof. Let $\langle \mathcal{C} \rangle = \langle C_1, C_2, \ldots, C_k \rangle$, be a solution derived by Algorithm GREEDY. Consider the color C_i and let e_j be the first edge inserted in C_i, i.e. $w_i = w(e_j)$. Let $E_i = \{e_1, e_2, \ldots, e_j\}$, G_i be the subgraph of G induced by the edges in E_i, and Δ_i be the maximum degree of G_i.

As the solution $\langle \mathcal{C} \rangle$ is a nice one and an optimal solution can be also considered to be nice, by Proposition 2, for general graphs, it follows that *(i)* $i \leq \left\lceil \frac{|E_i|}{b} \right\rceil - \left\lceil \frac{\Delta_i^2}{2b} \right\rceil + (2\Delta_i - 1)$, and *(ii)* in an optimal solution the edges of G_i appear in at least $i^* \geq \max\{\Delta_i, \left\lceil \frac{|E_i|}{b} \right\rceil\}$ colors, each one of weight at least w_i. Therefore, $\frac{i}{i^*} \leq \frac{\left\lceil \frac{|E_i|}{b} \right\rceil - \left\lceil \frac{\Delta_i^2}{2b} \right\rceil + (2\Delta_i - 1)}{\max\{\Delta_i, \left\lceil \frac{|E_i|}{b} \right\rceil\}}$. By distinguishing between $\Delta_i \geq \left\lceil \frac{|E_i|}{b} \right\rceil$ and $\Delta_i < \left\lceil \frac{|E_i|}{b} \right\rceil$ it follows that in either case $\frac{i}{i^*} \leq 3 - \frac{\Delta_i^2 + 2b}{2b\Delta_i}$. This bound is maximized when $\Delta_i = \sqrt{2b}$, that is $\frac{i}{i^*} \leq 3 - \frac{2}{\sqrt{2b}}$. Thus, $w_i \leq w_{i^*}^* \leq w_{\lceil i/(3 - \frac{2}{\sqrt{2b}}) \rceil}^*$. Summing up these inequalities for all i's, $1 \leq i \leq k$, we obtain the $(3 - \frac{2}{\sqrt{2b}})$ ratio for general graphs.

A similar analysis yields the $(3 - \frac{2}{\sqrt{b}})$ ratio for bipartite graphs.   □

Another approximation result for the $EC(w,b)$ problem is obtained by exploiting a general framework, presented in [12], which allows to convert a ρ-approximation algorithm for a coloring problem into an $e \cdot \rho$-approximation one for the corresponding max-coloring problem, for hereditary classes of graphs. In fact, this framework has been presented for such a conversion from the VC to the VC(w) problem, but it can be easily seen that this applies also for conversions from the

EC, VC(b) and EC(b) problems to the EC(w), VC(w,b) and EC(w,b) problems, respectively. However, this conversion leads to ratios greater than those shown in Table 1 for the EC(w) and VC(w,b) problems. For the EC(w,b) problem on general graphs this approach gives a ratio of at least $\frac{4}{3} \cdot e > 3$, as the EC, and hence the EC(b), problem cannot be approximated within a ratio less than $\frac{4}{3}$. On the other hand, the EC(w,b) problem on bipartite graphs can be approximated, this way, with a ratio of e, as the EC(b) problem is polynomial in this case (see Table 1).

Combining the discussion above with Lemma 2 and Proposition 1, it follows

Theorem 5. *The EC(w,b) problem can be approximated with ratio of* $\min\{3 - 2/\sqrt{2b}, H_b\}$ *for general graphs, and* $\min\{e, 3 - 2/\sqrt{b}, H_b\}$ *for bipartite graphs.*

Note that, the H_b ratio outperforms the other two ratios only for $b \leq 5$ for general graphs, for $b = 3$ in the case of bipartite graphs and hence, b can be considered as fixed.

4.2   NP-Completeness for Trees

We prove first that the bounded list edge-coloring, EC(ϕ,b), problem is NP-complete even if the graph $G = (V,E)$ is a set of chains, $|\phi(e)| = 2$ for all $e \in E$, and $b = 5$. We denote this problem as EC(chains, $|\phi(e)| = 2$, $b = 5$).

Proposition 3. *The EC(chains, $|\phi(e)| = 2$, $b = 5$) problem is NP-complete.*

Proof. By Theorem 1(i), the VC(chains, $|\phi(v)| \leq 2$, $b_i \leq 5$) problem is NP-complete. Given that the line-graph of a chain is also a chain, it follows that the EC(chains, $|\phi(e)| \leq 2$, $b_i \leq 5$) problem is also NP-complete. The later problem can be easily reduced to the EC(chains, $|\phi(e)| \leq 2$, $b = 5$) problem, where $b_i = b = 5$ for all colors: for every color C_i with $b_i < 5$, add $5 - b_i$ independent edges with just C_i in their lists. This last problem reduces to the EC(chains, $|\phi(e)| = 2$, $b = 5$) problem, where $|\phi(e)| = 2$ for all edges. This can be done by transforming an instance of EC(chains, $|\phi(v)| \leq 2, b = 5$) as follows: *(i)* add two new colors C_{k+1} and C_{k+2}, both with cardinality bound $b = 5$, *(ii)* add color C_{k+1} to the list of every edge e with $|\phi(e)| = 1$, *(iii)* add ten independent edges and put in their lists both colors C_{k+1} and C_{k+2}.   □

Theorem 6. *The EC(w,b) problem on trees is NP-complete.*

Proof. Our reduction is from EC(chains, $|\phi(e)| = 2$, $b = 5$) problem. We construct an instance of the EC(w,b) problem on a forest $G' = (V',E')$ as follows.

We replace every edge $e = (u,v) \in E$ with a chain of three edges: $e_1 = (u,u')$, $e_2 = (u',v')$ and $e_3 = (v',v)$, where $w(e_1) = w(e_2) = w(e_3) = 1$. Moreover, we create $k - |\phi(e)| = k - 2$ stars of $k - 1$ edges each. We add edges (u', s_t), $1 \leq t \leq k - 2$, between u' and the central vertex s_t of each of these $k - 2$ stars; thus every star has now exactly k edges. Let $\phi(e) = \{C_i, C_j\}$. The $k - 2$ edges (u', s_t) take different weights in $\{1, 2, \ldots, k\} \setminus \{i,j\}$. Let q be the weight taken by an edge (u', s_t). The remaining $k-1$ edges of the star t take different weights in $\{1, 2, \ldots, k\} \setminus \{q\}$. In the same way, we add $k-2$ stars connected to v'. In Figure 1, we show the u''s part of this edge-gadget for $e = (u, v)$. For every edge e of G, we add $2(k-2)$ stars and $2(k-2)k+2$ edges.

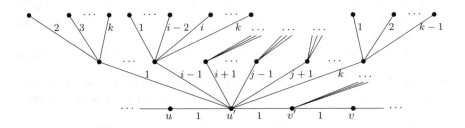

Fig. 1. The gadget for an edge $e = (u, v)$ with $\phi(e) = \{C_i, C_j\}$

To complete our construction we define f_i to be the number of edges that include C_i to their lists and $F = \max\{f_i | 1 \le i \le k\}$. For every color C_i we add $F - f_i$ disconnected copies of the color-gadget shown in Figure 2. Such a gadget consists of an edge $e = (x, y)$ and $k-1$ stars with $k-1$ edges each. There are also edges between one of the endpoints of e, say y, and the central vertices of all stars; thus every star has now exactly k edges. The edge e takes weight i and the edges in the stars of such a color-gadget take weights similar to those in the stars of an edge-gadget. For a color C_i we add $(F - f_i)(k - 1)$ stars and $(F - f_i)(k - 1)k + (F - f_i)$ edges.

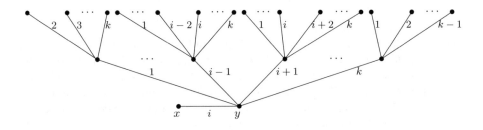

Fig. 2. A gadget for the color C_i

The number of stars in the forest G' we have constructed is $2|E|(k - 2) + \sum_{i=1}^{k}(F - f_i)(k - 1) = k(k - 1)F - 2|E|$, since $\sum_{i=1}^{k} f_i = 2|E|$. By setting $b' = k(k-1)F - 2|E| + 5 + F$, we prove that: "*There is a k-coloring for* EC(ϕ, b) *(chains,* $|\phi(e)| = 2$, $b = 5$*), if and only if,* G' *has a bounded max-edge-coloring of total weight* $\sum_{i=1}^{k} i$ *such that every color is used at most* b' *times*".

Consider, first, a solution $\mathcal{C}$ to the EC(ϕ, b) problem. We construct a solution $\mathcal{C}'$ for the EC(w, b) problem as follows. Let $e = (u, v) \in E$ be an edge with

$\phi(e) = \{C_i, C_j\}$, which, w.l.o.g., appears in the color C_i of $\mathcal{C}$. Put the edges e_1 and e_3 of the edge-gadget for e in color C_i', while the edge e_2 in color C_j'. After doing this for all edges in E, each color C_i' contains at most $2\cdot5+1\cdot(f_i-5) = f_i+5$ edges. Next, put the edges with weight i, $1 \leq i \leq k$, from the $k(k-1)F-2|E|$ stars in C_i'. Each color C_i' in $\mathcal{C}'$ constructed so far contains at most $k(k-1)F-2|E|+f_i+5 = b'-(F-f_i)$ edges and, by the construction of G', $\mathcal{C}'$ is a proper coloring. In the $F-f_i$ color-gadgets for C_i there are $F-f_i$ remaining (x,y) edges of weight i, which can still be inserted in color C_i'. Thus, we get a solution for the $\mathrm{EC}(w,b)$ problem of k colors, each one of at most b' edges, and total weight $\sum_{i=1}^{k} i$.

Conversely, consider a solution $\mathcal{C}'$ to the $\mathrm{EC}(w,b)$ problem. $\mathcal{C}'$ consists of exactly k colors of weights $1, 2, \ldots, k$, since each star in G' has k edges and each edge has a different weight in the range $\{1, 2, \ldots, k\}$. Thus, all edges of the same weight, say i, should belong in the same color C_i' of $\mathcal{C}'$. Therefore, C_i' contains one edge from each one of the $k(k-1)F-2|E|$ stars as well as the $F-f_i$ remaining (x,y) edges of the color-gadgets having weight i. Consider, now, the edges of G' corresponding to the edges e_1, e_2 and e_3 of the edge-gadget for an edge e with $\phi(e) = \{C_i, C_j\}$. By the construction of G' and the choice of edge weights, the edges e_1, e_2 and e_3 should appear in colors C_i' and C_j'. Thus, edges e_1 and e_3 should appear, w.l.o.g., in color C_i', while e_2 in color C_j'. Therefore, the edge $e \in E$ can be colored by color $C_i \in \phi(e)$. Finally, a color C_i' contains at most 5 edges of type e_1 (or e_3), corresponding to at most 5 edges of E; otherwise $|C_i'| \geq k(k-1)F-2|E|+(F-f_i)+(2\cdot6+1\cdot(f_i-6)) > b'$, a contradiction.

To complete our proof for the $\mathrm{EC}(w,b)$ problem on trees, let p be the number of trees in G'. We add a set of $p-1$ edges of weight $\epsilon < 1$ to transform the forest G' into a single tree T. This can be done as every tree of G' has at least two vertices. By keeping the same bound b', it is easy to see that there is a solution for the $\mathrm{EC}(w,b)$ problem on G' of weight $\sum_{i=1}^{k} i$, if and only if, there is a solution for the $\mathrm{EC}(w,b)$ problem on T whose weight is equal to $\sum_{i=1}^{k} i + \left\lceil \frac{p-1}{b'} \right\rceil \epsilon$. □

4.3   A 2-Approximation Algorithm for Trees

In [6] a 2-approximation algorithm for the $\mathrm{EC}(w)$ problem on trees has been presented, which is also exploited to derive a ratio of $3/2$ for that problem. This algorithm yields a solution of Δ colors, $\mathcal{M} = \{M_1, M_2, \ldots, M_\Delta\}$. Starting from this solution we obtain a solution to the $\mathrm{EC}(w,b)$ problem by finding the ordered b-partition of each color in $\mathcal{M}$. For the sake of completeness we give below the whole algorithm.

Algorithm Convert

```
1. Let T_r be the tree rooted in an arbitrary vertex r;
2. For each vertex v in pre-order traversal of T_r do
3.    Let ⟨E_v⟩ = ⟨e_1, e_2, ..., e_d(v)⟩ be the edges adjacent to v,
         and (v,p) be the edge from v, v ≠ r, to its parent;
4.    Using ordering ⟨E_v⟩, insert each edge in E_v, but (v,p),
         in the first matching which does not contain an edge in E_v;
5. Let M = {M_1, M_2, ..., M_Δ} be the colors constructed;
```

6. For $i = 1$ to Δ do
7.   Let $\mathcal{P}_{M_i} = \{M_1^i, M_2^i, \ldots, M_{k_i}^i\}$ be the ordered b-partition of $\langle M_i \rangle$;
8. Return a solution $\langle \mathcal{C} \rangle = \langle C_1, C_2, \ldots, C_k \rangle$, $\mathcal{C} = \bigcup_{i=1}^{\Delta} \mathcal{P}_{M_i}$;

The proof of the next theorem is based on similar arguments with those used in the analysis of Algorithm SPLIT.

Theorem 7. *Algorithm* CONVERT *is a 2-approximation one for the* EC(w, b) *problem on trees.*

References

1. Alon, N.: A note on the decomposition of graphs into isomorphic matchings. Acta Mathematica Hungarica 42, 221–223 (1983)
2. Baker, B.S., Coffman Jr., E.G.: Mutual exclusion scheduling. Theoretical Computer Science 162, 225–243 (1996)
3. Bodlaender, H.L., Jansen, K.: Restrictions of graph partition problems. Part I. Theoretical Computer Science 148, 93–109 (1995)
4. Bongiovanni, G., Coppersmith, D., Wong, C.K.: An optimum time slot assignment algorithm for an SS/TDMA system with variable number of transponders. IEEE Trans. on Communications 29, 721–726 (1981)
5. Boudhar, M., Finke, G.: Scheduling on a batch machine with job compatibilities. Belgian Journal of Oper. Res., Statistics and Computer Science 40, 69–80 (2000)
6. Bourgeois, N., Lucarelli, G., Milis, I., Paschos, V.T.: Approximating the max-edge-coloring problem. Theoretical Computer Science 411, 3055–3067 (2010)
7. Chvátal, V.: A greedy heuristic for the set-covering problem. Mathematics of Operations Research 4, 233–235 (1979)
8. de Werra, D., Demange, M., Escoffier, B., Monnot, J., Paschos, V.T.: Paschos. Weighted coloring on planar, bipartite and split graphs: Complexity and approximation. Discrete Applied Mathematics 157, 819–832 (2009)
9. de Werra, D., Hertz, A., Kobler, D., Mahadev, N.V.R.: Feasible edge coloring of trees with cardinality constraints. Discrete Mathematics 222, 61–72 (2000)
10. Demange, M., de Werra, D., Monnot, J., Paschos, V.T.: Paschos. Time slot scheduling of compatible jobs. Journal of Scheduling 10, 111–127 (2007)
11. Dror, M., Finke, G., Gravier, S., Kubiak, W.: On the complexity of a restricted list-coloring problem. Discrete Mathematics 195, 103–109 (1999)
12. Epstein, L., Levin, A.: On the max coloring problem. In: Kaklamanis, C., Skutella, M. (eds.) WAOA 2007. LNCS, vol. 4927, pp. 142–155. Springer, Heidelberg (2008)
13. Escoffier, B., Monnot, J., Paschos, V.T.: Weighted coloring: Further complexity and approximability results. Information Processing Letters 97, 98–103 (2006)
14. Gardi, F.: Mutual exclusion scheduling with interval graphs or related classes. Part II. Discrete Applied Mathematics 156, 794–812 (2008)
15. Gravier, S., Kobler, D., Kubiak, W.: Complexity of list coloring problems with a fixed total number of colors. Discrete Applied Mathematics 117, 65–79 (2002)
16. Holyer, I.: The NP-completeness of edge-coloring. SIAM Journal on Computing 10, 718–720 (1981)
17. Jarvis, M., Zhou, B.: Bounded vertex coloring of trees. Discrete Mathematics 232, 145–151 (2001)

18. Kavitha, T., Mestre, J.: Max-coloring paths: Tight bounds and extensions. In: Dong, Y., Du, D.-Z., Ibarra, O. (eds.) ISAAC 2009. LNCS, vol. 5878, pp. 87–96. Springer, Heidelberg (2009)
19. Kesselman, A., Kogan, K.: Nonpreemptive scheduling of optical switches. IEEE Trans. on Communications 55, 1212–1219 (2007)
20. Pemmaraju, S.V., Raman, R.: Approximation algorithms for the max-coloring problem. In: Caires, L., Italiano, G.F., Monteiro, L., Palamidessi, C., Yung, M. (eds.) ICALP 2005. LNCS, vol. 3580, pp. 1064–1075. Springer, Heidelberg (2005)
21. Pemmaraju, S.V., Raman, R., Varadarajan, K.R.: Buffer minimization using max-coloring. In: SODA 2004, pp. 562–571 (2004)
22. Zuckerman, D.: Linear degree extractors and the inapproximability of max clique and chromatic number. In: STOC 2006, pp. 681–690 (2006)

Parameterized Algorithms for Boxicity[*]

Abhijin Adiga[1], Rajesh Chitnis[2], and Saket Saurabh[3]

[1] Department of Computer Science and Automation, Indian Institute of Science,
Bangalore–560012, India
abhijin@csa.iisc.ernet.in
[2] Chennai Mathematical Institute, Siruseri–603103, India
rajesh@cmi.ac.in
[3] The Institute of Mathematical Sciences, Chennai–600113, India
saket@imsc.res.in

Abstract. In this paper we initiate an algorithmic study of BOXICITY, a combinatorially well studied graph invariant, from the viewpoint of parameterized algorithms. The boxicity of an arbitrary graph G with the vertex set $V(G)$ and the edge set $E(G)$, denoted by box(G), is the minimum number of interval graphs on the same set of vertices such that the intersection of the edge sets of the interval graphs is $E(G)$. In the BOXICITY problem we are given a graph G together with a positive integer k, and asked whether the box(G) is at most k. The problem is notoriously hard and is known to be NP-complete even to determine whether the boxicity of a graph is at most two. This rules out any possibility of having an algorithm with running time $|V(G)|^{O(f(k))}$, where f is an arbitrary function depending on k alone. Thus we look for other structural parameters like "vertex cover number" and "max leaf number" and see its effect on the problem complexity. In particular, we give an algorithm that given a vertex cover of size k finds box(G) in time $2^{O(2^k k^2)}|V(G)|$. We also give a faster additive one approximation algorithm for finding box(G) that given a graph with vertex cover of size k runs in time $2^{O(k^2 \log k)}|V(G)|$. Our next result is an additive two approximation algorithm for BOXICITY when parameterized by the max leaf number running in time $2^{O(k^3 \log k)}|V(G)|^{O(1)}$. Our results are based on structural relationships between boxicity and the corresponding parameter and could be of independent interest.

1 Introduction

Let $\mathcal{F} = \{S_1, S_2, \ldots, S_n\}$ be a family of sets. An intersection graph associated with $\mathcal{F}$ has $\mathcal{F}$ as the vertex set and we add an edge between S_i and S_j if and only if $i \neq j$ and $S_i \cap S_j \neq \varnothing$. Any graph can be represented as an intersection graph, but many important graph families can be described as intersection graphs of more restricted types of set families, for instance sets derived from some kind of geometric configuration, like interval graphs, circular arc graphs,

---

[*] Abhijin Adiga was supported by Infosys Technologies Ltd., Bangalore, under the "Infosys Fellowship Award".

O. Cheong, K.-Y. Chwa, and K. Park (Eds.): ISAAC 2010, Part I, LNCS 6506, pp. 366–377, 2010.

chordal graphs, grid-intersection graphs, string graphs and boxicity k-graphs. These graphs classes are not only interesting from a graph theoretic viewpoint and combinatorial perspective but are also useful in modeling many real life applications. In this paper our object of interest is boxicity k-graphs, an intersection graph obtained by a family of boxes in the k-dimensional Euclidean space.

A k-box is a Cartesian product of closed intervals $[a_1, b_1] \times [a_2, b_2] \times \cdots \times [a_k, b_k]$. A k-box representation of a graph G is a mapping of the vertices of G to k-boxes in the k-dimensional Euclidean space such that two vertices in G are adjacent if and only if their corresponding k-boxes have a non-empty intersection. The *boxicity* of a graph G, denoted box(G), is the minimum integer k such that G has a k-box representation. Boxicity was introduced by Roberts [23] in 1969 and it finds applications in modeling problems in social sciences and biology.

There has been significant amount of work done recently on finding lower and upper bounds on boxicity of different graph classes. Chandran and Sivadasan [8] showed that box$(G) \leq$ tree-width$(G) + 2$. Chandran et al. [7] proved that box$(G) \leq \chi(G^2)$ where, $\chi(G^2)$ is the chromatic number of G^2. In [14] Esperet proved that box$(G) \leq \Delta^2(G) + 2$, where $\Delta(G)$ is the maximum degree of G. Scheinerman [24] showed that the boxicity of outerplanar graphs is at most 2 while Thomassen [25] proved that the boxicity of planar graphs is at most 3. In [11], Cozzens and Roberts studied the boxicity of split graphs.

While there has been a lot of work on boxicity from graph theoretic view point, the problem remains hitherto unexplored in the light of algorithms and complexity; with exceptions that are few and far between. Cozzens [10] showed that computing the boxicity of a graph is NP-hard. This was later strengthened by Yannakakis [26] and finally by Kratochvíl [20] who showed that determining whether boxicity of a graph is at most two itself is NP-complete. Recently in [1,2], Adiga et al. showed that there exists no polynomial-time algorithm to approximate the boxicity of a bipartite graph on n vertices with a factor of $O(n^{0.5-\epsilon})$ for any $\epsilon > 0$, unless $NP = ZPP$. In this paper we study the Boxi-CITY problem – here we are given a graph G together with a positive integer k and asked whether the box(G) is at most k – from the parameterized complexity perspective.

Parameterized complexity is basically a two-dimensional generalization of "P vs. NP" where in addition to the overall input size n, one studies the effects on computational complexity of a secondary measurement that captures additional relevant information. This additional information can be, for example, a structural restriction on the input distribution considered, such as a bound on the treewidth of an input graph or the size of solution set. Parameterization can be deployed in many different ways; for general background on the theory see [12,18,22].

For decision problems with input size n, and a parameter k, the two dimensional analogue (or generalization) of P, is solvability within a time bound of $O(f(k)n^{O(1)})$, where f is a function of k alone, as contrasted with a trivial

$n^{k+O(1)}$ algorithm. Problems having such an algorithm is said to be fixed parameter tractable (FPT), and such algorithms are practical when small parameters cover practical ranges. The book by Downey and Fellows [12] provides a good introduction to the topic of parameterized complexity. For recent developments see the books by Flum and Grohe [18] and Niedermeier [22].

In the framework of parameterized complexity, an important aspect is the *choice of parameter* for a problem. Exploring how one parameter affects the complexity of different parameterized or unparameterized versions of the problem, often leads to non trivial combinatorics and better understanding of the problem. In general there are two kinds of parameterizations. In the first kind the parameter reflects the value of the objective function in question. The second kind, *structural parameterizations*, measure the structural properties of the input. A well developed structural parameter is the treewidth of the input graph. Other well established structural parameters include the *vertex cover number*, the size of the minimum vertex cover of graph [16,17] and the *max leaf number*, the maximum number of the leaves possible in a spanning tree of the input graph [15]. Observe that since determining whether the boxicity of the input graph is at most 2 is NP-complete we can not hope to have an algorithm to test whether the box(G) is at most k running in time $|V(G)|^{O(f(k))}$, where f is an arbitrary function depending on k alone. This initiates a study of BOXICITY from the structural parameterizations like treewidth, vertex cover number of the graph and the max leaf number. We parameterize the problem with vertex cover number and max leaf number of the input graph and obtain the following results:

1. an FPT algorithm for BOXICITY running in time $2^{O(2^k k^2)}|V(G)|$ when parameterized by the vertex cover number;

2. an additive one approximation for BOXICITY when parameterized by the vertex cover number running in time $2^{O(k^2 \log k)}|V(G)|$; and

3. an additive two approximation for BOXICITY when parameterized by the max leaf number running in time $2^{O(k^3 \log k)}|V(G)|^{O(1)}$.

Our other results include factor 2 approximation when parameterized by the feedback vertex set number of the input graph and a FPT algorithm for computing boxicity on co-bipartite graphs when parameterized by the vertex cover number of the "associated bipartite graph". Our results on parameterized approximation for BOXICITY are among the very few known results of such kind. These results contribute positively to the developing area of parameterized approximation and we refer to [5,9,13,21] for further details on parameterized approximation. All our results are based on structural relationships between boxicity and the corresponding parameter and they could be of independent interest. It is natural to ask why we do not consider parameterizing with the treewidth of the input graph. The reason for this is that, though we are not able to show it, we believe that BOXICITY is NP-hard even on graphs of constant treewidth. We leave this as an open problem.

2 Preliminaries

In this section we first give the known equivalent representation of boxicity k-graphs in terms of interval graphs. Then we show how to enumerate these graphs as they are useful for our algorithm. Finally we set up notations used throughout the paper.

2.1 Interval Graphs and Box Representations

Equivalent Characterization. It is easy to see that a graph has boxicity at most 1 if and only if it is an *interval graph*, that is, each vertex of the graph can be associated with a closed interval on the real line such that two intervals intersect if and only if the corresponding vertices are adjacent. By definition, boxicity of a complete graph is 0. Let G be any graph and G_i, $1 \leq i \leq k$ be graphs on the same vertex set as G such that $E(G) = E(G_1) \cap E(G_2) \cap \cdots \cap E(G_k)$. Then we say that G is the *intersection* of G_i s for $1 \leq i \leq k$ and denote it as $G = \bigcap_{i=1}^{k} G_i$. Boxicity can be stated in terms of intersection of interval graphs as follows:

Lemma 1. Roberts [23]: *The boxicity of a non-complete graph G is the minimum positive integer b such that G can be represented as the intersection of b interval graphs.*

We say that $\mathcal{B} = \{I_1, I_2, \ldots, I_b\}$ is a *b-box representation* of graph G if $G = \bigcap_{i=1}^{b} I_i$, where $I_1, I_2, \ldots I_b$ are interval graphs with fixed interval representations, i.e., $\mathcal{B}$ can be considered as a collection of b interval representations.

Interval Graphs and Box Representation of a Graph. Let I be an interval graph. Let f_I be an *interval representation* for I, that is, it is a mapping from the vertex set to closed intervals on the real line such that for any two vertices u and v, $\{u, v\} \in E(I)$ if and only if $f_I(u) \cap f_I(v) \neq \varnothing$. Let $l(u, f_I)$ and $r(u, f_I)$ denote the left and right end points of the interval corresponding to the vertex u respectively. In some sections we will never consider more than one interval representation for an interval graph, in which case we will simplify the notations to $l(u, I)$ and $r(u, I)$. Further, when there is no ambiguity about the graph under consideration and its interval representation, we simply denote the left and right end points as $l(u)$ and $r(u)$ respectively. For any interval graph there exists an interval representation with all end points distinct. Such a representation is called a *distinguishing* interval representation. It is an easy exercise to derive such a distinguishing interval representation starting from an arbitrary interval representation of the graph.

Enumeration of b-box representations. Consider a graph G on n vertices. The number of interval graphs on n vertices can be easily shown to be $2^{O(n \log n)}$. Therefore, a brute force enumeration of all distinct b-box representations of G will require time $\binom{2^{O(n \log n)}}{b} bn^2 = 2^{O(bn \log n)} bn^2$ time. The term bn^2 term is required to check the validity of the box representation. This results in the following proposition.

Proposition 1 ([4]). *There are at most $2^{O(nb\log n)}$ distinct b-box represen-tations of a graph G on n vertices and all these can be enumerated in time $2^{O(bn\log n)}bn^2$.*

2.2 Some Definitions and Notations

Let $[p]$ denote $\{1, 2, \ldots, p\}$ where p is a positive integer. For any graph G, let $V(G)$ and $E(G)$ denote its vertex set and edge set respectively. For $U \subseteq V$, let $G[U]$ be the subgraph of G induced by U. We use n to denote the number of vertices in the input graph. By $N(u)$ we denote (open) neighborhood of u that is the set of all vertices adjacent to u and by $N[u]$, the set $N(u) \cup \{u\}$. Similarly, for a subset $D \subseteq V$, we define $N[D] = \bigcup_{v \in D} N[v]$ and $N(D) = N[D] \setminus D$. The vertex cover number of a graph is the cardinality of the minimum sized subset of vertices of the graph that contains at least one end-point of every edge of the input graph. The max leaf number is the maximum number of the leaves possible in a spanning tree of the input graph. If $B_G = \{I_1, I_2, \ldots, I_k\}$ is a k-box representation, then we say that B_G has width k. A box representation of G of width box(G) is called as an *optimal box-representation*.

Property 1. Helly property of intervals: Suppose $A_1, A_2, \ldots, A_k$ is a finite set of intervals on the real line with pairwise non-empty intersection. Then there exists a common point of intersection for all the intervals, that is, $\bigcap_{i=1}^{k} A_i \neq \varnothing$.

3 Boxicity Parameterized by Vertex Cover

In this section we show that BOXICITY parameterized by the vertex cover number is fixed parameter tractable. This implies that given a graph G on n vertices we can find box(G) by finding the minimum $1 \leq k \leq n$ for which there exists a k-box representation for G.

We start with a few definitions. Let G be a graph with the vertex set $V(G)$. Let $U \subseteq V(G)$ be a vertex cover of G size k. A k-vertex cover can be computed in time $O(2^k + kn)$ (See [22] for references). Let $S = V(G) \setminus U$ be the independent set. We partition the vertices in the independent set S based on their neighborhoods in U. For every $A \subseteq U$, let $S_A = \{u \in S \mid N(u) = A\}$. Observe that in this way we can partition the vertices of S into at most 2^k parts – one for every subset of U. For $A \subseteq U$, if $|S_A| \geq 1$ then we retain an arbitrary vertex in S_A, say $v(A)$, which we call the *representative vertex* and remove other vertices of S_A from G. We call this step as *pruning* of the parts and denote the resulting graph obtained from this process by G' and let $S' = V(G') \setminus U$. The pruning step requires time $O(2^k n)$. Notice that every vertex in S' has a distinct neighborhood of U in G'. Throughout this section, G and G' represent the graphs defined above. Now we define the following.

Definition 1. *Let $B_{G'} = \{I'_1, I'_2, \ldots, I'_\ell\}$ be a box-representation of G' of width ℓ. We say that we can extend $B_{G'}$ to a representation, say B_G, of G of the same width if we can add intervals for vertices in $V(G) \setminus U$ while keeping the intervals of vertices from U as they are in each of the interval graphs $I'_1, I'_2, \ldots, I'_\ell$, and get an ℓ-box representation for G.*

Remark 1. Observe that according to the Definition 1 while extending the given box-representation for G' the only intervals that we are not allowed to change are that of the vertex set U. The intervals corresponding to $S' = V(G') \setminus U$ are allowed to be replaced/modified.

Now we characterize a relationship between box-representations of G' and G which is used crucially in the correctness of our algorithms later.

Lemma 2. *Let $B_{G'} = \{I'_1, I'_2, \ldots, I'_\ell\}$ be a box-representation of G'. Then $B_{G'}$ can be extended to a representation of G, say $B_G = \{I''_1, I''_2, \ldots, I''_\ell\}$ if and only if $\forall A \subseteq U$ such that $|S_A| > 1$ there exists a $j \in [\ell]$ such that A forms a clique in I'_j.*

Proof. First we prove the forward direction. Suppose that $B_{G'} = \{I'_1, I'_2, \ldots, I'_\ell\}$ can be extended to $B_G = \{I''_1, I''_2, \ldots, I''_\ell\}$. Recall that the intervals corresponding to the vertices from U remain unchanged by the definition. Let $A \subseteq U$ such that $|S_A| > 1$ and let a_1, a_2 be two arbitrary vertices in S_A. Since $\{a_1, a_2\} \notin E(G)$ there exists $j \in [\ell]$ such that intervals of a_1 and a_2 do not intersect in I''_j. Without loss of generality, let $r(a_1, I''_j) < l(a_2, I''_j)$. Then for every vertex in A, its interval in I''_j contains the interval $[r(a_1, I''_j), l(a_2, I''_j)]$ as they need to intersect intervals corresponding to both a_1 and a_2. Therefore, A forms a clique in I''_j. Hence A forms clique in I'_j also as the intervals of vertices from U remain unchanged while extending $B_{G'}$ to B_G.

Next we show the reverse direction of the lemma. By the arguments in Section 2.1 we can assume that each interval graph representation in $B_{G'}$ is a *distinguishing* interval representation. Let $A \subseteq U$ such that $|S_A| > 1$. We know that A forms a clique in some I'_j such that $j \in [\ell]$. By Property 1, the intervals in I'_j of vertices of A have a common intersection. Since I'_j is a distinguishing interval representation, this common intersection is not a point but rather a non-trivial interval, say J. Let the interval corresponding to $v(A)$ in I'_j be J_v. Now we assign all vertices from S_A including $v(A)$ to distinct point intervals in the common interval $J \cap J_v$. Notice that we can do this because $J \cap J_v$ is *not a point interval*. This follows from the arguments given in Section 2.1 that all the intervals have pairwise distinct end-points. In all other interval graphs from $B_{G'}$, we assign to all members of S_A the same interval as that of the representative element $v(A)$ of S_A. We do this for every $A \subseteq U$ for which $|S_A| > 1$. From the description above it is evident that the new interval graphs we get by above procedure is an ℓ-box representation for G. This concludes the proof of the lemma. ∎

Lemma 2 has a following important algorithmic consequence. This can be proved using the fact there are at most 2^k subsets of U such that $|S_A| > 1$.

Lemma 3. *Given a box representation for G' of width ℓ, in time $O(2^k k^2 \ell)$ we can determine whether it can be extended to a box representation for G and if so we can find an ℓ-box representation in time $O(2^k k^2 \ell n)$.*

Proof. Let $B_{G'} = \{I'_1, I'_2, \ldots, I'_\ell\}$ be a box-representation for G'. Lemma 2 provides a simple criteria to check whether $B_{G'}$ can be extended to an ℓ-box representation for G. The only thing we need to check is whether for every $A \subseteq U$ with

$|S_A| > 1$ there is at least one interval graph, say I'_t, in $B_{G'}$ such that the vertices corresponding to A forms a clique in I'_t. This can be done in $O(|A|^2 \ell) = O(k^2 \ell)$ time. Note that the list of As such that $|S_A| > 1$ can be assumed to be available from the pruning step. Since there are at most 2^k such subsets of U the total time required is $O(2^k k^2 \ell)$. We can find the explicit representation in the stated time using the construction given in the second half of the proof of Lemma 2. ■

Now we present a lemma which ensures an ℓ-box representation for G' which can be extended to an ℓ-box representation for G.

Lemma 4. *Let ℓ be the minimum integer such that there exists an ℓ-box representation of G' that can be extended to an ℓ-box representation of G. Then* $\mathrm{box}(G) = \ell$.

Proof. Let $\beta = \mathrm{box}(G)$. From the statement of the lemma we assume without loss of generality that $\beta \leq \ell$. Now consider a β-box-representation of G and look at its induced representation on vertices of G'. Clearly this induced representation on G' can be extended to a representation of G, in fact, it can be extended to the β-box-representation of G we started with. Thus $\beta \geq \ell$ and hence $\mathrm{box}(G) = \beta = \ell$. ■

Remark 2. Observe that the proof of Lemma 4 also implies that there exists a box representation for G' which can be extended to a box representation for G of the same width.

Observe that Lemmata 3 and 4 together with the above remark gives us an algorithm to find $\mathrm{box}(G)$. From [6] it is known that if G has vertex cover at most k, then, $\mathrm{box}(G) \leq \lfloor k/2 \rfloor + 1$. Hence we can enumerate all possible box-representation of G', where $|V(G')| \leq 2^k + k$, of width at most $\lfloor k/2 \rfloor + 1$, which by Proposition 1 takes time at most $2^{O(2^k k^2)}$, and check by Lemma 3 whether it can be extended to a box representation for G in time $O(2^k k^2)$. All this can be done in time $2^{O(2^k k^2)} n$. This results in the following theorem.

Theorem 1. *For graphs on n vertices with vertex cover bounded by k, the boxicity and an optimal box-representation can be computed in time* $2^{O(2^k k^2)} n$.

The running time obtained in Theorem 1 to compute the boxicity of a graph exactly is high. However if we are willing to accept an additive error of 1, that is, if we want an additive one approximation algorithm to compute boxicity of a graph parameterized by the vertex cover number then we can do much faster. We have the following result whose proof we omit due to lack of space:

Theorem 2. *Let G be a graph with vertex cover number at most k, then in time* $2^{O(k^2 \log k)} n$ *we can find a w-box-representation of G such that* $\mathrm{box}(G) \leq w \leq \mathrm{box}(G) + 1$.

3.1   On the Boxicity of Co-bipartite Graphs

In this section we give an algorithm to find the boxicity of co-bipartite graphs. A graph is called co-bipartite if it is the complement of a bipartite graph. In

[26], Yannakakis showed that it is NP-complete to determine if the boxicity of a co-bipartite graph is ≥ 3. Recently, Adiga, Bhowmick and Chandran [1] showed that it is hard to approximate the boxicity of a bipartite graph within $\sqrt{n}$ factor, where n is the order of the graph. A similar result can be derived for co-bipartite graphs too.

Observe that a co-bipartite graph on n vertices has a minimum vertex cover of size $n-2$. Therefore, Theorem 1 or parameterization by the vertex cover number of the input graph is not interesting for the class of co-bipartite graphs. However, interestingly, we show that given a co-bipartite graph G, finding box(G) is fixed parameter tractable when parameterized by the vertex cover of the following bipartite graph associated with it.

Definition 2. *Let H be an XY co-bipartite graph, that is, $V(H)$ is partitioned into cliques X and Y. The associated bipartite graph of H, denoted by H^* is the graph obtained by making the sets X and Y independent sets, but keeping the set of edges between vertices of X and Y identical to that of H, that is, $\forall u \in X, v \in Y, \{u,v\} \in E(H^*)$ if and only if $\{u,v\} \in E(H)$.*

We need the following relation between box(H^*) and box(H).

Lemma 5. *(Adiga, et al [1]) Let H be an XY co-bipartite graph and H^* its associated bipartite graph. If H is a non-interval graph, then box(H^*) $\leq$ box(H) $\leq$ 2box(H^*). If H is an interval graph, then box(H^*) ≤ 2.*

Our main theorem of this section is as follows.

Theorem 3. *Let G be a co-bipartite graph on n vertices and G^* be its associated bipartite graph. If the vertex cover of G^* is bounded by k, then the box(G) and an optimal box representation can be computed in time $2^{O(k^2 \log k)}n^2$.*

We skip the proof due to lack of space.

3.2   Boxicity Parameterized by Feedback Vertex Set Number

In this section we obtain a factor $2 + (2/\text{box}(G))$-approximation algorithm for finding the boxicity of a graph G running in time $f(k)|V(G)|^{O(1)}$ where k is the size of the minimum feedback vertex set of the input graph G. A feedback vertex set of a graph G is a subset $U \subseteq V(G)$ such that $G[V(G) \setminus U]$ is a forest. More precisely we have the following.

Theorem 4. *Let G be a graph with the minimum feedback vertex set size bounded by k. Then, there is a factor-$2 + (2/\text{box}(G))$-approximation algorithm to compute boxicity of G running in time $f(k)|V(G)|^{O(1)}$, where $f(k)$ is the exponential part of the running time of the algorithm to compute boxicity of a given graph having vertex cover of size at most k.*

Proof. Let $U \subseteq V(G)$ be a feedback vertex set of G. By definition, $S = V(G)\setminus U$ induces a forest in G and therefore, $G[S]$ is a bipartite graph with partite sets say X and Y. Let $G_1 = G[U \cup X]$ and $G_2 = G[U \cup Y]$. Clearly U is a vertex cover of G_1 and G_2. Using Theorem 1, we obtain a box representation $B_1 = \{I_{11}, I_{12}, \ldots, I_{1r}\}$ for G_1 and $B_2 = \{I_{21}, I_{22}, \ldots, I_{2s}\}$ for G_2 where $\text{box}(G_1) = r$ and $\text{box}(G_2) = s$. It is a well-known fact that the boxicity of a forest is at most 2 and can be constructed in polynomial time. Let $B_3 = \{I_{31}, I_{32}\}$ be a box representation for $G[S]$.

For each $I_{1j} \in B_1$, we construct I'_{1j} by introducing vertices of Y as universal vertices, that is, to every vertex $v \in Y$ we assign the following interval $l(v, I_{1,j}) = \min_{w \in V(G_1)} l(w, I_{1,j})$ and $r(v, I_{1,j}) = \max_{w \in V(G_1)} r(w, I_{1,j})$. Similarly, for each $I_{2j} \in B_2$, we construct I'_{2j} by introducing vertices of X as universal vertices. Finally, for each $I_{3j} \in B_3$, we construct I'_{3j} by introducing vertices in U as universal vertices. We call the new box representations B'_1, B'_2 and B'_3 respectively. It is easy to verify that the intersection of edge sets of interval graphs in B'_1, B'_2 and B'_3 is $E(G)$. This implies that the $\text{box}(G) \leq \text{box}(G_1) + \text{box}(G_2) + 2$. Since, $\text{box}(G_1), \text{box}(G_2) \leq \text{box}(G)$, we have a box representation of G comprising of at most $2\text{box}(G) + 2$ interval graphs. This gives us the desired approximation factor of the algorithm.   ∎

4   Boxicity Parameterized by Max Leaf Number

In this section we obtain an additive 2-approximation algorithm for finding the boxicity of a graph G running in time $f(k)|V(G)|^{O(1)}$ where k is the number of the maximum possible leaves in any spanning tree of the input graph G. The number of the maximum possible leaves in any spanning tree of the input graph G is called the *max-leaf number* of G.

Let us fix a connected graph G on n vertices and max-leaf number k. Let $V_{>2}$ and $V_{=1}$ denote the set of vertices in G with degree at least 3 and degree exactly 1, respectively. Let A be the set of vertices in G with degree at least 3, degree exactly 1 in G and their neighbors, that is, $A = N[V_{>2} \cup V_{=1}]$. Let H be $G[A]$. Our algorithm is based on the following lemma.

Lemma 6. *If G and H are as defined above then $\text{box}(H) \leq \text{box}(G) \leq \text{box}(H) + 2$. Furthermore given a b-box-representation of H it can be made into $(b + 2)$-box-representation of G in polynomial time.*

Proof. Since H is an induced subgraph of G, $\text{box}(H) \leq \text{box}(G)$. Now we will give a construction that shows that $\text{box}(G) \leq \text{box}(H) + 2$. Let $\text{box}(H) = b$ and $B = \{I_1, I_2, \ldots, I_b\}$ be a box representation of H. Consider $G[V(G)\backslash(V_{>2} \cup V_{=1})]$. It consists of vertex disjoint *excluded paths*. By excluded path we mean that the vertices of these paths have degree 2 in G. Their end points are adjacent to two vertices in $V(H)$. Let $P = xv_1v_2 \cdots v_py$ be one such path in $G[V(G) \backslash V_{>2} \cup V_{=1}]$ where $x, y \in V(H)$ have degree 2 in G and $v_i \in V(G) \setminus V(H)$. Note that x and y are associated with exactly one path. Let $G \setminus H = G[V(G) \setminus V(H)]$.

Now we order the paths of length at least 2 in $G[V(G) \setminus (V_{>2} \cup V_{=1})]$ and denote these paths as $P_1, P_2, \ldots, P_m$ where $P_i = x_iv_{i1}v_{i2} \cdots v_{ip(i)}y_i$ and m is the total number of paths. We only consider paths of length at least 2 as the vertices on the paths of length at most 1 are in the neighborhood of vertices in

$V_{>2} \cup V_{=1}$ and hence in $V(H)$. For each $I_a \in B$, we construct I'_a by introducing the vertices in $G \setminus H$ as follows: If x_i and y_i are not adjacent to each other in I_a, then without loss of generality we assume that $r(x_i, I_a) < l(y_i, I_a)$. Suppose $P_i = x_i v_{i1} v_{i2} \cdots v_{ip(i)} y_i$, then we consider $p(i) + 1$ distinct points $r(x, I_a) = c_0 < c_1 < c_2 < \cdots < c_{p(i)-1} < c_{p(i)} = l(y_i, I_a)$. We assign to v_{ij} the interval $[c_{j-1}, c_j]$. If x_i and y_i are adjacent, then we assign the same point interval within the region of intersection of the intervals of x_i and y_i to all vertices v_{ij}. We also construct two extra graphs I'_{b+1} and I'_{b+2} as follows:

$\mathbf{I'_{b+1}}$: All vertices in $V_{>2} \cup V_{=1}$ are assigned to the interval $[m+1, m+2]$. Consider the path P_i. The vertices x_i and y_i are assigned $[i, m+2]$ and other vertices in P_i are assigned to the point interval $[i + \frac{1}{2}, i + \frac{1}{2}]$.

$\mathbf{I'_{b+2}}$: All vertices in $V_{>2} \cup V_{=1}$ are assigned to the interval $[-1, 0]$. For a path P_i, the vertices x_i and y_i are assigned $[-1, i]$ and all other vertices in P_i are assigned the point interval $[i - \frac{1}{2}, i - \frac{1}{2}]$.

We skip the proof of $G = \bigcap_{a=1}^{b+2} I'_a$ due to lack of space.　　■

Our result in this section depends on Lemma 6 and the following known structural result.

Lemma 7. Kleitman and West [19] *For a graph G, if the max leaf number is equal to k, then G is a subdivision of a graph on at most $4k - 2$ vertices.*

From this we obtain the following theorem.

Theorem 5. *Let G be a connected graph on n vertices with max leaf number bounded by k. Then we can obtain an additive 2-approximation algorithm to compute the boxicity of the graph G running in time $2^{O(k^3 \log k)} n^{O(1)}$.*

5　Conclusion

In this paper we initiated a systematic study of computing the boxicity of a graph in the realm of parameterized complexity. The problem is notoriously hard and it is known to be NP-complete even to determine whether the boxicity of a graph is at most two. Hence we studied this problem by parameterizing with parameters that are FPT like the vertex cover number and the max-leaf number of the input graph. We showed that finding boxicity of a graph when parameterized by the vertex cover number is FPT, obtained a faster additive 1 approximation algorithm when parameterized by the vertex cover number and finally obtained an additive 2 approximation algorithm to boxicity of the graph when parameterized by the max leaf number of the graph. Our other results included factor 2 approximation when parameterized by the feedback vertex set number of the input graph and a FPT algorithm for computing boxicity on co-bipartite graphs when parameterized by the vertex cover number of the associated bipartite graph. Our results were based on structural relationships between boxicity and the corresponding parameter and could be of independent interest. We have not only obtained several algorithms for computing boxicity but also have opened up a plethora of interesting open problems. The main ones include.

- Is BOXICITY parameterized by the feedback vertex set or the max-leaf number FPT?
- Is computing the boxicity of a graph G, NP-hard on graphs of constant treewidth?

References

1. Adiga, A., Bhowmick, D., Chandran, L.S.: Boxicity and poset dimension. In: Thai, T. (ed.) COCOON 2010. LNCS, vol. 6196, pp. 3–12. Springer, Heidelberg (2010)
2. Adiga, A., Bhowmick, D., Chandran, L.S.: The hardness of approximating the threshold dimension, boxicity and cubicity of a graph. DAM 158(16), 1719–1726 (2010)
3. Booth, K.S., Lueker, G.S.: Testing for the consecutive ones property, interval graphs, and graph planarity using pq-tree algorithms. J. Comput. Syst. Sci. 13(3), 335–379 (1976)
4. Brandstädt, A., Le, V.B., Spinrad, J.P.: Graph classes: a survey. SIAM Monographs on Discrete Mathematics and Applications. SIAM, Philadelphia (1999)
5. Cai, L., Huang, X.: Fixed-parameter approximation: Conceptual framework and approximability results. In: Bodlaender, H.L., Langston, M.A. (eds.) IWPEC 2006. LNCS, vol. 4169, pp. 96–108. Springer, Heidelberg (2006)
6. Chandran, L.S., Das, A., Shah, C.D.: Cubicity, boxicity, and vertex cover. Disc. Math. 309, 2488–2496 (2009)
7. Chandran, L.S., Francis, M.C., Sivadasan, N.: Boxicity and maximum degree. J. Comb. Theory Ser. 98(2), 443–445 (2008)
8. Chandran, L.S., Sivadasan, N.: Boxicity and treewidth. J. Comb. Theory Ser. 97(5), 733–744 (2007)
9. Chen, Y., Grohe, M., Grüber, M.: On parameterized approximability. In: Bodlaender, H.L., Langston, M.A. (eds.) IWPEC 2006. LNCS, vol. 4169, pp. 175–183. Springer, Heidelberg (2006)
10. Cozzens, M.B.: Higher and multi-dimensional analogues of interval graphs, Ph.D. thesis. Department of Mathematics, Rutgers University, New Brunswick (1981)
11. Cozzens, M.B., Roberts, F.S.: Computing the boxicity of a graph by covering its complement by cointerval graphs. Disc. Appl. Math. 6, 217–228 (1983)
12. Downey, R.G., Fellows, M.R.: Parameterized complexity. Springer, New York (1999)
13. Downey, R.G., Fellows, M.R., McCartin, C.: Parameterized approximation problems. In: Bodlaender, H.L., Langston, M.A. (eds.) IWPEC 2006. LNCS, vol. 4169, pp. 121–129. Springer, Heidelberg (2006)
14. Esperet, L.: Boxicity of graphs with bounded degree. European J. Combin. 30(5), 1277–1280 (2009)
15. Fellows, M.R., Lokshtanov, D., Misra, N., Mnich, M., Rosamond, F.A., Saurabh, S.: The complexity ecology of parameters: An illustration using bounded max leaf number. Theory Comput. Syst. 45(4), 822–848 (2009)
16. Fellows, M.R., Lokshtanov, D., Misra, N., Rosamond, F.A., Saurabh, S.: Graph layout problems parameterized by vertex cover. In: Hong, S.-H., Nagamochi, H., Fukunaga, T. (eds.) ISAAC 2008. LNCS, vol. 5369, pp. 294–305. Springer, Heidelberg (2008)
17. Fiala, J., Golovach, P.A., Kratochvíl, J.: Parameterized complexity of coloring problems: Treewidth versus vertex cover. In: Chen, J., Cooper, S.B. (eds.) TAMC 2009. LNCS, vol. 5532, Springer, Heidelberg (2009)

18. Flum, J., Grohe, M.: Parameterized Complexity Theory. Texts in Theoretical Computer Science. An EATCS Series. Springer, Berlin (2006)
19. Kleitman, D.J., West, D.B.: Spanning trees with many leaves. SJDM 4, 99–106 (1991)
20. Kratochvíl, J.: A special planar satisfiability problem and a consequence of its NP-completeness. Disc. Appl. Math. 52, 233–252 (1994)
21. Marx, D., Razgon, I.: Constant ratio fixed-parameter approximation of the edge multicut problem. In: Fiat, A., Sanders, P. (eds.) ESA 2009. LNCS, vol. 5757, pp. 647–658. Springer, Heidelberg (2009)
22. Niedermeier, R.: Invitation to fixed-parameter algorithms. Oxford Lecture Series in Mathematics and its Applications, vol. 31. Oxford University Press, Oxford (2006)
23. Roberts, F.S.: On the boxicity and cubicity of a graph. In: Recent Progresses in Combinatorics, pp. 301–310. Academic Press, New York (1969)
24. Scheinerman, E.R.: Intersection classes and multiple intersection parameters, Ph.D. thesis, Princeton University (1984)
25. Thomassen, C.: Interval representations of planar graphs. J. Comb. Theory Ser. 40, 9–20 (1986)
26. Yannakakis, M.: The complexity of the partial order dimension problem. SIAM J. Alg. Disc. Math. 3(3), 351–358 (1982)

On Tractable Cases of Target Set Selection[*]

André Nichterlein, Rolf Niedermeier, Johannes Uhlmann, and Mathias Weller

Institut für Informatik, Friedrich-Schiller-Universität Jena,
Ernst-Abbe-Platz 2, D-07743 Jena, Germany
{andre.nichterlein,rolf.niedermeier,johannes.uhlmann,
mathias.weller}@uni-jena.de

Abstract. We study the NP-complete TARGET SET SELECTION (TSS) problem occurring in social network analysis. Complementing results on its approximability and extending results for its restriction to trees and bounded treewidth graphs, we classify the influence of the parameters "diameter", "cluster edge deletion number", "vertex cover number", and "feedback edge set number" of the underlying graph on the problem's complexity, revealing both tractable and intractable cases. For instance, even for diameter-two split graphs TSS remains very hard. TSS can be efficiently solved on graphs with small feedback edge set number and also turns out to be fixed-parameter tractable when parameterized by the vertex cover number, both results contrasting known parameterized intractability results for the parameter treewidth. While these tractability results are relevant for sparse networks, we also show efficient fixed-parameter algorithms for the parameter cluster edge deletion number, yielding tractability for certain dense networks.

1 Introduction

The NP-complete graph problem TARGET SET SELECTION (TSS) is defined as follows. Given an undirected graph $G = (V, E)$, a threshold function $\mathrm{thr} : V \to \mathbb{N}$, and an integer $k \geq 0$, is there a target set $S \subseteq V$ for G with $|S| \leq k$ activating all vertices in V, that is, for every $v \in V \setminus S$ eventually at least $\mathrm{thr}(v)$ of v's neighbors are activated? Note that *activation* is a dynamic process, where initially only the vertices in S are activated and the remaining vertices may become activated step by step during several rounds (see Section 2 for a formal definition). Roughly speaking, TSS offers a simple model to study the spread of influence, infection, or information in social networks; Kempe et al. [14] referred to it as influence maximization with a linear threshold model. In this work, we extend previous work [5,1] by studying the computational complexity of TSS for several special cases.

Domingos and Richardson [7] introduced TSS, studying it from the viewpoint of viral marketing and solving it heuristically. Next, Kempe et al. [14] formulated the problem using a threshold model (as we use it now), showed its NP-completeness, and presented a constant-factor approximation algorithm for a maximization variant. Next, Chen [5] showed the APX-hardness of a minimization variant, which holds even in case of some restricted threshold functions. He also provided a linear-time algorithm for trees. Most recently, Ben-Zwi et al. [1] generalized Chen's result for trees by showing

---

[*] Supported by the DFG, research projects PABI, NI 369/7, and DARE, NI 369/11.

O. Cheong, K.-Y. Chwa, and K. Park (Eds.): ISAAC 2010, Part I, LNCS 6506, pp. 378–389, 2010.

that TSS is polynomial-time solvable for graphs of constant treewidth; however, the degree of the polynomial of the running time depends on the treewidth (in other words, they showed that TSS is in the parameterized complexity class XP when parameterized by treewidth). They also proved that there is no $|V|^{o(\sqrt{w})}$ time algorithm (w denoting the treewidth) unless some unexpected complexity-theoretic collapse occurs. In particular, there is no hope for fixed-parameter tractability of TSS when parameterized by just treewidth.

Motivated by the mostly negative (or impractical) algorithmic results of Ben-Zwi et al. [1], we study further natural parameterizations of TSS. We obtain both hardness and tractability results. Since TSS resembles a dynamic variant of the well-known DOM-INATING SET problem, it seems unsurprising that the NP-hardness of DOMINATING SET on graphs with diameter two [15] carries over to TSS. We show that this is indeed the case. In contrast, we also observe that TSS can be solved in linear time for diameter-one graphs (that is, cliques).

The main part of the paper considers three different parameterizations of TSS exploiting different structural graph parameters: First, we look at the "cluster edge deletion number" ξ of the input graph denoting the minimum number of edges whose deletion transforms the graph into a disjoint union of cliques. We show that TSS can be solved in $O(4^\xi \cdot |E| + |V|^3)$ time and provide polynomial problem kernel with respect to ξ and the maximum threshold $t_{\max}$.

Following the spirit of previous work considering graph layout and coloring problems [9,10], we study the parameter "vertex cover number" τ of the underlying graph. This parameter imposes a stronger restriction than treewidth does. Indeed, we prove that TSS is fixed-parameter tractable when parameterized by τ. In addition, for constant thresholds we show a problem kernel consisting of $O(2^\tau)$ vertices and prove that there is little hope for a polynomial-size problem kernel in the general case.

Finally, as a third and (other than the previous parameters) easy-to-compute parameter also measuring tree-likeness, we study the feedback edge set number f (the minimum number of edges to delete to make a graph acyclic).[1] We develop polynomial-time data reduction rules that yield a linear-size problem kernel with respect to the parameter f. Moreover, we show that TSS can be solved in $4^f \cdot n^{O(1)}$ time, which again generalizes Chen's result [5] for trees.

Due to the lack of space, most proofs are deferred to a full version of the paper.

2   Preliminaries, Helpful Facts, and Small Diameter

Basic notation. Let $G = (V, E)$ be a graph and let $n := |V|$ and $m := |E|$ throughout this work. The *(open) neighborhood* of a vertex $v \in V$ in G is $N_G(v) := \{u : \{u, v\} \in E\}$ and the *degree* of v in G is $\deg_G(v) := |N_G(v)|$. The *closed neighborhood* of a vertex $v \in V$ in G is $N_G[v] := N_G(v) \cup \{v\}$. Moreover, for $V' \subseteq V$ let $N_G(V') := \bigcup_{v \in V'} N_G(v) \setminus V'$ and $N_G[V'] := \bigcup_{v \in V'} N_G[v]$. *Bypassing* a vertex $v \in V$ with $N(v) = \{u, w\}$ means to delete v from G and to insert the edge $\{u, w\}$. Note that, if u and w are already neighbors in G, then v must not be bypassed. We

---

[1] Graphs with small feedback edge set number are "almost trees"; such social networks occur in the context of e.g. sexually transmitted infections [19] and extremism propagation [12].

sometimes write $G-x$ as an abbreviation for the graph that results from G by deleting x, where x may be a vertex, edge, vertex set, or edge set. A *split graph* is a graph in which the vertices can be partitioned into a clique and an independent set. The class of split graphs is contained in the class of chordal graphs [4]. For a vertex set S, let $\mathcal{A}_G^i(S)$ denote the set of vertices of G that are *activated by S in the i'th round*, with $\mathcal{A}_G^0(S) := S$ and $\mathcal{A}_G^{j+1}(S) := \mathcal{A}_G^j(S) \cup \{v \in V : |N(v) \cap \mathcal{A}_G^j(S)| \geq \mathrm{thr}(v)\}$. For $S \subseteq V$, the uniquely determined positive integer r with $\mathcal{A}_G^{r-1}(S) \neq \mathcal{A}_G^r(S) = \mathcal{A}_G^{r+1}(S)$ is called the *number $r_G(S)$ of activation rounds*. Furthermore, we call $\mathcal{A}_G(S) := \mathcal{A}_G^{r_G(S)}(S)$ the set of vertices that are *activated by S*. If $\mathcal{A}_G(S) = V$, then S is called a *target set for G*. Thus, we arrive at the problem TARGET SET SELECTION: given an undirected graph $G = (V, E)$, a threshold function $\mathrm{thr} : V \to \mathbb{N}$ and an integer $k \geq 0$, is there a target set $S \subseteq V$ for G that contains at most k vertices?

Thresholds. Apart from arbitrary threshold functions, we consider two types of different threshold functions in this work.

Constant thresholds: All vertices have the same threshold $t_{\max}$. Since solving instances of TSS with $t_{\max} = 1$ is trivial (just select an arbitrary vertex in each connected component), we assume that all connected components of an input graph contain a vertex with threshold at least two.

Degree-dependent thresholds: The threshold of a vertex v depends on $\deg(v)$. In this context, note that, if $\mathrm{thr}(v) = \deg(v)$ for all vertices v, then TSS is identical to VERTEX COVER [5]. In this work, we particularly consider the "majority" threshold function, defined as $\mathrm{thr}(v) := \lceil \deg(v)/2 \rceil$.

In our work, sometimes thresholds of vertices are being decreased. However, thresholds can never be smaller than zero; further decreasing a threshold of zero has no effect.

Parameterized complexity. This is a two-dimensional framework for the analysis of computational complexity [8,11,17]. One dimension is the input size n, and the other one is the *parameter* (usually a positive integer). A problem is called *fixed-parameter tractable* (fpt) with respect to a parameter k if it can be solved in $f(k) \cdot n^{O(1)}$ time, where f is a computable function only depending on k. A core tool in the development of fixed-parameter algorithms is polynomial-time preprocessing by *data reduction* [2,13]. Here, the goal is to transform a given problem instance I with parameter k in polynomial time into an equivalent instance I' with parameter $k' \leq k$ such that the size of I' is upper-bounded by some function g only depending on k. If this is the case, we call I' a *kernel* of size $g(k)$. Usually, this is achieved by applying data reduction rules. We call a data reduction rule $\mathcal{R}$ *correct* if the new instance I' that results from applying $\mathcal{R}$ to I is a yes-instance if and only if I is a yes-instance. An instance is called *reduced* with respect to some data reduction rule if this rule has no further effect when applied to the instance. The whole process is called *kernelization*. Downey and Fellows [8] developed a parameterized theory of computational complexity to show fixed-parameter intractability by means of *parameterized reductions*. A parameterized reduction from a parameterized problem P to another parameterized problem P' is a function that, given an instance (x, k), computes in $f(k) \cdot n^{O(1)}$ time an instance (x', k') (with k' only depending on k) such that (x, k) is a yes-instance of P if and only if (x', k') is a yes-instance of P'. The basic complexity class for fixed-parameter intractability is called $W[1]$ and there is good reason to believe that $W[1]$-hard problems are not fpt [8,11,17]. Moreover, there is a whole hierarchy of classes $W[t]$, $t \geq 1$, where, intuitively, problems become harder with growing t.

Basic Facts. Based on the following observation, we present a data reduction rule to deal with vertices whose thresholds exceed their degrees.

Observation 1. *Let $G = (V, E)$ be a graph and let $v \in V$. If $\mathrm{thr}(v) > \deg(v)$, then v is contained in all target sets for G. If $\mathrm{thr}(v) = 0$, then v is not contained in any optimal target set for G.*

Reduction Rule 1. *Let $G = (V, E)$ and $v \in V$. If $\mathrm{thr}(v) > \deg(v)$, then delete v, decrease the threshold of all its neighbors by one and decrease k by one. If $\mathrm{thr}(v) = 0$, then delete v and reduce the thresholds of all its neighbors by one.*

Considering that changes propagate over each edge at most once, it is not hard to see that a graph can be reduced with respect to Reduction Rule 1 in $O(n + m)$ time. With Reduction Rule 1, we can define the graph that remains after an activation process.

Definition 1. *Let $(G = (V, E), \mathrm{thr}, k)$ be an instance of TSS, let $S \subseteq V$, and let $\mathrm{thr}_S : V \rightarrow \mathbb{N}$ with $\mathrm{thr}_S(v) := \infty$ for all $v \in S$ and $\mathrm{thr}_S(v) := \mathrm{thr}(v)$ for all $v \in V \setminus S$. Then we call the instance (G', thr', k') that results from exhaustively applying Reduction Rule 1 to (G, thr_S, k) the* reduced instance *of (G, thr, k) with respect to S.*

Observation 2. *Let $G = (V, E)$ be a graph reduced with respect to Reduction Rule 1. Then there is an optimal target set for G not containing vertices with threshold one.*

Small-Diameter Graphs. Since many real-world (social) networks have small diameter, it is natural to investigate the influence of the diameter on the complexity of TSS. If the diameter of a given graph G is one (that is, G is a clique), then a simple exchange argument allows us to observe that there is a minimum-size target set for G that contains the vertices with the highest thresholds. With Observation 1, we can sort the vertices by their thresholds in linear time using bucket sort and, once sorted, we can determine in linear time whether k vertices suffice to activate the clique. For graphs with diameter at least two, the close relationship to DOMINATING SET suggests that TSS is NP-hard [15]. We confirm this intuition by providing hardness results for several variants of TSS on graphs with constant diameter. Our hardness proofs use reductions from the $W[2]$-hard HITTING SET problem.

Theorem 1. TARGET SET SELECTION *is NP-hard and W[2]-hard for the parameter target set size k, even on*

1. *split graphs with diameter two,*
2. *bipartite graphs with diameter four and constant threshold two, and*
3. *bipartite graphs with diameter four and majority thresholds.*

3   Cluster Edge Deletion Number

Social networks may consist of almost cluster graphs. This motivates to study a parameterization measuring the distance from the input graph to a graph where all connected
components are cliques, a so-called "cluster graph" [20]. One such distance measure is
the size ξ of a minimum cluster edge deletion set, that is, a smallest set of edges whose
removal results in a cluster graph.

   For arbitrary thresholds, we show that TSS is fixed-parameter tractable with respect
to ξ by providing an algorithm running in $O(4^{\xi} \cdot m + n^3)$ time. For restricted thresholds,
we present a linear-time computable kernel whose number of vertices grows linearly
with ξ when the thresholds are bounded by some constant. In fact, we can bound the
number of vertices in this kernel by $2\xi(t_{\max} + 1)$, where $t_{\max}$ is the maximum threshold occurring in the input. Our elaborations highly depend on the notion of "critical
cliques": a clique K in a graph is a *critical clique* if all its vertices have the same closed
neighborhood and K is maximal with respect to this property. Computing all critical
cliques of a graph can be done in linear time [16].

Arbitrary Thresholds. To present a solving strategy for TSS in case of arbitrary thresholds, we first introduce a data reduction rule that shrinks clique-like substructures. The
key to showing fixed-parameter tractability is the observation that, in a graph reduced
with respect to this data reduction rule, there is an optimal target set consisting only
of vertices that are incident to the ξ edges of a given optimal cluster edge deletion set.
Since we can compute an optimal target set for a clique in linear time (see Section 2),
we assume that none of the connected components of the input graph is a clique.

   Now, we describe the data reduction rule that shrinks clique-like substructures. Consider a critical clique K in the input graph G and its neighbors $N_G(K)$ and note
that $N_G(K)$ separates K from the rest of G. If activating all vertices in $N_G(K)$ is
not enough to activate all vertices of K, then every target set of G has to contain some
vertices of K. Without loss of generality, we can assume those vertices to have the
highest threshold among all vertices of K. These considerations lead to the following.

Reduction Rule 2. *Let $I := (G, \mathrm{thr}, k)$ be an instance of TSS and let K be a critical
clique in G. Moreover, let (G', thr', k') be the reduced instance of $(G[N_G[K]], \mathrm{thr}, k)$
with respect to $N_G(K)$ and let S denote an optimal target set for (G', thr') (see
Definition 1). Then, reduce I with respect to S.*

Herein G' is a clique and, hence, an optimal target set S for (G', thr') can be computed
in linear time (see Section 2).

   Having reduced the input with respect to Reduction Rule 2, we claim that we can
limit our considerations to a small subset of vertices of the remaining graph. To see
this, consider a cluster edge deletion set E' of a graph G and let $V(E')$ denote the
vertices incident to the edges in E' (we call the vertices in $V(E')$ *affected* by E').

Lemma 1. *Let $I := (G, \mathrm{thr}, k)$ denote a yes-instance reduced with respect to
Reduction Rule 2 and let E' denote an optimal cluster edge deletion set of G. Then,
there exists an optimal target set S for I with $S \subseteq V(E')$.*

By Lemma 1, an optimal target set can be found by systematically checking every subset of the at most 2ξ vertices affected by an optimal cluster edge deletion set, leading directly to the following theorem.

Theorem 2. TARGET SET SELECTION *can be solved in* $O(4^\xi \cdot m + n^3)$ *time, where* ξ *denotes the cluster edge deletion number of G.*

Restricted Thresholds. In the following, we present a data reduction shrinking large critical cliques until their size can be bounded by the maximum threshold $t_{\max}$. Thereby, we obtain a problem kernel with $2\xi(t_{\max} + 1)$ vertices, which is linear in ξ if the thresholds of the input are bounded by a constant.

Consider a critical clique K containing at least $t_{\max} + 1$ vertices. Informally speaking, it is sufficient to keep the $t_{\max}$ vertices of K with the smallest thresholds, since if these are activated, then all vertices in $N_G[K]$ are activated.

Reduction Rule 3. *Let* (G, thr, k) *be an instance of* TSS *and let* $t_{\max}$ *denote the maximum threshold of this instance. Furthermore, let* K *be a critical clique in* G *with* $|K| > t_{\max}$ *and let* K^{high} *denote the* $|K| - t_{\max}$ *vertices with highest thresholds of* K. *Then, delete the vertices in* K^{high} *from* G.

In a reduced instance, the number of critical cliques is upper-bounded by 2ξ and each critical clique contains at most $t_{\max}$ vertices. Thus, the number of vertices in a reduced graph is at most $2\xi(t_{\max} + 1)$.

Theorem 3. TARGET SET SELECTION *admits a problem kernel with* $2\xi(t_{\max} + 1)$ *vertices. The kernelization runs in linear time.*

4   Vertex Cover Number

The vertex cover number of a graph G denotes the cardinality of an optimal vertex cover of G (that is, a minimum-size vertex set such that every edge has at least one endpoint in this set). Since deleting all vertices in a vertex cover results in a graph without edges, a vertex cover of a graph G is a feedback vertex set for G as well. Moreover, it is a well-known fact that the feedback vertex set number is an upper bound on the treewidth. Hence, the vertex cover number is an upper bound on the treewidth, too. Ben-Zwi et al. [1] have shown that TSS is W[1]-hard for the combined parameter treewidth and target set size. Indeed, the given reduction shows that TSS is W[1]-hard even for the combined parameter feedback vertex set number and target set size. These hardness results motivate the study of TSS parameterized by parameters larger than the feedback vertex set number as, for example, the vertex cover number [9,10].

Arbitrary thresholds. If the input graph G is reduced with respect to Reduction Rule 1, then the threshold of each vertex is bounded by its degree. Thus, it can be activated by activating all its neighbors. This implies that a vertex cover of G is also a target set for G and, hence, the target set size k is at most the vertex cover number τ of G. In this sense, the vertex cover number τ is a "weaker" parameter than the target set size k and allows for fixed-parameter algorithms with respect to parameter τ.

Let $G = (V, E)$ denote a graph and let Z denote a vertex cover of G. Then, $I :=$ $V \backslash Z$ is an independent set. The key to showing fixed-parameter tractability of TSS with respect to τ is to bound the number of vertices in I that are candidates for an optimal target set. To this end, vertices in I with an identical neighborhood are of particular interest. In this section, τ denotes the vertex cover number of G.

Definition 2. *Two pairwise non-adjacent vertices with the same open neighborhood are called* twins. *A set V' of vertices is called* critical independent set *if each two distinct vertices from V' are twins and V' is maximal with respect to this property.*

The vertices of I are contained in at most $2^{|Z|}$ critical independent sets (one for each subset of Z). The next observation allows us to focus on a restricted number of vertices for each critical independent set when looking for an optimal target set.

Observation 3. *Let $G = (V, E)$ denote a graph and let $u, w \in V$ denote two twins of G with $\mathrm{thr}(u) \geq \mathrm{thr}(w)$. In addition, let S denote a target set with $w \in S$ and $u \notin S$. Then, $S' := S \setminus \{w\} \cup \{u\}$ is a target set for G.*

In the following, we assume that the vertices of the input graph are ordered decreasingly by their thresholds. By Observation 3, we can directly conclude that there is an optimal target set that contains for each critical independent set I only vertices from the set of the $k \leq \tau$ vertices with highest threshold of I. Moreover, we can bound the number of critical independent sets by a function of τ, leading to fixed-parameter tractability of TSS with respect to τ.

Theorem 4. TARGET SET SELECTION *can be solved in $O(2^{(2^\tau+1)\cdot\tau} \cdot m)$ time, where τ denotes the vertex cover number of G.*

Proof. (Sketch) Let $(G = (V, E), \mathrm{thr}, k)$ denote the input instance for TSS. In the following, we assume that G is reduced with respect to Reduction Rule 1.

The algorithm works as follows. In a first phase, it computes an optimal vertex cover Z of G (this can be done in $O(1.3^\tau + \tau n)$ time [17]). If $k \geq |Z|$, then the algorithm returns Z. In a second phase it computes a set C of at most $(2^\tau + 1) \cdot \tau$ vertices for which there exists an optimal target set S with $S \subseteq C$. Then, in a third phase, the algorithm determines an optimal target set by systematically checking every subset of C. The set C in the second phase is computed as follows: For each critical independent set, C contains the k vertices with highest threshold (for a critical independent set of cardinality at most k all its vertices are in C). In the third phase, the algorithm systematically checks every size-k subset of C for being a target set. If no such target set is found, then it rejects the instance.

Next, we show that the presented algorithm finds a target set of size at most k for G, if it exists. For the correctness of the first phase note that, since G is reduced with respect to Reduction Rule 1, any vertex cover of G is a target set for G, as well. For the correctness of the other phases note that, by iteratively applying Observation 3, we can directly conclude that there exists an optimal target set S with $S \subseteq C$. Thus, the correctness of the algorithm follows by the fact that it checks every size-k subset of C.

We omit the running time analysis. $\square$

Restricted thresholds. Next, we show that TSS admits a problem kernel with $O(2^\tau \cdot t_{\max})$ vertices. Moreover, we show that TSS parameterized by the vertex cover number presumably does not admit a polynomial problem kernel even for majority thresholds.

First, we present a problem kernel for the combined parameter vertex cover number τ and maximum threshold $t_{\max}$. To bound the number of vertices in a reduced instance, we need (besides Reduction Rule 1) one further rule that reduces large critical independent sets. For every critical independent set of size at least $t_{\max} + 1$, this rule removes all but the $t_{\max}$ vertices with "smallest thresholds".

Reduction Rule 4. *Let (G, thr, k) denote a TSS-instance reduced with respect to Reduction Rule 1 and let I denote a critical independent set of G with $|I| > t_{\max} + 1$. Then delete the $|I| - (t_{\max} + 1)$ highest-threshold vertices of I.*

Theorem 5. TARGET SET SELECTION *admits a problem kernel with at most $O(2^\tau \cdot t_{\max})$ vertices, where $t_{\max}$ denotes the maximum threshold and τ denotes the vertex cover number of G. The kernelization runs in $O(n \cdot (n + m))$ time.*

Dom et al. [6] showed that HITTING SET does not admit a problem kernel of size $(|U| + k')^{O(1)}$ unless an unexpected complexity-theoretic collapse occurs. Here, k' denotes the solution size and $|U|$ denotes the size of the universe. Bodlaender et al. [3] introduced a refined concept of parameterized reduction (called polynomial time and parameter transformation) that allows to transfer such hardness results to new problems. By a reduction from HITTING SET to TSS, we can show that, under reasonable complexity-theoretic assumptions, TSS does not admit a problem kernel of size $(\tau + k)^{O(1)}$.

Theorem 6. TARGET SET SELECTION *on bipartite graphs with majority thresholds does not admit a problem kernel of size $(\tau + k)^{O(1)}$ unless coNP $\subseteq$ NP/poly, where k denotes the target set size and τ denotes the vertex cover number of G.*

5   Feedback Edge Set Number

Another structural parameter of a graph G that is lower bounded by its treewidth is the size f of a smallest feedback edge set of G, that is, a smallest set of edges of G whose deletion makes G acyclic. Recall from Section 4 that TSS is $W[1]$-hard with respect to the parameter feedback vertex set. Some real-world social networks are tree-like (for example sexual networks [19], see also [5]) or scale-free (see [12]). This often corresponds to small feedback edge sets and motivates parameterizing TSS with f. Like the vertex cover number τ, the feedback edge set number f is a "weaker" parameter than treewidth, raising hope for faster algorithms for TSS in cases where f is small. A clear advantage over treewidth (and tree decompositions) is that an optimal feedback edge set can be determined efficiently by computing a spanning tree.

First, we show a problem kernel of size $O(f)$ for TSS based on two data reduction rules. Second, we present an algorithm that solves TSS in $4^f \cdot n^{O(1)}$ time.

Kernelization for TSS. In this paragraph, we present two data reduction rules for TSS that can be applied exhaustively in linear time and leave a problem kernel of size $O(f)$. Reduction Rule 5 removes vertices v with $\mathrm{thr}(v) = \deg(v) = 1$ from G. Note that Reduction Rule 1 (see Section 2) removes all other degree-one vertices from G.

Reduction Rule 5. *Let* $(G = (V, E), \text{thr}, k)$ *be an instance for* TSS *reduced with respect to Reduction Rule 1 and let* $v \in V$ *with* $\text{thr}(v) = \deg(v) = 1$. *Then delete* v *from* G.

The correctness of Reduction Rule 5 follows immediately from Observation 2 and one easily verifies that Reduction Rule 5 can be exhaustively applied in linear time. In a graph reduced with respect to Reduction Rules 1 and 5, we can bound the number of vertices with degree at least three by $2f$.

Lemma 2. *A graph with feedback edge set number* f *reduced with respect to Reduction Rules 1 and 5 has at most* $2f$ *vertices with degree at least three.*

With Lemma 2 we bound the number of vertices with degree at least three in graphs that are reduced with respect to Reduction Rules 1 and 5. Since these graphs do not contain degree-one vertices, it remains to bound the degree-two vertices.

Reduction Rule 6. *Let* (G, thr, k) *be an instance for* TSS *reduced with respect to Reduction Rule 1 and let* (u, v, w) *be a path in* G *with* $\deg(u) = \deg(v) = \deg(w) = 2$ *where* u *and* w *are neither neighbors nor twins. If there is a vertex* $x \in \{u, v, w\}$ *with* $\text{thr}(x) = 1$, *then bypass* x, *otherwise, bypass* u *and* w *and decrease* k *by one.*

Theorem 7. TARGET SET SELECTION *admits a problem kernel of size* $O(f)$, *where* f *denotes the feedback edge set number. The kernelization runs in linear time.*

Proof. Let (G, thr, k) be an instance of TSS reduced with respect to Reduction Rules 1, 5, and 6. Furthermore, let F denote an optimal feedback edge set for G. Thus, $T := G - F$ is a forest. Consider the graph T^* that results from bypassing all vertices in T having degree two in both G and T. It is easy to see that all vertices in T^* have the same degree in T^* as they have in T. Hence, each vertex v with degree two in T^* is incident to an edge of F in G because otherwise it would have been bypassed. Furthermore, each leaf in T^* not incident to an edge of F is also a leaf in G, but since G is reduced with respect to Reduction Rules 1 and 5, each leaf of T^* is incident to an edge of F. Thus, the number of vertices of T^* with degree at most two is bounded by $2f$. Furthermore, by Lemma 2, the number of vertices of T^* with degree at least three is bounded by $2f$. Thus, the overall number of vertices in T^* is at most $4f$. Finally, since G is reduced with respect to Reduction Rule 6, for each edge $\{u, v\}$ of T^*, there are at most two vertices between u and v in T. Hence, the overall number of vertices of T can be bounded by $12f$ and since T is a forest, so can the overall number of edges of T. By construction of T, we can bound the overall number of vertices of G by $12f$ and the overall number of edges of G by $13f$.

It can be shown that any graph can be reduced with respect to Reduction Rules 1, 5, and 6 in linear time.   □

FPT algorithm for TSS. Theorem 7 already implies that TSS is fixed-parameter tractable by applying a brute-force search to the size-$O(f)$ problem kernel. Here, we develop a fixed-parameter algorithm with much better efficiency.

Our algorithm computes a solution S for an input instance (G, thr, k) for TSS if there is one. The algorithm runs in two phases. In Phase 1, we branch on degree-two vertices that are in a cycle of G. In Phase 2, we alternatingly try solutions for a cyclic subgraph and apply previous data reduction rules.

In the algorithm, we use the notion of branching rules. Here, we analyze an instance I and replace it with a set $\mathcal{I}$ of new instances. The creation of new instances is defined in branching rules, which we call *correct* if the original instance I is a yes-instance if and only if there is a yes-instance in $\mathcal{I}$. In analogy to data reduction rules, instances that are not subject to a specific branching rule $\mathcal{R}$ are called *reduced* with respect to $\mathcal{R}$. By considering each application of a branching rule as a vertex with parent I and children $\mathcal{I}$, we can define a *search tree* for each input instance.

In the following, we denote the set of all vertices of a graph G being part of cycles by $V_C(G)$. Let C be some two-edge connected component of G containing at least three vertices, that is, C is a non-singleton connected component that remains connected after removing all bridges from G.

Phase 1. In the first phase of our algorithm we branch on vertices of $V_C(G)$ with degree two and threshold two. To this end, we present a branching rule with branching number two that decreases the feedback edge set number f in every application. This lets us bound the search tree size by 2^f.

Branching Rule 1. *Let* $I := (G = (V, E), \mathrm{thr}, k)$ *be an instance of* TSS *and let* $v \in V \cap V_C(G)$ *and* $\mathrm{thr}(v) = \deg(v) = 2$. *Then, create the instance of* I *that is reduced with respect to* $\{v\}$ *and create the instance* $(G - v, \mathrm{thr}, k)$.

It is easy to see that Branching Rule 1 branches into at most two cases. After each application of Branching Rule 1, our algorithm exhaustively applies Reduction Rules 1 and 5 to the created instances. If none of Branching Rule 1 and Reduction Rules 1 and 5 applies to any of the created instances, then Phase 2 of the algorithm begins.

Phase 2. At the beginning of the second phase, no input instance is subject to Branching Rule 1 or Reduction Rules 1 and 5. In Phase 2, we branch on vertices with degree at least three that are contained in cycles, followed by the application of Reduction Rules 1 and 5. This process is iterated for all input instances until all input instances can trivially be determined to be yes- or no-instances.

Let $I_1 := (G_1, \mathrm{thr}_1, k_1)$ denote an input instance for Phase 2. For a two-edge connected component C of G_1, we denote the set of vertices of $V_C(C)$ that have degree two by $V_2(C)$. Note that, since G_1 is reduced with respect to Reduction Rule 1 and Branching Rule 1, all vertices in $V_2(C)$ have threshold one. Hence, by Observation 2, there is an optimal solution for I_1 that does not contain vertices of V_2. Note that, if each two-edge connected component of G_1 is contracted into a single vertex, then the remaining graph is a forest. In the following, we call this forest the *component forest* T_1 of G_1. Since G_1 is reduced with respect to Reduction Rules 1 and 5, every leaf of T_1 corresponds to a contracted non-singleton two-edge connected component of G.

Branching Rule 2. *Let* $I_1 := (G_1, \mathrm{thr}_1, k_1)$ *be an instance of* TSS *reduced with respect to Reduction Rules 1 and 5 and let* T_1 *be the component forest of* G_1. *Furthermore, let* C *denote a two-edge connected component of* G_1 *corresponding to a leaf in* T_1, *let* v *be the only vertex in* $N_{G_1}(V(C))$, *and let* w *be the only vertex in* $N(v) \cap V(C)$. *Then,* *for each $V' \subseteq V(C) \setminus V_2$, create a new instance by modifying I_1 in the following way. If V' is a target set for C, then create the instance of I_1 that is reduced with respect to V'. Else, if V' is a target set for C with $\mathrm{thr}_1(w)$ decreased by one, then delete $V(C)$ from G_1 and decrease k_1 by $|V'|$. Otherwise, reduce to a trivial no-instance.*

Note that one application of Branching Rule 2 can be pictured as branching into the cases $v \in S$ and $v \notin S$ for each vertex $v \in V_C(G_1) \setminus V_2$. By Lemma 2, $|V_C| \le 2f_1$, with f_1 denoting the feedback edge set number of G_1. Since each application of Branching Rule 1 decreases the feedback edge set number of the input, continuing the search tree of Phase 1 with Branching Rule 2 leads to a search tree of depth at most $2f$ and each node in the search tree has at most two children. Therefore, the overall search tree size is at most 4^f. After the exhaustive application of Branching Rule 2 and Reduction Rules 1 and 5, we can decide for each remaining instance whether it is a yes-instance of TSS or not. If some instance is a yes-instance, then the original instance I is also a yes-instance of TSS. Otherwise, I is a no-instance of TSS.

Theorem 8. TARGET SET SELECTION *can be solved in $4^f \cdot n^{O(1)}$ time, where f denotes the feedback edge set number.*

6 Conclusion and Open Problems

Following the spirit of multivariate algorithmics [18], we studied natural parameterizations of TSS. We confirmed the intuition deriving from the hardness of DOMINATING SET that TSS remains hard for the parameter diameter. However, for the structural parameters "cluster edge deletion number", "vertex cover number", and "feedback edge set number" we established tractability results by showing fixed-parameter algorithms and kernelizations. Since for several applications the parameter "number of activation rounds" appears to be relevant and should be small, this motivates to investigate the influence of this parameter on the complexity of TSS. As it turns out, however, TSS is already NP-hard for only two rounds; similarly to diameter, this calls for the combination with other parameters. Finally, note that we focused on activating *all* graph vertices; it is natural to extend our studies to cases when only a specified fraction of the graph vertices shall be activated at minimum cost.

References

1. Ben-Zwi, O., Hermelin, D., Lokshtanov, D., Newman, I.: An exact almost optimal algorithm for target set selection in social networks. In: Proc. 10th ACM EC, pp. 355–362. ACM Press, New York (2009)
2. Bodlaender, H.L.: Kernelization: New upper and lower bound techniques. In: IWPEC 2009. LNCS, vol. 5917, pp. 17–37. Springer, Heidelberg (2009)
3. Bodlaender, H.L., Thomassé, S., Yeo, A.: Analysis of data reduction: Transformations give evidence for non-existence of polynomial kernels. Technical Report UU-CS-2008-030, Department of Information and Computing Sciences, Utrecht University (2008)
4. Brandstädt, A., Le, V.B., Spinrad, J.P.: Graph Classes: a Survey. SIAM Monographs on Discrete Mathematics and Applications, vol. 3. SIAM, Philadelphia (1999)

5. Chen, N.: On the approximability of influence in social networks. SIAM Journal on Discrete Mathematics 23(3), 1400–1415 (2009)
6. Dom, M., Lokshtanov, D., Saurabh, S.: Incompressibility through colors and IDs. In: Albers, S., Marchetti-Spaccamela, A., Matias, Y., Nikoletseas, S., Thomas, W. (eds.) ICALP 2009. LNCS, vol. 5555, pp. 378–389. Springer, Heidelberg (2009)
7. Domingos, P., Richardson, M.: Mining the network value of customers. In: Proc. 7th ACM KDD, pp. 57–66. ACM Press, New York (2001)
8. Downey, R.G., Fellows, M.R.: Parameterized Complexity. Springer, Heidelberg (1999)
9. Fellows, M.R., Lokshtanov, D., Misra, N., Rosamond, F.A., Saurabh, S.: Graph layout problems parameterized by vertex cover. In: Hong, S.-H., Nagamochi, H., Fukunaga, T. (eds.) ISAAC 2008. LNCS, vol. 5369, pp. 294–305. Springer, Heidelberg (2008)
10. Fiala, J., Golovach, P.A., Kratochvíl, J.: Parameterized complexity of coloring problems: Treewidth versus vertex cover. In: TAMC 2009. LNCS, vol. 5532, pp. 221–230. Springer, Heidelberg (2009)
11. Flum, J., Grohe, M.: Parameterized Complexity Theory. Springer, Heidelberg (2006)
12. Franks, D.W., Noble, J., Kaufmann, P., Stagl, S.: Extremism propagation in social networks with hubs. Adaptive Behavior 16(4), 264–274 (2008)
13. Guo, J., Niedermeier, R.: Invitation to data reduction and problem kernelization. ACM SIGACT News 38(1), 31–45 (2007)
14. Kempe, D., Kleinberg, J., Tardos, E.: Maximizing the spread of influence through a social network. In: Proc. 9th ACM KDD, pp. 137–146. ACM Press, New York (2003)
15. McCartin, C., Rossmanith, P., Fellows, M.: Frontiers of intractability for dominating set (2007) (manuscript)
16. McConnell, R.M., Spinrad, J.: Linear-time modular decomposition and efficient transitive orientation of comparability graphs. In: Proc. 5th SODA, pp. 536–545. ACM/SIAM (1994)
17. Niedermeier, R.: Invitation to Fixed-Parameter Algorithms. Oxford University Press, Oxford (2006)
18. Niedermeier, R.: Reflections on multivariate algorithmics and problem parameterization. In: Proc. 27th STACS. LIPIcs, vol. 5, pp. 17–32. Schloss Dagstuhl–Leibniz-Zentrum fuer Informatik (2010)
19. Potterat, J.J., Phillips-Plummer, L., Muth, S.Q., Rothenberg, R.B., Woodhouse, D.E., Maldonado-Long, T.S., Zimmerman, H.P., Muth, J.B.: Risk network structure in the early epidemic phase of HIV transmission in Colorado Springs. Sexually Transmitted Infections 78, 159–163 (2002)
20. Shamir, R., Sharan, R., Tsur, D.: Cluster graph modification problems. Discrete Applied Mathematics 144(1-2), 173–182 (2004)

Combining Two Worlds:
Parameterised Approximation for Vertex Cover

Ljiljana Brankovic[1,*] and Henning Fernau[2]

[1] School of Electrical Engineering and Computer Science
The University of Newcastle, Callaghan, NSW 2308, Australia
Ljiljana.Brankovic@newcastle.edu.au
[2] Fachbereich 4, Abteilung Informatik
Universität Trier, 54286 Trier, Germany
fernau@uni-trier.de

Abstract. We explore opportunities for parameterising constant factor approximation algorithms for vertex cover. We provide a simple algorithm that works on any approximation ratio of the form $\frac{2l+1}{l+1}$ and has complexity that outperforms an algorithm by Bourgeois et al. derived from a sophisticated exact parameterised algorithm. In particular, for $l = 1$ (factor 1.5 approximation) our algorithm runs in time $\mathcal{O}^*(1.09^k)$. Additionally, we present an improved polynomial-time approximation algorithm for graphs of average degree four.

1 Introduction

MINIMUM VERTEX COVER: *A hard problem and how to deal with it.* Given a graph $G = (V, E)$, where V is the set of vertices and E the set of edges of G, a *vertex cover* C of G is a subset of V whose removal leaves only isolated vertices. The problem of finding a minimum vertex cover in an arbitrary graph, called MINIMUM VERTEX COVER or MINVC for short, has long attracted the attention of an army of theoretical computer scientists around the globe. MINVC is one of the most studied NP-hard graph problems in complexity theory. In particular, it has been the paradigmatic test-bed problem for the development of parameterised algorithms, being among the first problems presented in any introduction to that field. Minimum vertex cover can be formulated as a decision problem as follows:

Problem name: VERTEX COVER (VC)
Given: A graph $G = (V, E)$
Parameter: a positive integer k
Output: Is there a *vertex cover* $C \subseteq V$ such that $|C| \leq k$?

Both the polynomial-time approximation approach and the fixed-parameter approach to VC have their advantages and limitations. It is therefore of interest to combine both approaches, aiming at better approximation factors by allowing

* Supported by the RGC CEF grant G0189479 of The University of Newcastle.

O. Cheong, K.-Y. Chwa, and K. Park (Eds.): ISAAC 2010, Part I, LNCS 6506, pp. 390–402, 2010.
© Springer-Verlag Berlin Heidelberg 2010

FPT-time [4]. We refer the reader to the survey by Dániel Marx [19] for more details on how parameterised complexity can be extended to approximation algorithms. In this paper we follow the definition by Y.Chen et al. [10], which in our context can be phrased as follows. Given a graph G and a parameter k such that a minimum vertex cover C^* of G satisfies $|C^*| \leq k$, a *parameterised approximation algorithm with (constant) approximation ratio ρ* for MINIMUM VERTEX COVER produces a vertex cover C such that $|C| \leq \rho|C^*|$. Such an algorithm runs in time $\mathcal{O}^*(f(k))$ for some function f, where $\mathcal{O}^*$ notation suppresses polynomial factors and $\mathcal{O}^*(f(k))$ replaces $\mathcal{O}(f(k)poly(n))$.

Earlier results on VC. We report on three approaches to the problem: (1) polynomial-time approximation, (2) parameterised (exact) algorithms and (3) parameterised (and exponential-time) approximation.

(1) Despite all efforts, the best known constant factor approximation algorithm for general graphs is still (basically) a factor 2 approximation. More specifically, G. Karakostas [17] derived an approximation factor of $2 - \Theta(\frac{1}{\sqrt{\log n}})$. I. Dinur and S. Safra [13] showed that there is no polynomial-time approximation algorithm for MINVC achieving an approximation ratio better than $10\sqrt{5} - 21 \approx 1.36067$, unless P = NP. Moreover, assuming that the Unique Games Conjecture is true, no approximation factor of the form $2 - \varepsilon$ is possible for any $\varepsilon > 0$, as shown by S. Khot and O. Regev [18]. There are better results for some specific classes of graphs, most notably graphs with bounded degree:

Lemma 1. *[2] There exists a polynomial time factor 7/6 approximation algorithm for* MINIMUM VERTEX COVER *for any graph with maximum degree* 3.

Lemma 2. *[15] There exists a polynomial time factor 3/2 approximation algorithm for* MINIMUM VERTEX COVER *for any graph with maximum degree* 4.

Lemma 3. *[14] For an arbitrary graph with average degree d_{avg}, there exists a factor $(4d_{avg} + 1)/(2d_{avg} + 3)$ approximation algorithm for* MINIMUM VERTEX COVER, *providing that the size of a minimum vertex cover is at least $|V|/2$.*

(2) For general graphs, several parameterised algorithms offer run times of about $\mathcal{O}^*(1.28^k)$ [6,7,9,16,20]. For special classes of graphs, better run time bounds are known. For cubic graphs, a race has led to $\mathcal{O}^*(1.162^k)$ [8,21]. For graphs of bounded genus, $\mathcal{O}^*(c^{\sqrt{k}})$ can be obtained [12]. Such run times cannot be obtained for general graphs unless the Exponential Time Hypothesis fails [5].

(3) The benefits of allowing exponential time to improve the approximation factors have been studied by several authors. For example, E. Dantsin et al. [11] showed how to transform an existing approximation algorithm for MAX SAT into an algorithm with a better approximation factor by allowing exponential time. In relation to our paper, the article by N. Bourgeois et al. [3] is the most relevant one, as it shows that if there exists an exact exponential time algorithm for computing minimum vertex covers that runs in time $\mathcal{O}^*(\gamma^n)$, then an approximation factor of $2 - \rho$, for $\rho \in (0, 1]$ can be obtained with run time $\mathcal{O}^*(\gamma^{\rho n})$. Furthermore, any parameterised algorithm for VC running in time $\mathcal{O}^*(\delta^k)$ that

produces an associated feasible solution can be used to obtain a parameterised approximation algorithm with approximation ratio ρ and run time $\mathcal{O}^*(\delta^{\rho k})$.

Main contributions. In this paper we present the first genuine parameterised approximation algorithm for MINIMUM VERTEX COVER. Our $\mathcal{O}^*(1.09^k)$ $\frac{3}{2}$ approximation algorithm is simple, yet faster than the $\mathcal{O}^*(\delta^{.5k}) = \mathcal{O}^*(1.13^k)$ algorithm by Bourgeois et al. that incorporates a quite sophisticated exact parameterised $\mathcal{O}^*(1.28^k)$ algorithm known for VC. The main difference between our algorithm and earlier approaches lies in the fact that we exploit the target approximation factor in each branching step, as opposed to taking the exact (parameterised) algorithm as a black box. This way, we not only obtain far better running times, but our branching algorithms are distinguished by their simplicity. It is also the first such algorithm that introduces local-ratio techniques known from polynomial-time approximation, see [1], into parameterised algorithmics, combining them with the search tree analysis itself. We generalise this algorithm to any constant factor of the form $\frac{2l+1}{l+1}$. We also obtain improved polynomial-time approximation algorithms for graphs of average degree four.

2   A Parameterized $\frac{3}{2}$ Approximation Algorithm

We consider simple undirected graphs and we use $G = (V, E)$ to denote a graph G with the vertex set V and the edge set E. We use $\Delta(G)$ and $\delta(G)$ to denote the maximum and minimum degree in G, respectively, and $d_G(v)$ to denote the degree of the vertex $v \in V$ within G, suppressing G if understood from the context. The neighbourhood of the vertex v is denoted by $N_G(v)$. Given a graph G and two nonadjacent vertices in G, u and v, those neighbours of u that are not also neighbours of v are called *private neighbours of u with respect to v*, referring informally to the set $N(u) \backslash N(v)$.

Lemma 4. *Consider a connected graph G with maximum degree $\Delta(G)$. We delete one or more vertices from G together with their incident edges, and we denote the resulting graph by G'. If G' contains any edges at all, then there exists a vertex y in G' such that $0 < d_{G'}(y) < \Delta(G)$.*

Proof. Consider a connected graph $G = (V, E)$ with maximum degree $\Delta(G)$. Let $G' = (V', E')$ be a graph obtained from G by deleting $X \subseteq V$ (plus incident edges) and let Y be the set of neighbours of vertices from X in G that are not contained in X. Then $d_{G'}(y) < \Delta(G)$ for all $y \in Y$, as each vertex in Y has a neighbour in X in graph G. If $d_{G'}(y) = 0$ for all $y \in Y$, then the only neighbours of Y-vertices in G were from X. Since G is connected, there are no other vertices in G apart from vertices in X and Y, and thus $V' = Y$ and $E' = \emptyset$.   □

Main Strategy. The algorithm we present in this section is a combination local ratio techniques from polynomial time approximation, and two well-known techniques from parameterised algorithmics: reductions and bounded search trees. We associate with each node s of the search tree a VC instance $(G_s = (V_s, E_s), k_s)$, together with the set C_s, M_s and T_s of vertices and set D_s of vertex pairs. For the root r of the search tree we initially have $G_r = G$, $k_r = k$, $C_r = D_r = M_r = T_r = \emptyset$, before we apply any reduction rules, etc. A leaf node ℓ of the search tree is achieved when $E_\ell = \emptyset$, or $k_\ell < 0$. If $E_\ell = \emptyset$ and $k_\ell \geq 0$ for some leaf ℓ, then $D_\ell = \emptyset$ and an approximate cover is given by $A_\ell = C_\ell \cup M_\ell \cup T_\ell$. If for all leaves $k_\ell < 0$, then there does not exist an approximate cover for the given approximation factor.

The vertices within each C_s, D_s and T_s have been either determined by explicit branching or by reduction rules. Additionally, to each node s within the search tree designed for approximation, we associate a set M_s to intentionally "worsen" the vertex cover, in order to improve the run time of the algorithm. The vertices are always added to M_s two at the time, such that the two added vertices are adjacent in the current graph. If a vertex, or a pair of vertices is put into C_s, D_s, M_s and T_s then these vertices are deleted from the current graph, together with their incident edges and the parameter k_s is decremented accordingly. In particular, we add a vertex v to the set C_s in search for a minimum vertex cover (respecting previous choices) that contains v, and we decrement k_s by 1; we add a pair of vertices (v, u) to D_s in search for a minimum vertex cover (respecting previous choices) that contains exactly one of these two vertices, and we decrement k_s by 1; we select an edge in a current graph and we add its end vertices to M_s; note that any vertex cover contains at least one vertex from each pair added to M_s, therefore we decrement k_s by 1 for each vertex pair; finally, we always add three vertices at the time to the set T_s in search of a minimum vertex cover (respecting previous choices) that contains at least two of these three vertices; thus we decrement the parameter k_s by 2. Therefore,

$$k_s = k - (|C_s| + |D_s| + \frac{|M_s|}{2} + \frac{2|T_s|}{3}).$$

The partial approximative cover A_s at node s consists of $C_s \cup M_s \cup T_s$, and exactly one vertex from each pair of vertices in D_s. Thus the approximation factor of the partial vertex cover in the search tree node s is given by $\frac{|C_s|+|D_s|+|M_s|+|T_s|}{|C_s|+|D_s|+\frac{|M_s|}{2}+\frac{2|T_s|}{3}}$.
To achieve an approximation factor of 1.5, we choose $|M_s| = 2(|C_s| + |D_s|)$. Whenever we put a new element into the cover C_s or into the set D_s, we may select an edge and put its end vertices into M_s. To achieve approximation factor of $\frac{2l+1}{2l}$, we choose $|M_s| = 2l(|C_s| + |D_s|) + \frac{2(l-1)}{3}|T_s|$, where we put l pairs into M_s for each element in $|C_s|$ and $|D_s|$, and $l-1$ pairs for each 3 elements in $|T_s|$. More generally, whenever we put a vertex into the cover, we also add to M_s a certain (in general not necessarily integer) number of pairs of adjacent vertices to achieve a desired approximation factor.

Note that an exact parameterized algorithm, as well as a polynomial time factor 2 approximation can be seen as special cases of this strategy, if we choose $M_\ell = T_\ell = \emptyset$ or $C_\ell = D_\ell = T_\ell = \emptyset$, respectively. In the latter case, there is no branching at all and the only vertices in the cover are pairs of adjacent vertices. Although the following theorem was already shown in [3], our reasoning is distinctively different and allows for further extensions.

Theorem 1. *If* VC *can be solved in time* $\mathcal{O}^*(\delta^k)$ *by some branching algorithm (as formalised above), then a factor-α approximation, $\alpha \in \mathbb{Q}$, $1 \le \alpha < 2$, can be computed in time* $\mathcal{O}^*(\delta^{(2-\alpha)\tau(G)})$.

Reduction Rules. Some reduction rules that handle vertices of small degrees are given below. These rules have already been used extensively in the literature but we present them slightly differently in order to keep our main argument simple.
Degree-0 Rule: Delete isolated vertices; they do not go into the cover.
Degree-1 Rule: If v has a neighbour of degree one, put v into the cover and decrement the parameter.

Since we are designing a $\frac{3}{2}$ approximation algorithm, for every vertex of degree one we process in the above fashion, we can add a pair of adjacent vertices to M_s. Since at least two out of the three vertices added to the approximate cover must be included in any optimal cover, we are within the $\frac{3}{2}$ ratio.
Degree-2 Rule: [7] If $N(y) = \{x, z\}$, and x and z are adjacent, we add x and z to C_s and remove all three vertices x, y and z from the graph. If $N(y) = \{x, z\}$, and x and z are not adjacent, we add (x, y) to D_s, remove them from the graph and connect z to the remaining neighbours of x. Exactly one of the vertices x and y is in the cover and consequently we decrement the parameter k by 1. If later on, say in node t, z is put into the cover, then x is added the C_t and the pair (x, y) is deleted from D_t. If it is decided that z is not to go into the cover, then y is put into C_t and the pair (x, y) is deleted from D_t. Therefore, for each leaf ℓ such that $E_\ell = \emptyset$ and $k_\ell \ge 0$, $D_\ell = \emptyset$, as each pair of vertices added to D_s for some node s gets eventually deleted, say in node t, and one of its vertices gets added to the corresponding set C_t.
Triangles. [15] If there is a triangle we add all of its vertices to T_s (again, at least 2 of them must be in every vertex cover so we are within the 3/2 ratio).
Approximation Preserving (AP) Degree-2 Rule. We apply the AP Degree-2 Rule if there are no vertices of degree at most one but there is a vertex y of degree 2 with neighbours x and z such that there is no edge between x and z (otherwise we apply Degree-2 Rule). We choose one of the neighbours of y, say x, together with its neighbour $u \ne y$ (as there are no vertices of degree 1, there must exist such vertex u), and we add vertices x and u to M_s. This triggers the Degree-1 Rule, putting z into the cover.

Please note the difference between the Degree-2 Rule described in the previous section and the AP Degree-2 Rule described above. Importantly, while the Degree-0 Rule, Degree-1 Rule and AP Degree-2 Rule never increase degrees of vertices (as we only remove vertices and their incident edges), such a degree increase may happen with the Degree-2 Rule, as well as with the AP Degree-3 Rule described below.
Approximation Preserving (AP) Degree-3 Rule. If there are no vertices of degree at most 2, but at least one vertex v of degree 3, we select a neighbour u of the vertex v and a neighbour $x \ne v$ of u and add $\{x, u\}$ to M; in that way we have created a vertex of degree 2, namely v, and we apply the Degree-2 Rule to it; altogether we add 3 vertices to the vertex cover and some minimum cover must contain at least 2 of them, so we are within the 3/2 approximation ratio.

This provide a new way of dealing with vertices of degree three.

Consequences. We will always assume that the mentioned reduction rules have been carried out before any branching takes place. To explicitly refer to the approximation factor, we will call such a graph $\frac{3}{2}$-*reduced*. So, a $\frac{3}{2}$-reduced graph contains no vertices of degree ≤ 3 and no triangles. One can easily prove:

Lemma 5. *In a $\frac{3}{2}$-reduced graph, any vertex v has a vertex u at distance two.*

Proof. Let $G = (V, E)$ be a $\frac{3}{2}$-reduced graph. Consider some vertex $v \in V$. Since $d(v) > 3$, we have that $N(v) \neq \emptyset$. Consider any $x \in N(v)$. As $d(x) > 3$ it follows that $\exists u \in N(x) \setminus \{v\}$. Since there are no triangles in G, it follows that $u \notin N(v)$. Hence, $d(v, u) = 2$. $\qquad\square$

The Branching Algorithm. In what follows we assume that we have a connected graph without cut vertices, otherwise we branch on a cut vertex and/or process each graph component separately. We assume that the graph G is $\frac{3}{2}$-reduced, so it has no vertices of degree less than 4 and no triangles.

Consider a path xyz of length 2 and apply the following reasoning. 1) If at least one of the end vertices of the path, that is, x or z, is in some minimum cover (respecting the previous choices) then the cover remains unchanged if we add the edge xz and therefore we can treat xyz as a triangle and add all three vertices to T_s. 2) In the case that neither of the end vertices of the path (x and z) belongs to any minimum vertex cover (respecting the previous choices), then we remove x, z from the graph and add $N(x) \cup N(z)$ to the set C_s. Since we are not after a minimum vertex cover but rather a factor $\frac{3}{2}$ approximation, we select $|N(x) \cup N(z)|$ disjoint edges and add their end vertices to the set M_s.

The careful choice of the edges whose end vertices are added to M_s is crucial to the efficiency of the algorithm. In order to make such a choice we proceed as follows. We first apply repeatedly Degree-0, Degree-1, AP Degree-2 and the Triangle Rule, until no such rules can be applied. That leaves us with a triangle-free graph with minimum degree $\delta > 2$. We then select a vertex v with the minimum degree $\delta > 2$ and $\delta - 2$ neighbours of v, say $u_1, u_2, \ldots, u_{\delta-2}$, and a unique neighbour t_i for each vertex u_i, $1 \leq i \leq \delta - 2$ (note that this is always possible to do, as each of the vertices u_i, $1 \leq i \leq \delta - 2$ has degree at least δ). We put all vertices u_i and t_i, $1 \leq i \leq \delta - 2$, into M_s. Then, the vertex v has only 2 remaining neighbours and we apply the AP Degree-2 rule. Importantly, we did not increase any degrees in the graph (cf. Lemma 4) and we can add another $|N(x) \cup N(z)| - \delta + 2$ vertex pairs to M. We repeat this process as long as there are at least $\delta - 2$ pairs of vertices to add to M_s. As the very last step, we repeat the process described above and obtain a vertex with degree $d \geq 3$. In the case we have equality, that is, $d = 3$, we apply the AP Degree-3 Rule to it. The next step is a recursive call, so that the possible increase of degrees is now harmless.

The main advantage of our algorithm over branching on a single vertex is that in this way we effectively increase the degree by at least 2, which significantly improves the run time of the algorithm.

Algorithm 1. *1. Exhaustively apply the (approximation preserving) reduction rules.*
2. If $\Delta(G) < 5$, approximate in polynomial time, see Lemma 2.
3. Select a vertex v of maximum degree and a vertex u at distance two from v (see Lemma 5) so that $|N(u) \cup N(v)|$ is maximum.
4. If $N(u) \cup N(v) = N(v)$, the graph is bipartite; find exact cover in polynomial time.
5. Branch on two cases:
5a. At least one of u, v is in any minimum cover respecting previous choices.
⤳ Put u, v, x into T, where $x \in N(u) \cap N(v)$.
5b. None of u, v is is some minimum cover respecting previous choices.
⤳ Put $(N(u) \cup N(v))$ into C. We select $f = |N(u) \cup N(v)|$ disjoint edges and put their end vertices into C_s.

To formally reason about the correctness (and the claimed approximation factor), we recall that, when the procedure is started with $k \geq \tau(G)$, there is a leaf node ℓ with an approximate cover $A_\ell = C_\ell \cup M_\ell \cup T_\ell$ of size $|A_\ell| = |C_\ell| + |M_\ell| + |T_\ell|$, and a minimum cover C^* such that $|C^*| \geq |C_\ell| + \frac{1}{2}|M_\ell| + \frac{2}{3}|T_\ell|$. Since $|C_\ell| = \frac{1}{2}|M_\ell|$, we have $|C^*| \geq |M_\ell| + \frac{2}{3}|T_\ell| = \frac{2}{3}(\frac{3}{2}|M_\ell| + |T_\ell|) = \frac{2}{3}|A_\ell|$.

Theorem 2. *We have given an $\mathcal{O}^*(1.09^{\tau(G)})$ time, polynomial space algorithm for a factor $\frac{3}{2}$ approximation of MINIMUM VERTEX COVER.*

Proof. We consider a $\frac{3}{2}$-reduced connected graph G without cut vertices and we analyse branching in step 5. In the first branch, step 5a., three vertices are decided to be placed into the approximate vertex cover, i.e., into T_s, out of which at least two are also found in a minimum solution by assumption.

We now consider the step 5b. and the case when $|N(u) \setminus N(v)| \geq 2$. For the sake of simplicity, in what follows we refer to the maximum degree of the original graph G, $\Delta(G)$, simply as Δ. Since neither of the end vertices u and v are in any minimum vertex cover, we remove them from the graph and place all their neighbours (which is at least $\Delta + 2$ vertices) in the vertex cover, that is, in the set C_s. We can now select $\Delta + 2$ disjoint edges and put their end vertices into M_s as follows. We select a vertex v with the minimum degree δ, and select $\delta - 2$ disjoint edges in such a way to reduce the degree of v to 2. We then apply the AP Degree-2 Rule to v and add a vertex to C_s and a pair of vertices to M_s. By Lemma 4, the minimum degree $\delta(G)$ is no more than $\Delta - 1$ and thus we have used up at most $\Delta - 3$ edges, whose end vertices we added to the set M_s. Therefore, for each vertex v we process in this way, we add a pair of connected vertices to M_s and a vertex to C_s, which decreases the parameter k_s by 2, plus at most $\Delta - 3$ pairs of vertices to M_s. We can repeat this process $\lfloor \frac{\Delta+2}{\Delta-3} \rfloor$ times, each time reducing k_s by 2. If we reach a point where we do not have enough edges to reduce the degree of a vertex with minimum degree to 2, but we can reduce it to 3, we apply AP Degree-3 Rule, further reducing parameter k_s by 2. Altogether, in this way we process at least $\lfloor \frac{\Delta+3}{\Delta-3} \rfloor$ vertices. To summarise, in step 5b., we reduce k_s by at last $\Delta + 2$ vertices we put into C_s, at least $\Delta + 2$ pair of vertices we add to the set M_s, and by at least $2\lfloor \frac{\Delta+3}{\Delta-3} \rfloor$ for the vertices with minimum degree we process. So, we reduce k by $2(\Delta+2) + 2\lfloor \frac{\Delta+2}{\Delta-3} \rfloor$, which attains its minimum of 22 for $\Delta = 5, 6, 7$. This gives the branching vector $(2, 22)$.

We now consider the step 5b. and the case when $|N(u) \setminus N(v)| = 1$. We distinguish three cases. (1) There are no vertices at distance 3 from v — after removing the vertices v and u from the graph and putting all their neighbours in the set C_s, we are left with the graph with maximum degree 1 which we can process in linear time. (2) There is exactly 1 vertex at distance 3 from v — but then this is a cut vertex — a contradiction. (3) There are at least 2 vertices at distance 3 from v — after removing the vertices v and u from the graph and putting all their neighbours in the set C_s, we are left with at least one vertex of degree 1 — a vertex (other that u) at distance 2 from v, that has a neighbour at distance 3 from v. This allows us to put at least one extra vertex in the set C_s and at least one other vertex pair in M_s, which then brings us to the exactly same situation as in the case when $|N(u) \setminus N(v)| \geq 2$ ($\Delta + 2$ vertices added to set C_s and $\Delta + 2$ edges to add to M_s). This completes the proof. $\square$

3   Parameterised Approximation for a Factor of $\frac{2l+1}{l+1}$

In this section we present a generic, scalable algorithm for parameterised factor $(2l + 1)/(l + 1)$ approximation algorithm for MINIMUM VERTEX COVER. Since the case where $l = 1$ was dealt with in the previous section, we can now restrict our attention to the case when $l > 1$. We argue that this research direction is extremely fruitful when aiming at algorithms that could be practically competitive with polynomial-time, due to the quite small bases of the exponential part of the run-time estimate. Our reduction rules generalise as follows:

Small Odd Cycles. If there is an odd cycle of length $2i + 1 \leq 2l + 1$, we add all of its vertices to the vertex cover (at least $l + 1$ of them must be in every vertex cover, so we are within the $(2l + 1)/(l + 1)$ ratio). This rule is actually well-known, see, e.g., [15]. If $i < l$, we subsequently select $l - i$ disjoint edges and add their end vertices to the set M_s, while remaining within the target ratio.

Small Degree Rule. If v has degree $l + 2$ or less, select $d(v) - 2$ neighbours u_i of v and a neighbour $t_i \neq v$ of each u_i, and add all the verteces u_i and t_i, $1 \leq i \leq d(v) - 2$ to the set M_s. We then apply the Degree-2 Rule to the vertex v. If v has degree $l + 2$ or less, select $d(v) - 2$ neighbours u_i of v and a neighbour $t_i \neq v$ of each u_i, and add all the vertices u_i and t_i, $1 \leq i \leq d(v) - 2$ to the set M_s. We then apply the Degree-2 Rule to the vertex v. We then select $l + 2 - d(v)$ additional disjoint edges if $d(v) > 1$, or l if $d(v) = 1$ and place their end vertices in M_s. Altogether, we have placed $2l + 1$ vertices into M_s and C_s, out of which at least $l + 1$ are in a minimum vertex cover.

We term a graph $(2l + 1)/(l + 1)$- *reduced* if none of the above reduction rules apply for some specified $l > 1$. The following property is not hard to show.

Lemma 6. *If $l > 1$, then a $(2l + 1)/(l + 1)$-reduced graph is either bipartite or for each vertex v there exists a vertex u at distance at least 3 from v, such that there exists a path of length four between them.*

Proof. Let $G = (V, E)$ be a $(2l + 1)/(l + 1)$-reduced graph where $l > 1$. Let v be any vertex in G. We say that a vertex x is at level i if it is at distance i from

v. Then there cannot exist any edges between vertices of the same level i for $i < 3$, as any such edge would create a cycle of length 5 or less, which do not exist in a $(2l + 1)/(l + 1)$-reduced graph for $l > 1$. If there exists an edge uw between vertices u and v at level 3, then u is a vertex at distance at least 3 from v such that there exists a path of length 4 between them. If such an edge does not exist, we select any vertex u at distance 4 from v. If there are no vertices at distance 4, then G is bipartite, as there are no edges between the vertices at the same level i; the two partitions are formed by the vertices at odd and even levels, respectively.   □

Theorem 3. *There exists a polynomial factor $\frac{2l+1}{l+1}$ approximation algorithm for vertex cover for any graph with average degree $d_{avg} \leq 2.5l + 1$.*

Proof. The polynomial-time factor-$(2 - \frac{2}{\Delta})$ algorithm from [15] guarantees the $(2l+1)/(l+1)$ or better factor for $\Delta \leq 2l+2$. Further to this, by Lemma 3 there is a polynomial time algorithm for graphs with average degree d_{avg} and approximation factor $\frac{4d_{avg}+1}{2d_{avg}+3}$, assuming we are using a Nemhauser-Trotter reduction in each search tree node. It follows that for the approximation factor of the form $\frac{2l+1}{l+1}$, there is a polynomial time algorithm whenever $d_{avg} \leq 2.5l + 1$.   □

Therefore, we only need to consider $\Delta \geq \lfloor 2.5l + 1 \rfloor + 1$. We may also assume that the minimum degree in the graph G is $\delta = l+3$, as the vertices with degree less than $l + 3$ do not appear in a $(2l + 1)/(l + 1)$-reduced graph. Similarly, we assume that there are no odd cycles with length $2l + 1$ or less. Our algorithm for $l > 1$ is quite similar to the algorithm for $l = 1$, but there is one important difference. In the algorithm for $l = 1$ in the first branch (5.a.), we add to the vertex cover the three vertices on a path of length 3. We do not generalise the length of the path to $2l+1$ as it may be expected, as the algorithm is much more efficient for larger l if we always take the path of length 5. We consider a vertex of degree $\Delta(G)$ and a vertex u at distance at least 3 from v, such that there exists a path of length 4 between them, see Lemma 6. Note that if such vertex u does not exist, then the graph G is bipartite and there is an exact polynomial time algorithm for finding a minimum vertex cover. For a $(2l+1)/(l+1)$-reduced graph, we distinguish two branches: In the first one, we assume that at least one of the vertices v and u is in some minimum cover (respecting previous choices). Then we can add an edge vu and thus create a cycle of length 5, which we add to the vertex cover, together with end verteces of $l - i$ disjoint edges (see Small Odd Cycles reduction rule above). Altogether, we added $2l + 1$ vertices into the partial approximative cover, at least $l + 1$ of which must be in any minimum cover. In the second branch we assume that neither v nor u are in any minimum cover and we add their neighbours to the vertex cover, together with the end vertices of l carefully chosen disjoint edges for each neighbour.

Theorem 4. *The running time of the factor $(2l + 1)/(l + 1)$ approximation algorithm for $l > 1$ is described by the following recurrence relation: $T(k) = T(k - 2s - 1 - a) + T\left(k - (2s + 1 + a)\left(9s + 6 + 4a + \left\lfloor \frac{8s^2+10s+8as+7a}{3s+1+a}\right\rfloor\right)\right)$, where $l = 2s + a$, and $a = 0$ for l even and $a = 1$ for l odd.*

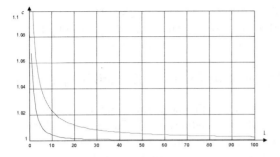

l	2	3	4	10
$1.28^{\frac{1}{l+1}}$	1.09	1.064	1.051	1.0420
c_l	1.04	1.024	1.017	1.0043

l	20	100	200
$1.28^{\frac{1}{l+1}}$	1.0118	1.00244	1.00122
c_l	1.0014	1.00008	1.00002

Fig. 1. With growing l, the base c_l quickly approaches one

Proof. We start from the recurrence relation $T(k) = T(k - (l + 1)) + T(k - y)$ and analyse the second branch. As we assume that neither v nor u are contained in any minimum vertex cover, we take $N(u) \cup N(v)$ into the partial current cover C_s; we refer to the remaining graph as G'. Recall that $d_G(v) = \Delta(G)$ and that the minimum degree in G is $\delta \geq l + 3$. As the distance between v and u is at least 3, they have no common neighbours and thus the total number of neighbours of v and u is $|N(v) \cup N(u)| \geq \Delta + l + 3$. Additionally, for each vertex $x \in N(v) \cup N(u)$ we select l disjoint edges and add their end vertices to M_s. Thus we have put altogether at least $(2l + 1)(\Delta + l + 3)$ vertices into the approximate cover. To improve the efficiency of the algorithm, we select the $l(\Delta + l + 3)$ disjoint edges in such a way to create as many as possible vertices of degree $l + 1$, as such vertices can be processed simply by removing vertices and edges and thus without increasing the degrees of the remaining vertices in G'. Hence, select a vertex z of minimum degree δ' in G'. We select $\delta' - l - 1$ disjoint edges such that after adding their end vertices to M_s we have $d(z) = l + 1$. We then process z by adding $2l + 1$ vertices to M_s. Since $\delta' \leq \Delta - 1$ (see Lemma 4), we can process $\left\lfloor \frac{l(\Delta + l + 3)}{\Delta - l - 2} \right\rfloor = l + \left\lfloor \frac{2l^2 + 5l}{\Delta - l - 2} \right\rfloor$ vertices this way. In total we put into the partial approximate vertex cover at least $(2l + 1)\left(\Delta + 2l + 3 + \left\lfloor \frac{2l^2 + 5l}{\Delta - l - 2} \right\rfloor\right)$ vertices, $y = (l + 1)\left(\Delta + 2l + 3 + \left\lfloor \frac{2l^2 + 5l}{\Delta - l - 2} \right\rfloor\right)$ of which must be in the minimum cover that can be found on the search tree path under consideration. For the worst case analysis of our algorithm we use the value of Δ for which y assumes the minimum value. Recall that we only need to consider $\Delta \geq \lfloor 2.5l + 1 \rfloor + 1$. We define $\Delta_{min} = \lfloor 2.5l + 1 \rfloor + 1$ and express Δ as $\Delta = \Delta_{min} + x$, where $x \in \mathbb{N}_0$. Then to perform the worst case analysis of our algorithm we need to find the value of x for which y assumes the minimum value. After considering several cases, we show that the minimum of $y(x)$ always occurs at $x = 1$ and the above recurrence relation follows. $\square$

These recurrences lead to branching numbers only slightly above 1 when l is growing, proving the claimed scalability of our approach. Figure 1. shows the constant appearing in $\mathcal{O}^*(constant^k)$ as a function of l. The upper curve corresponds to the $\mathcal{O}^*((1.28^{\frac{1}{l+1}})^k)$ algorithm derived from the Theorem 1 and the fastest known exact parameterised algorithm ; the lower curve corresponds to our $\mathcal{O}^*(c_l^k)$ parameterised approximation described in this section. Values of $1.28^{\frac{1}{l+1}}$ and constants c_l for some values of l are listed in a small table.

4   An Approximation Algorithm for Graphs with $d_{avg} \leq 4$

The seemingly unnatural condition of the result quoted in Lemma 3 can be satisfied by first applying a Nemhauser-Trotter reduction [7]. This might influence the average degree of the graph and hence the approximation factor. Lemma 7 describes situations when this type of reduction is not harmful to the average degree, and is crucial to Theorem 5, which gives a better approximation factor for the average degree 4 than the previously best known result [14].

Lemma 7. *In a graph G with average degree d_{avg}, if one deletes x vertices and at least $\frac{d_{avg}}{2}x$ edges, the remaining graph G' has average degree $d'_{avg} \leq d_{avg}$.*

Proof. Let $G = (V, E)$ be a graph and let $|V| = n$ and $|E| = m$. Then G has an average degree $d_{avg} = \frac{2m}{n}$. After deleting x vertices and at least $\frac{d_{avg}}{2}x$ edges, we have the the new average degree $d'_{avg} \leq \frac{2(m - \frac{d_{avg}}{2}x)}{n-x} = \frac{2m}{n} = d_{avg}$.   □

Theorem 5. *For any graph G with the average degree $d_{avg} \leq 4$, there exist a polynomial factor 1.5 approximation vertex cover algorithm.*

Proof. Let G be a connected graph (otherwise we process each component separately) with the average degree ≤ 4. Suppose that there is in G a vertex v of degree 1 or, if not, a vertex of degree 2. Then we apply the Degree-1 Rule or Degree-2 Rule from Section 2, respectively; then we removed 2 vertices and at least 2 edges from G. We add to the vertex cover a vertex v and its neighbour u that together have at least 6 adjacent edges. In total we removed 4 vertices and at least 8 edges and thus by Lemma 7 we have not increased the degree of the graph. If there are no such vertices v and u then the average degree of the remaining graph is < 3 and by Lemma 3 there is a polynomial time approximation with a factor less than $\frac{3}{2}$. In what follows we can assume that $\delta(G) \geq 3$, and that $d_{avg} > 3$. Suppose next that G contains one or more vertices with degree 3. Among all such vertices we select a vertex v that has a neighbour u with degree at least 4 (if there is no such vertex, then G is disconnected— a contradiction). Then we add the vertex u to the cover, together with any of its other neighbours (excluding v). Since there are no vertices of degree 1 or 2 in G, we thus removed 2 vertices and at least 6 edges. Then v has degree at most 2, and we can process it as above. In total we removed 4 vertices and at least 8 edges.

Suppose that all vertices in G have degree 4. Then we apply the algorithm from Lemma 2 to obtain the claimed approximation factor.   □

5   Conclusion and Future Directions

We presented the first genuine fixed-parameter approximation algorithms for MINIMUM VERTEX COVER, which combines local ratio arguments and the design of a search tree. We ask wether this idea be applied to other problems where both exact fixed-parameter algorithms are known as well as local-ratio arguments providing (in general) factor-two approximations (e.g., factor $3/2$ parameterised approximation algorithms for WEIGHTED VERTEX COVER, EDGE DOMINATING SET, or FEEDBACK VERTEX SET).

Acknowledgments. The authors would like to thank anonymous referees for their generous feedback.

References

1. Bar-Yehuda, R.: One for the price of two: a unified approach for approximating covering problems. Algorithmica 27, 131–144 (2000)
2. Berman, P., Fujito, T.: On approximation properties of the independent set problem for degree 3 graphs. In: Sack, J.-R., Akl, S.G., Dehne, F., Santoro, N. (eds.) WADS 1995. LNCS, vol. 955, pp. 449–460. Springer, Heidelberg (1995)
3. Bourgeois, N., Escoffier, B., Paschos, V.T.: Efficient approximation of combinatorial problems by moderately exponential algorithms. In: Dehne, F. (ed.) WADS 2009. LNCS, vol. 5664, pp. 507–518. Springer, Heidelberg (2009)
4. Cai, L., Huang, X.: Fixed-Parameter Approximation: Conceptual Framework and Approximability Results. Algorithmica 57(2), 398–412 (2010)
5. Cai, L., Juedes, D.: On the existence of subexponential parameterized algorithms. J. Computer and System Sciences 67, 789–807 (2003)
6. Sunil Chandran, L., Grandoni, F.: Refined memorization for vertex cover. Information Processing Letters 93, 125–131 (2005)
7. Chen, J., Kanj, I.A., Jia, W.: Vertex cover: further observations and further improvements. J. Algorithms 41, 280–301 (2001)
8. Chen, J., Kanj, I.A., Xia, G.: Labeled search trees and amortized analysis: improved upper bounds for NP-hard problems. Algorithmica 43, 245–273 (2005)
9. Chen, J., Kanj, I.A., Xia, G.: Improved parameterized upper bounds for vertex cover. In: Královič, R., Urzyczyn, P. (eds.) MFCS 2006. LNCS, vol. 4162, pp. 238–249. Springer, Heidelberg (2006)
10. Chen, Y., Grohe, M., Grüber, M.: On parameterized approximability. In: Bodlaender, H.L., Langston, M.A. (eds.) IWPEC 2006. LNCS, vol. 4169, pp. 109–120. Springer, Heidelberg (2006)
11. Dantsin, E., Gavrilovich, M., Hirsch, E.A., Konev, B.: MAX SAT approximation beyond the limits of polynomial-time approximation. Ann. Pure Appl. Logic 113(1-3), 81–94 (2001)
12. Demaine, E.D., Fomin, F.V., Hajiaghayi, M.T., Thilikos, D.M.: Subexponential parameterized algorithms on graphs of bounded genus and H-minor-free graphs. J. ACM 52(6), 866–893 (2005)
13. Dinur, I., Safra, S.: On the hardness of approximating minimum vertex cover. Annals of Mathematics 162, 439–485 (2005)
14. Halldórsson, M., Radhakrishnan, J.: Greed is Good: Approximating Independent Sets in Sparse and Bounded-Degree Graphs. Algorithmica 18(1), 145–163 (1997)
15. Hochbaum, D.S.: Efficient bounds for the stable set, vertex cover and set packing problems. Discrete Applied Mathematics 6, 243–254 (1983)
16. Kanj, I.: Vertex Cover: Exact and Approximation Algorithms and Applications. Phd Thesis, Texas A& M University (2001)

17. Karakostas, G.: A better approximation ratio for the vertex cover problem. In: Caires, L., Italiano, G.F., Monteiro, L., Palamidessi, C., Yung, M. (eds.) ICALP 2005. LNCS, vol. 3580, pp. 1043–1050. Springer, Heidelberg (2005)

18. Khot, S., Regev, O.: Vertex cover might be hard to approximate to within $2 - \varepsilon$. J. Computer and System Sciences 74, 335–349 (2008)

19. Marx, D.: Parameterized complexity and approximation algorithms. The Computer Journal 51(1), 60–78 (2008)

20. Niedermeier, R., Rossmanith, P.: Upper bounds for vertex cover further improved. In: Meinel, C., Tison, S. (eds.) STACS 1999. LNCS, vol. 1563, pp. 561–570. Springer, Heidelberg (1999)

21. Xiao, M.: A note on vertex cover in graphs with maximum degree 3. In: Thai, T. (ed.) COCOON 2010. LNCS, vol. 6196, pp. 150–159. Springer, Heidelberg (2010)

Listing All Maximal Cliques in Sparse Graphs in Near-Optimal Time

David Eppstein, Maarten Löffler, and Darren Strash

Department of Computer Science, University of California, Irvine, USA

Abstract. The *degeneracy* of an n-vertex graph G is the smallest number d such that every subgraph of G contains a vertex of degree at most d. We show that there exists a nearly-optimal fixed-parameter tractable algorithm for enumerating all maximal cliques, parametrized by degeneracy. To achieve this result, we modify the classic Bron–Kerbosch algorithm and show that it runs in time $O(dn3^{d/3})$. We also provide matching upper and lower bounds showing that the largest possible number of maximal cliques in an n-vertex graph with degeneracy d (when d is a multiple of 3 and $n \geq d+3$) is $(n-d)3^{d/3}$. Therefore, our algorithm matches the $\Theta(d(n-d)3^{d/3})$ worst-case output size of the problem whenever $n-d = \Omega(n)$.

Keywords: sparse graphs, d-degenerate graphs, maximal clique listing algorithms, Bron–Kerbosch algorithm, fixed-parameter tractability.

1 Introduction

Cliques, complete subgraphs of a graph, are of great importance in many applications. In social networks cliques may represent closely connected clusters of actors [6,14,28,40] and may be used as features in exponential random graph models for statistical analysis of social networks [17,19,20,44,48]. In bioinformatics, clique finding procedures have been used to detect structural motifs from protein similarities [26,35,36], to predict unknown protein structures [45], and to determine the docking regions where two biomolecules may connect to each other [22]. Clique finding problems also arise in document clustering [3], in the recovery of depth from stereoscopic image data [29], in computational topology [51], and in e-commerce, in the discovery of patterns of items that are frequently purchased together [50].

Often, it is important to find not just one large clique, but all *maximal cliques*. Many algorithms are now known for this problem [1,7,9,10,11,23,28,32,41,43,46] and for the complementary problem of finding maximal independent sets [16,31,37,39,47]. One of the most successful in practice is the *Bron–Kerbosch algorithm*, a simple backtracking procedure that recursively solves subproblems specified by three sets of vertices: the vertices that are required to be included in a partial clique, the vertices that are to be excluded from the clique, and some remaining vertices whose status still needs to be determined [7,9,32,35,46].

All maximal cliques can be listed in polynomial time per clique [37,47] or in total time proportional to the maximum possible number of cliques in an n-vertex graph, without additional polynomial factors [15,46]. In particular, a variant of the Bron–Kerbosch algorithm is known to be optimal in this sense [9,46]. Unfortunately

O. Cheong, K.-Y. Chwa, and K. Park (Eds.): ISAAC 2010, Part I, LNCS 6506, pp. 403–414, 2010.

this maximum possible number of cliques is exponential [42], so that all general-purpose algorithms for listing maximal cliques necessarily take exponential time.

We are faced with a dichotomy between theory, which states categorically that clique finding takes exponential time, and practice, according to which clique finding is useful and can be efficient in its areas of application. One standard way of resolving dilemmas such as this one is to apply *parametrized complexity* [13]: one seeks a parameter of instance complexity such that instances with small parameter values can be solved quickly. A parametrized problem is said to be *fixed-parameter tractable* if instances with size n and parameter value p can be solved in a time bound of the form $f(p)n^{O(1)}$, where f may grow exponentially or worse with p but is independent of n. With this style of analysis, instances with a small parameter value are used to model problems that can be solved quickly, while instances with a large parameter value represent a theoretical worst case that, one hopes, does not arise in practice.

The size of the largest clique does not work well as a parameter: the maximum clique problem, parametrized by clique size, is hard for $\mathbf{W}[1]$, implying that it is unlikely to have a fixed-parameter tractable algorithm [12], and Turán graphs $K_{\frac{n}{k},\frac{n}{k},\frac{n}{k},\dots}$ have $(n/k)^k$ maximal cliques of size k forcing any algorithm that lists them all to take time larger than any fixed-parameter-tractable bound. However, clique size is not the only parameter one can choose. In this paper, we study maximal clique finding parametrized by *degeneracy*, a frequently-used measure of the sparseness of a graph that is closely related to other common sparsity measures such as arboricity and thickness, and that has previously been used for other fixed-parameter problems [2,8,25,34]. We are motivated by the fact that sparse graphs often appear in practice. For instance, the World Wide Web graph, citation networks, and collaboration graphs have low arboricity [24], and therefore have low degeneracy. Empirical evidence also suggests that the h-index, a measure of sparsity that upper bounds degeneracy, is low for social networks [17]. Furthermore, planar graphs have degeneracy at most five [38], and the Barabási–Albert model of preferential attachment [4], frequently used as a model for large scale-free social networks, produces graphs with bounded degeneracy. We show that:

– A variant of the Bron–Kerbosch algorithm, when applied to n-vertex graphs with degeneracy d, lists all maximal cliques in time $O(dn3^{d/3})$.
– Every n-vertex graph with degeneracy d (where d is a multiple of three and $n \geq d+3$) has at most $(n-d)3^{d/3}$ maximal cliques, and there exists an n-vertex graph with degeneracy d that has exactly $(n-d)3^{d/3}$ maximal cliques. Therefore, our variant of the Bron–Kerbosch algorithm is optimal in the sense that its time is within a constant of the parametrized worst-case output size.

Our algorithms are fixed-parameter tractable, with a running time of the form $O(f(d)n)$ where $f(d) = d3^{d/3}$. Algorithms for listing all maximal cliques in graphs of constant degeneracy in time $O(n)$ were already known [10,11], but these algorithms had not been analyzed for their dependence on the degeneracy of the graph. We compare the parametrized running time bounds of the known alternative algorithms to the running time of our variant of the Bron–Kerbosch algorithm, and we show that the Bron–Kerbosch algorithm has a much smaller dependence on the parameter d. Thus we give theoretical evidence for the good performance for this algorithm that had previously been demonstrated empirically.

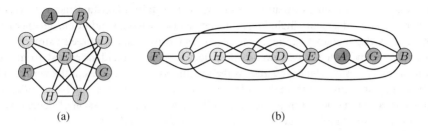

Fig. 1. (a) A graph with degeneracy 3. (b) A vertex ordering showing that the degeneracy is not larger than 3.

2 Preliminaries

We work with an undirected graph $G = (V,E)$, which we assume is stored in an adjacency list data structure. We let n and m be the number of vertices and edges of G, respectively. For a vertex v, we define $\Gamma(v)$ to be the set $\{w \mid (v,w) \in E\}$, which we call the *neighborhood* of v, and similarly for a subset $W \subset V$ we define $\Gamma(W)$ to be the set $\bigcap_{w \in W} \Gamma(w)$, which is the common neighborhood of all vertices in W.

2.1 Degeneracy

Our algorithm is parametrized on the *degeneracy* of a graph, a measure of its sparsity.

Definition 1 (degeneracy). *The degeneracy of a graph G is the smallest value d such that every nonempty subgraph of G contains a vertex of degree at most d [38].*

Figure 1(a) shows an example of a graph of degeneracy 3. Degeneracy is also known as the k-core number [5], width [21], and linkage [33] of a graph and is one less than the coloring number [18]. In a graph of degeneracy d, the maximum clique size can be at most $d + 1$, for any larger clique would form a subgraph in which all vertices have degree higher than d.

   If a graph has degeneracy d, then it has a *degeneracy ordering*, an ordering such that each vertex has d or fewer neighbors that come later in the ordering. Figure 1(b) shows a possible degeneracy ordering for the example. Such an ordering may be formed from G by repeatedly removing a vertex of degree d or less: by the assumption that G is d-degenerate, at least one such vertex exists at each step. Conversely, if G has an ordering with this property, then it is d-degenerate, because for any subgraph H of G, the vertex of H that comes first in the ordering has d or fewer neighbors in H. Thus, as Lick and White [38] showed, the degeneracy may equivalently be defined as the minimum d for which a degeneracy ordering exists. A third, equivalent definition is that d is the minimum value for which there exists an orientation of G as a directed acyclic graph in which all vertices have out-degree at most d [11]: such an orientation may be found by orienting each edge from its earlier endpoint to its later endpoint in a degeneracy ordering, and conversely if such an orientation is given then a degeneracy ordering may be found as a topological ordering of the oriented graph.

Degeneracy is a robust measure of sparsity: it is within a constant factor of other popular measures of sparsity including arboricity and thickness. In addition, degeneracy, along with a degeneracy ordering, can be computed by a simple greedy strategy of repeatedly removing a vertex with smallest degree (and its incident edges) from the graph until it is empty. The degeneracy is the maximum of the degrees of the vertices at the time they are removed from the graph, and the degeneracy ordering is the order in which vertices are removed from the graph [30]. The easy computation of degeneracy has made it a useful tool in algorithm design and analysis [11,16].

We can implement this algorithm in $O(n+m)$ time by maintaining an array D, where $D[i]$ stores a list of vertices of degree i in the graph [5]. To remove a vertex of minimum degree from the graph, we scan from the beginning of the array until we reach the first nonempty list, remove a vertex from this list, and then update its neighbors' degrees and move them to the correct lists. Each vertex removal step takes time proportional to the degree of the removed vertex, and therefore the algorithm takes linear time.

By counting the maximum possible number of edges from each vertex to later neighbors, we get the following bound on the number of edges of a d-degenerate graph:

Lemma 1 (Proposition 3 of [38]). *A graph $G = (V,E)$ with degeneracy d has at most $d(n - \frac{d+1}{2})$ edges.*

2.2   The Bron–Kerbosch Algorithm

The Bron–Kerbosch algorithm [7] is a widely used algorithm for finding all maximal cliques in a graph. It is a recursive backtracking algorithm which is easy to understand, easy to code, and has been shown to work well in practice.

A recursive call to the Bron–Kerbosch algorithm provides three disjoint sets of vertices R, P, and X as arguments, where R is a (possibly non-maximal) clique and $P \cup X = \Gamma(R)$ are the vertices that are adjacent to every vertex in R. The vertices in P will be considered to be added to clique R, while those in X must be excluded from the clique; thus, within the recursive call, the algorithm lists all cliques in $P \cup R$ that are maximal within the subgraph induced by $P \cup R \cup X$. The algorithm chooses a candidate v in P to add to the clique R, and makes a recursive call in which v has been moved from R to P; in this recursive call, it restricts X to the neighbors of v, since non-neighbors cannot affect the maximality of the resulting cliques. When the recursive call returns, v is moved to X to eliminate redundant work by further calls to the algorithm. When the recursion reaches a level at which P and X are empty, R is a maximal clique and is reported (see Fig. 2). To list all maximal cliques in the graph, this recursive algorithm is called with P equal to the set of all vertices in the graph and with R and X empty.

Bron and Kerbosch also describe a heuristic called *pivoting*, which limits the number of recursive calls made by their algorithm. The key observation is that for any vertex u in $P \cup X$, called a *pivot*, any maximal clique must contain one of u's non-neighbors (counting u itself as a non-neighbor). Therefore, we delay the vertices in $P \cap \Gamma(u)$ from being added to the clique until future recursive calls, with the benefit that we make fewer recursive calls. Tomita et al. [46] show that choosing the pivot u from $P \cup X$ in order to maximize $|P \cap \Gamma(u)|$ guarantees that the Bron–Kerbosch algorithm has worst-case running time $O(3^{n/3})$, excluding time to write the output, which is worst-case optimal.

proc BronKerbosch(P, R, X)

1: **if** $P \cup X = \emptyset$ **then**
2:     report R as a maximal clique
3: **end if**
4: **for each** vertex $v \in P$ **do**
5:     BronKerbosch($P \cap \Gamma(v)$, $R \cup \{v\}$, $X \cap \Gamma(v)$)
6:     $P \leftarrow P \setminus \{v\}$
7:     $X \leftarrow X \cup \{v\}$
8: **end for**

proc BronKerboschPivot(P, R, X)

1: **if** $P \cup X = \emptyset$ **then**
2:     report R as a maximal clique
3: **end if**
4: choose a pivot $u \in P \cup X$ {Tomita et al. choose u to maximize $|P \cap \Gamma(u)|$}
5: **for each** vertex $v \in P \setminus \Gamma(u)$ **do**
6:     BronKerboschPivot($P \cap \Gamma(v)$, $R \cup \{v\}$, $X \cap \Gamma(v)$)
7:     $P \leftarrow P \setminus \{v\}$
8:     $X \leftarrow X \cup \{v\}$
9: **end for**

Fig. 2. The Bron–Kerbosch algorithm without and with pivoting

3   The Algorithm

In this section, we show that apart from the pivoting strategy, the order in which the vertices of G are processed by the Bron–Kerbosch algorithm is also important. By choosing an ordering carefully, we develop a variant of the Bron–Kerbosch algorithm that correctly lists all maximal cliques in time $O(dn3^{d/3})$. Essentially, our algorithm performs the outer level of recursion of the Bron–Kerbosch algorithm without pivoting, using a degeneracy ordering to order the sequence of recursive calls made at this level, and then switches at inner levels of recursion to the pivoting rule of Tomita et al. [46].

In the original Bron–Kerbosch algorithm, in each recursive call the vertices in P are considered for expansion one by one (see line 4 of BronKerbosch in Figure 2). The order in which the vertices are treated is not specified. We first analyze what happens if we fix an order $v_1, v_2, \ldots, v_n$ on the vertices of V, and use the same order consistently to loop through the vertices of P in each recursive call of the non-pivoting version of BronKerbosch.

Observation 1. *When processing a clique R in the ordered variant of Bron–Kerbosch, the common neighbors of R can be partitioned into the set P of vertices that come after the last vertex of R, and the set X of remaining neighbors, as shown in Figure 3.*

Our algorithm computes a degeneracy ordering of the given graph, and performs the outermost recursive calls in the ordered variant of the Bron–Kerbosch algorithm (without pivoting) for this ordering. The sets P passed to each of these recursive calls will have at most d elements in them, leading to few recursive calls within each of these outer calls. Below the top level of the recursion we switch from the ordered non-pivoting version of

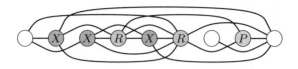

Fig. 3. Partitioning the common neighbors of a clique R into the set P of later vertices and the set X of remaining neighbors

the Bron–Kerbosch algorithm to the pivoting algorithm (with the same choice of pivots as Tomita et al. [46]) to further control the number of recursive calls.

proc BronKerboschDegeneracy(V, E)
  1: **for** each vertex v_i in a degeneracy ordering $v_0, v_1, v_2, \ldots$ of (V,E) **do**
  2:     $P \leftarrow \Gamma(v_i) \cap \{v_{i+1}, \ldots, v_{n-1}\}$
  3:     $X \leftarrow \Gamma(v_i) \cap \{v_0, \ldots, v_{i-1}\}$
  4:     BronKerboschPivot($P, \{v_i\}, X$)
  5: **end for**

Fig. 4. Our algorithm

Lemma 2. *The Bron–Kerbosch algorithm using the Tomita et al. pivoting strategy generates all and only maximal cliques containing all vertices in R, some vertices in P, and no vertices in X, without duplication.*

Proof. See Tomita et al. [46]. ☐

Theorem 1. *Algorithm* BronKerboschDegeneracy *generates all and only maximal cliques without duplication.*

Proof. Let C be a maximal clique, and v its earliest vertex in the degeneracy order. By Lemma 2, C will be reported (once) when processing v. When processing any other vertex of C, v will be in X, so C will not be reported. ☐

To make pivot selection fast we pass as an additional argument to BronKerboschPivot a subgraph $H_{P,X}$ of G that has $P \cup X$ as its vertices; an edge (u, v) of G is kept as an edge in $H_{P,X}$ whenever at least one of u or v belongs to P and both of them belong to $P \cup X$. The pivot chosen according to the pivot rule of Tomita et al. [46] is then just the vertex in this graph with the most neighbors in P.

Lemma 3. *Whenever* BronKerboschDegeneracy *calls* BronKerboschPivot *it can form* $H_{P,X}$ *in time* $O(d(|P| + |X|))$.

Proof. The vertex set of $H_{P,X}$ is known from P and X. Each edge is among the d outgoing edges from each of its vertices. Therefore, we can achieve the stated time bound by looping over each of the d outgoing edges from each vertex in $P \cup X$ and testing whether each edge meets the criterion for inclusion in $H_{P,X}$. ☐

The factor of d in the time bound of Lemma 3 makes it too slow for the recursive part of our algorithm. Instead we show that the subgraph to be passed to each recursive call can be computed quickly from the subgraph given to its parent in the recursion tree.

Lemma 4. *In a recursive call to* BronKerboschPivot *that is passed the graph* $H_{P,X}$ *as an auxiliary argument, the sequence of graphs* $H_{P\cap\Gamma(v),X\cap\Gamma(v)}$ *to be passed to lower-level recursive calls can be computed in total time* $O(|P|^2(|P|+|X|))$.

Proof. It takes $O(|P|+|X|)$ time to identify the subsets $P\cap\Gamma(v)$ and $X\cap\Gamma(v)$ by examining the neighbors of v in $H_{P,X}$. Once these sets are identified, $H_{P\cap\Gamma(v),X\cap\Gamma(v)}$ may be constructed as a subgraph of $H_{P,X}$ in time $O(|P|(|P|+|X|))$ by testing for each edge of $H_{P,X}$ whether its endpoints belong to these sets. There are $O(|P|)$ graphs to construct, one for each recursive call, hence the total time bound.   □

Lemma 5 (Theorem 3 of [46]). *Let T be a function which satisfies the following recurrence relation:*

$$T(p) \le \begin{cases} \max_k\{kT(p-k)\}+dp^2 & \text{if } p>0 \\ e & \text{if } p=0 \end{cases}$$

where p and k are integers, such that $p \ge k$, and d,e are constants greater than zero. Then, $T(p) \le \max_k\{kT(p-k)\}+dp^2 = O(3^{p/3})$.

Lemma 6. *Let v be a vertex, P_v, be v's later neighbors, and X_v be v's earlier neighbors. Then* BronKerboschPivot$(P_v, \{v\}, X_v)$ *executes in time $O((d+|X_v|)3^{|P_v|/3})$, excluding the time to report the discovered maximal cliques.*

Proof. Define $D(p,x)$ to be the running time of BronKerboschPivot$(P_v, \{v\}, X_v)$, where $p = |P_v|$, and $x = |X_v|$. We show that $D(p,x) = O((d+x)3^{p/3})$. By the description of BronKerboschPivot, D satisfies the following recurrence relation:

$$D(p,x) \le \begin{cases} \max_k\{kD(p-k,x)\}+c_1 p^2(p+x) & \text{if } p>0 \\ c_2 & \text{if } p=0 \end{cases}$$

where c_1 and c_2 are constants greater than 0.

Since our graph has degeneracy d, the inequality $p+x \le d+x$ always holds. Thus,

$$D(p,x) \le \max_k\{kD(p-k,x)\}+c_1 p^2(p+x)$$
$$\le (d+x)\left(\max_k\left\{\frac{kD(p-k,x)}{d+x}\right\}+c_1 p^2\right)$$
$$\le (d+x)\left(\max_k\{kT(p-k)\}+c_1 p^2\right)$$
$$= O((d+x)3^{p/3}) \quad \text{by letting } d=c_1, e=c_2 \text{ in Lemma 5}\qquad \square$$

Theorem 2. *Given an n-vertex graph G with degeneracy d, our algorithm reports all maximal cliques of G in time $O(dn3^{d/3})$.*

Proof. For each initial call to BronKerboschPivot for each vertex v, we first spend time $O(d(|P_v| + |X_v|))$ to set up subgraph H_{P_v,X_v}. Over the entire algorithm we spend time

$$\sum_v O(d(|P_v| + |X_v|)) = O(dm) = O(d^2 n)$$

setting up these subgraphs. The time spent performing the recursive calls is

$$\sum_v O((d + |X_v|)3^{|P_v|/3}) = O((dn + m)3^{d/3}) = O(dn3^{d/3}),$$

and the time to report all cliques is $O(d\mu)$, where μ is the number of maximal cliques. We show in the next section that $\mu = (n - d)3^{d/3}$ in the worst case, and therefore we take time $O(d(n - d)3^{d/3})$ reporting cliques in the worst case. Therefore, the algorithm executes in time $O(dn3^{d/3})$.   □

This running time is nearly worst-case optimal, since there may be $\Theta((n - d)3^{d/3})$ maximal cliques in the worst case.

4   Worst-Case Bounds on the Number of Maximal Cliques

Theorem 3. *Let d be a multiple of 3 and $n \geq d + 3$. Then the largest possible number of maximal cliques in an n-vertex graph with degeneracy d is $(n - d)3^{d/3}$.*

Proof. We first show that there cannot be more than $(n - d)3^{d/3}$ maximal cliques. We then show that there exists a graph that has $(n - d)3^{d/3}$ maximal cliques.

An Upper Bound. Consider a degeneracy ordering of the vertices, in which each vertex has at most d neighbors that come later in the ordering. For each vertex v that is placed among the first $n - d - 3$ vertices of the degeneracy ordering, we count the number of maximal cliques such that v is the clique vertex that comes first in the ordering.

Since vertex v has at most d later neighbors, the Moon–Moser bound [42] applied to the subgraph induced by these later neighbors shows that they can form at most $3^{d/3}$ maximal cliques with each other. Since these vertices are all neighbors of v, v participates in at most $3^{d/3}$ maximal cliques with its later neighbors. Thus, the first $n - d - 3$ vertices contribute to at most $(n - d - 3)3^{d/3}$ maximal cliques total.

By the Moon–Moser bound, the remaining $d + 3$ vertices in the ordering can form at most $3^{(d+3)/3}$ maximal cliques. Therefore, a graph with degeneracy d can have no more than $(n - d - 3)3^{d/3} + 3^{(d+3)/3} = (n - d)3^{d/3}$ maximal cliques.

A Lower Bound. By a simple counting argument, we can see that the graph $K_{n-d,3,3,3,3,\ldots}$ contains $(n - d)3^{d/3}$ maximal cliques: Each maximal clique must contain exactly one vertex from each disjoint independent set of vertices, and there are $(n - d)3^{d/3}$ ways of forming a maximal clique by choosing one vertex from each independent set. Figure 5 shows this construction for $d = 6$. We can also see that this graph is d-degenerate since in any ordering of the vertices, the first vertex must have d or more later neighbors, and in any ordering where the $n - d$ disjoint vertices come first, these first $n - d$ vertices have exactly d later neighbors, and the last d vertices have fewer later neighbors.   □

Relatedly, a bound of $(n - d + 1)2^d$ on the number of cliques (without assumption of maximality) in n-vertex d-degenerate graphs was already known [49].

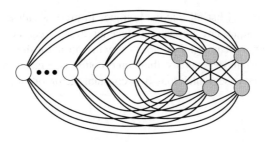

Fig. 5. The lower bound construction for $d = 6$, consisting of a Moon–Moser graph of size d on the right (blue vertices) and an independent set of $n - d$ remaining vertices that are each connected to all of the last d vertices

5 Comparison with Other Algorithms

Chiba and Nishizeki [10] describe two algorithms for finding cliques in sparse graphs. The first of these two algorithms reports all maximal cliques using $O(am)$ time per clique, where a is the arboricity of the graph, and m is the number of edges in G. The *arboricity* is the minimum number of edge-disjoint spanning forests into which the graph can be decomposed [27]. The degeneracy of a graph is closely related to arboricity: $a \leq d \leq 2a - 1$. In terms of degeneracy, Chiba and Nishizeki's algorithm uses $O(d^2n)$ time per clique. Combining this with the bound on the number of cliques derived in Section 4 results in a worst-case time bound of $O(d^2n(n-d)3^{d/3})$. For constant d, this is a quadratic time bound, in contrast to the linear time of our algorithm.

Another algorithm of Chiba and Nishizeki [10] lists cliques of order l in time $O(la^{l-2}m)$. It can be adapted to enumerate all maximal cliques in a graph with degeneracy d by first enumerating all cliques of order $d + 1, d, \ldots$ down to 1, and removing cliques that are not maximal. Applying their algorithm directly to a d-degenerate graph takes time $O(ld^{l-1}n)$. Therefore, the running time to find all maximal cliques is $\sum_{1 \leq i \leq d+1} O(ind^{i-1}) = O(nd^{d+1})$. Like our algorithm, this is linear when d is constant, but with a much worse dependence on the parameter d.

Chrobak and Eppstein [11] list triangles and 4-cliques in graphs of bounded degeneracy by testing all sets of two or three later neighbors of each vertex according to a degeneracy ordering. The same idea extends in an obvious way to finding maximal cliques of size greater than four, by testing all subsets of later neighbors of each vertex. For each vertex v, there are at most 2^d subsets to test; each subset may be tested for being a clique in time $O(d^2)$, by checking whether each of its vertices has all the later vertices in the subset among its later neighbors, giving a total time of $O(nd^22^d)$ to list all the cliques in the graph. However, although this singly-exponential time bound is considerably faster than Chiba and Nishizeki, and is close to known bounds on the number of (possibly non-maximal) cliques in d-degenerate graphs [49], it is slower than our algorithm by a factor that is exponential in d. Our new algorithm uses this same idea of searching among the later neighbors in a degeneracy order but achieves much greater efficiency by combining it with the Bron–Kerbosch algorithm.

Makino and Uno [41] list all maximal cliques in graphs with maximum degree Δ in time $O(\Delta^4)$ per clique. For graphs with degeneracy d, Δ can be any value between d and $n-1$, making a meaningful comparison with our algorithm difficult. Therefore, for graphs with constant degeneracy, their time bound is a factor Δ^4 slower than our algorithm in the worst case, which may be a constant factor, or much worse.

6 Conclusion

We have presented theoretical evidence for the fast performance of the Bron–Kerbosch algorithm for finding cliques in graphs, as has been observed in practice. We observe that the problem is fixed-parameter tractable in terms of the degeneracy of the graph, a parameter that is expected to be low in many real-world applications, and that a slight modification of the Bron–Kerbosch algorithm performs optimally in terms of the degeneracy.

We explicitly prescribe the order in which the Bron–Kerbosch algorithm processes the vertices of the graph, something that has not been considered before. Without this particular order, we do not have a bound on the running time. It would be interesting to determine whether a random order gives similar results, as this would further explain the observed performance of implementations of Bron–Kerbosch that do not use the degeneracy order.

It would also be of interest to determine the largest possible number of maximal cliques among all n-vertex graphs with degeneracy d (or, slightly differently, at most d), under less restrictive assumptions than those made in Theorem 3.

Acknowledgments. This research was supported in part by the National Science Foundation under grant 0830403, and by the Office of Naval Research under MURI grant N00014-08-1-1015.

References

1. Akkoyunlu, E.A.: The enumeration of maximal cliques of large graphs. SIAM J. Comput. 2(1), 1–6 (1973)
2. Alon, N., Gutner, S.: Linear time algorithms for finding a dominating set of fixed size in degenerated graphs. Algorithmica 54(4), 544–556 (2009)
3. Augustson, J.G., Minker, J.: An analysis of some graph theoretical cluster techniques. J. ACM 17(4), 571–588 (1970)
4. Barabási, A.L., Albert, R.: Emergence of scaling in random networks. Science 286, 509–512 (1999)
5. Batagelj, V., Zaveršnik, M.: An O(m) algorithm for cores decomposition of networks (2003)
6. Berry, N.M., Ko, T.H., Moy, T., Smrcka, J., Turnley, J., Wu, B.: Emergent clique formation in terrorist recruitment. In: Dignum, V., Corkill, D., Jonker, C., Dignum, F. (eds.) Proc. AAAI 2004 Worksh. Agent Organizations. AAAI Press, Menlo Park (2004), http://www.aaai.org/Papers/Workshops/2004/WS-04-02/WS04-02-005.pdf
7. Bron, C., Kerbosch, J.: Algorithm 457: finding all cliques of an undirected graph. Commun. ACM 16(9), 575–577 (1973)

8. Cai, L., Chan, S., Chan, S.: Random separation: A new method for solving fixed-cardinality optimization problems. In: Bodlaender, H.L., Langston, M.A. (eds.) IWPEC 2006. LNCS, vol. 4169, pp. 239–250. Springer, Heidelberg (2006)
9. Cazals, F., Karande, C.: A note on the problem of reporting maximal cliques. Theor. Comput. Sci. 407(1-3), 564–568 (2008)
10. Chiba, N., Nishizeki, T.: Arboricity and subgraph listing algorithms. SIAM J. Comput. 14(1), 210–223 (1985)
11. Chrobak, M., Eppstein, D.: Planar orientations with low out-degree and compaction of adjacency matrices. Theor. Comput. Sci. 86(2), 243–266 (1991)
12. Downey, R.G., Fellows, M.R.: Fixed-parameter tractability and completeness II: On completeness for W[1]. Theor. Comput. Sci. 141(1-2), 109–131 (1995)
13. Downey, R.G., Fellows, M.R.: Parameterized Complexity. Springer, Heidelberg (1999)
14. Du, N., Wu, B., Pei, X., Wang, B., Xu, L.: Community detection in large-scale social networks. In: Proc. 9th WebKDD and 1st SNA-KDD 2007 Workshop on Web Mining and Social Network Analysis, pp. 16–25 (2007)
15. Eppstein, D.: Small maximal independent sets and faster exact graph coloring. J. Graph Algorithms & Applications 7(2), 131–140 (2003)
16. Eppstein, D.: All maximal independent sets and dynamic dominance for sparse graphs. ACM Trans. Algorithms 5(4), A38 (2009)
17. Eppstein, D., Spiro, E.S.: The h-index of a graph and its application to dynamic subgraph statistics. In: WADS 2009. LNCS, vol. 5664, pp. 278–289. Springer, Heidelberg (2009)
18. Erdős, P., Hajnal, A.: On chromatic number of graphs and set-systems. Acta Mathematica Hungarica 17(1-2), 61–99 (1966)
19. Frank, O.: Statistical analysis of change in networks. Statistica Neerlandica 45(3), 283–293 (1991)
20. Frank, O., Strauss, D.: Markov graphs. J. Am. Stat. Assoc. 81(395), 832–842 (1986)
21. Freuder, E.C.: A sufficient condition for backtrack-free search. J. ACM 29(1), 24–32 (1982)
22. Gardiner, E.J., Willett, P., Artymiuk, P.J.: Graph-theoretic techniques for macromolecular docking. J. Chem. Inf. Comput. Sci. 40(2), 273–279 (2000)
23. Gerhards, L., Lindenberg, W.: Clique detection for nondirected graphs: Two new algorithms. Computing 21(4), 295–322 (1979)
24. Goel, G., Gustedt, J.: Bounded arboricity to determine the local structure of sparse graphs. In: Fomin, F.V. (ed.) WG 2006. LNCS, vol. 4271, pp. 159–167. Springer, Heidelberg (2006)
25. Golovach, P.A., Villanger, Y.: Parameterized complexity for domination problems on degenerate graphs. In: Broersma, H., Erlebach, T., Friedetzky, T., Paulusma, D. (eds.) WG 2008. LNCS, vol. 5344, pp. 195–205. Springer, Heidelberg (2008)
26. Grindley, H.M., Artymiuk, P.J., Rice, D.W., Willett, P.: Identification of tertiary structure resemblance in proteins using a maximal common subgraph isomorphism algorithm. J. Mol. Biol. 229(3), 707–721 (1993)
27. Harary, F.: Graph Theory. Addison-Wesley, Reading (1972)
28. Harary, F., Ross, I.C.: A procedure for clique detection using the group matrix. Sociometry 20(3), 205–215 (1957)
29. Horaud, R., Skordas, T.: Stereo correspondence through feature grouping and maximal cliques. IEEE Trans. Patt. An. Mach. Int. 11(11), 1168–1180 (1989)
30. Jensen, T.R., Toft, B.: Graph Coloring Problems. Wiley Interscience, New York (1995)
31. Johnson, D.S., Yannakakis, M., Papadimitriou, C.H.: On generating all maximal in- dependent sets. Inf. Proc. Lett. 27(3), 119–123 (1988)
32. Johnston, H.C.: Cliques of a graph—variations on the Bron–Kerbosch algorithm. Int. J. Parallel Programming 5(3), 209–238 (1976)
33. Kirousis, L., Thilikos, D.: The linkage of a graph. SIAM J. Comput. 25(3), 626–647 (1996)

34. Kloks, T., Cai, L.: Parameterized tractability of some (efficient) Y-domination variants for planar graphs and t-degenerate graphs. In: Proc. International Computer Symposium (2000), http://hdl.handle.net/2377/2482

35. Koch, I.: Enumerating all connected maximal common subgraphs in two graphs. Theor. Comput. Sci. 250(1-2), 1–30 (2001)

36. Koch, I., Lengauer, T., Wanke, E.: An algorithm for finding maximal common subtopologies in a set of protein structures. J. Comput. Biol. 3(2), 289–306 (1996)

37. Lawler, E.L., Lenstra, J.K., Rinnooy Kan, A.H.G.: Generating all maximal independent sets: NP-hardness and polynomial-time algorithms. SIAM J. Comput. 9(3), 558–565 (1980)

38. Lick, D.R., White, A.T.: k-degenerate graphs. Canad. J. Math. 22, 1082–1096 (1970), http://www.smc.math.ca/cjm/v22/p1082

39. Loukakis, E., Tsouros, C.: A depth first search algorithm to generate the family of maximal independent sets of a graph lexicographically. Computing 27(4), 349–366 (1981)

40. Luce, R.D., Perry, A.D.: A method of matrix analysis of group structure. Psychometrika 14(2), 95–116 (1949)

41. Makino, K., Uno, T.: New algorithms for enumerating all maximal cliques. In: Hagerup, T., Katajainen, J. (eds.) SWAT 2004. LNCS, vol. 3111, pp. 260–272. Springer, Heidelberg (2004)

42. Moon, J.W., Moser, L.: On cliques in graphs. Israel J. Math. 3(1), 23–28 (1965)

43. Mulligan, G.D., Corneil, D.G.: Corrections to Bierstone's algorithm for generating cliques. J. ACM 19(2), 244–247 (1972)

44. Robins, G., Morris, M.: Advances in exponential random graph (p*) models. Social Networks 29(2), 169–172 (2007)

45. Samudrala, R., Moult, J.: A graph-theoretic algorithm for comparative modeling of protein structure. J. Mol. Biol. 279(1), 287–302 (1998)

46. Tomita, E., Tanaka, A., Takahashi, H.: The worst-case time complexity for generating all maximal cliques and computational experiments. Theor. Comput. Sci. 363(1), 28–42 (2006)

47. Tsukiyama, S., Ide, M., Ariyoshi, H., Shirakawa, I.: A new algorithm for generating all the maximal independent sets. SIAM J. Comput. 6(3), 505–517 (1977)

48. Wasserman, S., Pattison, P.: Logit models and logistic regressions for social networks: I. An introduction to Markov graphs and p*. Psychometrika 61(3), 401–425 (1996)

49. Wood, D.R.: On the maximum number of cliques in a graph. Graphs and Combinatorics 23(3), 337–352 (2007)

50. Zaki, M.J., Parthasarathy, S., Ogihara, M., Li, W.: New algorithms for fast discovery of association rules. In: Proc. 3rd Int. Conf. Knowledge Discovery and Data Mining, pp. 283–286. AAAI Press, Menlo Park (1997), http://www.aaai.org/Papers/KDD/1997/KDD97-060.pdf

51. Zomorodian, A.: The tidy set: a minimal simplicial set for computing homology of clique complexes. In: Proc. 26th ACM Symp. Computational Geometry, pp. 257–266 (2010), http://www.cs.dartmouth.edu/~afra/papers/socg10/tidy-socg.pdf

Lower Bounds for Howard's Algorithm for Finding Minimum Mean-Cost Cycles

Thomas Dueholm Hansen[1,*] and Uri Zwick[2]

[1] Department of Computer Science, Aarhus University
[2] School of Computer Science, Tel Aviv University, Tel Aviv 69978, Israel

Abstract. Howard's policy iteration algorithm is one of the most widely used algorithms for finding optimal policies for controlling *Markov Decision Processes* (MDPs). When applied to weighted directed graphs, which may be viewed as *Deterministic* MDPs (DMDPs), Howard's algorithm can be used to find Minimum Mean-Cost cycles (MMCC). Experimental studies suggest that Howard's algorithm works extremely well in this context. The theoretical complexity of Howard's algorithm for finding MMCCs is a mystery. No polynomial time bound is known on its running time. Prior to this work, there were only linear lower bounds on the number of iterations performed by Howard's algorithm. We provide the first weighted graphs on which Howard's algorithm performs $\Omega(n^2)$ iterations, where n is the number of vertices in the graph.

1 Introduction

Howard's policy iteration algorithm [11] is one of the most widely used algorithms for solving Markov decision processes (MDPs). The complexity of Howard's algorithm in this setting was unresolved for almost 50 years. Very recently, Fearnley [5], building on results of Friedmann [7], showed that there are MDPs on which Howard's algorithm requires exponential time. In another recent breakthrough, Ye [17] showed that Howard's algorithm is strongly polynomial when applied to *discounted* MDPs, with a fixed discount ratio. Hansen *et al.* [10] recently improved some of the bounds of Ye and extended them to the 2-player case.

Weighted directed graphs may be viewed as *Deterministic* MDPs (DMDPs) and solving such DMDPs is essentially equivalent to finding minimum mean-cost cycles (MMCCs) in such graphs. Howard's algorithm can thus be used to solve this purely combinatorial problem. The complexity of Howard's algorithm in this setting is an intriguing open problem. Fearnley's [5] exponential lower bound seems to depend in an essential way on the use of stochastic actions, so it does not extend to the deterministic setting. Similarly, Ye's [17] polynomial upper bound depends in an essential way on the MDPs being *discounted* and does not extend to the non-discounted case.

---

* Supported by the Center for Algorithmic Game Theory at Aarhus University, funded by the Carlsberg Foundation.

O. Cheong, K.-Y. Chwa, and K. Park (Eds.): ISAAC 2010, Part I, LNCS 6506, pp. 415–426, 2010.

The MMCC problem is an interesting problem that has various applications. It generalizes the problem of finding a negative cost cycle in a graph. It is also used as a subroutine in algorithms for solving other problems, such as min-cost flow algorithms, (See, e.g., Goldberg and Tarjan [9].)

There are several polynomial time algorithms for solving the MMCC problem. Karp [12] gave an $O(mn)$-time algorithm for the problem, where m is the number of edges and n is the number of vertices in the input graph. Young *et al.* [18] gave an algorithm whose complexity is $O(mn+n^2 \log n)$. Although this is slightly worse, in some cases, than the running time of Karp's algorithm, the algorithm of Young *et al.* [18] behaves much better in practice.

Dasdan [3] experimented with many different algorithms for the MMCC problem, including Howard's algorithm. He reports that Howard's algorithm usually runs much faster than Karp's algorithm, and is usually almost as fast as the algorithm of Young *et al.* [18]. A more thorough experimental study of MMCC algorithms was recently conducted by Georgiadis *et al.* [8].[1]

Understanding the complexity of Howard's algorithm for MMCCs is interesting from both the applied and theoretical points of view. Howard's algorithm for MMCC is an extremely simple and natural combinatorial algorithm, similar in flavor to the Bellman-Ford algorithm for finding shortest paths [1,2],[6] and to Karp's [12] algorithm. Yet, its analysis seems to be elusive. Howard's algorithm also has the advantage that it can be applied to the more general problem of finding a cycle with a *minimum cost-to-time ratio* (see, e.g., Megiddo [14,15]).

Howard's algorithm works in iteration. Each iteration takes $O(m)$ time. It is trivial to construct instances on which Howard's algorithm performs n iterations. (Recall that n and m are the number of vertices and edges in the input graph.) Madani [13] constructed instances on which the algorithm performs $2n - O(1)$ iterations. No graphs were known, however, on which Howard's algorithm performed more than a *linear* number of iterations. We construct the first graphs on which Howard's algorithm performs $\Omega(n^2)$ iterations, showing, in particular, that there are instances on which its running time is $\Omega(n^4)$, an order of magnitude *slower* than the running times of the algorithms of Karp [12] and Young *et al.* [18].

We also construct n-vertex outdegree-2 graphs on which Howard's algorithm performs $2n - O(1)$ iterations. (Madani's [13] examples used $\Theta(n^2)$ edges.) This example is interesting as it shows that the number of iterations performed may differ from the number of edges in the graph by only an additive constant. It also sheds some more light on the non-trivial, and perhaps non-intuitive behavior of Howard's algorithm.

Our examples still leave open the possibility that the number of iterations performed by Howard's algorithm is always at most m, the number of edges. (The graphs on which the algorithm performs $\Omega(n^2)$ iterations also have $\Omega(n^2)$ edges.) We conjecture that this is always the case.

---

[1] Georgiadis *et al.* [8] claim that Howard's algorithm is not robust. From personal conversations with the authors of [8] it turns out, however, that the version they used is substantially different from Howard's algorithm [11].

2   Howard's Algorithm for Minimum Mean-Cost Cycles

We next describe the specialization of Howard's algorithm for *deterministic*
MDPs, i.e., for finding Minimum Mean-Cost Cycles. For Howard's algorithm
for general MDPs, see Howard [11], Derman [4] or Puterman [16].

Let $G = (V, E, c)$, where $c : E \to \mathbb{R}$, be a weighted directed graph. We assume
that each vertex has a unique *serial number* associated with. We also assume,
without loss of generality, that each vertex $v \in V$ has at least one outgoing edge.

If $C = v_0 v_1 \ldots v_{k-1} v_0$ is a cycle in G, we let $val(C) = \frac{1}{k} \sum_{i=0}^{k-1} c(v_i, v_{i+1})$,
where $v_k = v_0$, be its *mean cost*. The vertex on C with the smallest serial
number is said to be the *head* of the cycle. Our goal is to find a cycle C that
minimizes $val(C)$.

A *policy* π is a mapping $\pi : V \to V$ such that $(v, \pi(v)) \in E$, for every $v \in V$.
A policy π, defines a subgraph $G_\pi = (V, E_\pi)$, where $E_\pi = \{(v, \pi(v)) \mid v \in V\}$.
As the outdegree of each vertex in G_π is 1, we get that G_π is composed of a
collection of disjoint directed cycles with directed paths leading into them.

Given a policy π, we assign to each vertex $v_0 \in V$ a value $val_\pi(v_0)$ and a
potential $pot_\pi(v_0)$ in the following way. Let $P_\pi(v_0) = v_0 v_1, \ldots$ be the infinite
path defined by $v_i = \pi(v_{i-1})$, for $i > 0$. This infinite path is composed of a finite
path P leading to a cycle C which is repeated indefinitely. If $v_r = v_{r+k}$ is the first
vertex visited for the second time, then $P = v_0 v_1 \ldots v_r$ and $C = v_r v_{r+1} \ldots v_{r+k}$.
We let v_ℓ be the head of the cycle C. We now define

$$val_\pi(v_0) = val(C) = \tfrac{1}{k} \sum_{i=0}^{k-1} c(v_{r+i}, v_{r+i+1}),$$
$$pot_\pi(v_0) = \sum_{i=0}^{\ell-1} (c(v_i, v_{i+1}) - val(C)).$$

In other words, $val_\pi(v_0)$ is the mean cost of C, the cycle into which $P_\pi(v_0)$ is
absorbed, while $pot_\pi(v_0)$ is the *distance* from v_0 to v_ℓ, the head of this cycle,
when the mean cost of the cycle is subtracted from the cost of each edge. It is
easy to check that values and potentials satisfy the following equations:

$$val_\pi(v) = val_\pi(\pi(v)),$$
$$pot_\pi(v) = c(v, \pi(v)) - val_\pi(v) + pot_\pi(\pi(v)).$$

The *appraisal* of an edge $(u, v) \in E$ is defined as the pair:

$$A_\pi(u, v) = (\, val_\pi(v) \,,\ c(u, v) - val_\pi(v) + pot_\pi(v) \,).$$

Howard's algorithm starts with an arbitrary policy π and keeps *improving* it.
If π is the current policy, then the next policy π' produced by the algorithm is
defined by

$$\pi'(u) = \operatorname*{arg\,min}_{v:(u,v)\in E} A_\pi(u, v).$$

In other words, for every vertex the algorithm selects the outgoing edge with the
lowest appraisal. (In case of ties, the algorithm favors edges in the current policy.)
As appraisals are pairs, they are compared lexicographically, i.e., (u, v_1) is better

than (u, v_2) if and only if $A_\pi(u, v_1) \prec A_\pi(u, v_2)$, where $(x_1, y_1) \prec (x_2, y_2)$ if and only if $x_1 < x_2$, or $x_1 = x_2$ and $y_1 < y_2$. When $\pi' = \pi$, the algorithm stops. The correctness of the algorithm follows from the following two lemmas whose proofs can be found in Howard [11], Derman [4] and Puterman [16].

Lemma 1. *Suppose that π' is obtained from π by a policy improvement step. Then, for every $v \in V$ we have $(val_{\pi'}(v), pot_{\pi'}(v)) \preceq (val_\pi(v), pot_\pi(v))$. Furthermore, if $\pi'(v) \neq \pi(v)$, then $(val_{\pi'}(v), pot_{\pi'}(v)) \prec (val_\pi(v), pot_\pi(v))$.*

Lemma 2. *If policy π is not modified by an improvement step, then $val_\pi(v)$ is the minimum mean weight of a cycle reachable from v in G. Furthermore by following edges of π from v we get into a cycle of this minimum mean weight.*

Each iteration of Howard's algorithm takes only $O(m)$ time and is not much more complicated than an iteration of the Bellman-Ford algorithm.

3   A Quadratic Lower Bound

We next construct a family of weighted directed graphs for which the number of iterations performed by Howard's algorithm is *quadratic* in the number of vertices. More precisely, we prove the following theorem:

Theorem 1. *Let n and m be even integers, with $2n \leq m \leq \frac{n^2}{4} + \frac{3n}{2}$. There exists a weighted directed graph with n vertices and m edges on which Howard's algorithm performs $m - n + 1$ iterations.*

All policies generated by Howard's algorithm, when run on the instances of Theorem 1, contain a *single* cycle, and hence all vertices have the same *value*. Edges are therefore selected for inclusion in the improved policies based on *potentials*. (Recall that potentials are essentially adjusted distances.) The main idea behind our construction, which we refer to as the *dancing cycles* construction, is the use of cycles of very large costs, so that Howard's algorithm favors long, i.e., containing many edges, paths to the cycle of the current policy attractive, delaying the discovery of better cycles.

Given a graph and a sequence of policies, it is possible check, by solving an appropriate *linear program*, whether there exist costs for which Howard's algorithm generates the given sequence of policies. Experiments with a program that implements this idea helped us obtain the construction presented below.

For simplicity, we first prove Theorem 1 for $m = \frac{n^2}{4} + \frac{3n}{2}$. We later note that the same construction works when removing pairs of edges, which gives the statement of the theorem.

3.1   The Construction

For every n we construct a weighted directed graph $G_n = (V, E, c)$, on $|V| = 2n$ vertices and $|E| = n^2 + 3n$ edges, and an initial policy π_0 such that Howard's algorithm performs $n^2 + n + 1$ iterations on G_n when it starts with π_0.

The graph G_n itself is fairly simple. (See Figure 1.) Most of the intricacy goes into the definition of the cost function $c : E \rightarrow \mathbb{R}$. The graph G_n is composed of two symmetric parts. To highlight the symmetry we let $V = \{v_n^0, \ldots, v_1^0, v_1^1, \ldots, v_n^1\}$. Note that the set of vertices is split in two, according to whether the superscript is 0 or 1. In order to simplify notation when dealing with vertices with different superscripts, we sometimes refer to v_1^0 as v_0^1 and to v_1^1 as v_0^0. The set of edges is:

$$E = \{(v_i^0, v_j^0), (v_i^1, v_j^1) \mid 1 \le i \le n,\ i - 1 \le j \le n\}.$$

We next describe a sequence of policies Π_n of length $n^2 + n + 1$. We then construct a cost function that causes Howard's algorithm to generate this long sequence of policies. For $1 \le \ell \le r \le n$ and $s \in \{0,1\}$, and for $\ell = 1$, $r = 0$ and $s = 0$, we define a policy $\pi_{\ell,r}^s$:

$$\pi_{\ell,r}^s(v_i^t) = \begin{cases} v_{i-1}^t & \text{for } t \ne s \text{ or } i > \ell \\ v_r^t & \text{for } t = s \text{ and } i = \ell \\ v_n^t & \text{for } t = s \text{ and } i < \ell \end{cases}$$

The policy $\pi_{\ell,r}^s$ contains a single cycle $v_r^s v_{r-1}^s \ldots v_\ell^s v_r^s$ which is determined by its *defining* edge $e_{\ell,r}^s = (v_\ell^s, v_r^s)$. As shown in Figure 1, all vertices to the left of v_ℓ^s, the *head* of the cycle, choose an edge leading furthest to the right, while all the vertices to the right of v_ℓ^s choose an edge leading furthest to the left.

The sequence Π_n is composed of the policies $\pi_{\ell,r}^s$, where $1 \le \ell \le r \le n$ and $s \in \{0,1\}$, or $\ell = 1$, $r = 0$ and $s = 0$, with the following ordering. Policy $\pi_{\ell_1,r_1}^{s_1}$ precedes policy $\pi_{\ell_2,r_2}^{s_2}$ in Π_n if and only if $\ell_1 > \ell_2$, or $\ell_1 = \ell_2$ and $r_1 > r_2$, or $\ell_1 = \ell_2$ and $r_1 = r_2$ and $s_1 < s_2$. (Note that this is a reversed lexicographical ordering on the triplets $(\ell_1, r_1, 1 - s_1)$ and $(\ell_2, r_2, 1 - s_2)$.) For every $1 \le \ell \le r \le n$ and $s \in \{0,1\}$, or $\ell = 1$, $r = 0$ and $s = 0$, we let $f(\ell, r, s)$ be the *index* of $\pi_{\ell,r}^s$ in Π_n, where indices start at 0. We can now write:

$$\Pi_n = (\pi_k)_{k=0}^{n^2+n} = \left(\left(\left((\pi_{n-\ell,n-r}^s)_{s=0}^1 \right)_{r=0}^\ell \right)_{\ell=0}^{n-1}, \pi_{1,0}^0 \right)$$

We refer to Figure 1 for an illustration of G_4 and the corresponding sequence Π_4.

3.2   The Edge Costs

Recall that each policy $\pi_k = \pi_{\ell,r}^s$ is determined by an edge $e_k = e_{\ell,r}^s = (v_\ell^s, v_r^s)$, where $k = f(\ell, r, s)$. Let $N = n^2 + n$. We assign the edges the following *exponential* costs:

$$\begin{aligned} c(e_k) &= c(v_\ell^s, v_r^s) = n^{N-k} &,& \quad 0 \le k < N, \\ c(v_1^0, v_1^1) &= c(v_1^1, v_1^0) = -n^N &,& \\ c(v_i^s, v_{i-1}^s) &= 0 &,& \quad 2 \le i \le n,\ s \in \{0,1\}. \end{aligned}$$

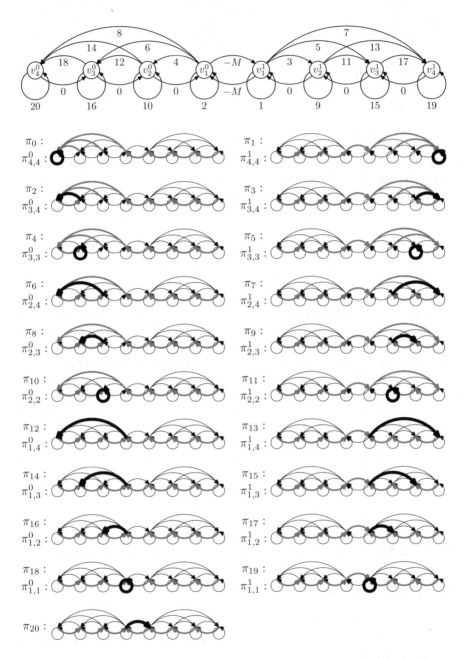

Fig. 1. G_4 and the corresponding sequence Π_4. Π_4 is shown in left-to-right order. Policies $\pi_{f(\ell,r,s)} = \pi^s_{\ell,r}$ are shown in bold, with $e^s_{\ell,r}$ being highlighted. Numbers below edges define costs. 0 means 0, $k > 0$ means n^k, and $-M$ means $-n^N$.

We claim that with these exponential edge costs Howard's algorithm does indeed produce the sequence Π_n. To show that π_{k+1} is indeed the policy that Howard's algorithm obtains by improving π_k, we have to show that

$$\pi_{k+1}(u) = \underset{v:(u,v)\in E}{\arg\min}\ A_{\pi_k}(u,v)\ ,\quad \forall u \in V. \tag{1}$$

For brevity, we let $c(v_i^s, v_j^s) = c_{i,j}^s$. The only cycle in $\pi_k = \pi_{\ell,r}^s$ is $C_{\ell,r}^s = v_r^s v_{r-1}^s \ldots v_\ell^s v_r^s$. As $c_{i,i-1}^s = 0$, for $2 \le i \le n$ and $s \in \{0,1\}$, we have

$$\mu_{\ell,r}^s\ =\ val(C_{\ell,r}^s)\ =\ \frac{c_{\ell,r}^s}{r-\ell+1}.$$

As $c_{\ell,r}^s = n^{N-k}$ and all cycles in our construction are of size at most n we have

$$n^{N-k-1}\ \le\ \mu_{\ell,r}^s\ \le\ n^{N-k}.$$

As all vertices have the same value $\mu_{\ell,r}^s$ under $\pi_k = \pi_{\ell,r}^s$, edges are compared based on the second component of their appraisals $A_{\pi_k}(u,v)$. Hence, (1) becomes:

$$\pi_{k+1}(u) = \underset{v:(u,v)\in E}{\arg\min}\ c(u,v) + pot_{\pi_k}(v)\ ,\quad \forall u \in V. \tag{2}$$

Note that an edge (u, v_1) is preferred over (u, v_2) if and only if

$$c(u,v_1) - c(u,v_2) < pot_{\pi_k}(v_2) - pot_{\pi_k}(v_1).$$

Let v_ℓ^s be the head of the cycle $C_{\ell,r}^s$. Keeping in mind that $c_{i,i-1}^s = 0$, for $2 \le i \le n$ and $s \in \{0,1\}$, it is not difficult to see that the potentials of the vertices under policy $\pi_{\ell,r}^s$ are given by the following expression:

$$pot_{\pi_{\ell,r}^s}(v_i^t) = \begin{cases} c_{i,n}^s - (n-\ell+1)\mu_{\ell,r}^s & \text{if } t = s \text{ and } i < \ell, \\ -(i-\ell)\mu_{\ell,r}^s & \text{if } t = s \text{ and } i \ge \ell, \\ c_{1,0}^t - i\mu_{\ell,r}^s + pot_{\pi_{\ell,r}^s}(v_1^s) & \text{if } t \ne s. \end{cases}$$

It is convenient to note that we have $pot_{\pi_{\ell,r}^s}(v_i^t) \le 0$, for every $1 \le \ell \le r \le n$, $1 \le i \le n$ and $s, t \in \{0,1\}$. In the first case ($t = s$ and $i < \ell$), this follows from the fact that (v_i^s, v_n^s) has a larger index than (v_ℓ^s, v_r^s) (note that $i < \ell$). In the third case ($t = s$), it follows from the fact that $c_{1,0}^t = -n^N < 0$.

3.3   The Dynamics

The proof that Howard's algorithm produces the sequence Π_n is composed of three main cases, shown in Figure 2. Each case is broken into several subcases. Each subcase on its own is fairly simple and intuitive.

Case 1. Suppose that $\pi_k = \pi_{\ell,r}^0$. We need to show that $\pi_{k+1} = \pi_{\ell,r}^1$. We have the following four subcases, shown at the top of Figure 2.

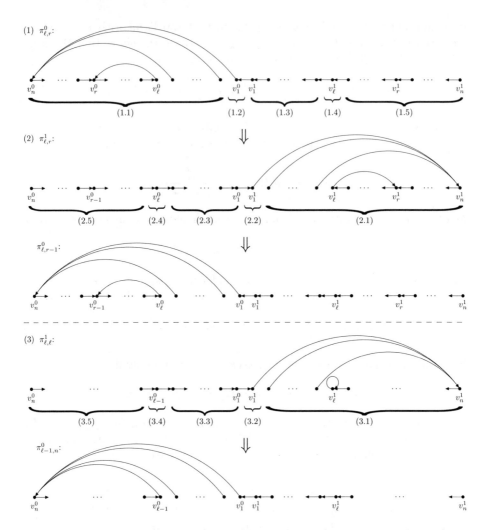

Fig. 2. Policies of transitions (1) $\pi^0_{\ell,r}$ to $\pi^1_{\ell,r}$, (2) $\pi^1_{\ell,r}$ to $\pi^0_{\ell,r-1}$, and (3) $\pi^1_{\ell,\ell}$ to $\pi^0_{\ell-1,n}$. Vertices of the corresponding subcases have been annotated accordingly.

Case 1.1. We show that $\pi_{k+1}(v^0_i) = v^0_{i-1}$, for $2 \le i \le n$. We have to show that (v^0_i, v^0_{i-1}) beats (v^0_i, v^0_j), for every $2 \le i \le j \le n$, or in other words that

$$c^0_{i,i-1} + pot_{\pi^0_{\ell,r}}(v^0_{i-1}) < c^0_{i,j} + pot_{\pi^0_{\ell,r}}(v^0_j) \quad , \quad 2 \le i \le j \le n.$$

Case 1.1.1. Assume that $j < \ell$. We then have

$$pot_{\pi^0_{\ell,r}}(v^0_{i-1}) = c^0_{i-1,n} - (n - \ell + 1)\mu^0_{\ell,r},$$
$$pot_{\pi^0_{\ell,r}}(v^0_j) = c^0_{j,n} - (n - \ell + 1)\mu^0_{\ell,r}.$$

Recalling that $c^0_{i,i-1} = 0$, the inequality that we have to show becomes

$$c^0_{i-1,n} < c^0_{i,j} + c^0_{j,n} \quad , \quad 2 \le i \le j < \ell \le n.$$

As the edge (v^0_{i-1}, v^0_n) comes *after* (v^0_i, v^0_j) in our ordering, we have $c^0_{i-1,n} < c^0_{i,j}$. The other term on the right is non-negative and the inequality follows easily.

Case 1.1.2. Assume that $i - 1 < \ell \le j$. We then have

$$pot_{\pi^0_{\ell,r}}(v^0_{i-1}) = c^0_{i-1,n} - (n - \ell + 1)\mu^0_{\ell,r},$$
$$pot_{\pi^0_{\ell,r}}(v^0_j) = -(j - \ell)\mu^0_{\ell,r},$$

and the required inequality becomes

$$c^0_{i-1,n} < c^0_{i,j} + (n - j + 1)\mu^0_{\ell,r} \quad , \quad 1 \le i - 1 < \ell \le j \le n.$$

As $j \le n$, the inequality again follows from the fact that $c^0_{i-1,n} < c^0_{i,j}$.

Case 1.1.3. Assume that $\ell \le i - 1 < j$. We then have

$$pot_{\pi^0_{\ell,r}}(v^0_j) - pot_{\pi^0_{\ell,r}}(v^0_{i-1}) = (i - j - 1)\mu^0_{\ell,r},$$

and the required inequality becomes

$$(j - i + 1)\mu^0_{\ell,r} < c^0_{i,j} \quad , \quad 1 \le \ell \le i - 1 < j \le n.$$

This inequality holds as $(j - i + 1)\mu^0_{\ell,r} < n\, c^0_{\ell,r} \le c^0_{i,j}$. The last inequality follows as (v^0_ℓ, v^0_r) appears after (v^0_i, v^0_j) in our ordering. (Note that we are using here, for the first time, the fact that the weights are exponential.)

Case 1.2. We show that $\pi_{k+1}(v^0_1) = v^1_1$. We have to show that

$$c^0_{1,0} + pot_{\pi^0_{\ell,r}}(v^1_1) < c^0_{1,j} + pot_{\pi^0_{\ell,r}}(v^0_j) \quad , \quad 1 \le j \le n.$$

This inequality is easy. Note that $c^0_{1,0} = -n^N$, $pot^{\pi^0_{\ell,r}}(v^1_1) \le 0$, while $c^0_{1,j} > 0$ and $pot_{\pi^0_{\ell,r}}(v^0_j) > -n^N$.

Case 1.3. We show that $\pi_{k+1}(v^1_i) = v^1_n$, for $1 \le i < \ell$. We have to show that

$$c^1_{i,n} - c^1_{i,j} < pot_{\pi^0_{\ell,r}}(v^1_j) - pot_{\pi^0_{\ell,r}}(v^1_n) \quad , \quad 1 \le i < \ell, \, i - 1 \le j < n.$$

Case 1.3.1. Suppose that $i = 1$ and $j = 0$. We need to verify that

$$c^1_{1,n} - c^1_{1,0} < pot_{\pi^0_{\ell,r}}(v^1_0) - pot_{\pi^0_{\ell,r}}(v^1_n).$$

As $pot_{\pi^0_{\ell,r}}(v^1_0) = pot_{\pi^0_{\ell,r}}(v^1_1)$ and $pot_{\pi^0_{\ell,r}}(v^1_n) = c^1_{1,0} - n\mu^0_{\ell,r} + pot_{\pi^0_{\ell,r}}(v^1_1)$, we have to verify that $c^1_{1,n} < n\mu^0_{\ell,r}$, which follows from the fact that $\ell > 1$ and that (v^0_ℓ, v^0_r) has a smaller index than (v^1_1, v^n_1).

Case 1.3.2. Suppose that $j \geq 1$. We have to verify that

$$c^1_{i,n} - c^1_{i,j} < pot_{\pi^0_{\ell,r}}(v^1_j) - pot_{\pi^0_{\ell,r}}(v^1_n) = (n-j)\mu^0_{\ell,r} \quad , \quad 1 \leq i < \ell \,,\, i-1 \leq j < n.$$

As in Case 1.3.1 we have $c^1_{i,n} \leq \mu^0_{\ell,r}$ while $c^1_{i,j} > 0$.

Case 1.4. We show that $\pi_{k+1}(v^1_\ell) = v^1_r$. We have to show that

$$c^1_{\ell,r} - c^1_{\ell,j} < pot_{\pi^0_{\ell,r}}(v^1_j) - pot_{\pi^0_{\ell,r}}(v^1_r) \quad , \quad \ell-1 \leq j \leq n \,,\, j \neq r.$$

Case 1.4.1. Suppose that $\ell = 1$ and $j = 0$. As in case 1.3.1, the inequality becomes $c^1_{1,r} < r\mu^0_{\ell,r}$ which is easily seen to hold.

Case 1.4.2. Suppose that $\ell - 1 \leq j < r$ and $0 < j$. We need to show that

$$c^1_{\ell,r} - c^1_{\ell,j} < pot_{\pi^0_{\ell,r}}(v^1_j) - pot_{\pi^0_{\ell,r}}(v^1_r) = (r-j)\mu^0_{\ell,r} \quad , \quad \ell-1 \leq j \leq n \,,\, 0 < j < r.$$

As (v^1_ℓ, v^1_r) immediately follows (v^0_ℓ, v^0_r) in our ordering, we have $c^1_{\ell,r} = n^{-1}c^0_{\ell,r}$. Thus $c^1_{\ell,r} \leq \mu^0_{\ell,r} \leq (r-j)\mu^0_{\ell,r}$. As $c^1_{\ell,j} > 0$, the inequality follows.

Case 1.4.3. Suppose that $r < j$. We need to show that

$$c^1_{\ell,r} - c^1_{\ell,j} < pot_{\pi^0_{\ell,r}}(v^1_j) - pot_{\pi^0_{\ell,r}}(v^1_r) = (r-j)\mu^0_{\ell,r} \quad , \quad r < j \leq n,$$

or equivalently that $c^1_{\ell,j} - c^1_{\ell,r} < (j-r)\mu^0_{\ell,r}$, for $r < j \leq n,$. This follows from the fact that (v^1_ℓ, v^1_r) comes after (v^1_ℓ, v^1_j) in the ordering and that $c^1_{\ell,r} > 0$.

Case 1.5. We show that $\pi_{k+1}(v^1_i) = v^1_{i-1}$, for $\ell < i \leq n$. We have to show that

$$c^1_{i,i-1} - c^1_{i,j} < pot_{\pi^s_{\ell,r}}(v^1_j) - pot_{\pi^s_{\ell,r}}(v^1_{i-1}) = (i-j-1)\mu^s_{\ell,r} \quad , \quad \ell < i \leq j \leq n.$$

This is identical to case 1.1.3.

Case 2. Suppose that $\pi_k = \pi^1_{\ell,r}$ and $\ell < r$. We need to show that $\pi_{k+1} = \pi^0_{\ell,r-1}$. The proof is very similar to Case 1 and is omitted.

Case 3. Suppose that $\pi_k = \pi^1_{\ell,\ell}$. We need to show that $\pi_{k+1} = \pi^0_{\ell-1,n}$. The proof is very similar to Cases 1 and 2 and is omitted.

3.4   Remarks

For any $0 \leq \ell \leq r < n$, if the edges (v^0_ℓ, v^0_r) and (v^1_ℓ, v^1_r) are removed from G_n, then Howard's algorithm skips $\pi^0_{\ell,r}$ and $\pi^1_{\ell,r}$, but otherwise Π_n remains the same. This can be repeated any number of times, essentially without modifying the proof given in Section 3.3, thus giving us the statement of Theorem 1.

Let us also note that the costs presented here have been chosen to simplify the analysis. It is possible to define smaller costs, but assuming $c^s_{i,i-1} = 0$ for $s \in \{0,1\}$ and $2 \leq i \leq n$, which can always be enforced using a potential transformation, one can show that integral costs must be exponential in n. Details will appear in the full version of the paper.

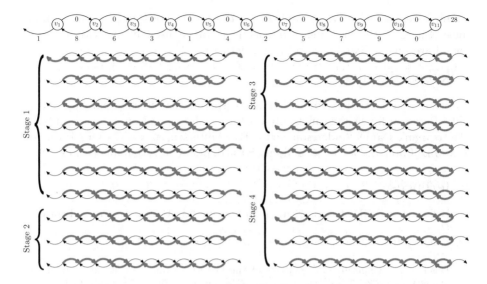

Fig. 3. G_5 and the corresponding sequence of policies

4   A $2n - O(1)$ Lower Bounds for Outdegree-2 Graphs

In this section we briefly mention a construction of a sequence of outdegree-2 DMDPs on which the number of iterations of Howard's algorithm is only two less than the total number of edges.

Theorem 2. *For every $n \geq 3$ there exists a weighted directed graph $G_n = (V, E, c)$, where $c : E \rightarrow \mathbb{R}$, with $|V| = 2n + 1$ and $|E| = 2|V|$, on which Howard's algorithm performs $|E| - 2 = 4n$ iterations.*

The graph used in the proof of Theorem 2 is simply a *bidirected cycle* on $2n + 1$ vertices. (The graph G_5 is depicted at the top of Figure 3.) The proof of Theorem 2 will appear in the full version of the paper.

5   Concluding Remarks

We presented a quadratic lower bound on the number of iterations performed by Howard's algorithm for finding Minimum Mean-Cost Cycles (MMCCs). Our lower bound is quadratic in the number of *vertices*, but is only linear in the number of *edges*. We conjecture that this is best possible:

Conjecture. *The number of iterations performed by Howard's algorithm, when applied to a weighted directed graph, is at most the number of edges in the graph.*

Proving (or disproving) our conjecture is a major open problem. Our lower bounds shed some light on the non-trivial behavior of Howard's algorithm, even on deterministic DMDPs, and expose some of the difficulties that need to be overcome to obtain non-trivial upper bounds on its complexity.

Our lower bounds on the complexity of Howard's algorithm do not undermine the usefulness of Howard's algorithm, as the instances used in our quadratic lower bound are very unlikely to appear in practice.

Acknowledgement

We would like to thank Omid Madani for sending us his example [13], and to Mike Paterson for helping us to obtain the results of Section 4. We would also like to thank Daniel Andersson, Peter Bro Miltersen, as well as Omid Madani and Mike Paterson, for helpful discussions on policy iteration algorithms.

References

1. Bellman, R.E.: Dynamic programming. Princeton University Press, Princeton (1957)
2. Bellman, R.E.: On a routing problem. Quarterly of Applied Mathematics 16, 87–90 (1958)
3. Dasdan, A.: Experimental analysis of the fastest optimum cycle ratio and mean algorithms. ACM Trans. Des. Autom. Electron. Syst. 9(4), 385–418 (2004)
4. Derman, C.: Finite state Markov decision processes. Academic Press, London (1972)
5. Fearnley, J.: Exponential lower bounds for policy iteration. In: Proc. of 37th ICALP (2010), Preliminaey version available at http://arxiv.org/abs/1003.3418v1
6. Ford Jr., L.R., Fulkerson, D.R.: Maximal flow through a network. Canadian Journal of Mathematics 8, 399–404 (1956)
7. Friedmann, O.: An exponential lower bound for the parity game strategy improvement algorithm as we know it. In: Proc. of 24th LICS, pp. 145–156 (2009)
8. Georgiadis, L., Goldberg, A.V., Tarjan, R.E., Werneck, R.F.F.: An experimental study of minimum mean cycle algorithms. In: Proc. of 11th ALENEX, pp. 1–13 (2009)
9. Goldberg, A.V., Tarjan, R.E.: Finding minimum-cost circulations by canceling negative cycles. Journal of the ACM 36(4), 873–886 (1989)
10. Hansen, T.D., Miltersen, P.B., Zwick, U.: Strategy iteration is strongly polynomial for 2-player turn-based stochastic games with a constant discount factor. CoRR, abs/1008.0530 (2010)
11. Howard, R.A.: Dynamic programming and Markov processes. MIT Press, Cambridge (1960)
12. Karp, R.M.: A characterization of the minimum cycle mean in a digraph. Discrete Mathematics 23(3), 309–311 (1978)
13. Madani, O.: Personal communication (2008)
14. Megiddo, N.: Combinatorial optimization with rational objective functions. Mathematics of Operations Research 4(4), 414–424 (1979)
15. Megiddo, N.: Applying parallel computation algorithms in the design of serial algorithms. Journal of the ACM 30(4), 852–865 (1983)
16. Puterman, M.L.: Markov decision processes. Wiley, Chichester (1994)
17. Ye, Y.: The simplex method is strongly polynomial for the Markov decision problem with a fixed discount rate (2010),
   http://www.stanford.edu/~yyye/simplexmdp1.pdf
18. Young, N.E., Tarjan, R.E., Orlin, J.B.: Faster parametric shortest path and minimum-balance algorithms. Networks 21, 205–221 (1991)

Solving Two-Stage Stochastic Steiner Tree Problems by Two-Stage Branch-and-Cut

Immanuel Bomze[1], Markus Chimani[2,*], Michael Jünger[3,**],
Ivana Ljubić[1,***], Petra Mutzel[4,**], and Bernd Zey[4]

[1] Faculty of Business, Economics and Statistics, University of Vienna
{immanuel.bomze,ivana.ljubic}@univie.ac.at
[2] Institute of Computer Science, Friedrich-Schiller-University of Jena
markus.chimani@uni-jena.de
[3] Department of Computer Science, University of Cologne
mjuenger@informatik.uni-koeln.de
[4] Department of Computer Science, TU Dortmund
{petra.mutzel,bernd.zey}@tu-dortmund.de

Abstract. We consider the Steiner tree problem under a 2-stage stochastic model with recourse and finitely many scenarios (SSTP). Thereby, edges are purchased in the first stage when only probabilistic information on the set of terminals and the future edge costs is known. In the second stage, one of the given scenarios is realized and additional edges are purchased to interconnect the set of (now known) terminals. The goal is to choose an edge set to be purchased in the first stage while minimizing the overall expected cost of the solution.

We provide a new semi-directed cut-set based integer programming formulation that is stronger than the previously known undirected model. To solve the formulation to provable optimality, we suggest a two-stage branch-and-cut framework, facilitating (integer) L-shaped cuts. The framework itself is also applicable to a range of other stochastic problems.

As SSTP has yet been investigated only from the theoretical point of view, we also present the first computational study for SSTP, showcasing the applicability of our approach and its benefits over solving the extensive form of the deterministic equivalent directly.

1 Introduction

Motivation. The classical Steiner tree problem in graphs is a quite well-studied combinatorial optimization problem and has a wide range of applications: from the planning of various infrastructure networks (e.g., communication or energy supply) to the study of protein-interaction networks in bioinformatics. Given an

* Funded via a juniorprofessorship by the Carl-Zeiss-Foundation.
** M.J. and P.M. gratefully acknowledge the hospitality they enjoyed during their stay as visiting research professors at Univ. Vienna when much of this research was done.
*** Supported by the project T334 of the Austrian Science Fund (FWF).

O. Cheong, K.-Y. Chwa, and K. Park (Eds.): ISAAC 2010, Part I, LNCS 6506, pp. 427–439, 2010.

undirected graph $G = (V, E)$ with edge weights (*costs*) $c_e \geq 0$, for all $e \in E$, and a subset of required vertices (*terminals*) $R \subseteq V$, the problem consists of finding a subset of edges that interconnects all the terminals at minimum (edge installation) cost.

In practice, however, network planners are often faced with uncertainty with respect to the input data. The actual demand patterns become known only after the network has been built. In that case, networks found by solving an instance in which it is assumed that the complete knowledge of the input is known up-front, might not provide appropriate solutions if deviations from the assumed scenario are encountered. Stochastic optimization is a promising way to take uncertainties into account.

Our problem. We consider the two-stage stochastic Steiner tree problem (SSTP) with fixed recourse and finitely many scenarios in which the terminal set and the edge installation costs are subject to uncertainty. This means: We consider two points in times (*stages*): In the first stage, we only have probabilistic information in terms of possible *scenarios*; a scenario thereby specifies a terminal set, edge costs for the second stage, and a probablity for it to be realized. Based on this information, we may buy some edges at a price lower than their expected second stage costs. Then, in the second stage, one of the given scenarios is realized and we have to purchase additional edges in order to interconnect the (now known) terminal nodes. Our goal is to make a decision about edges to be purchased in the first stage, while minimizing the *expected cost* of the full solution (i.e., after the second stage).

The SSTP has obvious applications in the design of various communication, distribution or transportation networks. For example, a telecommunication company wants to expand its broadband infrastructure without a precise knowledge about the demand patterns and link costs. These values are therefore estimated using collected statistical data and realistic forecast models to generate various scenarios to model possible outcomes (probably using sampling methods [16,17]). Due to economic and/or geographical factors, the future link costs may vary between different scenarios and they are usually more expensive than the current costs. Hence, a company is interested in building a cost minimal subnetwork today that takes all possible future outcomes into account.

Previous work. SSTP is one of the fundamental network design problems under uncertainty. Gupta et al. started a series of papers on approximation algorithms for the SSTP. E.g., they provided a constant factor approximation for the SSTP when the second stage costs are determined from the first stage costs by multiplication with a fixed inflation factor [8]. The algorithm is based on a primal-dual scheme, guided by a relaxed integer linear programming (ILP) solution. For the general case that we consider, Gupta et al. [7] have shown that the problem becomes as hard as Label Cover (which is $\Omega(2^{\log^{1-\epsilon} n})$-hard). Shmoys and Swamy [16] presented a 4-approximation for the SSTP with cost-sharing properties. We are not aware of any computational study concerning the SSTP.

Our Contribution consists of three parts:

1) The ILP model used in [8] is based on an undirected cut-set formulation for Steiner trees. We propose a new semi-directed ILP model and show that it is provably stronger than the undirected one. This may in turn lead to stronger approximation algorithms, as this has been recently achieved in the context of traditional Steiner trees [4]. Furthermore, we show that the recourse function decomposes into a set of independent *restricted* Steiner arborescence problems.

2) To solve the problem, we present a Benders-like decomposition [1]. In our case, the subproblems are themselves again NP-hard and we propose the concept of *2-stage branch-and-cut* (2-B&C), which may be of further interest beyond the SSTP (cf. Sect. 5). Thereby we nest a branch-and-cut framework for solving the subproblems into a branch-and-cut master approach in which (integer) L-shaped cuts are generated, while guaranteeing overall convergence of the process. We know from the traditional STP that cut-based formulations outperform compact formulations (based on multicommodity flow) by orders of magnitudes. To our knowledge, this is the first time that a non-compact formulation is considered as a second stage subproblem.

3) While there is a series of approximation results for the SSTP (e.g., [16,8,7]), this is the first time that it is studied computationally. In our experiments, we investigate the behaviour of our 2-B&C algorithm for two different decompositions of the new semi-directed ILP model and compare it to solving the deterministic equivalent directly. We report optimal results for SSTP instances with up to 165 vertices, 274 edges, and 5–200 scenarios.

2   ILP Models

2.1   Problem Definition

We consider the following two-stage stochastic Steiner tree problem. Let $G = (V, E)$ be an undirected network with a selected root r and with known first-stage edge costs $c_e \geq 0$, for all $e \in E$; let $V_r := V \setminus \{r\}$. The set of terminals, as well as the costs of edges to be purchased in the second stage, is known only in the second stage. These values together form a random variable ξ, for which we assume that it has a finite support. It can therefore be modeled using a finite set of scenarios $\mathcal{K} = \{1, \ldots, K\}$, $K \geq 1$. The realization probability of each scenario is given by $p_k > 0$, $k \in \mathcal{K}$; we have $\sum_{k \in \mathcal{K}} p_k = 1$. Denote by $q_e^k \geq 0$ the cost of an edge $e \in E$ if it is bought in the second stage, under scenario $k \in \mathcal{K}$. Denote the *expected second stage cost* of an edge $e \in E$ by $q_e^* := \sum_{k \in \mathcal{K}} p_k q_e^k$. We assume that $q_e^* > c_e$, for all $e \in E$. Furthermore, let $R_k \subseteq V_r$ be the set of terminals under the k-th scenario. We denote by E_0 the set of edges purchased in the first-stage, and by E_k the set of additional edges purchased under scenario k, $k \in \mathcal{K}$.

The *Stochastic Steiner Tree problem (SSTP)* can then be formulated as follows: Determine the subset of edges $E_0 \subseteq E$ to be purchased in the first stage, so that the overall cost defined as $\sum_{e \in E_0} c_e + \sum_{k \in \mathcal{K}} p_k \sum_{e \in E_k} q_e^k$ is minimized, while $E_0 \cup E_k$ spans R_k for all $k \in \mathcal{K}$.

Obviously, the optimal first-stage solution of the SSTP is not necessarily a tree [8]. In fact, the optimal solution might contain several disjoint fragments, depending on the subsets of terminals throughout different scenarios, or depending on the second-stage cost structure.

2.2   Undirected Model, Deterministic Equivalent

A deterministic equivalent (in extensive form) of the stochastic Steiner tree problem has been originally proposed in [8]. The authors developed an undirected ILP formulation as a natural extension of the undirected cut-set model for Steiner trees. We briefly recall this model here. Binary variables x_e indicate whether an edge $e \in E$ belongs to E_0, and binary second-stage variables y_e^k indicate whether e belongs to E_k, for all $k \in \mathcal{K}$. For $D \subseteq E$, let $(x + y^k)(D) = \sum_{e \in D}(x_e + y_e^k)$. For $S \subseteq V$, let $\delta(S) = \{\{i, j\} \in E \mid i \in S \text{ and } j \notin S\}$. A *deterministic equivalent* of the SSTP can then be written using *undirected* cuts:

$$(UD) \quad \min_{x \in \{0,1\}^{|E|}, y \in \{0,1\}^{|K||E|}} \left\{ \sum_{e \in E} c_e x_e + \sum_{k \in \mathcal{K}} p_k \sum_{e \in E} q_e^k y_e^k \mid \right.$$

$$(x + y^k)(\delta(S)) \geq 1, \forall S \subseteq V_r, S \cap R_k \neq \emptyset, \ \forall k \in \mathcal{K}\}$$

Gupta et al. [8] have shown that the solution of the canonical LP-relaxation of the above model can be rounded to a feasible solution with value of at most 40 times that of the optimal solution, if the edge costs in the second stage are given by $q_e^k = \sigma_k c_e$, for all $e \in E$, $k \in \mathcal{K}$, for some fixed scalar σ_k.

2.3   Semi-directed Model, Deterministic Equivalent

It is well known that directed models for Steiner trees provide better lower LP-bounds, and therefore the natural question arises whether we can extend the model (UD) by bi-directing the given graph G and replacing edge- by arc-variables in the same model. The main difficulty with the stochastic Steiner tree problem is that the arcs of the first-stage solution cannot be derived using this technique. It is not difficult to imagine an instance in which an edge $\{i, j\} \in E$ is used in direction (i, j) for one scenario, and in the opposite direction (j, i) for another scenario.

Cut-set formulation. Despite the difficulty mentioned above, we can model SSTP using oriented edges to describe the second stage solutions. In other words, we are looking for the optimal first-stage solution (an undirected subgraph of G) such that each solution of scenario k represents a Steiner arborescence rooted at r, whose *arcs* are built upon all the (already installed) first stage *edges* and additional second-stage arcs. In order to derive the new model, we first bi-direct graph G by defining the set of arcs $A = \{(i, j) \cup (j, i) \mid \{i, j\} \in E, i, j \neq r\} \cup \{(r, i) \mid \{r, i\} \in E\}$. Denote by A_k the arcs of the optimal solution of scenario k, $k \in \mathcal{K}$. For each scenario $k \in \mathcal{K}$, we now introduce binary arc-variables z_{ij}^k, for all $(i, j) \in A$. A variable z_{ij}^k is set to 1 iff the final solution after the second stage in scenario k uses the arc (i, j). Note that for edges bought in the first stage, each scenario solution has to select one of its corresponding arcs.

We can then write a new *semi-directed deterministic equivalent* of the SSTP as follows; we denote the formulation by (EF) for *extensive form*, to distinguish it from its decomposed variants (Sect. 3):

$$(EF) \quad \min \sum_{e \in E} c_e x_e + \sum_{k \in \mathcal{K}} p_k \sum_{e=\{i,j\} \in E} q_e^k (z_{ij}^k + z_{ji}^k - x_e)$$

$$\text{s.t.} \quad z^k(\delta^-(S)) \geq 1, \qquad \forall S \subseteq V_r, S \cap R_k \neq \emptyset, \forall k \in \mathcal{K} \tag{1}$$

$$z_{ij}^k + z_{ji}^k \geq x_e, \qquad \forall e = \{i,j\} \in E, \forall k \in \mathcal{K} \tag{2}$$

$$z_{ij}^k \in \{0,1\}, \quad \forall (i,j) \in A, \forall k \in \mathcal{K} \tag{3}$$

$$0 \leq x_e \leq 1, \qquad \forall e \in E, \forall k \in \mathcal{K} \tag{4}$$

Here, $\delta^-(S) = \{(i,j) \in A \mid i \notin S, j \in S\}$. Constraints (1) ensure that for each terminal $v \in R_k$, there is a directed path (using the second stage arcs) from r to v. Inequalities (2) are capacity constraints ensuring that at least one second stage arc is installed for every edge purchased in the first stage.

Lemma 1. *Formulation (EF) models the deterministic equivalent of the stochastic Steiner tree problem correctly. In particular, in every optimal solution of the model, variables x_e take value 0 or 1.*

Proof. It should be clear from the above description that the formulation is correct when restricting $x \in \{0,1\}^{|E|}$. Assume there exists an optimal solution $(\bar{x}, \bar{z})$ with $0 < \bar{x}_e < 1$, for some $e = \{i,j\} \in E$. Inequalities (2), together with (3) and the fact that $q_e^k > 0$, imply that for all scenarios $k \in \mathcal{K}$ we have $\bar{z}_{ij}^k + \bar{z}_{ji}^k = 1$. The term in the objective function corresponding to e then is:

$$c_e \bar{x}_e + \sum_{k \in \mathcal{K}} p_k q_e^k (1 - \bar{x}_e) = c_e \bar{x}_e + q_e^*(1 - \bar{x}_e) = (c_e - q_e^*)\bar{x}_e + q_e^*.$$

Since $c_e - q_e^* < 0$, we could reduce the value of the objective function by setting $\bar{x}_e := 1$, which is a contradiction to $\bar{x}$ being an optimal solution. $\qquad \square$

Clearly, the semi-directed formulation (EF) for the stochastic Steiner tree problem is at least as strong as the undirected formulation. We can show that the new formulation is even strictly stronger.

Lemma 2. *Denote by $Proj_{x,y}(EF)$ the projection of the polytope defined by the LP-relaxation of (EF) onto the space of x and y variables in which $y_e^k = z_{ij}^k + z_{ji}^k - x_e$, for all $e = \{i,j\} \in E$, for all $k \in \mathcal{K}$. Let $\mathcal{P}_u$ be the polytope defined by the LP-relaxation of (UD). Then for any instance of SSTP we have $Proj_{x,y}(EF) \subseteq \mathcal{P}_u$ and there are instances for which strict inequality holds and the optimal LP-relaxation value of (EF) is strictly larger than the corresponding LP-relaxation value of (UD).*

Proof. It is not difficult to see that the $\subseteq$-relationship holds. To show the strict inequality, consider the following example.

For the network given in Figure 1, we assume that scenarios are assigned a constant inflation factor, σ_k, for all $k \in \mathcal{K}$, so that $q_e^k = \sigma_k c_e$, for all $e \in E$. The following scenario values are given:

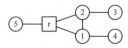

Fig. 1. Problem instance for the proof of Lemma 2. All edge costs are equal to one.

$$\text{Scenario 1: } \sigma_1 = 3/2, \quad p_1 = 1/4, \quad R_1 = \{1,2,3\},$$
$$\text{Scenario 2: } \sigma_2 = 3/2, \quad p_2 = 1/4, \quad R_2 = \{1,2,4\},$$
$$\text{Scenario 3: } \sigma_3 = 3, \quad\;\; p_3 = 1/2, \quad R_3 = \{5\}.$$

The optimal LP-solution of (UD) sets $x_{r5} = y_{23}^1 = y_{14}^2 = 1$ and $y_{r2}^\ell = y_{r1}^\ell = y_{12}^\ell = 1/2$, for $\ell = 1, 2$. The other variables are set to zero. Therefore, the solution value of (UD) is 2.875. On the other hand, there exists no feasible solution in (EF) with the same objective value, which proves the strict inequality.   □

3   Algorithmic Framework

3.1   Decomposition of the (EF) Model

The large number of decision variables makes the extensive form (EF) very difficult to solve when considering many scenarios. However, we can rewrite the (EF) formulation as

$$\min_{x \in \{0,1\}^{|E|}} c^t x + Q(x),$$

in which the so-called *recourse function* $Q(x)$ decomposes into K independent problems, i.e., $Q(x) = \mathbb{E}Q(x, \xi) = \sum_{k \in \mathcal{K}} p_k Q(x, k)$. For a fixed vector $\tilde{x}$, the k-th subproblem related to $Q(\tilde{x}, k)$ is the following NP-hard *restricted Steiner arborescence problem*:

$$(RSAP^k) \quad \min \sum_{e=\{i,j\} \in E} q_e^k (z_{ij}^k + z_{ji}^k - \tilde{x}_e)$$

$$\text{s.t.} \quad z^k(\delta^-(S)) \geq 1, \qquad \forall S \subseteq V_r, S \cap R_k \neq \emptyset \tag{5}$$

$$z_{ij}^k + z_{ji}^k \geq \tilde{x}_e, \qquad \forall e = \{i,j\} \in E \tag{6}$$

$$z_{ij}^k \in \{0,1\}, \quad \forall (i,j) \in A \tag{7}$$

Due to the integrality restrictions on the second stage variables, the recourse function $Q(x)$ is non-convex and discontinuous. Let $R(x)$ and $R(x, k)$ denote the relaxation of the function $Q(x)$ and $Q(x, k)$, in which the second stage variables z^k are continuous, respectively.

3.2   Two-Stage Branch-and-Cut

The key idea is to apply a *nested* Branch-and-Cut approach: a Benders-like decomposition method determines the *Master Branch-and-Cut Framework*. Let the following problem be the relaxed master problem (RMP):

$$(RMP) \quad \min_{x \in [0,1]^{|E|}, \Theta_k \geq 0} \{c^t x + \Theta \mid \Theta = \sum_{k \in \mathcal{K}} p_k \Theta_k,$$

a set of L-shaped cuts and integer L-shaped cuts}.

For a given first stage solution in x, the variables Θ_k are estimated second stage costs of scenario k needed for purchasing *additional* arcs in the second stage in order to interconnect the terminals from R_k. As *optimality cuts* we use *L-shaped* and *integer L-shaped* cuts [3,10] to guarantee the convergence of the algorithm as described below. Observe that no *feasibility cuts* are needed, since we are dealing with the problem with *complete recourse*, i.e., every first-stage solution is feasible.

Step 0: Initialization. $UB = +\infty$ (global upper bound, corresponding to a feasible solution), $\nu = 0$. Create the first pendant node. In the initial (RMP), the set of (integer) L-shaped cuts is empty.

Step 1: Selection. Select a pendant node from the B&C tree, if such a node exists, otherwise STOP.

Step 2: Separation. Solve (RMP) at the current node. $\nu = \nu + 1$. Let $(x^\nu, \Theta_1^\nu, \dots, \Theta_K^\nu)$ be the current optimal solution, $\Theta^\nu = \sum_{k \in \mathcal{K}} p_k \Theta_k^\nu$.

   (2.1) If $c^t x^\nu + \Theta^\nu > UB$ fathom the current node and goto Step 1.

   (2.2) Search for **violated L-shaped cuts**:

   For all $k \in \mathcal{K}$, compute the LP-relaxation value $R(x^\nu, k)$ of $(RSAP^k)$. If $R(x^\nu, k) > \Theta_k^\nu$: insert L-shaped cut (10) into (RMP).

   If at least one L-shaped cut was inserted goto Step 2.

   (2.3) If x is binary, search for **violated integer L-shaped cuts**:

   (2.3.1) For all $k \in \mathcal{K}$ s.t. z^k is not binary in the previously computed LP-relaxation, solve $(RSAP^k)$ to optimality. Let $Q(x^\nu, k)$ be the optimal $(RSAP^k)$ value. If $\sum_{k \in \mathcal{K}} p_k Q(x^\nu, k) > \Theta^\nu$ insert integer L-shaped cut (11) into (RMP). Goto Step 2.

   (2.3.2) $UB = \min(UB, c^t x^\nu + \Theta^\nu)$. Fathom the current node and goto Step 1.

Step 3: Branching. Using a branching criterion, create two nodes, append them to the list of pendant nodes, goto Step 1.

The algorithm described above is a B&C approach in which each of the subproblems $(RSAP^k)$ is solved to optimality using another B&C. This explains the name *two-stage branch-and-cut*.

L-shaped cuts. To solve the LP-relaxation of the (EF) formulation via the models $(RSAP^k)$ given above, we will relax the integrality constraints (7) to $0 \leq z_{ij}^k$, for all $(i, j) \in A$, for all $k \in \mathcal{K}$. Only a small number among the exponential number of cuts will be needed to solve the LP-relaxations (cf. cutting plane method). Therefore, in the corresponding dual problems only those dual variables associated to cuts found in the cutting plane phase will be of interest. We associate dual variables α_S^k to constraints (5) and β_e^k to (6).

The dual of the LP-relaxation of $(RSAP^k)$ then reads:

$$(DRSAP^k) \quad \max \sum_{S \subseteq V_r : S \cap R_k \neq \emptyset} \alpha_S^k + \sum_{e \in E} \tilde{x}_e \beta_e^k - \sum_{e \in E} q_e^k \tilde{x}_e$$

s.t. $\quad \sum_{S:(i,j)\in\delta^-(S)} \alpha_S^k + \beta_e^k \le q_{ij}^k, \quad \forall (i,j) \in A, e = \{i,j\} \quad$ (8)

$$\alpha_S^k \ge 0, \quad \beta_e^k \ge 0, \quad \forall S \subseteq V_r, S \cap R_k \ne \emptyset, \quad \forall e \in E \quad (9)$$

Denote by $(\tilde{\alpha}^k, \tilde{\beta}^k)$ the optimal solutions of the dual of the k-th subproblem. Since our relaxed recourse function is convex, continuous and bounded, we know:

Lemma 3. *Let $\partial f(x)$ denote the subdifferential of a function f at the point x. We have $\tilde{\beta}^k - q^k \in \partial R(\tilde{x}, k)$ and hence (see [13]) $\sum_{k\in\mathcal{K}} p_k(\tilde{\beta}^k - q^k) \in \partial R(\tilde{x})$.*

Instead of inserting one optimality cut per iteration, we can consider a *multicut* version [2] of the L-shaped-type cuts for this problem: We can apply a disaggregation of optimality cuts per each single scenario. Thereby, the number of master iterations may be significantly reduced, which is of great importance if the number of scenarios is large and/or the recourse function $Q(\tilde{x}, k)$ is difficult to solve. For a fixed first-stage solution $(\tilde{x}, \tilde{\Theta}_1, \ldots, \tilde{\Theta}_K)$, we will solve LP-relaxations of all K scenarios, and can insert L-shaped-type cuts

$$\Theta_k + \sum_{e\in E}(q_e^k - \tilde{\beta}_e^k)x_e \ge \sum_{S\subseteq V_r:S\cap R_k\ne\emptyset} \tilde{\alpha}_S^k, \quad (10)$$

for all $k \in \mathcal{K}$ where $\tilde{\Theta}_k < R(\tilde{x}, k)$. Observe that due to the cutting plane method, L-shaped cuts can be found in polynomial time.

Integer L-shaped cuts. Let x^ν be a binary first stage solution with its corresponding optimal second stage value $Q(x^\nu) = \sum_{k\in\mathcal{K}} p_k Q(x^\nu, k)$. Let $\mathcal{I}^\nu := \{e \in E : x_e^\nu = 1\}$ be the index set of the edge variables chosen in the first stage, and the constant L be a known lower bound of the recourse function (before branching: $L = 0$). We want to explicitly cut off the solution (x^ν, Θ^ν). In our case, the general integer optimality cuts of the L-shaped scheme [10] already suffice, as we never need to generate them in our experiments (see below):

$$\Theta \ge (Q(x^\nu) - L)\left(\sum_{e\in\mathcal{I}^\nu} x_e - \sum_{e\in E\setminus\mathcal{I}^\nu} x_e - |\mathcal{I}^\nu| + 1\right) + L. \quad (11)$$

Solving the subproblems. Each of the K subproblems is solved using a *Subproblem Branch-and-Cut Framework* for the restricted Steiner arborescence problem. The subproblems are solved using the algorithm given in [11], augmented with (6). Cuts found during the separation of one subproblem are then stored in a pool where they can be reused by other subproblems (if applicable).

3.3   Reformulation with Negative Edge Costs in the First Stage

Alternatively to above, we can consider the following two objective functions when decomposing the problem: $\min \sum_{e\in E}(c_e - q_e^*)x_e + \sum_{k\in\mathcal{K}} p_k\Theta_k$ for the (RMP) formulation. The second stage subproblem is then decomposable into the following subproblems:

$$(RSAP_*^k) \quad Q^*(\tilde{x}, k) = \min\left\{\sum_{(i,j)\in A} q_{\{i,j\}}^k z_{ij}^k \mid z_{ij}^k \text{ satisfies (5)–(7)}\right\}.$$

In this formulation, variables Θ_k denote the expected costs for interconnecting terminals from R_k plus purchasing all edges from $\tilde{x}$ in the second stage. The difference in using this decomposition, rather than the one described before, is that the edge costs in the first stage become negative and the initial iterations of the master B&C will therefore select many instead of few edges.

Moreover, the dual of $(RSAP_*^k)$ then reads as follows

$$(DRSAP_*^k) \quad \max\{\sum_{S \subseteq V_r : S \cap R_k \neq \emptyset} \alpha_S^k + \sum_{e \in E} \tilde{x}_e \beta_e^k \mid (\alpha^k, \beta^k) \text{ satisfies (8), (9)}\}$$

and, since $\tilde{\beta}^k \in \partial R^*(\tilde{x}, k)$ where $R^*(x, k)$ is the relaxation of $Q^*(x, k)$, the generated L-shaped cuts are written as

$$\Theta_k - \sum_{e \in E} \tilde{\beta}_e^k x_e \geq \sum_{S \subseteq V_r : S \cap R_k \neq \emptyset} \tilde{\alpha}_S^k. \tag{12}$$

We will see that, from the computational point of view, this second approach significantly outperforms the previous one, presumably for sparsity reasons.

4   Computational Results

All experiments were performed on an Intel Core-i7 2.67GHz Quad Core machine with 12 GB RAM, under Ubuntu 9.04. Each run was performed on a single core. We used ABACUS 3.0 as a generic B&C framework; for solving the LP relaxations we used the commercial package IBM CPLEX (version 10.1) via COIN-Osi 0.102.

Depending on the used decompositions $(RSAP^k)$ and $(RSAP_*^k)$, we denote the implementations of the two-stage B&C algorithms by $2BC$ and $2BC^*$, respectively. Thereby, we use the following primal heuristic at the root node of the B&C tree (after each iteration, until we obtain the first upper bound): Round the fractional solution x' to a binary solution x''. If x'' is cycle free, solve all K subproblems to optimality and obtain a valid upper bound $UB = c^t x'' + \sum_{k \in K} p_k Q(x'', k)$. For solving (EF) directly, we implemented a branch-and-cut approach analogous to [11]; we denote the algorithm by EF.

4.1   Benchmark Instances: SSTPLib

As of now, there seems to be no established benchmark set for SSTP. Herein we propose the SSTPLib [14] (available online), which is derived from well-known benchmark sets for the deterministic (prize-collecting) Steiner tree problem. We use the underlying graph and cost structures, and generate multiple SSTP instances with varying parameterizations for the stochastic properties. This allows us to experimental deduce dependencies on these.

– **Instance groups K and P.** These prize-collecting Steiner tree instances were originally proposed in [9]. Our inputs are graphs obtained by applying several valid reduction procedures as described in [11] and contain up to 91 nodes and 237 edges (available online [12]).

– **Instance group** lin. These instances are borrowed from the well-known SteinLib [15]. The graphs contain up to 165 nodes and 274 edges with up to 14 terminals. Although these instances appear to be solvable by preprocessing or by dual ascent heuristics for the deterministic Steiner tree problem, the same techniques cannot be applied straight-forwardly to the corresponding SSTPs.

Converting Deterministic into Stochastic Inputs. Deterministic Steiner tree input graphs $G = (V, E)$ with edge costs c_e, $e \in E$ are transformed into the SSTP instances as follows:

1. We generate K scenarios. To obtain scenario probabilities p_k, we distribute 1000 points (corresponding to the probability of 1‰, each) among these scenarios randomly (ensuring that each scenario has at least probability 1‰).
2. For each scenario k, we construct R_k by independently picking each terminal or Steiner node with probability 0.3 or 0.05, respectively.
3. Each second stage edge cost q_e^k is randomly (independent, uniform) drawn from $[1.1c_e, 1.3c_e]$.

4.2   Comparing the Deterministic Equivalent vs. Two-Stage Branch-and-Cut Approaches

For the K and P instance groups, we focus on comparing the time to obtain provably optimal solutions, required by our two decomposition-based algorithms $2BC$, $2BC^*$ and the standard approach EF. Figure 2 shows the running times in seconds, averaged over all instances of the corresponding group. We observe that decomposing the problem is not worthwhile for instances with less than 20 scenarios. However, as the number of scenarios increases, the benefit of decomposing is obvious: already with 100 scenarios, EF needs 10 times the running time of the two-stage B&C approaches. In additional experiments with 500 scenarios, EF is not able to solve 6 out of 11 instances within two hours, whereas the two-stage approach $2BC^*$ needs only 510 seconds on average.

We also observe that $2BC^*$ always outperforms $2BC$. In particular for the group K instances with 100–500 scenarios, it is 1.8 times faster. This is because the L-shaped cuts generated by $2BC^*$ are sparser ($\tilde{\beta}_e^k$ are often 0) and numerically more stable than the corresponding cuts generated by $2BC$ (cf. Section 3.1).

Table 1 shows the comparison between EF and the two-stage approach $2BC^*$. Instances lin01–lin06 were used to generate inputs with $K \in \{5, 10, 20, 50\}$ scenarios. Column $|R_{avg}|$ gives the average number of terminals in each scenario; OPT* gives the optimal values (or the best upper bound, if the time limit of 2 hours is reached). We compare the running time in seconds ($t[s]$), the number of branch-and-bound nodes (b&b), the final gap obtained after the time limit of two hours, as well as the overall number of iterations in the B&C (#iter). We observe that, as the number of scenarios increases, the number of iterations decreases for $2BC^*$. This is due to the larger number of multi-cuts inserted in each primal iteration. In contrast to this, the number of iterations for EF increases drastically with the number of scenarios, which explains why instances with more than 20 scenarios are not solvable within the time limit.

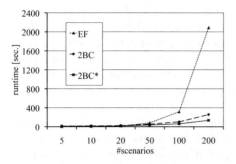 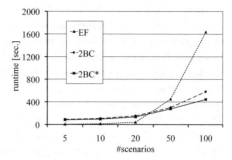

Fig. 2. Average running times in seconds for both two-stage branch-and-cut algorithms $2BC$ and $2BC^*$, and for the extensive formulation of the deterministic equivalent EF. Left: K group, 11 instances, right: P group, 5 instances.

Table 1. Results for lin instances: within the time limit of two hours, EF was not able to solve most of the instances with 50 scenarios

Instance	K	$|R_{avg}|$	OPT*	EF $t[s]$	b&b	gap	#iter	$2BC^*$ $t[s]$	b&b	gap	#iter
lin01_53_80	5	4.6	797.0	**0.2**	1	—	34	2.2	1	—	61
lin01_53_80	10	4.2	633.2	**0.7**	3	—	59	2.5	3	—	50
lin01_53_80	20	4.6	753.9	**5.7**	3	—	63	6.9	3	—	52
lin01_53_80	50	4.7	768.9	33.4	3	—	70	**10.4**	3	—	36
lin02_55_82	5	4.6	476.2	**0.1**	1	—	24	1.1	1	—	45
lin02_55_82	10	5.3	739.1	**1.0**	1	—	33	3.0	1	—	47
lin02_55_82	20	5.3	752.2	4.9	1	—	69	**4.3**	1	—	37
lin02_55_82	50	5.1	732.6	31.2	1	—	70	**10.7**	1	—	35
lin03_57_84	5	4.4	653.0	**0.5**	1	—	80	1.9	1	—	55
lin03_57_84	10	5.2	834.7	**3.8**	7	—	90	8.7	7	—	91
lin03_57_84	20	5.8	854.9	10.8	1	—	92	**7.3**	1	—	41
lin03_57_84	50	5.5	895.7	103.1	3	—	106	**21.3**	3	—	43
lin04_157_266	5	10.4	1922.1	**140.4**	3	—	315	959.2	47	—	567
lin04_157_266	10	9.8	1959.1	**415.8**	7	—	244	989.2	7	—	339
lin04_157_266	20	9.3	1954.9	5498.7	11	—	833	**3016.7**	13	—	575
lin04_157_266	50	9.8	2097.7	(2h)	1	19.5	185	**5330.2**	11	—	269
lin05_160_269	5	10.2	2215.5	**282.0**	53	—	722	2681.2	35	—	1558
lin05_160_269	10	11.4	2210.2	**1866.7**	5	—	1130	4096.0	35	—	1502
lin05_160_269	20	11.1	2412.2	(2h)	11	5.6	1060	(2h)	17	**4.7**	890
lin05_160_269	50	11.6	2297.0	(2h)	1	21.3	210	**3627.4**	1	—	159
lin06_165_274	5	11.0	1975.8	**212.8**	53	—	797	760.9	19	—	834
lin06_165_274	10	10.6	1918.7	**501.7**	5	—	260	808.4	3	—	306
lin06_165_274	20	14.0	2457.6	(2h)	11	—	1099	**3222.9**	11	—	459
lin06_165_274	50	12.6	2186.8	(2h)	1	22.5	221	**2795.5**	11	—	215

5   Extensions and Future Work

Gupta et al. [8] also consider the SSTP in which the first stage solution is a tree. Using our above ideas and bi-directing G already for the first stage, we can deduce an even stronger fully directed model that ensures that the first-stage solution is a rooted Steiner arborescence as well. It will be interesting to evaluate the potentially arising benefits.

Along the lines of the *algorithm engineering cycle*, our above approach leaves multiple areas for further improvements: The integration of stronger primal heuristics may lead to further significant speed-ups. A broader set of specifically designed benchmark instances may allow a better insight in the dependencies between input properties and solvability; e.g., it seems to be hard to generate SSTP instances that require integer L-shaped cuts in practice. It is also an open question how to integrate further known strong arborescence constraint classes like flow-balance constraints, as they are not directly valid in our SSTP setting.

Our 2-B&C framework is extendable to theoretical concepts as, e.g., the ones of Carøe and Tind [5] where strengthening inequalities of an integer second-stage polytope are used to generate additional L-shaped cuts. Furthermore, our framework can be applied to other stochastic network design models whose deterministic counterpart is of Goemans-Williamson-type [6] (e.g., shortest path, MST, generalized STP, network survivability, etc.), in order to facilitate strong and efficient non-compact models.

References

1. Benders, J.F.: Partitioning procedures for solving mixed-variables programming problems. Numerische Mathematik 4, 238–252 (1962)
2. Birge, J.R., Louveaux, F.: A multicut algorithm for two-stage stochastic linear programs. European Journal of Operational Research 34, 384–392 (1988)
3. Birge, J.R., Louveaux, F.: Introduction to Stochastic Programming. Springer, New York (1997)
4. Byrka, J., Grandoni, F., Rothvoß, T., Sanità, L.: An improved LP-based approximation for Steiner tree. In: ACM STOC (to appear, 2010)
5. Carøe, C.C., Tind, J.: L-shaped decomposition of two-stage stochastic programs with integer recourse. Mathematical Programming 83(3), 451–464 (1998)
6. Goemans, M.X., Williamson, D.P.: The primal-dual method for approximation algorithms and its application to network design problems. In: Approximation algorithms for NP-hard problems, pp. 144–191 (1996)
7. Gupta, A., Hajiaghayi, M., Kumar, A.: Stochastic Steiner tree with non-uniform inflation. In: Charikar, M., Jansen, K., Reingold, O., Rolim, J.D.P. (eds.) RANDOM 2007 and APPROX 2007. LNCS, vol. 4627, pp. 134–148. Springer, Heidelberg (2007)
8. Gupta, A., Ravi, R., Sinha, A.: LP rounding approximation algorithms for stochastic network design. Math. of Operations Research 32(2), 345–364 (2007)
9. Johnson, D.S., Minkoff, M., Phillips, S.: The prize-collecting Steiner tree problem: Theory and practice. In: ACM-SIAM SODA, pp. 760–769. SIAM, Philadelphia (2000)
10. Laporte, G., Louveaux, F.: The integer L-shaped method for stochastic integer programs with complete recourse. Oper. Res. Lett. 13, 133–142 (1993)
11. Ljubić, I., Weiskircher, R., Pferschy, U., Klau, G., Mutzel, P., Fischetti, M.: An algorithmic framework for the exact solution of the prize-collecting Steiner tree problem. Mathematical Programming 105(2-3), 427–449 (2006)
12. PCSTP Benchmark, http://homepage.univie.ac.at/ivana.ljubic/research/pcstp/

13. Shapiro, A., Ruszczynski, A.P., Dentcheva, D.: Lectures on Stochastic Programming: Modeling and Theory. SIAM, Philadelphia (2009)
14. SSTPLib, http://ls11-www.cs.tu-dortmund.de/staff/zey/sstp
15. SteinLib, http://steinlib.zib.de/steinlib.php
16. Swamy, C., Shmoys, D.: Approximation algorithms for 2-stage stochastic optimization problems 1(37), 33–46 (2006)
17. Verweij, B., Ahmed, S., Kleywegt, A.J., Nemhauser, G., Shapiro, A.: The sample average approximation method applied to stochastic routing problems: A computational study. Computational Optimization and Appl. 24(2-3), 289–333 (2003)

An Optimal Algorithm for Single Maximum Coverage Location on Trees and Related Problems

Joachim Spoerhase

Lehrstuhl für Informatik I, Universität Würzburg
Am Hubland, 97074 Würzburg, Germany
joachim.spoerhase@uni-wuerzburg.de

Abstract. The single maximum coverage location problem is as follows. We are given an edge-weighted tree with customers located at the nodes. Each node u is associated with a demand $w(u)$ and a radius $r(u)$. The goal is to find, for some facility, a node x such that the total demand of customers u whose distance to x is at most $r(u)$ is maximized.

We give a simple $O(n \log n)$ algorithm for this problem which improves upon the previously fastest algorithms. We complement this result by an $\Omega(n \log n)$ lower bound showing that our algorithm is optimal.

We observe that our algorithm leads also to improved time bounds for several other location problems such as indirect covering subtree and certain competitive location problems. Finally, we outline how our algorithm can be extended to a large class of distance-based location problems.

Keywords: graph algorithm, coverage, tree, efficient algorithm.

1 Introduction and Preliminaries

Suppose that a company wants to open a facility (for example, a shop, a warehouse, a plant, or a computer server) in a given network, modeled by an edge-weighted graph. A potential customer u, located at some node of the graph, is willing to use the service provided by that company if the cost thereby incurred is limited by some bound $r(u)$. The costs are modeled by shortest-path distances in the underlying graph, that is, the cost for customer u equals the distance to the closest server of the company. For example, the distances may represent transportation costs or travel times in a logistical network, but also response times in a communication network. If the company serves customer u—that is, his distance to the closest server does not exceed $r(u)$—it earns a profit of $w(u)$, which generally corresponds to the demand of the customer. The goal of the company is to identify a location (that is, a node of the graph) for its facility such that the total profit is maximized.

1.1 Problem Definition

Megiddo et al. [9] introduced the maximum coverage location problem which allows for the location of r servers. We consider the following single-location

O. Cheong, K.-Y. Chwa, and K. Park (Eds.): ISAAC 2010, Part I, LNCS 6506, pp. 440–450, 2010.

variant. The input of the *single maximum coverage location problem* is an undirected tree $T = (V, E)$ with non-negative edge weights $c \colon E \to \mathbb{R}_{\geq 0}$. The edge weights induce a distance function $d \colon V \times V \to \mathbb{R}_{\geq 0}$ on the node set. Each node is associated with a radius $r(u)$ and a non-negative demand $w(u)$. The goal is to find a node x (location) that maximizes the total demand $W(x) := \sum \{ w(u) \mid u \in V$ and $d(u, x) \leq r(u) \}$ served by x. For a set $S \subseteq V$ of nodes we define $W(x, S) := \sum \{ w(u) \mid u \in S$ and $d(u, x) \leq r(u) \}$. Clearly $W(x) = W(x, V)$.

1.2   Related Work and Previous Results

The maximum coverage location problem (allowing the placement of an arbitrary set of r nodes) is NP-hard on general graphs [9] while it can be solved in time $O(rn^2)$ on trees [15]. This leads to an $O(n^2)$ algorithm for the *single* maximum coverage location problem on trees by setting $r = 1$. Kim et al. [7] provide a faster algorithm running in $O(n \log^2 n)$. Their algorithm works even for the more general indirect covering subtree problem (confer Section 4.2). Recently a slightly faster $O(n \log^2 n / \log \log n)$-time algorithm for single maximum coverage location has been reported [11].

1.3   Contribution and Outline of This Paper

We propose an $O(n \log n)$ algorithm for single maximum coverage location, which is faster than the previously best algorithms for that problem. We complement this result with a matching lower bound on the running time showing that our algorithm is optimal. We show that our approach also leads to an $O(n \log n)$ algorithm for indirect covering subtree beating the running time of $O(n \log^2 n)$ of Kim et al. [7]. Finally, we outline how to generalize our technique to a large class of distance-based location problems.

Our algorithm relies on a simple technique to subdivide trees, which we call *two-terminal subtree subdivision*. This technique is a simplification of the recursive coarsening strategy [11]. The source of our speedup is that we manage to avoid explicitly sorting the nodes according to their distances and radii during the recursion, which has been necessary in the coarsening approach and also in the algorithm of Kim et al. [7]. One further advantage of our algorithm is that it is a lot simpler than the recursive coarsening algorithm.

The two-terminal subtree technique has proved successful also for other location problems [14,12]. We believe that there are further problem classes where it can be applied.

We remark that there are other methods of recursively subdividing trees such as the so-called spine decomposition [2]. It is possible that our algorithm can be made work also with this subdivision technique. However, we feel that the concept of two-terminal subtrees is simpler.

2   The Algorithm

The algorithms of Kim et al. [7] relies on computing $W(x)$ for each node x by divide-and-conquer. It partitions the node set V into two sets V_1, V_2 of bounded size such that both induce subtrees and have exactly one node (called centroid [6]) in common. It sorts the sets V_i according to their radii and their distances to the centroid, respectively. Then it computes, by means of a clever merge-and-scan procedure, for all $x \in V_i$ the demands $W(x, V_j)$ of the users in V_j where $j \neq i$. Applying the routine recursively to the subtrees induced by V_1, V_2 one can determine for $i = 1, 2$ and for each $x \in V_i$ the value $W(x, V_i)$. Finally, one obtains the total demand $W(x, V)$ of any node $x \in V$ by adding $W(x, V_1)$ and $W(x, V_2)$. The algorithm has a total running time of $O(n \log^2 n)$.

The algorithm of Spoerhase and Wirth [11] decomposes at each stage into $O(\log n)$ subproblems achieving the slightly better running time of $O(n \log^2 n$ $/ \log \log n)$.

Our approach proceeds in a similar way but uses a more suitable subdivision technique, which we call two-terminal subtree subdivision (2TS for short). This decomposition allows us to avoid the explicit sorting thereby suppressing the additional log-factor. Spoerhase and Wirth [14,12] used the two-terminal technique for solving competitive location problems on undirected trees.

Consider the input tree $T = (V, E)$. We may assume that T has maximum degree three. Otherwise, we can split nodes of larger degree by introducing suitable zero-length edges and zero-weighted nodes [15].

If s and t are distinct nodes then T_{st} denotes the maximal subtree of T having s and t as leaves. Let V_{st} be the node set of T_{st}. We call s and t *terminals* and T_{st} *two-terminal subtree* (2TS).

Our algorithm divides the input tree recursively into 2TSs. Since we are dealing with a degree-bounded tree we can subdivide any 2TS S into at most five 2TSs, called *child 2TSs*, of bounded size (confer Figure 1). We make use of the following lemma.

Lemma 1 (2TS Subdivision [12]). *Let S be a 2TS with maximum degree three. Then S can be partitioned into at most five edge-disjoint 2TS each of which has at most $\frac{1}{2}|S| + 1$ nodes. This subdivision can be computed in $O(|S|)$ time.* □

Consider a 2TS T_{st}. We introduce the lists $L_{d,s}(T_{st})$ and $L_{r,s}(T_{st})$. Both lists contain all nodes v of T_{st} sorted in increasing order with respect to the values $d(s, v)$ and $r(v) - d(v, s)$, respectively. The lists $L_{d,t}(T_{st})$ and $L_{r,t}(T_{st})$ are defined symmetrically.

The algorithm computes $W(v, T_{st})$ for all $v \in T_{st}$ as well as the four lists $L_{d,s}(T_{st})$, $L_{d,t}(T_{st})$, $L_{r,s}(T_{st})$ and $L_{r,t}(T_{st})$ for any 2TS T_{st} occurring during the recursion. We shall see that these information can be propagated inductively from child towards parent 2TSs such that we will have computed $W(\cdot, V) = W(\cdot)$ at the top of the recursion.

To this end consider an arbitrary 2TS $S = T_{st}$ being subdivided into at most five child 2TSs S_i with terminals s_i, t_i. Moreover assume that we have already computed $W(\cdot, S_i)$ and the four lists corresponding to S_i for all child 2TSs S_i.

We start with computing $L_{r,s}(S)$. Confer Algorithm 1. We perform the following operations for all child 2TSs S_i where $1 \leq i \leq 5$. Assume that s_i is the terminal of S_i closest to s. Then the list $L_{r,s_i}(S_i)$ contains all nodes $u \in S_i$ with

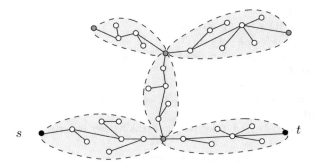

Fig. 1. An example of a subdivision of a 2TS with terminals s, t into five child 2TSs. The child 2TSs are marked by the gray-shaded areas. The gray nodes are terminals of child 2TSs.

associated sorting keys $r(u) - d(s_i, u)$. Now we create a copy L'_i of this list and subtract for all nodes $u \in L'_i$ the value $d(s, s_i)$ from the sorting key $r(u) - d(u, s_i)$, which does not affect the order of L'_i. As a result L'_i contains all nodes u of S_i sorted with respect to $r(u) - d(s, u)$ as desired. Finally, we obtain $L_{r,s}(S)$ as the union of all L'_i where $1 \leq i \leq 5$. The computation of this union can be carried out in a merge-sort-like manner in order to ensure $O(|S|)$ time.

> $L_{r,s}(S) \leftarrow \emptyset$
> **for** $i = 1, \ldots, 5$ **do**
> > let s_i be the terminal of S_i closest to s
> > $L'_i \leftarrow L_{r,s_i}(S_i)$
> > **for** $u \in L'_i$ **do**
> > > decrease key $r(u) - d(u, s_i)$ of u by $d(s, s_i)$ yielding $r(u) - d(u, s)$
> >
> > **end**
> > $L_{r,s}(S) \leftarrow L_{r,s}(S) \cup L'_i$
>
> **end**

Algorithm 1. Computation of the list $L_{r,s}(S)$

The computation of the list $L_{d,s}(S)$ is even simpler. We perform the following operations for all child 2TSs S_i where $1 \leq i \leq 5$. Assume that s_i is the terminal of S_i closest to s. Then the list $L_{d,s_i}(S_i)$ contains all nodes $u \in S_i$ with associated sorting keys $d(s_i, u)$. We create a copy L'_i of this list and add for all nodes $u \in L'_i$ the value $d(s, s_i)$ to the sorting key $d(s_i, u)$, which does not affect the order of L'_i. As a result L'_i contains all nodes u of S_i sorted with respect to $d(s, u)$ as desired. Finally, we obtain $L_{d,s}(S)$ as the union of all L'_i where $1 \leq i \leq 5$.

The respective lists for terminal t are computed symmetrically. The total running time for computing the four lists associated with S is $O(|S|)$ since we handle a constant number of child 2TSs.

We are now going to explain how $W(x, S)$ can be determined for all $x \in S$ (confer Algorithm 2). To this end assume that x is contained in some S_i. Since we already know $W(x, S_i)$ by the inductive hypothesis it suffices to determine $W(x, S_j)$ for all $S_j \neq S_i$ and to add these values to $W(x, S_i)$.

```
for i = 1, ..., 5 do
 │ W(x, S) ← W(x, Sᵢ) for all x ∈ Sᵢ
 │ for j = 1, ..., 5 and j ≠ i do
 │  │ let sᵢ, sⱼ be the terminals of Sᵢ, Sⱼ closest to each other
 │  │ L' ← L_{d,sᵢ}(Sᵢ)
 │  │ for x ∈ L' do
 │  │  │ increase key d(x, sᵢ) by d(sᵢ, sⱼ) yielding key d(x, sⱼ)
 │  │ end
 │  │ L' ← L' ∪ L_{r,sⱼ}(Sⱼ) decreasing, breaking ties in favor of nodes in Sⱼ
 │  │ W ← 0
 │  │ for v ∈ L' in decreasing order do
 │  │  │ if v ∈ Sⱼ then
 │  │  │  │ W ← W + w(v)
 │  │  │ else
 │  │  │  │ W(v, S) ← W(v, S) + W
 │  │  │ end
 │  │ end
 │ end
 end
end
```

Algorithm 2. Computation of $W(x, S)$ for all $x \in S$

Consider an arbitrary $S_j \neq S_i$ and assume that s_i, s_j are the terminals of these 2TSs closest to each other, respectively. We create a copy L' of list $L_{d,s_i}(S_i)$ and add the distance $d(s_j, s_i)$ to all sorting keys in this list. As a result L' contains all nodes x of S_i sorted with respect to the key $d(x, s_j)$.

At this point we can compute $W(x, S_j)$ for *all* $x \in S_i$ in linear time $O(|S_i| + |S_j|)$. To this end we merge the sorted list L' with the sorted list $L_{r,s_j}(S_j)$ and store the result in L'. We assume that the nodes in L' are sorted in *decreasing* order with respect to their numerical sorting keys. Ties are broken in favor of nodes in S_j. This can be carried out in linear time $O(|S_i| + |S_j|)$.

Now recall that a node $u \in S_j$ generates a profit $w(u)$ for $x \in S_i$ if $d(x, s_j) + d(s_j, u) = d(x, u) \leq r(u)$ or equivalently $d(x, s_j) \leq r(u) - d(u, s_j)$. This is tantamount to that u precedes x in L' since the left-hand side of the inequality contains the sorting key of x in L' and the right-hand side the one of u in L'. This allows us to compute $W(x, S_j)$ for all $x \in S_i$ in one single traversal of L'. During this traversal we maintain the total demand of all nodes $u \in S_j$ encountered so far, which equals the profit $W(x, S_j)$ whenever a node $x \in S_i$ is reached.

The total running time of this routine (Algorithm 2) is $O(|S|)$ since the number of child 2TSs is constant and the necessary lists and $W(\cdot)$-values have already been computed for all child 2TSs in the preceding recursion levels.

Note that the bottom of the recursion, that is, when T_{st} consists merely of a single edge (s, t) can easily be handled in constant time.

To sum up, this leads us to an algorithm whose running time $T(|S|)$ can be described by the following recurrence

$$T(|S|) = O(|S|) + \sum_{i=1}^{k} T(|S_i|),$$

where $k \leq 5$, $\sum_{i=1}^{k} |S_i| = |S| + 4$ and $|S_i| \leq \frac{1}{2}|S| + 1$. This implies that $T(n)$ is $O(n \log n)$.

Theorem 1. *The single maximum coverage location can be solved in time $O(n \log n)$.* □

3   A Matching Lower Bound

In this section we complement our algorithm with a lower bound $\Omega(n \log n)$ on the running time for solving single maximum coverage location on a tree. This shows that (for certain computational models) our algorithm is optimal.

We make use of a recent result which is summarized in the following theorem.

Theorem 2 ([1]). *Let $W \subseteq \mathbb{R}^n$. If W is recognized in time $T(n)$ on a real-number RAM that supports direct assignments, memory access, flow control, and arithmetic instructions $\{+, -, \times, /\}$ then $T(n) = \Omega(\log \beta(W^\circ))$.* □

Here, W° denotes the interior of W and $\beta(W')$ denotes the number of connected components of some set $W' \subseteq \mathbb{R}^n$.

To prove our lower bound we introduce a variant of the set disjointness problem. To this end let $n \in \mathbb{N}$. The set $W_n \subseteq \mathbb{R}_+^{2n}$ contains all tuples $(x_1, \ldots, x_n, y_1, \ldots, y_n)$ such that $x_1 < \ldots < x_n$ and $x_i \neq y_j$ for all pairs i, j. Consider a permutation π on the set $\{1, \ldots, n\}$ and some tuple $x_1 < y_{\pi(1)} < x_2 < y_{\pi(2)} < \ldots < x_n < y_{\pi(n)}$ in W_n. It is easy to see that for different permutations such tuples lie in different connected components of W_n° so W_n° contains at least $n!$ connected components. Hence any RAM of the above described type takes time $\Omega(n \log n)$ to recognize W_n.

We establish a linear time reduction from the problem to recognize W_n to the single maximum coverage location problem on a tree with $O(n)$ nodes. To this end we consider a tuple $(x_1, \ldots, x_n, y_1, \ldots, y_n)$ for which we want to decide whether or not it is contained in W_n.

First we check if $x_1 < \ldots < x_n$. Then we create an edge (u, v) of some length $c(u, v) > \max\{x_i, y_i \mid i = 1, \ldots, n\}$ and choose some radius r such that $r > c(u, v)$. For any y_i we create two edges (u, u_i) and (v, v_i) of lengths $r - y_i$ and $y_i + r - c(u, v)$, respectively. Finally, we create for each x_i a node $\widetilde{x}_i$ on edge (u, v) with distance $d(u, \widetilde{x}_i) := x_i$. For each node z in the node set $V := \{u_i, v_i, \widetilde{x}_i \mid i = 1, \ldots, n\} \cup \{u, v\}$ we set $w(z) := 1$ and $r(z) := r$, which completes the reduction.

First suppose that we locate a facility outside the path $P(u, v)$. Assume that the facility is located at some node u_i. Then the distance of u_i to u is positive and $d(u, v_j) \geq r$ for any j. Hence, none of the nodes v_j is covered by u_i and a demand of at least n remains uncovered by u_i. The case where the facility is placed at some node v_i is treated analogously.

Now suppose for a moment that we can locate a facility everywhere at the path $P(u, v)$, that is, also at interior points of edges on $P(u, v)$. The point x

where the facility is located can then be identified with the distance $d(u, x)$. First, all nodes on $P(u, v)$ are covered by x since r was chosen to exceed $d(u, v)$. Due to our construction x covers all nodes u_i where $x \leq y_i$ and all nodes v_j where $x \geq y_j$. Thus, exactly n nodes and thus a demand of n remains uncovered by x if x is not contained in the set $\{y_1, \ldots, y_n\}$. If $x = y_j$ then x covers both u_j and v_j and the uncovered demand is bounded by $n - 1$. Since the facility can only be placed at nodes $\widetilde{x}_i$, that is, at distances x_i from u we conclude that the uncovered demand is n if the input tuple $(x_1, \ldots, x_n, y_1, \ldots, y_n)$ lies in W_n and $n - 1$ otherwise.

Theorem 3. *Any real-number RAM that complies with Theorem 2 takes at least $\Omega(n \log n)$ time to solve the single maximum coverage location problem on a tree even for uniform demands and uniform radii.* $\square$

4   Implications for Related Problems

We point to some problems that can be reduced to the maximum coverage location problem and for which our result leads to algorithms that are faster than the existing ones. Finally, we outline a more general framework of distance-based functions to which our technique can be applied.

4.1   Absolute Variant of Maximum Coverage Location

The variant of single maximum coverage location where the facility can be placed not only at the nodes but also at interior points of edges is called the *absolute* maximum coverage location problem. Kim et al. show [7] that a set of $O(n)$ critical points (that is, a set of point which is guaranteed to contain an optimal point) for absolute single maximum coverage location on trees can be found in time $O(n \log n)$. We infer that also the absolute variant can be solved in $O(n \log n)$ on a tree.

4.2   An Optimal Algorithm for Indirect Covering Subtree

Kim et al. [7] introduce the *indirect covering subtree problem*. Instead of assigning to each customer a demand $w(u)$ they assume that a *penalty* $\pi(u)$ is imposed on the company if customer u is *not* served. Moreover they allow the facility to occupy a *subtree* Y of the input tree and not only a single node.

   The input of the indirect covering subtree problem is an undirected tree $T = (V, E)$ with non-negative edge weights $c \colon E \to \mathbb{R}_{\geq 0}$. The edge weights induce a distance function $d \colon V \times V \to \mathbb{R}_{\geq 0}$ on the node set. Each node is associated with a radius $r(u)$ and a non-negative penalty $\pi(u)$. Consider a subtree Y of T. A node u is said to be *covered* by Y if $d(u, Y) \leq r(u)$, that is, if u lies within distance $r(u)$ from Y. If u is *not* covered by Y, then u imposes a penalty $\pi(u)$ on Y. If $U \subseteq V$ is a set of nodes then $p(U, Y) := \sum \{ \pi(u) \mid u \in U \text{ and } d(u, Y) > r(u) \}$ denotes the penalty imposed on Y by U. The total penalty imposed on Y is given by $p(Y) := p(V, Y)$. It is further assumed that establishing a tree-shaped facility Y leads to a setup cost of $c(Y)$. The indirect covering subtree problem asks for a subtree Y of T such that the total cost $c(Y) + p(Y)$, given by the sum of setup cost $c(Y)$ and penalty cost $p(Y)$, is minimum among all subtrees of T.

Kim et al. [7] show that the indirect covering subtree can be reduced to the single maximum coverage location problem on a so-called *bitree*. A bitree is a directed graph obtained from an undirected tree by replacing each edge with a pair of anti-parallel arcs. The arc lengths are allowed to be negative and asymmetric. Kim et al. [7] show that there is a linear time reduction from indirect covering subtree on an undirected tree to single maximum coverage location on a bitree or more precisely, to the problem of computing the profit $W(x)$ for every node x. Kim et al. show how this information can be computed in total time $O(n \log^2 n)$. Our Algorithm of Section 2 accomplishes the same at least for undirected trees. But it is not hard to verify that our algorithm works also on bitrees without noteworthy changes.

Theorem 4. *The indirect covering subtree problem on a tree can be solved in* $O(n \log n)$ *time.*   □

Consider an instance of single maximum coverage location. Now scale all edge-lengths and radii by the ratio of the total demand of all customers in the tree and the length of the shortest tree edge. If we now set $\pi(u) := w(u)$ for nodes u we obtain an instance of indirect covering subtree, which is equivalent to the original single maximum coverage location instance since it is now too expensive for a tree-shaped facility to occupy more than a single node. In other words this is a linear time reduction from single maximum coverage location to indirect covering subtree showing that our algorithm is also optimal for the latter problem.

4.3   Competitive Location

Another implication of our result leads us to the realm of *competitive location*. Let a graph $G = (V, E)$ and $r, p \leq n$ be given. We assume that the graph is edge and node weighted. Let $X, Y \subseteq G$ be sets of nodes or interior points of edges. Then $w(Y \prec X)$ denotes the total weight $\sum\{w(u) \mid u \in V \text{ and } d(u, Y) < d(u, X)\}$ of nodes that are closer to Y than to X. Given some point set X the goal of the (r, X)-*medianoid problem* [4] is to identify a set Y of r points such that $w(Y \prec X)$ is maximized. This maximum weight is denoted by $w_r(X)$. The goal of the (r, p)-*centroid problem* is to find a p-element point set X such that $w_r(X)$ is minimized. By setting $r(u) := d(u, X) - \varepsilon$ (where ε is a suitably small constant) one can easily verify that (r, X)-medianoid is a special case of the multiple maximum coverage location problem with r servers. On general graphs the problem is NP-hard [4]. It can be solved efficiently in $O(rn^2)$ on trees [15]. Our result leads to an $O(n \log n)$ algorithm for the absolute and the discrete version of $(1, X)$-medianoid on trees. The previously best result is $O(n \log^2 / \log \log n)$ [11].

Corollary 1. *The discrete and the absolute* $(1, X)$-*medianoid problem can be solved in* $O(n \log n)$ *on trees.*   □

Now let's turn our view to the (r, p)-centroid problem. The problem is known to be Σ_2^P-complete on general graphs [10] and NP-hard even on path graphs [13]. However, both the absolute and the discrete variant of $(1, p)$-centroid on trees can be solved in polynomial time $O(n^3 \log^2 n / \log \log n)$ and $O(n^3 \log^3 n / \log \log n)$, respectively [8]. Those algorithms rely on $O(n^2 \log n)$ (resp. $O(n^2 \log^2 n)$) calls to a subroutine solving $(1, X)$-medianoid on a tree. The algorithm provided here allows us to solve $(1, X)$-medianoid in $O(n \log n)$ which yields.

Corollary 2. *The discrete and the absolute $(1, p)$-centroid problem for trees can be solved in $O(n^3 \log n)$ and $O(n^3 \log^2 n)$, respectively.*   □

4.4 Distance-Based Location Problems

We outline how our technique can be generalized to a large class of distance-based location problems.

Let u be a customer and let x be a facility. We assume that u generates a cost which is a customer-specific function $f_u(d(u, x))$ that depends solely on the distance of u and x. The total cost for x is given by the aggregation of all customer costs. The aggregation could, for example, be summation or maximization. We write $\text{cost}(x) := \text{agg}_{u \in V} f_u(d(u, x))$. The goal is to find a node x minimizing $\text{cost}(x)$. Note that a similar model has been considered by Tamir [15] for p-facility location models on trees. However, in his model the customer-specific cost functions need to be increasing while ours are unrestricted and may even attain negative values.

Consider the following examples:

- $f_u(\delta) = w(u) \cdot \delta$ with agg $= +$ gives rise to the 1-median problem. If the demands are allowed to be negative, we obtain the 1-median with so-called pos/neg weights.
- $f_u(\delta) = w(u) \cdot \delta$ with agg $= \max$ amounts to weighted 1-center.
- $f_u(\delta) = 0$ if $\delta \le r(u)$ and $f_u(\delta) = w(u)$ otherwise, yields single maximum coverage location (more precisely, an equivalent minimization formulation).

Consider now a 2TS $S := T_{st}$ we define the aggregated function $f_s^S(\delta)$ as $\text{agg}_{u \in S} f_u(\delta + d(s, u))$. One can picture $f_s^S(\delta)$ as the total cost caused by the customers in S for a virtual node outside S at distance δ from terminal s.

Analogously to the algorithm of Section 2, we compute for each 2TS $S = T_{st}$ that occurs during the recursive subdivision the lists $L_{d,s}(S)$, $L_{d,t}(S)$ and the functions $f_s^S(\cdot)$ and $f_t^S(\cdot)$. Moreover, we compute $\text{cost}(x, S)$ for each node $x \in S$ where $\text{cost}(x, S)$ denotes $\text{agg}_{u \in S} f_u(d(u, x))$.

Consider the child 2TSs $S_1, \dots, S_5$ of S with terminals s_i, t_i. Assume that for each S_i the terminal s_i is closest to s among all nodes in S_i. We outline in an abstract way how to compute $L_{d,s}(S)$, $f_s^S(\cdot)$ and $\text{cost}(\cdot, S)$ if $L_{d,s_i}(S_i)$, $f_{s_i}^{S_i}(\cdot)$ and $\text{cost}(\cdot, S_i)$ are known for $i = 1, \dots, 5$.

The list $L_{d,s}(S)$ can be computed exactly as in Section 2.

The function $f_s^S(\cdot)$ can be computed by means of the relation $f_s^S(\delta) = \text{agg}_i f_{s_i}^{S_i}(\delta + d(s, s_i))$. That is we have to determine for each of the functions $f_{s_i}^{S_i}(\cdot)$ the *translation* by the additive constant $d(s, s_i)$. Afterward we have to *aggregate* those translated functions. If this can be performed in linear time $O(|S|)$ we say that the customer-specific cost functions allow *quick translation and aggregation*.

The cost $c(x, S)$ can be computed for each $x \in S$ as follows. Suppose that $x \in S_i$. By inductive hypothesis we already know $\text{cost}(x, S_i)$. In order to determine $\text{cost}(x, S)$ we need to compute $\text{cost}(x, S_j)$ for all $j \neq i$ and aggregate those values with $\text{cost}(x, S_i)$. The critical step is to *evaluate* $f_{s_j}^{S_j}(d(u, s_i) + d(s_i, s_j))$. Note that we find the distances $d(u, s_i)$ sorted in the list $L_{d,s_i}(S_i)$ and that $d(s_i, s_j)$ is a constant. Hence all we have to do is to evaluate the function $f_{s_j}^{S_j}(\cdot)$ for a list of $O(|S|)$ increasing distances. If this can be done in linear time $O(|S|)$ we say the the customer-specific cost functions allow *quick evaluation*.

We conclude

Theorem 5. *A node with minimum cost $\text{cost}(x)$ can be determined in time $O(n \log n)$ on a tree if the customer-specific cost functions allow quick translation, aggregation and evaluation.*   □

One class of such customer-specific cost functions for which this is possible is the class of piece-wise linear functions with a constant number breakpoints. The aggregated functions $f_s^S(\cdot)$ for 2TSs S that occur during the recursion are still piecewise linear with $O(|S|)$ breakpoints. It is not hard to verify that such functions allow quick translation, aggregation and evaluation if they are represented by line segments and rays in the distance-cost coordinate system.

Note that all of the above examples (1-median, 1-center, and single maximum coverage location) fit into this framework. For 1-median and 1-center the customer-specific cost functions are linear (without breakpoints) and admit linear time algorithms for trees [3,5]. Already one breakpoint makes the problem more difficult as it needs time $\Omega(n \log n)$ to be solved. (Note that maximum coverage location corresponds to a step function with a single step and needs time $\Omega(n \log n)$ as stated in Theorem 3). Our technique provides an algorithm that achieves this optimal time bound of $O(n \log n)$ even for arbitrarily (but constant) many breakpoints.

References

1. Ben-Amram, A.M., Galil, Z.: Topological lower bounds on algebraic random access machines. SIAM Journal on Computing 31(3), 722–761 (2001)
2. Benkoczi, R., Bhattacharya, B.K.: A new template for solving p-median problems for trees in sub-quadratic time. In: Brodal, G.S., Leonardi, S. (eds.) ESA 2005. LNCS, vol. 3669, pp. 271–282. Springer, Heidelberg (2005)
3. Goldman, A.J.: Optimal center location in simple networks. Transportation Science 5, 212–221 (1971)
4. Hakimi, S.L.: On locating new facilities in a competitive environment. European Journal of Operational Research 12, 29–35 (1983)
5. Handler, G.Y.: Minimax location of a facility in an undirected tree graph. Transportation Science 7, 287–293 (1973)

6. Kang, A.N.C., Ault, D.A.: Some properties of a free centroid of a free tree. Information Processing Letters 4, 18–20 (1975)
7. Kim, T.U., Lowe, T.J., Tamir, A., Ward, J.E.: On the location of a tree-shaped facility. Networks 28(3), 167–175 (1996)
8. Lazar, A., Tamir, A.: Improved algorithms for some competitive location centroid problems on paths, trees and graphs. Tech. rep., Tel Aviv University (submitted, 2010)
9. Megiddo, N., Zemel, E., Hakimi, S.: The maximum coverage location problem. SIAM Journal on Algebraic and Discrete Methods 4(2), 253–261 (1983)
10. Noltemeier, H., Spoerhase, J., Wirth, H.C.: Multiple voting location and single voting location on trees. European Journal of Operational Research 181(2), 654–667 (2007), http://dx.doi.org/10.1016/j.ejor.2006.06.039
11. Spoerhase, J., Wirth, H.C.: An $O(n(\log n)^2/loglogn)$ algorithm for the single maximum coverage location or the $(1, X_p)$-medianoid problem on trees. Information Processing Letters 109(8), 391–394 (2009), http://dx.doi.org/10.1016/j.ipl.2008.12.009
12. Spoerhase, J., Wirth, H.C.: Optimally computing all solutions of Stackelberg with parametric prices and of general monotonous gain functions on a tree. Journal of Discrete Algorithms 7(2), 256–266 (2009), http://dx.doi.org/10.1016/j.jda.2009.02.004
13. Spoerhase, J., Wirth, H.C.: (r, p)-centroid problems on paths and trees. Theoretical Computer Science 410(47-49), 5128–5137 (2009), http://dx.doi.org/10.1016/j.tcs.2009.08.020
14. Spoerhase, J., Wirth, H.C.: Relaxed voting and competitive location under monotonous gain functions on trees. Discrete Applied Mathematics 158, 361–373 (2010), http://dx.doi.org/10.1016/j.dam.2009.05.006
15. Tamir, A.: An $O(pn^2)$ algorithm for the p-median and related problems on tree graphs. Operations Research Letters 19, 59–64 (1996)

A Faster Algorithm for the Maximum Even Factor Problem

Maxim A. Babenko*

Moscow State University
max@adde.math.msu.su

Abstract. Given a digraph $G = (VG, AG)$, an *even factor* $M \subseteq AG$ is a subset of arcs that decomposes into a collection of node-disjoint paths and even cycles. Even factors in digraphs were introduced by Geelen and Cunningham and generalize path matchings in undirected graphs.

Finding an even factor of maximum cardinality in a general digraph is known to be NP-hard but for the class of *odd-cycle symmetric* digraphs the problem is polynomially solvable. So far, the only combinatorial algorithm known for this task is due to Pap; its running time is $O(n^4)$ (hereinafter n stands for the number of nodes in G).

In this paper we present a novel *sparse recovery* technique and devise an $O(n^3 \log n)$-time algorithm for finding a maximum cardinality even factor in an odd-cycle symmetric digraph. This technique also applies to a wide variety of related problems.

1 Introduction

In [3] Cunningham and Geelen introduced the notion of *independent path matchings* and investigated their connection with the separation algorithms for the matchable set polytope, which was previously studied by Balas and Pulleyblank [1]. Finding an independent path matching of maximum size was recognized as an intriguing example of a graph-theoretic optimization problem that is difficult to tackle by the purely combinatorial means. Two algorithms were given by Cunningham and Geelen: one relies on the ellipsoid method [3], and the other is based on deterministic evaluations of the Tutte matrix [4]. Later, a rather complicated combinatorial algorithm was proposed by Spille and Weismantel [13].

The notion of an *even factor* was introduced as a further generalization of path matchings in a manuscript of Cunningham and Geelen [5] (see also [2]). An even factor is a set of arcs that decomposes into a node-disjoint collection of simple paths and simple cycles of even lengths. Since cycles of length 2 are allowed, it is not difficult to see that finding a maximum matching in an undirected graph G reduces to computing a maximum even factor in the digraph obtained from G by replacing each edge with a pair of oppositely directed arcs. On the other hand, no reduction from even factors to non-bipartite matchings is known.

---

* Supported by RFBR grant 09-01-00709-a.

O. Cheong, K.-Y. Chwa, and K. Park (Eds.): ISAAC 2010, Part I, LNCS 6506, pp. 451–462, 2010.

Finding a maximum cardinality even factor in a general digraph is known to be NP-hard [2]. For the class of *weakly symmetric digraphs* a min-max relation and an Edmonds–Gallai-type structure were established by Cunningham and Geelen [5] and by Pap and Szegő [12]. Later it was noted by Pap [10] that these arguments hold for a slightly broader class of *odd-cycle symmetric digraphs*. Kobayashi and Takazawa [8] pointed out a relation between even factors and jump systems and showed that the requirement for a digraph to be odd-cycle symmetric is natural, in a sense.

The question of finding a combinatorial solution to the maximum even factor problem in an odd-cycle symmetric digraph had been open for quite a while until Pap gave a direct $O(n^4)$-time algorithm [10,11]. His method can be slightly sped up to $O(n^2(m + n \log n))$, as explained in Section 3. (Hereinafter n denotes for the number of nodes in G and m denotes the number of arcs.) To compare: the classical algorithm of Micali and Vazirani for finding a maximum non-bipartite matching, which is a special case of the maximum even factor problem, runs in $O(mn^{1/2})$ time [9]. Thus, it is tempting to design a faster algorithm for the maximum even factor problem by applying the ideas developed for matchings (e.g. blocking augmentation [6]). Howerver, there are certain complications that make even the bound of $O(mn)$ nontrivial.

To explain the nature of these difficulties let us briefly review Pap's approach (a more detailed exposition will be given in Section 3). It resembles Edmonds' non-bipartite matching algorithm and executes a series of iterations each trying to increase the size of the current even factor M by one. At each such iteration, a search for an augmenting path P is performed. If no such path is found then M is maximum. Otherwise, the algorithm tries to apply P to M. If no odd cycle appears after the augmentation then the iteration completes. Otherwise, a certain contracting reduction is applied to G and M.

Hence, each iteration consists of *phases* and the number of nodes in the current digraph decreases with each phase. Totally there are $O(n)$ iterations and $O(n)$ phases during each iteration, which is quite similar to the usual blossom-shrinking method. The difference is that during a phase the reduction may change the alternating reachability structure completely so the next phase is forced to start looking for P from scratch. (Compare this with Edmonds' algorithm where a blossom contraction changes the alternating forest in a predicable and consistent way thus allowing this forest to be reused over the phases.)

In this paper we present a novel $O(n^3 \log n)$-time algorithm for solving the maximum even factor problem. It is based on Pap's method but grows the alternating forest in a more careful fashion. When a contraction is made in the current digraph the forest gets destroyed. However, we are able to restore it by running a *sparse recovery* procedure that carries out a reachability search in a specially crafted digraph with $O(n)$ arcs in $O(n \log n)$ time (where the $\log n$ factor comes from manipulations with balanced trees used for odd-cycle testing).

Our method extends to a large variety of related problems. In particular, the $O(mn^3)$-time algorithm of Takazawa [14] solves the weighted even factor problem in $O(mn^3)$ time and also involves recomputing the alternating forest from scratch on each phase. Same applies to the maximum C_4-free 2-factor problem [11]. Considering matroidal structures one gets a *maximum independent even factor problem*, which is solvable by the methods similar to the discussed above, see [7]. All these problems benefit from the sparse recovery technique. Due to the lack of space we omit the details on these extensions.

2 Preliminaries

We employ some standard graph-theoretic notation throughout the paper. For an undirected graph G we denote its sets of nodes and edges by VG and EG, respectively. For a directed graph we speak of arcs rather than edges and denote the arc set of G by AG. A similar notation is used for paths, trees, and etc. We allow parallel edges and arcs but not loops. As long as this leads to no confusion, an arc from u to v is denoted by (u, v).

A path or a cycle is called *even* (respectively *odd*) if is consists of an even (respectively *odd*) number of arcs or edges. For a digraph G a *path-cycle matching* is a subset of arcs M that is a union of node-disjoint simple paths and cycles in G. When M contains no odd cycle it is called an *even factor*. The *size* of M is its cardinality and the *maximum even factor problem* prompts for constructing an even factor of maximum size.

An arc (u, v) in a digraph G is called *symmetric* if (v, u) is also present in G. Following the terminology from [10], we call G *odd-cycle symmetric* (respectively *weakly symmetric*) if for each odd (respectively any) cycle C all the arcs of C are symmetric. As observed by Kobayashi and Takazawa [8], this symmetry is essential for the tractability of the problem. Also, it provides a link between odd cycles (in digraphs) and factor critical subgraphs (in undirected graphs) and makes contractions possible, see Lemma 1 below.

For a digraph G and $U \subseteq VG$, the set of arcs entering (respectively leaving) U is denoted by $\delta_G^{in}(U)$ and $\delta_G^{out}(U)$. We write $\gamma_G(U)$ to denote the set of arcs with both endpoints in U and $G[U]$ to denote the subgraph of G induced by U, i.e. $G[U] = (U, \gamma_G(U))$. When the digraph G is clear from the context it is omitted from notation.

To *contract* a set $U \subseteq VG$ in a digraph G one has to replace nodes in U by a single *complex node*. The arcs in $\gamma(VG - U)$ are not affected, arcs in $\gamma(U)$ are dropped, and the arcs in $\delta^{in}(U)$ (respectively $\delta^{out}(U)$) are redirected so as to enter (respectively leave) the complex node. The resulting graph is denoted by G/U. We identify the arcs in G/U with their pre-images in G. Note that G/U may contain multiple parallel arcs but not loops. If G' is obtained from G by an arbitrary series of contractions then $G' = G/U_1/ \ldots /U_k$ for a certain family of disjoint subsets $U_1, \ldots, U_k \subseteq VG$ (called the *maximum contracted sets*).

Maximizing the size of an even factor M in a digraph G is equivalent to minimizing its *deficiency* $\mathrm{def}(G, M) := |VG| - |M|$. The minimum deficiency of an even factor in G is called the *deficiency* of G and is denoted by $\mathrm{def}(G)$.

3   Pap's Algorithm

Consider an odd-cycle symmetric digraph G. The algorithm for finding a maximum even factor in G follows the standard scheme of cardinality augmentation. Namely, we initially start with the empty even factor M and execute a series of *iterations* each aiming to increase $|M|$ by one. Iterations call SIMPLE-AUGMENT routine that, given an odd-cycle symmetric digraph G and an even factor M in G either returns a larger even factor M^+ or NULL indicating that the maximum size is reached.

3.1   Augmentations

Let us temporarily allow odd cycles and focus on path-cycle matchings in G. The latter are easily characterized as follows. Construct two disjoint copies of VG: $V^1 := \{v^1 \mid v \in VG\}$ and $V^2 := \{v^2 \mid v \in VG\}$. For each arc $a = (u, v) \in AG$ add the edge $\{u_1, v_2\}$ (*corresponding* to a). Denote the resulting undirected bipartite graph by $\widetilde{G}$.

Clearly, a path-cycle matching M in G is characterized by the following properties: for each node $v \in VG$, M has at most one arc entering v and also at most one arc leaving v. Translating this to $\widetilde{G}$ one readily sees that M generates a matching $\widetilde{M}$ in $\widetilde{G}$. Moreover, this correspondence between matchings in $\widetilde{G}$ and path-cycle matchings in G is one-to-one. A node u^1 (respectively u^2) in $\widetilde{G}$ not covered by $\widetilde{M}$ is called a *source* (a *sink*, respectively).

Given a digraph G and a path-cycle matching M in G we turn $\widetilde{G}$ into a digraph $\overrightarrow{G}(M)$ by directing the edges $\{u^1, v^2\}$ corresponding to arcs $(u, v) \in M$ from v^2 to u^1 and the other edges from u^1 to v^2.

Definition 1. *A simple path in $\overrightarrow{G}(M)$ starting in a source node is called* alternating. *An alternating path ending in a sink node is called* augmenting.

For an alternating path P let $A(P)$ denote the set of arcs in G corresponding to the arcs of P in $\overrightarrow{G}(M)$. Hereinafter $A \bigtriangleup B$ denotes the symmetric difference of sets A and B. The next statements are well-known.

Claim 1. *If $\overrightarrow{G}(M)$ admits no augmenting path then M is a path-cycle matching of maximum size.*

Claim 2. *If P is an augmenting (respectively an even alternating) path in $\overrightarrow{G}(M)$ then $M' := M \bigtriangleup A(P)$ is path-cycle matching obeying $|M'| = |M| + 1$ (respectively $|M'| = |M|$).*

The augmentation procedure (see Algorithm 1) constructs $\overrightarrow{G}(M)$ and searches for an augmenting path P there (line 1). In case no such path exists, the current even factor M is maximum by Claim 1 (even in the broader class of path-cycle matchings), hence the algorithm terminates (line 3). Next, let $\overrightarrow{G}(M)$ contain an augmenting path P. Claim 2 indicates how a larger path-cycle matching M' can be formed from M, however M' may contain an odd cycle. The next definition focuses on this issue.

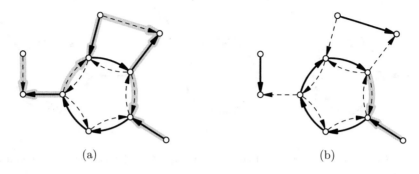

Fig. 1. Preparing for a contraction. Subfigure (a): the arcs of M are bold and the grayed arcs correspond to path P_{i+1}. Subfigure (b): path P_i is applied, the arcs of M_i are bold, and the grayed arcs indicate the remaining part of P_{i+1}.

Algorithm 1. SIMPLE-AUGMENT(G, M)

1: Search for an augmenting path P in $\overrightarrow{G}(M)$
2: **if** P does not exist **then**
3:     **return** NULL
4: **else if** P exists and is feasible **then**
5:     **return** $M \bigtriangleup A(P)$
6: **else** {P exists but is not feasible}
7:     For $i = 0, \ldots, k+1$ put $M_i \Leftarrow M \bigtriangleup A(P_i)$, where P_i $(0 \leq i \leq k)$ is the prefix of P of length $2i$ and $P_{k+1} = P$
8:     Find an index i $(0 \leq i \leq k)$ such that P_i is feasible while P_{i+1} is not
9:     Find the unique odd cycle C in M_{i+1}
10:     $G' \Leftarrow G/C$, $M' \Leftarrow M_i/C$
11:     $\overline{M}' \Leftarrow$ SIMPLE-AUGMENT(G', M')
12:     **if** $\overline{M}' =$ NULL **then** {M' is maximum in G'}
13:         **return** NULL
14:     **else** {M' is augmented in G' to a larger even factor $\overline{M}'$}
15:         Undo the contractions and transform $\overline{M}'$ into an even factor M^+ in G
16:         **return** M^+
17:     **end if**
18: **end if**

Definition 2. *Let P be an augmenting or an even alternating path. Then P is called* feasible *if $M' := M \bigtriangleup A(P)$ is again an even factor.*

If P is feasible then SIMPLE-AUGMENT exits with the updated even factor $M \bigtriangleup A(P)$ (line 5).

Consider the contrary, i.e. P is not feasible. Clearly, P is odd, say it consists of $2k + 1$ arcs. Construct a series of even alternating paths $P_0, \ldots, P_k$ where P_i is formed by taking the first $2i$ arcs of P $(0 \leq i \leq k)$. Also, put $P_{k+1} := P$ and $M_i := M \bigtriangleup A(P_i)$ $(0 \leq i \leq k+1)$.

Then there exists an index i $(0 \leq i \leq k)$ such that P_i is feasible while P_{i+1} is not feasible. In other words, M_i is an even factor obeying $\mathrm{def}(G, M_i) =$

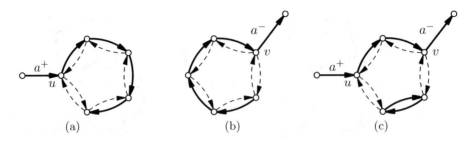

Fig. 2. Some cases appearing in Lemma 1. Digraph G and even factor N (bold arcs) are depicted.

$\operatorname{def}(G, M)$ and M_{i+1} contains an odd cycle. Since M_i and M_{i+1} differ by at most two arcs, it can be easily shown that an odd cycle in M_{i+1}, call it C, is unique (see [10]). Moreover, M_i *fits* C, that is, $|M_i \cap AC| = |VC| - 1$ and $\delta^{\text{out}}(VC) \cap M_i = \emptyset$. See Fig. 1 for an example. It turns out that when an even factor fits an odd cycle then a certain optimality-preserving contraction is possible. As long as no confusion is possible, for a digraph H and a cycle K we abbreviate H/VK to H/K. Also, for $X \subseteq AH$ we write X/K to denote $X \setminus \gamma_H(VK)$.

Claim 3 (Pap [10]). *Let K be an odd cycle in H and N be an even factor that fits K. Put $H' := H/K$ and $N' := N/K$. Then H' is an odd-cycle symmetric digraph and N' is an even factor in H'. Moreover, if N' is maximum in H' then N is maximum in H.*

Note that M is maximum in G if and only if M_i is maximum in G. The algorithm contracts C in G. Let $G' := G/C$ and $M' := M_i/C$ denote the resulting digraph and the even factor. To check if M' is maximum in G' a recursive call Simple-Augment(G', M') is made (line 11). If NULL is returned then M' is a maximum even factor in G', which by Claim 3 implies that the initial even factor M was maximum in G. In this case Simple-Augment terminates returning NULL (line 13).

Otherwise, the recursive call has augmented M' to a larger even factor $\overline{M}'$ in G'. The algorithm transforms $\overline{M}'$ into an even factor in G and terminates (lines 15–16). The following statement from [10] is applied (see Fig. 2 for examples):

Lemma 1. *Let K be an odd cycle in an odd-cycle symmetric digraph H. Put $H' := H/K$ and let N' be an even factor in H'. Then there exists an even factor N in H obeying $\operatorname{def}(H, N) = \operatorname{def}(H', N')$.*

3.2 Complexity

There are $O(n)$ iterations each consisting of $O(n)$ phases. To bound the complexity of a single phase note that it takes $O(m)$ time to find an augmenting path P (if it exists). We may construct all the path-cycle matchings $M_0, \ldots, M_{k+1}$ and decompose each of them into node-disjoint paths and cycles in $O(n^2)$ time. Hence, finding the index i and the cycle C takes $O(n^2)$ time. Contracting C in G takes $O(m)$ time. (We have already spent $O(m)$ time looking for P, so it is feasible to spend another $O(m)$ time to construct the new digraph $G' = G/C$ explicitly.) An obvious bookkeeping allows to undo all the contractions performed during the iteration and convert the final even factor in the contracted digraph into an even factor the initial digraph in $O(m)$ time. Totally, the algorithm runs in $O(n^4)$ time.

The above bound can be slightly improved as follows. Note that the algorithm needs an arbitrary index i such that M_i is an even factor and M_{i+1} is not, i.e. i is not required to be minimum. Hence, we may carry out a binary search over the range $[0, k+1]$. At each step we keep a pair of indices (l, r) such that M_l is an even factor while M_r is not. Replacing the current interval $[l, r)$ by a twice smaller one takes $O(n)$ time and requires constructing and testing a single path-cycle matching M_t, $t := \lfloor (l + r)/2 \rfloor$. This way, the $O(n^2)$ term reduces to $O(n \log n)$ and the total running time decreases to $O(n^2(m + n \log n))$. The ultimate goal of this paper is to get rid of the $O(m)$ term.

4   A Faster Algorithm

4.1   Augmentations

The bottleneck of SIMPLE-AUGMENT are the augmenting path computations. To obtain an improvement we need better understanding of how these paths are calculated. Similarly to the usual path-finding algorithms we maintain a directed out-forest $\mathcal{F}$ rooted at the source nodes. The nodes belonging to this forest are called $\mathcal{F}$-reachable. At each step a new arc (u, v) leaving an $\mathcal{F}$-reachable node u is scanned and either gets added to $\mathcal{F}$ (thus making v $\mathcal{F}$-reachable) or skipped because v is already $\mathcal{F}$-reachable. This process continues until a sink node is reached or no unscanned arcs remain in the digraph.

Definition 3.   *Let G be a digraph and M be an even factor in G. An* alternating forest $\mathcal{F}$ *for M is a directed out-forest in $\overrightarrow{G}(M)$ such that: (i) the roots of $\mathcal{F}$ are all the source nodes in $\overrightarrow{G}(M)$; (ii) every path from a root of $\mathcal{F}$ to a leaf of $\mathcal{F}$ is even.*

The intuition behind the suggested improvement is to grow $\mathcal{F}$ carefully and to avoid exploring infeasible alternating paths.

Definition 4.   *An alternating forest $\mathcal{F}$ is called* feasible *if every even alternating or augmenting path in $\mathcal{F}$ is feasible. An alternating forest $\mathcal{F}$ is called* complete *if it contains no sink node and for each arc (u, v) in $\overrightarrow{G}(M)$ if u is $\mathcal{F}$-reachable then so is v.*

We replace SIMPLE-AUGMENT by a more sophisticated recursive FAST-AUGMENT procedure. It takes an odd-cycle symmetric digraph G, an even factor M in G, and an additional flag named *sparsify*. The procedure returns a digraph $\overline{G}$ obtained from G by a number of contractions and an even factor $\overline{M}$ in $\overline{G}$. Additionally, it may return an alternating forest $\overline{\mathcal{F}}$ for $\overline{M}$ in $\overline{G}$. Exactly one of the following two cases applies:

(1) $\mathrm{def}(\overline{G},\overline{M}) = \mathrm{def}(G,M) - 1$ and $\overline{\mathcal{F}}$ is undefined;

(2) $\mathrm{def}(\overline{G},\overline{M}) = \mathrm{def}(G,M)$, $\overline{M}$ is maximum in $\overline{G}$, M is maximum in G, and $\overline{\mathcal{F}}$ is a complete feasible alternating forest for $\overline{M}$ in $\overline{G}$.

Assuming the above properties are true, let us explain how FAST-AUGMENT can be used to perform a single augmenting iteration. Given a current even factor M in G the algorithm calls FAST-AUGMENT(G, M, TRUE) and examines the result. If $\mathrm{def}(\overline{G},\overline{M}) = \mathrm{def}(G,M)$ then by (2) M is a maximum even factor in G, the algorithm stops. (Note that the forest $\overline{\mathcal{F}}$, which is also returned by FAST-AUGMENT, is not used here. This forest is needed due to the recursive nature of FAST-AUGMENT.) Otherwise $\mathrm{def}(\overline{G},\overline{M}) = \mathrm{def}(G,M) - 1$ by (1); this case will be referred to as a *breakthrough*. Applying Lemma 1, $\overline{M}$ is transformed to an even factor M^+ in G such that $\mathrm{def}(G,M^+) = \mathrm{def}(\overline{G},\overline{M}) = \mathrm{def}(G,M) - 1$. This completes the current iteration.

Clearly, the algorithm constructs a maximum even factor correctly provided that FAST-AUGMENT obeys the contract. Let us focus on the latter procedure. It starts growing a feasible alternating forest $\mathcal{F}$ rooted at the source nodes (line 1). During the course of the execution, FAST-AUGMENT *scans* the arcs of G in a certain order. For each node u in G we keep the list $L(u)$ of all unscanned arcs leaving u. The following invariant is maintained:

(3) if $a = (u,v)$ is a scanned arc then either $a \in M$ or both u^1 and v^2 are $\mathcal{F}$-reachable.

Consider an $\mathcal{F}$-reachable node u^1. To enumerate the arcs leaving u^1 in $\overrightarrow{G}(M)$ we fetch an unscanned arc $a = (u,v)$ from $L(u)$. If $a \in M$ or v^2 is $\mathcal{F}$-reachable then a is skipped and another arc is fetched. (In the former case a does not generate an arc leaving u^1 in $\overrightarrow{G}(M)$, in the latter case v^2 is already $\mathcal{F}$-reachable so a can be made scanned according to (3).)

Otherwise, consider the arc $a_1 := (u^1, v^2)$ in $\overrightarrow{G}(M)$ and let P_0 denote the even alternating feasible path from a root of $\mathcal{F}$ to u^1. Note that each node x^2 in $\overrightarrow{G}(M)$ (for $x \in VG$) is either a sink or has a unique arc leaving it. A *single* step occurs when v^2 is a sink (lines 10–13). The algorithm constructs an augmenting path $P_1 = P_0 \circ a_1$ leading to v^2. (Here $L_1 \circ L_2$ stands for the concatenation of L_1 and L_2.) If P_1 is feasible, the current even factor gets augmented according to Claim 2 and FAST-AUGMENT terminates. Otherwise, forest growing stops and the algorithm proceeds to line 22 to deal with a contraction.

A *double* step is executed when v^2 is not a sink (lines 15–20). To keep the leafs of $\mathcal{F}$ on even distances from the roots, $\mathcal{F}$ is extended by adding pairs of arcs. Namely, there is a unique arc leaving v^2 in $\overrightarrow{G}(M)$, say $a_2 = (v^2, w^1)$ (evidently $(w,v) \in M$). Moreover, w^1 is not a source node and (v^2, w^1) is the only arc entering w^1. Hence, w^1 is not $\mathcal{F}$-reachable. If $P_1 := P_0 \circ a_1 \circ a_2$ is feasible then a_1 and a_2 are added to $\mathcal{F}$ thus making v^2 and w^1 $\mathcal{F}$-reachable. Otherwise, a contraction is necessary.

Algorithm 2. FAST-AUGMENT$(G, M, sparsify)$

1: Initialize forest $\mathcal{F}$
2: **for all** unscanned arcs $a = (u, v)$ such that $u^1 \in V\mathcal{F}$ **do**
3:     Mark a as scanned
4:     **if** $a \in M$ or $v^2 \in V\mathcal{F}$ **then**
5:         **continue for** {to line 2}
6:     **end if**
7:     $a_1 \Leftarrow (u^1, v^2)$
8:     Let P_0 be the even alternating path to u^1 in $\mathcal{F}$ {P_0 is feasible}
9:     **if** v^2 is a sink **then** {single step}
10:         $P_1 \Leftarrow P_0 \circ a_1$ {P_1 is augmenting}
11:         **if** P_1 is feasible **then**
12:             **return** $(G, M \triangle A(P_1), \text{NULL})$
13:         **end if**
14:     **else** {double step}
15:         Let $a_2 = (v^2, w^1)$ be the unique arc leaving v^2 {$w^1 \notin V\mathcal{F}$}
16:         $P_1 \Leftarrow P_0 \circ a_1 \circ a_2$ {P_1 is even alternating}
17:         **if** P_1 is feasible **then**
18:             Add nodes v^2 and w^1 and arcs a_1, a_2 to $\mathcal{F}$
19:             **continue for** {to line 2}
20:         **end if**
21:     **end if**
22:     $M_0 \Leftarrow M \triangle A(P_0)$, $M_1 \Leftarrow M \triangle A(P_1)$
23:     Find a unique odd cycle C in M_1
24:     $G' \Leftarrow G/C$, $M' \Leftarrow M_0/C$
25:     **if** $sparsify = \text{FALSE}$ **then**
26:         **return** FAST-AUGMENT(G', M', FALSE)
27:     **end if**
28:     Construct the digraph H'
29:     $(\overline{H}', \overline{M}', \overline{\mathcal{F}}) \Leftarrow$ FAST-AUGMENT(H', M', FALSE)
30:     Compare $V\overline{H}'$ and VG: let $Z_1, \ldots, Z_k$ be the maximum contracted sets and $z_1, \ldots, z_k$ be the corresponding complex nodes in $\overline{H}'$
31:     $\overline{G}' \Leftarrow G/Z_1/\ldots/Z_k$
32:     **if** $\text{def}(\overline{G}', \overline{M}') < \text{def}(G', M')$ **then**
33:         **return** $(\overline{G}', \overline{M}', \text{NULL})$
34:     **end if**
35:     Unscan the arcs in $\overline{G}'$ that belong to M and the arcs that enter $z_1, \ldots, z_k$
36:     $G \Leftarrow \overline{G}'$, $M \Leftarrow \overline{M}'$, $\mathcal{F} \Leftarrow \overline{\mathcal{F}}$
37: **end for**
38: **return** $(G, M, \mathcal{F})$

Now we explain how the algorithm deals with contractions at line 22. One has an even alternating feasible path P_0 and an infeasible augmenting or even alternating path P_1 (obtained by extending P_0 by one or two arcs). Put $M_0 := M \triangle A(P_0)$ and $M_1 := M \triangle A(P_1)$. Let C denote the unique odd cycle in M_1. Put $G' := G/C$, $M' := M_0/C$. If $sparsify = \text{FALSE}$ then FAST-AUGMENT acts like SIMPLE-AUGMENT, namely, it makes a recursive call passing G' and M' as an input and, hence, restarting the whole path computation.

Next, suppose *sparsify* = TRUE. In this case the algorithm tries to *recover* some feasible alternating forest for the contracted digraph G' and the updated even factor M'. To accomplish this, a sparse digraph H' is constructed and FAST-AUGMENT is called for it recursively (with *sparsify* = FALSE). The latter nested call may result into a breakthrough, that is, find an even factor of smaller deficiency. In this case, the outer call terminates immediately. Otherwise, the nested call returns a complete feasible alternating forest $\overline{\mathcal{F}}$ for an even factor $\overline{M}'$ in a digraph $\overline{H}'$ (obtained from H' by contractions). This forest is used by the outer call to continue the path-searching process. It turns out that almost all of the arcs that were earlier fetched by the outer call need no additional processing and may remain scanned w.r.t. the new, recovered forest. This way, the algorithm amortizes arc scans during the outer call.

More formally, the algorithm first constructs a spanning subgraph H of G as follows. Take the node set of G, add all the arcs of M and all the arcs $(u, v) \in AG$ such that (u^1, v^2) is present in $\mathcal{F}$. We need to ensure that H is odd-cycle symmetric: if some arc (u, v) is already added to H and the reverse arc (v, u) exists in G then add (v, u) to H. Obviously, H is an odd-cycle symmetric spanning subgraph of G and M is an even factor in H.

Secondly, put $H' := H/C$. Note that H' is an odd-cycle symmetric spanning subgraph of G' and $|AH'| = O(n)$. Also, M' is an even factor in H'.

Finally, make the recursive call FAST-AUGMENT(H', M', FALSE) and let $\overline{H}'$ and $\overline{M}'$ be the resulting digraph and the even factor, respectively.

Compare the node sets of $\overline{H}'$ and H. Clearly, $\overline{H}'$ may be viewed as obtained from H by a number of contractions. Let $Z_1, \ldots, Z_k$ be the maximum contracted subsets in VH (= VG), i.e. $Z_1, \ldots, Z_k$ are disjoint and $\overline{H}' = H/Z_1/\ldots/Z_k$. (Note that $k \geq 1$ since C is already contracted in H'.) Let $z_1, \ldots, z_k$ be the composite nodes corresponding to $Z_1, \ldots, Z_k$. The algorithm applies these contractions to G and constructs the digraph $\overline{G}' := G/Z_1/\ldots/Z_k$. Clearly $\overline{M}'$ is an even factor in both $\overline{G}'$ and $\overline{H}'$. If $\mathrm{def}(\overline{H}', \overline{M}') < \mathrm{def}(H', M') = \mathrm{def}(G, M)$, then one has a breakthrough, FAST-AUGMENT terminates yielding $\overline{G}'$ and $\overline{M}'$.

Otherwise, the recursive call in line 29 also returns a complete feasible forest $\overline{\mathcal{F}}$ for $\overline{H}'$ and $\overline{M}'$. Recall that some arcs in G are marked as *scanned*. Since we identify the arcs of $\overline{G}'$ with their pre-images in G, one may speak of scanned arcs in $\overline{G}'$. The algorithm "unscans" certain arcs $a = (u, v) \in A\overline{G}'$ by adding them back to their corresponding lists $L(u)$ to ensure (3). Namely, the arcs that belong to M and are present in $\overline{G}'$ and the arcs that enter any of the complex nodes $z_1, \ldots, z_k$ in $\overline{G}'$ are unscanned. After this, the algorithm puts $G := \overline{G}'$, $M := \overline{M}'$, $\mathcal{F} := \overline{\mathcal{F}}$ and proceeds with growing $\mathcal{F}$ (using the adjusted set of the scanned arcs).

Finally, all the arcs of G are scanned by FAST-AUGMENT and no sink is reached, the resulting forest $\mathcal{F}$ is both complete and feasible. In particular, by (3) no augmenting path for M exists. By Claim 3 this implies the maximality of M in G. The algorithm returns the current digraph G, the current (maximum) even factor M, and also the forest $\mathcal{F}$, which certifies the maximality of M.

The correctness of FAST-AUGMENT is evident except for the case when it tries to recover $\mathcal{F}$ and alters the set of scanned arcs. One has to prove that (3) holds for the updated forest and the updated set of the scanned arcs. Due to the lack of space the proof of this statement will be given in the full version of the paper.

4.2   Complexity

We employ arc lists to represent digraphs. When a subset U in a digraph Γ is contracted we enumerate the arcs incident to U and update the lists accordingly. If a pair of parallel arcs appears after a contraction, these arcs are merged, so all our digraphs remain simple. The above contraction of U takes $O(|V\Gamma| \cdot |U|)$ time. During FAST-AUGMENT the sum of sizes of the contracted subsets telescopes to $O(n)$, so graph contractions take $O(n^2)$ time in total. The usual bookkeeping allows to undo the contractions and recover a maximum even factor in the original digraph in $O(m)$ time.

Consider an invocation FAST-AUGMENT$(\Gamma, N, \text{FALSE})$ and let us bound its complexity (together with the recursive calls). The outer loop of the algorithm (lines 2–37) enumerates the unscanned arcs. Since $sparsify = \text{FALSE}$, each arc can be scanned at most once, so the bound of $O(|A\Gamma|)$ for the number of arc scans follows. Using the balanced search trees to represent even factors the reachability checks in lines 11 and 17 can be carried out in $O(\log|V\Gamma|)$ time. Constructing M_0, M_1, and C takes $O(|V\Gamma|)$ time. This way, FAST-AUGMENT$(\Gamma, N, \text{FALSE})$ takes $O((k+1)|A\Gamma|\log|V\Gamma|)$ time, where k denotes the number of graph contractions performed during the invocation.

Next, we focus on FAST-AUGMENT(Γ, N, TRUE) call. Now one may need to perform more than $|A\Gamma|$ arc scans since forest recovery may produce new unscanned arcs (line 35). Note that forest recovery totally occurs $O(|V\Gamma|)$ times (since each such occurrence leads to a contraction). During each recovery M generates $O(|V\Gamma|)$ unscanned arcs or, in total, $O(|V\Gamma|^2)$ such arcs for the duration of FAST-AUGMENT. Also, each node z_i generates $O(|V\Gamma|)$ unscanned arcs (recall that we merge parallel arcs and keep the current digraph simple). The total number of these nodes processed during FAST-AUGMENT is $O(|V\Gamma|)$ (since each such node corresponds to a contraction). Totally these nodes produce $O(|V\Gamma|^2)$ unscanned arcs. Hence, the total number of arc scans is $O(|A\Gamma| + |V\Gamma|^2) = O(|V\Gamma|^2)$.

Each feasibility check costs $O(\log|V\Gamma|)$ time, or $O(|V\Gamma|^2\log|V\Gamma|)$ in total. Finally, we must account for the time spent in the recursive invocations during FAST-AUGMENT(Γ, N, TRUE). Each such invocation deals with a sparse digraph and hence takes $O((k+1)|V\Gamma|\log|V\Gamma|)$ time (where, as earlier, k denotes the number of contractions performed by the recursive invocation). Since the total number of contractions is $O(|V\Gamma|)$, the sum over all recursive invocations telescopes to $O(|V\Gamma|^2\log|V\Gamma|)$.

The total time bound for FAST-AUGMENT(Γ, N, TRUE) (including the recursive calls) is also $O(|V\Gamma|^2\log|V\Gamma|)$. Therefore a maximum even factor in an odd-cycle symmetric digraph can be found in $O(n^3\log n)$ time, as claimed.

Acknowledgements

The author is thankful to anonymous referees for providing valuable comments.

References

1. Balas, E., Pulleyblank, W.: The perfectly matchable subgraph polytope of an arbitrary graph. Combinatorica 9, 321–337 (1989)
2. Cunningham, W.H.: Matching, matroids, and extensions. Mathematical Programming 91(3), 515–542 (2002)
3. Cunningham, W.H., Geelen, J.F.: The optimal path-matching problem. Combinatorica 17, 315–337 (1997)
4. Cunningham, W.H., Geelen, J.F.: Combinatorial algorithms for path-matching (2000) (manuscript)
5. Cunningham, W.H., Geelen, J.F.: Vertex-disjoint dipaths and even dicircuits (2001) (manuscript)
6. Hopcroft, J.E., Karp, R.M.: An $n^{5/2}$ algorithm for maximum matchings in bipartite graphs. SIAM Journal on Computing 2(4), 225–231 (1973)
7. Iwata, S., Takazawa, K.: The independent even factor problem. In: Proceeinds of the 18th Annual ACM-SIAM Symposium on Discrete algorithms, pp. 1171–1180 (2007)
8. Kobayashi, Y., Takazawa, K.: Even factors, jump systems, and discrete convexity. J. Comb. Theory Ser. B 99(1), 139–161 (2009)
9. Micali, S., Vazirani, V.: An $O(\sqrt{|V|}.|E|)$ algorithm for finding maximum matching in general graphs. In: Proc. 45th IEEE Symp. Foundations of Computer Science, pp. 248–255 (1980)
10. Pap, G.: A combinatorial algorithm to find a maximum even factor. In: Proceedings of the 11th Integer International IPCO Conference on Programming and Combinatorial Optimization, pp. 66–80 (2005)
11. Pap, G.: Combinatorial algorithms for matchings, even factors and square-free 2-factors. Math. Program. 110(1), 57–69 (2007)
12. Pap, G., Szegö, L.: On the maximum even factor in weakly symmetric graphs. J. Comb. Theory Ser. B 91(2), 201–213 (2004)
13. Spille, B., Weismantel, R.: A generalization of Edmonds' matching and matroid intersection algorithms. In: Proceedings of the 9th International IPCO Conference on Integer Programming and Combinatorial Optimization, pp. 9–20 (2002)
14. Takazawa, K.: A weighted even factor algorithm. Mathematical Programming 115(2), 223–237 (2008)

Author Index

Printing: Mercedes-Druck, Berlin
Binding: Stein+Lehmann, Berlin